PHYTOTOXINS IN PLANT DISEASES

Proceedings of the NATO Advanced Study Institute
Pugnochiuso, Italy, June 1970

Edited by

R. K. S. WOOD

Department of Botany and Plant Technology,
Imperial College of Science and Technology,
London, England

A. BALLIO

Instituto Chimico dell'Università, Naples, Italy

and

A. GRANITI

Istituto di Patologia vegetale dell'Università
Bari, Italy

1972

ACADEMIC PRESS · LONDON · NEW YORK

ACADEMIC PRESS INC. (LONDON) LTD
24/28 Oval Road
London NW1

U.S. Edition published by
ACADEMIC PRESS INC.
111 Fifth Avenue,
New York, New York 10003

Library of Congress Catalog Card Number: 76-175835
ISBN: 0-12-762850-9

PRINTED IN GREAT BRITAIN BY
WILLIAM CLOWES AND SONS LIMITED
LONDON, COLCHESTER AND BECCLES

PHYTOTOXINS IN PLANT DISEASES

List of Participants

ANDEL, Miss Dr. O. M. van. Instituut voor Biologisch en Scheikundig Onderzoek van Landbouwgewassen, Bornsesteeg 65/67, Postbus 14, Wageningen, Netherlands.

ARNTH-JENSEN, Dr. H. Scientific Affairs Division, North Atlantic Treaty Organization, 1110 Bruxelles, Belgium.

BACKMAN, Dr. Paul A. U.S.D.A.-A.R.S. Department of Plant Pathology and Physiology, Clemson University, Clemson, South Carolina 29631, U.S.A.

BALLIO, Prof. Alessandro. Laboratorio di Chimica delle Sostanze Naturali, Istituto Chimico dell'Università, Via Mezzocannone 4, 80134 Napoli, Italy.

BARBIER, Prof. Michel. Institut de Chimie des Substances Naturelles, C.N.R.S., 91-Gif-Sur-Yvette, France.

BISIACH, Dr. Marco. Istituto di Patologia Vegetale dell'Università, Via Celoria 2, 20133 Milano, Italy.

BOTTALICO, Dr. Antonio. Istituto di Patologia Vegetale dell'Università, Via Amendola 165/A, 70126 Bari, Italy.

BOUHOT, Dr. Daniel. I.N.R.A., Département de Pathologie Végétale, Station de Recherches sur la Flore Pathogène dans le Sol, 7 Rue Sully, 21-Dijon, France.

CANONICA, Prof. Luigi. Istituto di Chimica Organica dell'Università, Via C. Saldini 50, 20133 Milano, Italy.

CASINOVI, Prof. Carlo G. Istituto Superiore di Sanità, Viale Regina Elena 299, 00161 Roma, Italy.

CATARA, Dr. Antonino. Istituto di Patologia Vegetale dell'Università, Via Valdisavoia 5, 95123 Catania, Italy.

CICCARONE Prof. Antonio. Istituto di Patologia Vegetale dell'Università, Via Amendola 165/A, 70126 Bari, Italy.

CIRULLI Dr. Matteo. Istituto di Patologia vegetale dell'Università, Via Amendola 165/A, 70126 Bari, Italy.

DÉFAGO, Miss Dr. Geneviève. Institut für Spezielle Botanik, Eidg. Technische Hochschule, Universitätstrasse 2, 8006 Zürich, Switzerland.

DIMOND, Dr. Albert E. Department of Plant Pathology and Botany, The Connecticut Agricultural Experiment Station, 123 Huntington Street, Box 1106, New Haven, Connecticut 06504, U.S.A.

DURBIN, Dr. Richard D. U.S.D.A. Pioneering Research Laboratory, Department of Plant Pathology, University of Wisconsin, Madison, Wisconsin 53706, U.S.A.

ELGERSMA, Dr. Doekle M. Phytopathologisch Laboratorium "Willie Commelin Scholten", Javalaan 20, Baarn, Netherlands.

ERCOLANI, Dr. Gian Luigi. Istituto di Patologia Vegetale dell'Università, Via Amendola 165/A, 70126 Bari, Italy.

FELDMAN, Prof. Albert W. Citrus Experiment Station, IFAS, University of Florida, P.O. Box 1088, Lake Alfred, Florida 33850, U.S.A.

FINDLAY, Dr. John A. Laboratorio di Chimica Biologica, Istituto Superiore di Sanità, Viale Regina Elena 299, 00161 Roma, Italy. (*Permanent Address:* Department of Chemistry, University of New Brunswick, Fredericton, N.B., Canada).

FRAMONDINO, Dr. Michele. Laboratorio di Chimica delle Sostanze Naturali, Istituto Chimico dell'Università, Via Mezzocannone 4, 80134 Napoli, Italy.

GEUKES FOPPEN, Dr. Fredrik H. Laboratorio di Chimica Biologica, Istituto Superiore di Sanità, Viale Regina Elena 299, 00161 Roma, Italy.

GIACOBINI, Miss Dr. Clelia. Istituto Centrale del Restauro, Piazza S. Francesco di Paola 9, 00184 Roma, Italy.

GIANANI, Miss Dr. Luciana. Laboratorio di Chimica delle Sostanze Naturali, Istituto Chimico dell'Università, Via Mezzocannone 4, 80134 Napoli, Italy.

GOODMAN, Prof. Robert N. Department of Plant Pathology, University of Missouri, Columbia, Missouri 65201, U.S.A.

GOTTLIEB, Prof. David. Department of Plant Pathology, University of Illinois, Urbana, Illinois 61801, U.S.A.

GRANITI, Prof. Antonio. Istituto di Patologia Vegetale dell'Università, Via Amendola 165/A, 70126 Bari, Italy.

GRASSO, Prof. Vincenzo. Istituto di Patologia Vegetale dell'Università, Via Amendola 165/A, 70126 Bari, Italy.

GROSSMANN, Prof. Friedrich. Institut für Phytopathologie, Justus-Liebig-Universität, Ludwigstrasse 23, 6300 Giessen, Germany.

HESS, Dr. Wilford M. Department of Botany and Range Science, Brigham Young University, Provo, Utah 84601, U.S.A.

HUBBELING, Dr. Nicolaas. Instituut voor Plantenziektenkundig Onderzoek, Binnenhaven 12, Wageningen, Postbus 42, Netherlands.

KERN, Prof. Heinz. Institut für Spezielle Botanik, Eidg. Technische Hochschule, Universitätstrasse 2, 8006 Zürich, Switzerland.

KÚC, Prof. Joseph. Department of Biochemistry, Purdue University, Lafayette, Indiana 47907, U.S.A.

LAMBERTI, Dr. Franco. Istituto di Patologia Vegetale dell'Università, Via Amendola 165/A, 70126 Bari, Italy.

LAVIOLA, Dr. Cesare. Istituto di Patologia Vegetale dell'Università, Via Amendola 165/A, 70126 Bari, Italy.

LEBEAU, Dr. Jack B. Canada Department of Agriculture, Research Station, Lethbridge, Alberta, Canada.

LINNASALMI, Miss Dr. Annikki. Maatalouden Tutkimuskeskus, Kasvitautien Tutkimuslaitos, Tikkurila, Finland.

LOCCI, Dr. Romano. Istituto di Patologia Vegetale dell'Università, Via Celoria 2, 20133 Milano, Italy.

LOUSBERGE, Dr. Robert J. J. C. Organisch Chemisch Laboratorium. Rijksuniversiteit te Utrecht, Croesestraat 79, Utrecht, Netherlands.

LUKEZIC, Dr. Felix L. Department of Plant Pathology, The Pennsylvania State University, 211 Buckhout Laboratory, University Park, Pennsylvania 16802, U.S.A.

MAIN, Dr. Charles E. Department of Plant Pathology, North Carolina State University, Box 5397, Raleigh, North Carolina 27607, U.S.A.

MAJORANA, Prof. Giuseppe. Istituto di Patologia Vegetale dell'Università, Via Amendola 165/A, 70126 Bari, Italy.

MANTLE, Dr. Peter G. Department of Biochemistry, Imperial College of Science and Technology, London, S.W.7, England.

MARAITE, Dr. Henri M. M. Laboratoire de Phytopathologie et de Mycologie Générale, Université de Louvain, De Croylaan 42, Heverlee, Belgium.

MARRÈ, Prof. Erasmo. Cattedra di Fisiologia Vegetale, Istituto di Scienze Botaniche dell'Università, Via Giuseppe Colombo 60, 20133 Milano, Italy.

MARTELLI, Prof. Giovanni. Istituto di Patologia Vegetale dell'Università, Via Amendola 165/A, 70126 Bari, Italy.

MATTA, Dr. Alberto. Istituto di Patologia Vegetale dell'Università, Via Pietro Giuria 15, 10126 Torino, Italy.

MAZZUCCHI, Dr. Umberto. Istituto di Patologia Vegetale dell'Università, Via Filippo Re 8, 40126 Bologna, Italy.

MEYER, Prof. J.-A. Laboratoire de Phytopathologie et de Mycologie Générale, Université de Louvain, De Croylaan 42, Heverlee, Belgium.

MIROCHA, Prof. Chester J. Department of Plant Pathology, University of Minnesota, St. Paul, Minnesota 55101, U.S.A.

MOLINAS DE FERRER, Miss, Dr. Luisa M. Laboratorio de Fisiologia Vegetal, Facultad de Ciencias, Universidad de Barcelona, Spain.

MOREAU, Dr. Claude. Laboratoire de Biologie Végétale, Faculté des Sciences, Avenue Le Gorgeu, 29N-Brest, France.

MOREAU, Mrs. Prof. Mireille. Laboratoire de Biologie Végétale, Faculté des Sciences, Avenue Le Gorgeu, 29N-Brest, France.

MORPURGO, Prof. Giorgio. Istituto Superiore di Sanità, Viale Regina Elena 299, 00161 Roma, Italy.

MUSSELL, Dr. Harry W. Boyce Thompson Institute for Plant Research, Inc., 1086 North Broadway, Yonkers, N.Y. 10701, U.S.A.

NAEF-ROTH, Mrs. Dr. Stephi. Institut für Spezielle Botanik, Eidg. Technische Hochschule, Universitätstrasse 2, 8006 Zürich, Switzerland.

PAGE, Prof. Orville T. Department of Biology, The University of New Brunswick, Fredericton, N.B., Canada.

PATIL, Dr. Suresh S. Department of Plant Pathology, University of Hawaii, 3190 Maile Way, Honolulu, Hawaii 96822, U.S.A.

PAXTON, Dr. Jack. Department of Plant Pathology, University of Illinois, Urbana, Illinois 61801, U.S.A.

PIGLIONICA, Dr. Vito. Istituto di Patologia Vegetale dell'Università, Via Amendola 165/A, 70126 Bari, Italy.

PRINGLE, Dr. Ross B. Canada Department of Agriculture, Physiology Section, Cell Biology Research Institute, Ottawa 3, Ontario, Canada.

RAGGI, Dr. Vittorio. Istituto di Patologia Vegetale dell'Università, S. Pietro, 06100 Perugia, Italy.

RANDAZZO, Dr. Giacomino. Laboratorio di Chimica delle Sostanze Naturali, Istituto Chimico dell'Università, Via Mezzocannone 4, 80134 Napoli, Italy.

RODRIGUES, Dr. Carlos J. Jr. Centro de Investigação das Ferrugens do Cafeeiro, Estação Agronómica Nacional, Oeiras, Portugal.

ROSSI, Dr. Carlo. Laboratorio di Chimica Biologica, Istituto Superiore di Sanità, Viale Regina Elena 299, 00161 Roma, Italy.

RUDOLPH, Dr. Klaus. Institut für Pflanzenpathologie und Pflanzenschutz, Georg-August-Universität, Nikolausberger Weg 5a, 34 Göttingen, Germany.

RUFFNER, Dr. Fortunat. Institut für Spezielle Botanik, Eidg. Technische Hochschule, Universitätstrasse 2, 8006 Zürich, Switzerland.

RUSSI, Mrs. Dr. Serena. Laboratorio di Chimica Biologica, Istituto Superiore di Sanità, Viale Regina Elena 299, 00161 Roma, Italy.

SALEMINK, Prof. C. A. Organisch Chemisch Laboratorium, Rijksuniversiteit te Utrecht, Croesestraat 79, Utrecht, Netherlands.

SCALA, Dr. Antonio. Istituto di Chimica Organica dell'Università, Via C. Saldini 50, 20133 Milano, Italy.

SCHEFFER, Prof. Robert P. Department of Botany and Plant Pathology, Michigan State University, 166 Plant Biology Building, East Lansing, Michigan 48823, U.S.A.

SCHIPPERS, Dr. Bob. Phytopathologisch Laboratorium "Willie Commelin Scholten", Javalaan 20, Baarn, Netherlands.

SCHLOSSER, Dr. Wolf Eckart. Division of Crop Production and Protection, Faculty of Agricultural Sciences, American University of Beirut, Beirut, Lebanon.

SILANO, Dr. Vittorio. Laboratorio di Chimica Biologica, Istituto Superiore di Sanità, Viale Regina Elena 299, 00161 Roma, Italy.

SMEDEGÅRD-PETERSEN, Dr. Vigo. Plantepatologisk Afdeling, Kgl. Veterinaerog Landbohojskole, Rolighedsvej 23, 1958 Cobenhavn V., Denmark.

SPARAPANO, Dr. Lorenzo. Istituto di Patologia Vegetale dell'Università, Via Amendola 165/A, 70126 Bari, Italy.

STOESSL, Dr. Albert. Laboratorium für Organische Chemie, Eidg. Technische Hochschule, Universitätstrasse 6, 8006 Zürich, Switzerland. (*Permanent Address*: Canada Department of Agriculture, Research Institute, University Sub Post Office, London 72, Ontario, Canada).

STROBEL, Dr. Gary A. Department of Botany and Microbiology, Montana State University, Bozeman, Montana 59715, U.S.A.

SUNDHEIM, Dr. Leif. The Agricultural College of Norway, 1432 Vollebekk, Norway.

TORRACCA, Mrs. Dr. Anna Maria. Laboratorio di Chimica Biologica, Istituto Superiore di Sanità, Viale Regina Elena 299, 00161 Roma, Italy.

TROUVELOT, Dr. Alain. I.N.R.A., Département de Pathologie Végétale, Station de Recherches sur la Flore Pathogène dans le Sol, 7 Rue Sully, 21-Dijon, France.

TURNER, Dr. Neil C. Department of Ecology and Climatology, The Connecticut Agricultural Experiment Station, 123 Huntington Street, Box 1106, New Haven, Connecticut 06504, U.S.A.

VURRO, Miss Dr. Antonietta. Istituto di Patologia Vegetale dell'Università, Via Amendola 165/A, 70126 Bari, Italy.

WARD, Dr. Edmund W. B. Canada Department of Agriculture, Research Institute, University Sub Post Office, London 72, Ontario, Canada.

WOLTZ, Dr. Shreve S. Gulf Coast Experiment Station, IFAS, University of Florida, 5007 60th Street East, Bradenton, Florida 33505, U.S.A.

WOOD, Prof. Ronald K. S. Department of Botany, Imperial College of Science and Technology, Prince Consort Road, London, S.W.7, England.

YODER, Dr. Olen C. Department of Botany and Plant Pathology, Michigan State University, 166 Plant Biology Building, East Lansing, Michigan 48823, U.S.A.

Preface

The idea that the toxic effects of an infectious disease may be caused by substances elaborated by the pathogen dates back to the last century when the aetiology of certain human and animal diseases was still under discussion. Much later, the discovery of the endotoxins of *Clostridium tetani*, *C. botulinum* and *Corynebacterium diphtheriae* emphasized the importance in disease of substances elaborated by the pathogen during pathogenesis. Many other toxins are now known, or are believed to be implicated in a variety of diseases of humans and animals.

In plant pathology, the identification of phytotoxins as microbial products, and the systematic study of their effects on higher plants is more recent but progress in this field has been rapid and there is now good evidence that demonstrates a role for these substances in a wide range of plant diseases. However, even for the best studied of these diseases much remains to be learned about how phytotoxins damage plants and about their significance in the development of different syndromes. Indeed, for many diseases that have been studied in this respect not much more is known than that the pathogen can produce one or more substances that, applied to plants, reproduce some or many of the symptoms of the natural disease. Much more is necessary before a role in pathogenesis can unequivocally be ascribed to a particular metabolite.

As this important field of study in plant pathology has developed during the past decade, so has the need for a rational and critical approach to the study of phytotoxins and for an agreed terminology. It was fortunate, therefore, that the idea of an international, interdisciplinary summer school on phytotoxins arose naturally from the important cooperative work that has been in progress on these substances in Italy since 1963. This work brought together from different laboratories research scientists with different backgrounds and expertise. The results have been profitable and fruitful. This highly successful collaboration between plant pathologist, physiologist, microbiologist and chemist, and the growing interest in phytotoxins on an international scale encouraged Professor A. Graniti, University of Bari, and Professor A. Ballio, University of Naples, Italy to submit an application for a NATO Advanced Study Institute on Phytotoxins in Plant Diseases after consulting with and receiving enthusiastic support from Professor A. Ciccarone, University of Bari, Italy, Dr. A. E.

Dimond, Connecticut Agricultural Experiment Station, New Haven, Connecticut, U.S.A. and Professor R. K. S. Wood, Imperial College, London, England.

The NATO Scientific Affairs Division formally approved the Institute in 1969. Professor R. K. S. Wood accepted an invitation to be its Scientific Director and Dr. G. L. Ercolani, University of Bari, Italy, agreed to be the Secretary.

The Institute was most generously supported by the North Atlantic Treaty Organization, Bruxelles, Belgium, and also received a financial contribution from the Italian Ministry of Agriculture and Forestry, Rome. Grants were also awarded to participants by the National Science Foundation, Washington, D.C., U.S.A., the French Ministry of Foreign Affairs, Paris, France, the European Branch of Richardson-Merrel Co., Paris, France, and a number of Italian Universities. Certain needs of the Institute were met by the Universities of Bari and Naples and facilities for the meetings were provided by the Ente Nazionale Idrocarburi, Rome.

The Institute was held at Pugnochiuso, near Foggia, Italy from 7 to 21 June, 1970. The subject matter was considered under three main headings, general concepts, chemistry and mode of action of toxins produced by plant pathogens. In addition there were six discussion groups on special topics and another on future trends in research on phytotoxins.

The Institute was opened and addressed by Professor A. Ciccarone who represented the University of Bari and the Italian National Council for Research (C.N.R.). Each day there were two main lectures and a number of seminars, followed by discussions almost all of which continued beyond the time allocated to them. The twenty main lectures were given by scientists from nine countries and there were fifty-four official participants from fourteen countries each of whom had been invited to attend because of their current interest in phytotoxins, and to give accounts of their work in this field. In addition, twelve observers and visitors from several countries attended part, or the whole of the Institute.

The organizers of the Institute take this opportunity to express their gratitude to the North Atlantic Treaty Organization and to all other bodies whose financial help ensured the success of the Institute. They also thank Dr. G. L. Ercolani most warmly for the excellence of his work before and during the meeting, and Dr. H. Arnth-Jensen of the North Atlantic Treaty Organization for his most generous help and advice and his personal interest in the Institute itself.

The Editors thank, too, all those who in many different ways have contributed to the publication of this book which aims to give a full account of the Proceedings of the Institute and which is, we believe, the first given entirely to this large and rapidly expanding field of research in plant pathology.

September, 1971

A.B.
A.G.
R.K.S.W.

Address of Welcome

by

PROFESSOR ANTONIO CICCARONE

Istituto di Patologia vegetale, University of Bari, Italy

It is a great pleasure for me to convey to you the warmest welcome from Professor Vincenzo Caglioti, President of the Italian Council for Research, and the kindest greetings from the Rector of the University and of the Agricultural College of Bari.

Your field is one of the most active and promising in plant pathology and it is also one of the most complex, one where biochemistry, ultrastructure, physiology, genetics and ecology of the host plant and of the pathogen are brought together and investigated in their natural interrelationships, with the aim of a deeper insight of what disease and life processes are. These studies of attack by pathogens and defence by host must have important consequences for disease control and crop production and, in the end, for human welfare.

The complexity of the different disciplines involved in modern research is such that interdisciplinary discussions are more and more required, and it is advantageous for these discussions to take place in isolated spots where attention can be continuously focused on the topics under review. It is fitting that the Institute will be held in this secluded bay of the Gargano peninsula, far from large towns, not too well endowed with transport facilities, quiet and serene, between the sea and near the rocks of Monte S. Angelo, where in the fifth century the Archangel Michael appeared to men.

You will forgive me if I mention the rich history of this region, recently considered somewhat peripheral to our world. And yet Apulia and the neighbouring regions have made many valuable contributions over the centuries to the histories of Magna Graecia, Rome, the Normans, the Holy Roman Empire, and the Crusades, contributions to our culture, and to the maintenance and advancement of our civilization. Now, at last, we are gratified in realizing that Apulia and other parts of the warm Mediterranean region are enjoying a revival, of which we are more than a little proud and which I know you will cordially welcome.

I sincerely wish you a fruitful and happy stay in the Gargano peninsula and I wish all success to your Institute. I would like finally to express my kindest thanks to the North Atlantic Treaty Organization for the financial support so generously granted for this meeting.

Contents

SHORTER PAPERS

A. BACTERIAL PHYTOTOXINS

B. FUNGAL PHYTOTOXINS

C. MISCELLANEOUS

1. The Evolution of the Toxin Concept in Plant Pathology

ANTONIO GRANITI

Istituto di Patologia vegetale dell'Università degli Studi, Bari, Italy

Toxins, Poisons, and Toxic Substances

The meaning of the term "toxin" is related not so much to the idea of death as suggested by its Greek etymology ("*toxon*" = bow), as to the poison on the head of arrows shot from the bow. Since this implies aggression and a vector, "toxin", as a word, should differ substantially from "venom" and similar terms used in other languages (Italian "*veleno*", and Spanish "*veneno*"), which originally meant "love potion",[1] and later were used to designate any substance injurious to man and animals if ingested. This idea of ingestion is quite apparent in the French and English word "poison", which is again derived from the Latin ("*potio*" = a drink, a potion). Plants and animals from which such substances originate are called "poisonous". Furthermore, the current connotation of poisons refers to substances having a rapid action at high dilutions.

Usually, however, any material which, independently of its nature and origin, damages or kills when introduced into plants or animals is generally termed a "toxic substance". The toxicity depends on the dose and usually increases with concentration.

These terms have been used in very different and often contradictory ways to designate one and the same substance.

In animal pathology, the meaning of the term "toxin" is somewhat more restricted than in toxicology because of the tendency to differentiate chemical substances harmful through respiration, ingestion and contact, from those produced by a pathogen in close association with its host. In pathology "toxins" usually means substances produced by organisms, especially micro-organisms, that are injurious, or lethal, to macro-organisms (man, animals, plants) in very low concentrations. Hence, it is the suscept, a macro-organism instead of a micro-organism, that distinguishes toxins from

[1] The Latin word "*venenum*" originates from "*Venus*", the Goddess of Love, and is indirectly related to the idea of gift, offering (see the Latin verb "*venerari*"). Also the German word "*Gift*" = poison derives from "*geben*" = to give.

antibiotics. Substances that are toxic only in excessive doses, although they are excreted by micro-organisms, are not considered toxins.

The concepts and definitions of "toxin" given by various authors in different scientific fields differ widely. They have been influenced by the characteristics of the microbial toxins affecting man and animals, such as diffusibility and solubility, protein or protein-like nature, high toxicity, antigenicity[2] and specificity of the toxic effect. However, few or none of these characteristics are usable if a definition valid for all microbial toxins and for all pathogens that produce them, is to be devised. Moreover, the strict application of any definition of toxins prevents discussion of closely related substances which are proved to be physiologically active in diseases. There is, however, general agreement that toxins should have certain characteristics. Thus, they are usually regarded as diffusible and translocatable substances produced by living organisms, active in small amounts, and injurious to certain animals or plants. Mineral substances are consequently excluded.

Toxins are readily classified according to their origin. In toxicology, where only one suscept, Man, is considered, toxins are currently distinguished in three classes:

1. animal toxins, or zootoxins, produced by animals, such as the poison of snakes, fish, or arthropods
2. plant toxins or phytotoxins, produced by mushrooms or higher plants, such as alkaloids and glucosides
3. microbial toxins, produced by micro-organisms, including micromycetes, and further differentiated into "bacterial toxins", such as the diphtherial and botulin toxins, and "mycotoxins", such as the aflatoxins

In other applied sciences, toxins are not classified in this way, and the same term often refers to the organism injured by the toxin rather than to the organism producing it. This results in ambiguity. For example, toxins produced by *Mycota* (Micromycetes) are called "mycotoxins" in human and animal pathology, "phytotoxins" in plant pathology, and "antibiotics" in pharmacology. In another sense, the same toxin may affect entirely different classes of organisms, in different circumstances acting as a phytotoxin, a mycotoxin, or an antibiotic.

Of the two meanings given to the term "phytotoxin", namely plant-produced toxins, and toxins harmful to plants, the latter is the one generally used in plant pathology. This ambiguity would be eliminated if the classification of phytotoxins were established on the basis of the organisms producing them as given below.

[2] The strikingly antigenic properties of the exotoxins produced by pathogenic bacteria causing diseases in man and animals have led to the tacit agreement to consider them as antigenic poisons of microbial origin. "Soluble bacterial antigens" has been suggested as a more proper term for these substances (Oakley, 1970).

Phytotoxins (toxins injurious to plants):

1. microbial phytotoxins (from mycoplasma, bacteria, algae, fungi)
2. higher plant phytotoxins
3. animal phytotoxins (entomo-, acaro-, nemato- toxins)

This obviously could be confusing but unfortunately clarity in this matter is not easily achieved. An alternative solution might be to use the term "toxin" in plant pathology *sic et simpliciter* as many actually do and "define toxins in terms of effects they produce and not on the basis of their origin and nature" (Wood, 1967).

Classification and nomenclature of phytotoxins defined as substances toxic to higher plants have been mainly based on their specificity and mode of action. Most toxins are not specific in their action, that is, they harm many species belonging to different genera or families with no relation between the host range of the pathogen and the plants affected by the toxins it produces. A few toxins are species- or cultivar-specific. They only affect plants susceptible to the pathogen; resistant plants are not damaged.

Some toxins affect various kinds of plant tissue whereas others damage a single type of tissue. They may induce a particular syndrome when applied to plants in suitable concentrations. Wilting, for example, is a common effect of many phytotoxins which cause loss of cell turgor and leaf flaccidity, and eventually a general wilting of the whole shoot. Such toxins have been called "marasmins" or "wilting toxins". Lycomarasmin and fusaric acid were the first representatives of this class to be purified and identified. Wilting toxins are produced not only by vascular parasites such as *Fusarium oxysporum* Schl. f. sp. *lycopersici* (Sacc.) Sny. *et* Hans. but also by localized parasites that cause systemic effects, such as species of *Colletotrichum* and *Fusicoccum*. Toxins causing necrosis, the so-called "necrotoxins" are mostly produced by other types of localized parasites such as species of *Alternaria*, *Phytophthora* and *Pseudomonas*. In infected plants parenchyma cells are damaged or killed in advance of these pathogens. Necrotic spots are a typical symptom.

The categories given above were based on the effects that the phytotoxins produce on sensitive hosts. Plants, however, can show damage only in a limited number of ways so that a nomenclature based on symptoms is likely to be inadequate. A classification and nomenclature based on the relationship between chemical structure and toxicity will be preferable once sites and mode of action are defined at the histological and cellular level and in functional and molecular terms.

Toxins in Plant Diseases

Information on toxigenic diseases of plants caused by insects, especially aphids, has been extensively reviewed by Leach (1940), Carter (1962) and

Miles (1968). However, little is known about the substances involved in these diseases. Still less is known about the action of toxins on plants damaged by mites and nematodes or by parasitic phanerogams. More is known about substances produced by certain higher plants that are toxic to other plants. Juglone produced by walnut trees is an example of such toxins. The release of cyanide from decomposing roots of old peach trees is said to be responsible in part for the peach re-plant problem. Such compounds can be considered as phytotoxins which are comparable in function to the phytotoxic metabolites produced in soil by saprophytic micro-organisms.

Because of their nature, it is unlikely that viruses produce toxins. Toxic effects in virus-infected plants are undoubtedly a consequence of metabolic changes induced by viruses in plant cells, which are expressed in various ways such as necrosis or premature senescence of infected tissues. Ethylene, for instance, is produced in local lesions but not in systemic infections caused by TMV and CMV on tobacco (Balázs *et al.*, 1969).

Although there is, so far, little evidence, it seems likely that phytotoxins are involved in diseases of plants caused by mycoplasma.

Toxins are readily produced by many species of plant pathogenic bacteria and fungi in a wide variety of environments. When produced outside the plant, as for example in soil, they may cause disorders of plants or predispose them to attack by weak pathogens. When produced inside the plant, they may contribute to the development of a disease during both its determinative and expressive phases.

The idea that in infectious diseases of plants harmful effects could be caused by substances produced by the pathogen began to be formulated when the aetiology of many plant diseases was still uncertain, when the plant pathology of the Tulasne Brothers changed into the plant pathology of de Bary. As early as 1886, de Bary succeeded in reproducing soft rot by applying a sterile extract from rotten carrots to healthy carrot tissues. This and the results of other early physiological investigations of plant diseases were the beginning of the toxin concept in plant pathology. But for a long time the concept was applied only to diseases where although the pathogen is localized, the symptoms are widespread in the infected plant.

In 1954 Gäumann stated that "Micro-organisms are pathogenic only if they are toxigenic: in other words, the agents responsible for diseases can damage their hosts only if they form toxins—microbial poisons—that penetrate into the host tissue", a provocative statement at that time although, in fact, the term toxin was used by Gäumann for all substances produced by the pathogen including enzymes. This statement summarized the findings of a number of investigations starting with Brown's work on macerating enzymes, published in 1915 and then developed over 40 years to include the study of the toxins associated with vascular wilts, soft rots and other necrotic diseases.

The essential characteristic of a phytotoxin is that it acts directly upon the protoplasts of plant cells, damaging or killing them.

Clearly the effects of any toxic substance will depend upon its concentration. When a toxin is continuously produced in small amounts by a pathogen in an infected host, its effect may be first stimulatory, then inhibitory, and eventually lethal. Unstable products of the pathogen may, however, fail to accumulate in plant tissues in concentrations high enough to cause serious damage. Also, the effect on the host may depend on where the toxin is produced. Thus, a toxin distributed via the vascular stream may cause wilting or general blighting of leaves and shoots, but a toxin produced in parenchyma will usually cause local lesions and chlorotic or necrotic spots.

In vascular diseases, toxins can move easily along the conducting elements provided the flow in the elements is not materially decreased by vascular dysfunction. If this does occur, then the movement of toxins to leaves and other organs is reduced and wilt is caused more by cell wall occlusions than by the action of the toxins. These, however, may move freely through the vascular system when the parasite invades adjacent tissues (bark or wood) rather than the functioning xylem. Then the toxic metabolites may rapidly accumulate in leaves and induce a sudden and severe wilt.

In infections of parenchyma such as occur in leaf spots and cankers on stems, toxins move mainly through middle lamellae and cell wall routes obviously more slowly than through the vascular elements. Chlorosis, necrosis, and death of the affected tissues follow the injury of cells which usually occurs in advance of the pathogen.

In hypersensitive hosts, the effects of toxins may eventually benefit the plant if they lead to reactions around the infection court that prevent further growth of the pathogen. Thus in apple scab, toxins produced by the pathogen during the early phases of infection are thought to trigger a chain of metabolic changes in the host tissue that lead to resistance.

The role of toxins produced by the pathogen in hypersensitivity and resistance is likely to be studied increasingly in the future.

Toxins as Chemical Determinants of Plant Diseases

Most research on this subject has been done with substances produced by parasites or saprophytes in various environments and known to be injurious to plants.

The presence of toxic substances in cell-free culture filtrates of many plant pathogens is often easily demonstrated. Unfortunately, it is unusual for a toxin to be readily produced by a pathogen both *in vivo* and *in vitro.* Also, the amount of a toxin in culture filtrates may be too low for biological assay, unless the filtrates are concentrated. Furthermore, examples are known in which toxicity depends on the simultaneous action of two or more synergistic components. Within a plant the concentration of a toxin will be the result of deactivation processes and the production, movement and accumulation of the metabolite. Because of these and other conditions it is usually difficult to show that these substances are present in the infected plant in an active form and in concentrations high enough to be responsible for the symptoms.

Dimond and Waggoner (1953) discriminated between "toxins" and "vivotoxins". A toxin was defined as "any compound produced by a micro-organism which is toxic to plants", and a vivotoxin as "a substance produced in the infected host by the pathogen and/or its host, which functions in the production of disease, but is not itself the initial inciting agent of disease".

Vivotoxins were considered as toxic compounds producing a portion of the symptoms of disease in naturally infected plants. The definition of vivotoxin was probably devised to include the different substances of plant or pathogen origin which were thought to be responsible for symptoms in vascular wilts. This definition has undoubtedly influenced the current broad concept of toxin in plant pathology. If we omit "and/or its host" from the definition the class of active substances which results is homogeneous, for they are all chemopathogens of microbial origin.

Since a vivotoxin is a disease-producing entity and therefore a pathogenic agent, it has to satisfy rules similar to Koch's postulates for establishing the pathogenicity of a micro-organism. Three criteria were proposed by Dimond and Waggoner (1953) as minimal to establish vivotoxicity, namely: (1) reproducible separation from the sick plant; (2) purification; and (3) reproduction of at least a portion of the syndrome when the toxin is placed in a healthy plant.

Later, Dimond (1955) slightly modified these criteria stating that before a substance can be regarded as a vivotoxin, it must be: (1) isolated from the diseased plant but not be present in the healthy host; (2) characterized chemically; (3) introduced in pure form into a healthy host, it must produce the symptoms of disease or a portion of the syndrome.

The concept of vivotoxin has served well to clarify the ideas of plant pathologists about toxins and plant diseases. Since its introduction, it has come to be widely used in the literature. But the concept, the definitions and the criteria have not been wholly accepted.

Ludwig (1960) considered the concept too restrictive, because it excludes phytotoxic metabolites produced by a plant pathogen outside the plant, for example, in soil. These can predispose plants to infection by a pathogen. Wood (1967) has observed that if such metabolites are produced in close association with the host and act by damaging plant tissues, they can appropriately be regarded as vivotoxins.

Other substantial points have been outlined by Deverall (1964) and Wood (1967). In a diseased plant, toxins must not only be associated with symptoms, but also be present in concentrations sufficient to cause the symptoms. Absence of a toxin does not necessarily mean that they have not been present because degradation could have occurred in the plant tissues or during extraction. Also, a toxin could be continuously produced by the pathogen in very low amounts. The toxic effect on plant tissue of this slow poisoning may not reflect the concentration of toxin in the tissue at a particular time; this, in fact, may be misleadingly low. Because plants can show damage only in a limited number of ways, reproduction of symptoms of a disease by introducing a purified toxin into a healthy plant is not necessarily sufficient evidence to implicate it in the disease. Physiological and biochemical changes induced by the toxin must also be considered in this context.

Braun and Pringle (1959) expressed the view that Dimond and Waggoner's criteria for vivotoxins are valid but would be impractical for certain types of disease.

Wheeler and Luke (1963) did not consider it necessary for a toxin to be isolated from a diseased plant and identified before its involvement in disease can be accepted. Thus, it may be difficult or impossible to isolate a highly potent, unstable toxin present only in traces in an infected plant, and chemical characterization of a toxin that has been purified may also be a slow and difficult process.

Nevertheless it is worth emphasizing at this point the importance of chemical identification in showing whether the toxin is a plant or a microbial product. Unfortunately, serological methods are, or will be, of little or no value for most of the toxins involved in plant diseases. Difficulties similar to those mentioned above have long been encountered by microbiologists aiming to satisfy Koch's postulates for viruses, rickettsiae, mycoplasma and certain still unidentified causes of disease.

Wheeler and Luke considered the type of disease in which all symptoms are caused by a single toxin produced by the pathogen involved in that disease. This is the case with victorin, the toxin of *Helminthosporium victoriae* Meeh. *et* Murphy as well as a few other host-specific toxins. Wheeler and Luke proposed the term "pathotoxin" for such toxins and listed four criteria that must be satisfied before a pathotoxin can be implicated in this type of disease: (a) the toxin, applied at concentrations which could be reasonably expected in or around the diseased plant, produces in a susceptible host all the symptoms characteristic of the disease; (b) the pathogen and the toxin exhibit similar suscept specificity; (c) the ability of the pathogen to produce the toxin varies directly with its ability to cause disease; (d) a single toxin is involved.

These criteria are, however, unlikely to be satisfied in most diseases in which toxins are likely to be implicated. Wheeler and Luke were aware of this because they also listed a series of modifications of the criteria. These can be applied to diseases other than those in which a pathogen produces one host-specific toxin that causes all symptoms of the disease. These "modifications" are somewhat confusing because selective toxins producing only a part of the symptoms, and non-host-specific toxins produced by highly specialized pathogens may also be considered pathotoxins. This compelled Goodman *et al.* (1967) to call a toxin such as colletotin a "semipathotoxin".

Restriction of vivotoxins to "phytotoxic compounds isolated from diseased plants and identified chemically", as proposed by Wheeler and Luke misrepresents the meaning given by Dimond and Waggoner to vivotoxins.

Wheeler and Luke also used the term "phytotoxins" for "all products of living organisms toxic to plants" without reference to their role in plant diseases. This definition is substantially that given by Dimond and Waggoner for "toxin" as "any compound produced by a micro-organism which is toxic to plants".

In 1964, Pringle and Scheffer commented on the confusion in the literature on the terminology applied to toxins. They stressed the advantages of considering toxins as metabolic products of plant pathogens in distinguishing them from other naturally occurring or synthetic compounds which damage plants.

They emphasized that a significant role for toxins in plant diseases was established only with the discovery of the "host-specific toxins" defining such a toxin as "a metabolic product of a pathogenic micro-organism which is toxic only to the host of that pathogen".

The known host-specific toxins produce all the symptoms of the diseases caused by the pathogens so that such toxins can be used as a substitute for pathogens in biochemical and physiological studies of diseases.

In 1967, Scheffer and Pringle divided the substances produced by the pathogen and involved in diseases into two classes: (a) primary determinants essential for pathogenicity; (b) secondary determinants not required for pathogenicity. Primary determinants are largely the "host-specific plant toxins", but may also be non-specific toxins. Loss of ability to produce these toxins leads to loss of pathogenicity. Secondary determinants are substances that contribute to virulence but do not control pathogenicity.

Classification of Substances Involved in Plant Diseases

Goodman *et al.* (1967) have pointed out that there are doubts as to the appropriateness of classifying toxins solely on their capacity to evoke disease symptoms in specific host plants.

In my opinion the main reason for the present confusion in the nomenclature of chemical factors involved in plant diseases arises from the inclusion in one class of metabolites of the pathogens, substances produced by the host, and substances formed in diseased tissues, merely because they may be associated with the production of symptoms.

Buddenhagen and Kelman (1964), Goodman *et al.* (1967), Wood (1967) and other authors have already recognized the need to distinguish between enzymes destroying the integrity of plant tissues and microbial metabolites that alter or disrupt cell metabolism.

The scheme I now submit for comment and criticism (Table I) is an attempt to discriminate between and classify the various substances that are known or likely to be involved in plant diseases.

The term "toxin" should be used only in its original sense, for microbial metabolites excreted (exotoxins) or released by lysis of cells (endotoxins), which in low concentrations are directly toxic to cells of suscepts.

The term "phytotoxin" should be retained, *sensu stricto*, for microbial metabolites that interfere with or alter the metabolism of plant cells and so injure the plant symplast. Phytotoxins are produced either in the host, on its surface or even outside the host for example in soil.

The question may arise whether or not phytotoxins include enzymes. Some authors (Owens, 1969) consider phytotoxins as non-enzymic. I am inclined not to exclude enzymes from phytotoxins if they are proved to possess inherent primary toxicity.

"Vivotoxin" should be confined to phytotoxins whose role in the pathogenesis has been demonstrated. The definition given by Dimond and Waggoner (1953) should thus be amended to include another criterion, "production by the casual organism(s) of the disease".

Host-specific plant toxins as defined by Pringle and Scheffer (1964) are vivotoxins with host-selective toxicity.

"Pathotoxin" as a substitute for vivotoxin should be rejected for priority reasons. However, this term might be used for host-specific vivotoxins fulfilling the four criteria given by Wheeler and Luke (1963), i.e. necessarily linked to virulence of the pathogen and causing all disease symptoms.

For microbial products which are not toxins as defined above but related entities having a function in pathogenesis and symptom expression, the term "phytoaggressins" is proposed.

TABLE I. Substances involved in plant diseases

Main classes		Subclasses according to:	
Pathogen-derived substances	Phytotoxins (and vivotoxins)	chemical nature	(low mocular and high molecular weight compounds; inorganic compounds; organic compounds: aliphatic, aromatic, alicyclic, heterocyclic; ...)
		suscept specificity	(host-specific; non-host-specific; taxaspecific; ...)
		substrate specificity	(active on meristems, parenchyma, leaves, stems, roots, flowers, ...)
		virulence of the pathogen	[primary determinants (pathotoxins); secondary determinants]
		mode of action	(antimetabolites; enzyme inhibitors; chelating agents; pseudoplasmolysing agents; ...)
		host reaction	[chlorotic toxins; necrotic toxins (necrotoxins); wilting toxins (marasmins); ...]
	Phytoaggressins (and vivoaggressins)	toxicity and chemical nature	[compounds toxic in excessive concentrations (alcohols, organic acids, ...); non-toxic compounds (pigments, polysaccharides, polypeptides, ...); ...]
		mode of action	[enzymes (cutinases, macerating enzymes, phenoloxidases, lignin-degrading enzymes, ...); growth regulators (growth promoters; growth inhibitors); metabolic determinants (nucleic acids; tumour inducing principle; ...); ...]

Plant-derived substances	Products of tissue alteration or degradation by the pathogen	toxicity	(toxic and non-toxic compounds)
		chemical nature and origin	(byproducts of cellulose, pectic substances, lignin, ...)
		mode of action	...
	Products of plant defence processes	toxicity	(toxic and non-toxic compounds)
		chemical nature	(phenolic compounds, gums, resins, mucilages, pigments, ...)
		mode of action	(enzyme inhibitors; antimetabolites; phytoalexins; growth regulators; ...)
	Products of cell or tissue degeneration and death	toxicity	(toxic and non-toxic compounds)
		chemical nature	(ethylene, ammonia, amines, organic acids, melanoid substances, ...)
		mode of action	...

The high molecular weight metabolites of microbial origin such as polysaccharides and related compounds forming occlusions in cell walls or plugs in conducting elements with harmful effects on the water economy of plants are not toxic *per se* and, consequently are not toxins. Degradative enzymes such as the cellulases and pectolytic enzymes are not directly toxic to cells. In fact, they are used to produce wall-free, living protoplasts in hypertonic solutions. These and other enzymes such as polyphenoloxidases primarily affect the plant apoplast. However, certain phytoaggressins may alter metabolism without killing the cells. Examples are growth regulating substances and other metabolic determinants such as nucleic acids, and the "tumour inducing principle" of *Agrobacterium tumefaciens* (Sm. et Town.) Conn. Other microbial metabolites such as certain organic acids may accumulate in host tissues sufficiently to damage or kill cells but the large amounts needed to cause damage do not allow us to call these substances toxins which I conceive as compounds harmful in low concentrations. They are better regarded as phytoaggressins.

"Aggressin" as used in mammal microbiology (Bonventre, 1970) relates to microbial metabolites which enhance virulence of the pathogen but are not directly toxic to the host. Essentially, aggressins promote the invasion of the host by parasites by overcoming host barriers or depressing defence reactions. The term aggressin was used by Scott (1931) for microbe-associated substances which do not qualify as toxins. Van Heyningen (1951, 1970) included these substances among toxins and Miles (1955) called them "auxiliary virulence factors".

To implicate phytoaggressins in plant diseases I suggest criteria similar to those proposed for vivotoxins. Phytoaggressins satisfying these criteria would be called "vivoaggressins".

Three other classes of substances involved in plant diseases should now be tentatively classified.

The first class includes substances that accumulate in infected plants as a result of the action of enzymes or other entities produced by the pathogen on host constituents. These products may affect the symplast, the apoplast, or both. Examples are the products of cell wall degradation which form plugs in xylem elements or occlude cell wall micropores. Certain phenolic compounds, liberated by the pathogen from host tissues and then oxidized or polymerized, may also be placed in this class.

The second class includes substances produced by the plant in response to infection. These usually appear in tissues surrounding the infected area and include macromolecular compounds such as gums, gels and resins, phenolic compounds liberated by hydrolysis of glycosides, and substances arising from the altered metabolism of plant tissues such as the phytoalexins. Products of defense reactions such as phloridzin derivatives in apple leaves or ipomeamarone in sweet potato roots are well known examples. These substances may contribute to hypersensitivity phenomena.

The third class includes substances that often accumulate in the late stages of diseases and which are mainly products liberated during or following death of plant cells both in invaded and adjacent tissues. These compounds are comparable to those originated in plants by senescence, wounds or by abiotic agents.

In conclusion, definitions and classification of phenomena are not academic exercises in science. Good terminology requires clarity of thought and this contributes to progress in our knowledge of the phenomena we attempt to define.

As Northrop (1961) has said, biologists use the language of semantics and are often satisfied with descriptive and qualitative theories; chemists use the language of mathematics based on quantitative data.

In our study of the role of toxins in plant diseases, our approach is still largely in the semantic, qualitative phase. With the help of our colleagues who are chemists we must now aim at defining and resolving our problems in more quantitative terms. The Proceedings of this NATO Advanced Study Institute will, undoubtedly, catalyse interactions between the methods and viewpoints of the biologist and the chemist that are needed before the significance of toxins in disease can be properly assessed.

References

Balázs, E., Gáborjányi, R., Tóth, A. and Király, Z. (1969). *Acta Phytopath. Acad. Sci. Hung.* **4**, 355–358.

Bonventre, P. F. (1970). *In* "Microbial Toxins", Vol. I, "Bacterial Protein Toxins" (S. J. Ajl, S. Kadis and T. C. Montie, Eds) Chapter 2, pp. 29–66. Academic Press, New York and London.

Braun, A. C. and Pringle, R. B. (1959). *In* "Plant Pathology. Problems and Progress 1908–1958" (C. S. Holton, G. W. Fisher, R. W. Fulton, H. Hart and S. E. A. McCallan, Eds), Chapter 9, pp. 88–99. University of Wisconsin Press, Madison.

Brown, W., 1915. *Ann. Bot.* **29**, 313–348.

Buddenhagen, I. and Kelman, A. (1964). *A. Rev. Phytopath.* **2**, 203–230.

Carter, W. (1962). "Insects in relation to plant disease." Interscience Publishers, New York and London, 705 pp.

Deverall, B. J. (1964). *In* "Microbial behaviour *in vivo* and *in vitro*", 14th Symposium Soc. General Microbiology. Cambridge University Press, pp. 165–186.

Dimond, A. E. (1955). *A. Rev. Pl. Physiol.* **6**, 329–350.

Dimond, A. E. and Waggoner, P. E. (1953). *Phytopathology* **43**, 229–235.

Gäumann, E. (1954). *Endeavour* **13**, 198–204.

Goodman, R. N., Király, Z. and Zaitlin, M. (1967). "The Biochemistry and

Physiology of Infectious Plant Disease." Van Nostrand Co. Inc., Princeton, 354 pp.

Leach, J. G. (1940). "Insect transmission of plant diseases." McGraw-Hill Book Co., Inc., New York and London, 615 pp.

Ludwig, R. A. (1960). *In* "Plant Pathology. An Advanced Treatise" (J. G. Horsfall and A. E. Dimond, Eds), Vol. 2, "The Pathogen", Chapter 9, pp. 315–357. Academic Press, New York and London.

Miles, A. A. (1955). *In* "Mechanisms of microbial pathogenicity", 5th Symp. Soc. Gen. Microbiology. Cambridge University Press, pp. 1–16.

Miles, P. W. (1968). *A. Rev. Phytopath.* **6**, 137–164.

Northrop, J. H. (1961). *A. Rev. Biochem.* **30**, 1–10.

Oakley, C. L. (1970). *In* "Microbial Toxins", Vol. I, "Bacterial Protein Toxins" (S. J. Ajl, S. Kadis, and T. C. Montie, Eds), Chapter 10, pp. 355–387. Academic Press, New York and London.

Owens, L. D. (1969). *Science, N.Y.* **165**, 18–25.

Pringle, R. B. and Scheffer, R. P. (1964). *A. Rev. Phytopath.* **2**, 133–156.

Scheffer, R. P. and Pringle, R. B. (1967). *In* "The Dynamic Role of Molecular Constituents in Plant-Parasite Interaction" (C. J. Mirocha and I. Uritani, Eds), The United States–Japan Cooperative Science Program. The Amer. Phytopath. Soc., Inc., St. Paul, Minnesota, pp. 217–236.

Scott, J. P. (1931). *J. Bact.* **22**, 323–337.

van Heyningen, W. E. (1951). "Bacterial toxins." Blackwell, Oxford, 1033 pp.

van Heyningen, W. E. (1970). *In* "Microbial Toxins", Vol. I, "Bacterial Protein Toxins" (S. J. Ajl, S. Kadis, and T. C. Montie, Eds), Chapter 1, 1–28. Academic Press, New York and London.

Wheeler, H. and Luke, H. H. (1963). *A. Rev. Phytopath.* **17**, 223–242.

Wood, R. K. S. (1967). "Physiological Plant Pathology." Blackwell Scientific Publications, Oxford and Edinburgh, 570 pp.

Additional Selected References on Toxins and Phytotoxins

Agrios, G. N. (1969). "Plant Pathology." Academic Press, New York and London, 629 pp.

Beckman, C. H. (1964). *A. Rev. Phytopath.* **2**, 231–252.

Bollard, E. G. and Mattews, R. E. F. (1966). *In* "Plant Physiology. A treatise", Vol. 4B, "Metabolism: Intermediary Metabolism and Pathology" (F. C. Steward, Ed.), Chapter 8, pp. 417–477. Academic Press, New York and London.

Braun, R. (1963). *In* "Moderne Methoden der Pflanzenanalyse" (H. F. Linskens and M. V. Tracey, Eds), Vol. 6, pp. 219–243. Springer-Verlag, Berlin.

Brian, P. W. (1955). *In* "Mechanisms of microbial pathogenicity", 5th Symp. Soc. Gen. Microbiology. Cambridge University Press, pp. 294–319.

Dimond, A. E. (1967). *In* "The Dynamic Role of Molecular Constituents in Plant–Parasite Interaction" (C. J. Mirocha and I. Uritani, Eds). The United States–Japan Cooperative Science Program. The Amer. Phytopath. Soc., Inc., St. Paul, Minnesota, pp. 100–120.

Dubos, R. J. (1954). "Biochemical Determinants of Microbial Diseases." Harvard University Press, Cambridge, Massachusetts, 152 pp.
Gäumann, E. (1950). "Principles of Plant Infection. A text-book of general plant pathology for biologists, agriculturists, foresters and plant breeders." Crosby Lockwood & Son, Ltd., London, 543 pp.
Gäumann, E. (1951). *Adv. Enzymol.* **11**, 401–437. Interscience Publications, New York.
Gäumann, E. (1951). *Experientia* **6**, 441–447.
Graniti, A. and Matta, A. (1969). *Ann. Phytopath.* **1**, n. hors-série, 77–120.
Grümmer, G. (1961). *In* "Mechanisms in biological competition", *Symp. Soc. Experim. Biology*, **15**, 219–228. Cambridge, University Press.
Oku, H. (1967). *In* "The Dynamic Role of Molecular Constituents in Plant–Parasite Interaction" (C. J. Mirocha and I. Uritani, Eds), The United States–Japan Cooperative Science Program. The Amer. Phytopath. Soc., Inc., St. Paul, Minnesota. 237–255.
Orsenigo, M. (1960). *Annal. Microbiol.* **10**, 146–173.
Rubin, B. A. and Artsikhovskaya, Y. V. (1963). "Biochemistry and Physiology of Plant Immunity." Pergamon Press, Oxford etc., 358 pp.
Standfast, A. F. B. (1964). *In* "Microbial behaviour, *in vivo* and *in vitro*", 14th Symp. Soc. Gen. Microbiology. Cambridge University Press, pp. 64–164.
Strobel, G. A. and Mathre, D. E. (1970). "Outlines of Plant Pathology." Van Nostrand Reinhold Co., New York etc., 405 pp.
van Heyningen, W. E. (1955). *In* "Mechanisms of microbial pathogenicity", 5th Symp. Soc. Gen. Microbiology. Cambridge University Press, pp. 17–39.
Wright, D. E. (1968). *A. Rev. Microbiol.* **22**, 269–282.

Discussion

Scheffer: Several years ago I suggested that pathogen-produced substances may be thought of as primary or secondary determinants of disease. I would now amend and possibly simplify this by using instead, primary or secondary determinants of pathogenicity. Disease development is complicated and I prefer to limit myself to a study of the ability to induce disease. The difference is one of emphasis. I suggest this not as a basis for classification, but simply to clarify our thinking on this subject.

Goodman: I disagree with Scheffer's concepts in that I consider pathogenicity to be an inherent and stable condition of a given genus and species. A pathogen is always a pathogen. On the other hand, the variable frequently referred to as pathogenicity is in reality *virulence* and this is only one of the many thousands of characteristics that are gene controlled and which may be lost or modified.

Scheffer: Goodman and I appear to be following different definitions of pathogenicity. I follow the usual concept used in plant pathology, as is stated in plant pathology textbooks.

Dimond: Is it worthwhile to attempt to define the word toxin? As our biochemical and biophysical knowledge becomes more complete the need for the term may disappear. Our interest lies in knowing what the fungus and the plant in response to the fungus produce that causes symptoms of disease to develop.

Durbin: Would you exclude compounds that occur in the healthy plant?

Dimond: No. For example indole acetic acid occurs naturally, yet plays an important role in pathogenesis. The work of Matta has shown how polyphenol oxidase becomes activated in *Fusarium* wilt of tomato. It, too, is a normal component of the healthy plant. Matta has shown how its activity and the phenols contribute to the higher levels of indole acetic acid found in diseased plants and cause symptoms of disease.

Kuć: When we know more about the mode of action and chemical structure of phytotoxins we can place them into appropriate chemical or physiological classes. The most descriptive name is the chemical one while we argue about the definitions of "toxins". I remain confused by the term "pathogen" because I have apparently worked mainly with fungi that do cause symptoms (disease) on plants but which are usually classified as non-pathogens!

Page: Would it be appropriate at this time to mention the term *mycotoxin*? There are a number of examples in the literature where a substance classified as a mycotoxin also causes plant disease. It would thus appear that the prefix *myco-* should not only imply a capacity to cause disease in animals but should have wider implications. Comments would be appreciated.

Mirocha: The use of the term *mycotoxin* is useful and exact in the sense that it relates to a group of specific diseases in animals caused by the ingestion of food contaminated by fungi and their products. It is distinct from a *phytotoxin* involved in a plant disease and communicates this difference well. As Page has pointed out, some mycotoxins like diacetoxyscerpenol and aflatoxin have been shown to be toxic to plants. But this does not mean that they are involved in a plant disease nor would it detract from their importance as mycotoxins in diseases of animals if they were.

Graniti: The term mycotoxin has been extensively used for metabolites injurious to man and animals. It has now become ordinary terminology in medicine and pharmacy in this limited sense. Widening it to the field of plants would add confusion to that already existing in the terminology of toxins. It would thus be wiser to consider it as a term "conservandum" with its present meaning. However, the fact remains that, just as there are known examples of mycotoxins active towards plants, there are also phytotoxins injurious to man and animals.

Strobel: I should like us to think about, and comment on the nomenclature that should be applied to specific groups of toxins that have and will be isolated from plant parasites. What guide lines should be followed in giving trivial names to toxins?

Wood: Surely we must be guided by the chemists before we introduce and use trivial names for substances. They must be the arbiters in the naming of all chemical substances.

Mirocha: Let me emphasize what has been said by Strobel so that we may all understand his proposition. Strobel accepts the broad classification of phytotoxins or mycotoxins but within it he would like to see a general chemical classification of the toxins. In answer to Durbin's objection to the trivial names presently being used by some investigators, e.g. T-2 or HT-2 toxins, most of the examples he quoted could be placed in the chemical group called trichothecenes; this would accord with Strobel's suggestion.

Gottlieb: Certainly classification of toxins is important, but this should be based on known properties or structures and not on all-defined interactions about which very little is known. I would favour a classification on broad, basic structural, groupings. Further separation of compounds would depend on the variations in the substituents of the basic chemical groups. Generally, compounds with similar basic structures tend to react on cells in the same general manner but with differences that depend on minor structural variations. The name of each class of compound should be that of the common central chemical grouping. Only a knowledge of the presence of this grouping is needed to place a new toxin in the proper group. Later studies will define it more accurately within the group.

Smedegaard-Petersen: Many people are concerned about the confusion in naming toxins. For example *mycotoxin* refers to toxins that affect animal but its name derives from its origin, whereas *phytotoxin* differs in that it relates to the organism affected.

Would it not be better to classify toxins according to the organisms producing them, for example (1) mycotoxins, toxins produced by fungi, (2) bacterial toxins, toxins produced by bacteria, (3) virus toxins, toxins produced by viruses and so on. These broad classes could then be subdivided by using other criteria.

Mantle: So far as the use of the terms *phytotoxin* and *mycotoxin* is concerned I think that the terms of reference of this Institute preclude it from legislating for those who work in other fields for example, on fungal toxins as animal poisons, and who have their own understanding of the term *mycotoxin*.

Page: I would like to caution against the use of species names in the naming of toxin compounds. For example, in the genus *Fusarium* I am aware of a compound which has been isolated from three species and chemically characterized. Later it was decided that the three species were the same. Also, where the taxonomy of a fungal genus is in a state of flux, use of a species name can lead to confusion.

Wood: For me and in the context of this discussion a pathogen is a micro-organism that damages another plant. Many mechanisms are involved in pathogenicity so we should be careful in assuming that pathogenesis is other than a complex sequence of events in most if not all diseases. In particular I should like to warn against the argument that runs: organism A produces system X which could produce symptoms that are observed and A is pathogenic; organism B produces system X, but is not pathogenic; therefore, system X is not involved in pathogenesis. For obvious reasons this could be an erroneous argument though it is one that has been put forward by some plant pathologists.

2. Bacterial Phytotoxins: a Survey of Occurrence, Mode of Action and Composition

RICHARD D. DURBIN

Pioneering Research Laboratory, Crops Research Division, Agricultural Research Service, U.S. Dept. of Agriculture and Department of Plant Pathology, University of Wisconsin, Madison, Wisconsin, 53706, *U.S.A.*

Introduction

The concept that phytotoxins play a role in the development of diseases incited by bacteria has been widely accepted. Indeed, selected species within all genera of phytopathogenic bacteria have been reported to produce phytotoxins considered important in pathogenesis. In general, however, unequivocal evidence for such a causative role for these compounds is still lacking.

In the context of this presentation a "bacterial phytotoxin" will be defined in rather broad terms as any non-enzymatic compound synthesized by a phytopathogenic bacterium that at physiological concentrations adversely affects a plant. As long as a compound meets these requirements, the fact that other organisms, e.g., higher plants, can also synthesize the same compound is not sufficient reason for not considering it a phytotoxin. Because of bacterial synthesis it may be present in diseased tissue in such vastly greater concentrations than normal that the imbalance thus created contributes to the syndrome. From this viewpoint such compounds are as important for our understanding of disease development as are those whose synthesis is restricted to the pathogen.

Within such a broad definition several distinctions should be made: first, between phytotoxins produced by bacteria only in culture and phytotoxins produced in the diseased plant; second, this latter group should include only compounds produced by the bacterium which, *per se*, are phytotoxic and exclude those which are not directly phytotoxic but which secondarily cause a toxic response, e.g., inducers of oxidase synthesis by the host.

Dimond and Waggoner (1953) have presented criteria designed to distinguish toxins produced both in diseased plants and *in vitro* from those which are only produced *in vitro*. In essence the criteria involve: (i) isolation of the compound from diseased but not healthy plants; (ii) chemical purification; and (iii) reproduction of a portion of the syndrome by introduction of the compound into a healthy plant. Compounds satisfying conditions (i) and (iii) would be called "vivotoxins".

These criteria were subsequently redefined by Wheeler and Luke (1963), who after reducing the toxin hypothesis to its most elemental form—a form which they admit would probably hold for only a small number of cases—presented the following requirements for establishing that a toxin plays a causal role in disease: (i) a single toxin is involved; (ii) application of the toxin at physiological concentrations reproduces the characteristic disease symptoms; (iii) the pathogen and toxin exhibit similar host specificity; and (iv) toxin production is directly correlated with virulence. Compounds satisfying these requirements would be called "pathotoxins".

Effects of Bacterial Phytotoxins on Plants

If we examine bacterial phytotoxins on the basis of the kinds of gross symptoms they elicit, we see that the plant responds in only a few ways, mainly by the development of chlorosis, growth abnormalities, necrosis, water soaking or wilting. One response may grade into another, e.g., water soaking or chlorosis to necrosis. Also, very commonly the response will depend upon phytotoxin concentration, environmental conditions and/or physiological state of the host. These responses by the plant are not only limited in kind but are also non-specific in that different agents acting in different ways can induce the same response. This is both an advantage, in that it allows one wider latitude in selecting a bioassay, and a disadvantage, in that several substances in a mixture being purified may affect the bioassay. Furthermore, as will be discussed later, the fact that different compounds can elicit the same response demands that proof of complicity in disease development does not rest solely on the identification of the phytotoxin from an *in vitro* source.

Mode of Action and Composition of Bacterial Phytotoxins

This section will briefly consider the mode of action and composition of the phytotoxins synthesized by plant pathogenic bacteria. For convenience of presentation the various phytotoxins are grouped under one of the above five headings that describe the primary response of the suscept to the phytotoxin.

Chlorosis

Pseudomonas coronafaciens (*Ell.*) *Stev. and* P. tabaci (*Wolf* et *Fost.*) *Stev.*

Pseudomonas tabaci produces a phytotoxin called "wildfire toxin", the trivial name being derived from a description of the disease on tobacco. Recent evidence strongly suggests that *P. coronafaciens*, which attacks oats, and an undescribed pseudomonad attacking timothy (*Phleum pratense* L.), also synthesize the same compound (Sinden and Durbin, 1970; Taylor *et al.*, 1970). Braun (1955) proposed that the toxin interferes with methionine utilization because: (i) there was thought to be a structural resemblance between it and L-methionine; (ii) inhibition of algal growth by the toxin was reversed by L-methionine; (iii) methionine sulfoximine (MSO), a reported antagonist of methionine, produced chlorotic lesions on tobacco similar to those induced by the toxin; (iv) MSO toxicity to algae was also reversed by L-methionine; and (v) algal mutants resistant to the toxin were also resistant to MSO. However, the failure of L-methionine to prevent chlorosis induced by either the phytotoxin or MSO, as well as several other objections (Owens, 1969), have not been satisfactorily explained. In addition, more recent evidence indicates that MSO affects glutamate rather than methionine metabolism, and that the reason why methionine protects against MSO in intact organisms is that there is competition for passage of these metabolites across membranes rather than for metabolic sites within the cell (Lamar and Sellinger, 1965).

Lovrekovich *et al.* (1964) suggested that the wildfire toxin may directly affect RNA metabolism. In diseased leaves the concentrations of RNA and soluble proteins decreased concomitantly with increases in free amino acids and ammonia, an effect counteracted by addition of kinetin (Lovrekovich and Farkas, 1963). The toxin does not seem to affect protein synthesis directly. Vanderhoef and Durbin (1970) have shown that *in vitro* a protein-synthesizing system using mung bean embryo axes as a source of active polyribosomes and enzymes was not inhibited by purified wildfire toxin (*c.* 1·25 mM), whereas cycloheximide (0·25 mM) and puromycin (0·1 mM) inhibited amino acid incorporation.

The wildfire toxin may act by inhibiting glutamine synthetase, chlorosis being due to either: (i) the accumulation of toxic amounts of ammonia, and/or (ii) a depletion of various critical molecules the synthesis of which requires glutamine, e.g., adenylic and cytidylic acids, histidine and carbamyl-phosphate (Sinden and Durbin, 1968). At present the experimental evidence does not allow us to distinguish between these two proposed modes of action and the possibility that inhibition of some other enzyme system is also involved. However, the fact that leaves can be protected against chlorosis if glutamine is applied at the same time as the toxin, suggests that the site of action does involve glutamate metabolism.

Woolley *et al.* (1952, 1955) proposed that the wildfire toxin is composed of the amino acid tabtoxinine (α,ϵ-diamino-β-hydroxypimelic acid) linked to lactic acid by a lactone ring. Stewart (1961) however, synthesized the four possible DL-diastereoisomers of tabtoxinine, and found that none matched the properties of the compound derived from the wildfire toxin. Woolley (1959) later indicated that tabtoxinine may be branched. More recently, working with toxin preparations from both *P. coronafaciens* and *P. tabaci*, Sinden and Durbin (1970) failed to detect lactic acid, but did find that threonine and serine were present in molar ratios of 1 and 0·1, respectively, to 1 of tabtoxinine. Similar results were found by Taylor *et al.* (1970). From these data it has been suggested that the bacterium synthesizes two closely related compounds, one containing serine and the other threonine. Whether only one or both are phytotoxic is unknown at present.

Pseudomonas phaseolicola (*Burkh.*) *Dows. and* P. glycinea *Coerp.* Preliminary evidence, based upon isolation characteristics and chromatographic mobilities, indicates that both species produce the same toxin in culture and in infected plants (Hoitink and Sinden, 1970).

Rudolph (1969) partially purified a phytotoxin from culture filtrates of *P. phaseolicola* which is predominantly composed of hexose units; hydrolysis (3% HNO_3 for 2 h at 100°C) yielded glucose, rhamnose and fructose in molar ratios of 11:1:1. About 2% of the toxin is nitrogen, probably as amino groups. The toxin was estimated to have a molecular weight of 2100. It induces chlorosis, causes accumulation of ornithine and inhibits the growth of beet leaves.

Patil (1970a) has also partially purified a phytotoxin from filtrates of cultures of *P. phaseolicola*. However, there are several discrepancies between his results and those of Rudolph. Patil (1970b) finds that his purified material is ninhydrin-positive, and on acid hydrolysis yields four components two of which correspond to known amino acids. Results from molecular sieving experiments using Sephadex G-10 indicate that the molecular weight of the toxin is much smaller than that reported by Rudolph.

Bajaj *et al.* (1969) reported that in chloroplasts of naturally infected bean leaves or of leaves treated with toxin-containing culture filtrates the grana were disrupted and large starch grains accumulated. Other cellular organelles appeared normal. However, Lesemann and Rudolph (1969) reported that the toxin indirectly induces a helix-like aggregation of the ribosomes, and that other organelles are only slightly affected if at all.

Acting on the earlier evidence that there is an accumulation of ornithine in chlorotic tissues (Patel and Walker, 1963), Patil (1969) has shown that the phytotoxin inhibits ornithine carbamyltransferase. Chlorosis could be prevented by repeated applications of citrulline, the product of the reaction, to toxin-treated tissue (Patil, 1970b).

Rhizobium japonicum (*Kirch.*) *Buch.* Some strains of this root nodule bacterium synthesize a phytotoxin, designated rhizobitoxine, which induces chlorosis in certain soybean varieties (Owens, 1969). The phytotoxin has been isolated from culture media, and from chlorotic leaves as well as from the nodules of diseased plants.

Using nutritional mutants of *Salmonella typhimurium* (Loeffl.) Cast. *et* Chalm. Owens *et al.* (1968) found that rhizobitoxine inhibited β-cystathionase. Growth inhibition could be prevented by homocysteine or L-methionine, but not by cystathionine or homoserine which are earlier precursors in the methionine pathway. The toxin has a very high affinity for bacterial β-cystathionase, i.e., a K_i of 0·02 μM compared with a K_m of 0·36 mM, and also inhibits the enzyme from spinach.

Although the chemical structure of the toxin has not been fully elucidated, it is thought to be a basic, sulfur-containing amino acid, presumably comprised of two molecules of an ether derivative of homoserine linked through a thioether bridge (Owens, 1969).

Xanthomonas campestris (*Pamm.*) *Dows.* Blackrot of crucifers aptly demonstrates the succession of symptoms exhibited by a host. The pathogen first induces a chlorotic, V-shaped lesion in the leaf, veins in the affected area then darken, there is a localized wilting of the chlorotic area, and finally the lesion becomes necrotic.

In culture this bacterium copiously produces an extracellular polysaccharide that is chemically and serologically similar to material obtained from xylem vessels of infected cabbage leaves (Sutton, 1969). Mechanical plugging of the xylem precedes the first visible symptom of the disease and, according to Sutton and Williams (1970), is primarily due to the polysaccharide. They also found that an avirulent strain did not produce the polysaccharide.

Because of possible industrial use, this polysaccharide has been extensively characterized. It is composed of D-glucose, D-mannose and D-glucuronic acid (2·8:3:2), together with small amounts of pyruvic and acetic acids (Sloneker *et al.*, 1964; Orentas *et al.*, 1963).

Growth Abnormalities

Agrobacterium rhizogenes (*Riker* et al.) *Conn.* This bacterium induces its host to form adventitious roots. We have found that cell-free culture filtrates of this bacterium will significantly increase root formation in a bioassay using sections of mung bean hypocotyl (Hopkins and Durbin, 1971). A portion of this effect is ascribable to IAA synthesized by the bacterium. However, the major part of the rooting activity is due to some other factor(s) having chemical properties different from those of the indole plant growth hormones.

Agrobacterium tumefaciens (*Sm.* et *Towns.*) *Conn.* Many workers have tried to determine whether this bacterium produces a phytotoxin that can induce neoplastic growth. Smith *et al.* (1911) first suggested that such a compound might be produced. Thung (1929) obtained from culture filtrates a toxin which caused necrosis, but not tumor formation. Other workers eventually concluded that toxins extracted from the bacterium or from culture filtrates do not induce tumors. Attempts to correlate production by the bacterium of IAA, a gibberellin-like substance, and the bacteriocin, agrobacteriocin I, with tumor induction have also failed (Schilperoort, 1969).

Bacillus cereus *Frankl.* et *Frankl.* This bacterium induces an interesting and unusual effect on tobacco. It apparently does not parasitize the plant, but grows in the rhizosphere producing a metabolite which when absorbed by the plant causes a disease called "frenching". The initial symptom is a chlorosis which begins at the leaf margins of the young leaves and gradually spreads until all of the interveinal regions become affected. At a later stage, laminae of the new leaves fail to develop so that long, narrow strap-shaped leaves are produced. Apical dominance is lost and in extreme instances axillary buds develop.

Shear and Ussery (1940) proposed that frenching was caused by a phytotoxin produced by some widely distributed soil micro-organism. Subsequently, Steinberg (1950) showed that, of some 60 species of common soil bacteria tested, only *B. cereus* was unequivocally able to cause the symptoms of frenching on tobacco. He found large populations of *B. cereus* in the root zone of naturally affected plants (Steinberg, 1951). Sterile filtrates from cultures of the bacterium could also cause chlorosis and strap-shaped tobacco leaves (Steinberg, 1952). Microbiological analysis of affected leaves revealed that there were large amounts of free "isoleucine" present, and indeed, the application of L-isoleucine (1 mM) or L-alloisoleucine (10 μM) to tobacco seedlings produced symptoms closely resembling those of frenching (Steinberg, 1952). Unfortunately, this work has not been continued, and it is not known whether certain amino acids are produced by the bacterium in sufficient quantities to cause frenching, or whether the primary factor is some other compound(s) which leads to an imbalance in amino acid metabolism.

Corynebacterium fascians (*Tilf.*) *Dows.* Thimann and Sachs (1966) showed that fasciations of pea plants could be mimicked by treatments with kinetin, and that *C. fascians* produced a substance in culture that had cytokinin-like activity and chemical properties of a purine. Infected but not healthy pea tissue yielded a similar compound which was subsequently identified as 6-(γ,γ-diamethylallylamino)purine (Helgeson and Leonard, 1966).

The mode of action of cytokinins is still unknown but recently there has been speculation that they may act via tRNAs (Helgeson, 1968).

Pseudomonas solanacearum (*E. Sm.*) *E. Sm.* Grieve (1939) first pointed out that epinasty, the earliest visual symptom of infection on tomato, could also be elicited by ethylene. Similarly, the premature ripening of diseased banana fruit can also be induced by ethylene. Freebairn and Buddenhagen (1964) showed that isolates of both race 1 and 2 of *P. solanacearum* produced ethylene in culture. Race 1 isolates produced more ethylene than did those of race 2. However, an insect-transmitted (SFR) strain of race 2 did not produce ethylene.

Necrosis

Erwinia amylovora (*Burr.*) *Winsl.* et al. Lovrekovich *et al.* (1970) found that infected apple shoots evolve ammonia. When they exposed healthy apple shoots to ammonia vapors until the shoots became discolored, their content of ammonia was about the same as that in infected shoots. From this and the fact that *E. amylovora* produces ammonia in culture, they concluded that ammonia is probably the phytotoxin responsible for the twig blight phase of fire blight.

The mechanism(s) by which ammonia might act is uncertain. A number of hypotheses have been proposed, but none has been shown to apply to diseased plants (Gilpatrick, 1969; Katunuma *et al.*, 1966; Puritch and Barker, 1967; Visek, 1968).

Pseudomonas syringae *V. Hall.* Erikson and Montgomery (1945) suggested that a phytotoxin was involved in necrosis caused by the *mors-prunorum* ecotype (virulent to plum and cherry). They isolated a toxic fraction from dried bacterial cells with weak acid, and suggested on the basis of physical and chemical properties that the toxin in this fraction was proteinaceous.

The *syringae* ecotype (virulent to peach) produces a polypeptide antibiotic, syringomycin, in culture that has properties different from the phytotoxin described from the *mors-prunorum* ecotype (Sinden and DeVay, 1967). Pathogenicity is correlated with antibiosis; highly pathogenic isolates are very inhibitory to a wide range of micro-organisms, weakly virulent isolates are less inhibitory and mutants that had lost pathogenicity are not inhibitory (DeVay *et al.*, 1968). Syringomycin induces shoot cankers and necrotic leaf spots in peach identical in appearance to those produced by virulent bacteria. Furthermore, a compound has been isolated from diseased host tissue with the same properties as syringomycin (Sinden and DeVay, 1967).

Syringomycin was initially shown to inhibit malonate oxidation by sweet potato mitochondria (Sinden and DeVay, 1967). More recent evidence indicates that its primary effect is on cellular membranes (see Backman, p. 383).

Pseudomonas tabaci. Lovrekovich *et al.* (1969) have presented evidence, similar to that cited for *E. amylovora*, that undissociated ammonia formed exogenously by the bacteria is responsible for the necrotic symptom of the wildfire disease.

Water Soaking

The water-soaked appearance of leaves is a characteristic early symptom of infection by many bacteria. Early workers suggested that proteolytic and lipolytic enzymes might play a role in producing the water soaking symptom by altering plasmalemma permeability (Kelman, 1969). However, thus far critical evidence has not been brought forth to support this suggestion. More recent work has shown that phytotoxins may be partially involved.

It is not clear whether the phytotoxins reportedly causing water soaking are basically different from some of those included in the section on wilting since semi-purified preparations of toxins causing water congestion in leaves can cause cuttings to wilt.

Pseudomonas lachrymans (*Sm.* et *Br.*) *Cars.* A fraction containing lipid, protein and polysaccharide, called lipomucopolysaccharide (LMP) has been isolated from culture filtrates (Keen *et al.*, 1969). It causes non-enzymatic solubilization of proteins. Its possible role in water soaking of cucumber leaves, however, is unknown. LMP has been extracted from diseased leaves, but the yield was 2–5 times lower than that required to produce water congestion. But in the diseased leaf it may act in conjunction with proteolytic enzymes produced by the pathogen (Keen and Williams, 1970).

Wilting

Wilting is caused primarily by bacteria which colonize the xylem early in the course of the disease, and only later, when the host is in the advanced stages of infection, spread to other tissues. Early workers, noting the presence of bacterial masses, gums and tyloses in the vessels of infected plants, suggested that mechanical plugging was responsible for wilting. Later evidence suggested that toxins produced by the bacterium or by its action on the host, ultimately induce water stress. More recently, the participation of various pathogen-synthesized carbohydrases has become more fully appreciated. As wilt inducers, phytotoxins are most probably only one of several agents acting together to cause symptoms. The physiological aspects of this problem have recently been reviewed (Talboys, 1968; see also Dimond, p. 289).

A number of phytopathogenic bacteria produce substances in culture, notably high molecular weight polysaccharides, that can cause cuttings to wilt, presumably by mechanical means. But, as has been pointed out, many other high molecular weight compounds have a similar effect. Bacteria often produce wilt-inducing substances although wilting is not a part of the syndrome (Feder and Ark, 1951; Hodgson *et al.*, 1945; Hodgson *et al.*, 1949). Thus, in certain cases, one should accept with caution the interpretation that high molecular weight substances produced by the pathogen are directly causing the diseased plant to wilt.

Corynebacterium insidiosum (*McCull.*) *Jens.* Spencer and Gorin (1969) reported that this bacterium produces a "gum" composed of D-glucose, D-galactose and L-fucose. When placed in dilute solutions of the partially purified gum, alfalfa cuttings wilted in 10–15 min. They proposed that this substance may be involved in the disease since they were able to isolate fucose from an acid hydrolysate of infected but not healthy alfalfa. More recent results indicate that the phytotoxin contains only galactose and fucose (Ries and Strobel, 1970). It also has a blue chromophore based on chelated copper which is not required for biological activity.

Corynebacterium michiganense (*E. Sm.*) *Jens.* Rai and Strobel (1969a) have isolated an extracellular glycopeptide that causes cuttings to wilt. They suggest that this phytotoxin directly affects cellular membranes, disorganization of which leads to water imbalance (Rai and Strobel, 1969b).

Corynebacterium sepedonicum (*Spieck.* et *Kotlp.*) *Skapt.* et *Burkh.* Originally it was reported that *C. sepedonicum* synthesized a polysaccharide similar to that from *C. insidiosum* (Spencer and Gorin, 1969). This compound has now been more fully characterized and found to be a glycopeptide (Strobel, 1970). Rather than acting mechanically, it appears directly to affect plasma membrane permeability as does the phytotoxin from *C. michiganense* (Strobel and Hess, 1968). Affected cells do not plasmolyse and lose electrolytes rapidly. Those from naturally infected plants behave similarly. Such plants contain sufficient glycopeptide to account for disease symptoms (Strobel, 1970).

The peptide portion of the phytotoxin (mol.wt, 21,400) is composed of one mole of aspartic acid, threonine, serine, glycine, lysine, alanine, methionine and histidine to 2 mol of glutamic acid/mol of toxin; alanine and glycine occupy the N-terminal position in a molar ratio of 3:2. The carbohydrate portion is composed of glucose (48%), mannose (13%), L-fucose (1%) and 2-keto-3-deoxy-gluconic acid (6%).

Erwinia tracheiphila (*E. Sm.*) *Holl.* One of the most diagnostic characteristics of cucumber wilt is a viscose, extremely stringy exudate that forms in the xylem. Harris (1940) found what is presumably the same substance in autoclaved, three-week-old culture filtrates, and demonstrated that it would cause wilting. Since wilting was always accompanied by a decrease in transpiration rate, he suggested that the xylem had been mechanically occluded.

Pseudomonas caryophylli (*Burkh.*) *Starr* et *Burkh.* Carnation cuttings wilt when placed in filtrates from cultures of *P. caryophylli* (Holtzmann and Thomas, 1953). However, when "large molecular weight compounds" were removed from the filtrates, wilting did not occur. The extent to which a phytotoxin may be responsible for the wilt phase of this disease has not been further clarified.

Pseudomonas solanacearum. Hutchinson (1913) reported that wilting of infected plants results from the action of a toxin which alters cellular permeability. In further studies, Kunz (1952) obtained two wilt-inducing fractions from 2- to 5-month-old cultures. One, a polysaccharide complex, induced wilting by plugging the vessels. The other fraction presumably acted by altering membrane permeability. Husain and Kelman (1958) demonstrated that an extracellular, wilt-inducing polysaccharide was synthesized *in vitro* only by virulent strains, and that a substance with similar properties could be obtained from the vessels of infected plants. They proposed that wilting occurs because the polysaccharide causes an increase in the viscosity of the xylem fluid, thereby reducing the rate of water movement (Kelman, 1969). Synthesis of IAA by the bacterium may also be a contributing factor (Phelps and Sequeira, 1967).

Xanthomonas phaseoli (*E. Sm.*) *Dows.* Leach *et al.* (1957) found that this bacterium produces an extracellular, high molecular weight heteropolysaccharide which causes wilting when introduced into the vascular system of tomato cuttings. Corey and Starr (1957), studying what was apparently the same compound, found that production in culture by four colony types was directly correlated with lesion production but not with bacterial growth in the host. Introduction of the semi-purified material into healthy bean leaves produced lesions similar to those induced by *X. phaseoli.*

Using a cell-free but otherwise crude culture filtrate, Leach *et al.* (1957) estimated from light-scattering data that the phytotoxin had a molecular weight of about $19{\cdot}5 \times 10^6$. Acid hydrolysis yielded glucose and mannose as well as some other unidentified reducing substances; a Lassaigne test for nitrogen was negative.

Xanthomonas stewartii (*E. Sm.*) *Dows.* Stewart (1897) suggested from observations of the disease that wilting was due to mechanical plugging of the xylem by the bacterium, and to a lesser extent by gum and tyloses. Later, Harris (1940) found a heat stable, wilt-inducing substance in culture filtrates.

The exact nature of this wilt-inducing substance and its occurrence in naturally-infected tissue have not been reported. Presumably however, the substance may be the same as the extracellular polysaccharide which Gorin and Spencer (1961) isolated from cultures, and characterized as a highly branched polymer of glucose, galactose and an aldobiouronic acid linked to the 4 position of glucose.

Xanthomonas *spp.* Many pathogenic xanthomonads produce heteropolysaccharides similar to that formed by *X. campestris*, except that they contain galactose in place of the mannose (Orentas *et al.*, 1963). Such bacteria include: *X. carotae* (Rend.) Dows., *X. hederae* (Arn.) Dows., *X. malvacearum* (E. Sm.) Dows., *papavericola* (Bryan *et* McWhor.) Dows., *translucens* (Jones *et al.*) Dows. f. sp. *hordei-avenae* Hagb. and *X. translucens* f. sp. *undulosa* (Sm. *et al.*) Hagb., and *X. vesicatoria* (Doidge) Dows. What role these polysaccharides play in disease production is as yet unknown.

Discussion

It is evident from the foregoing that our overall knowledge of bacterial phytotoxins is still fragmentary, though in selected cases it is quite extensive. Several useful generalizations can be made however: (i) phytotoxins inducing the same plant response may have different mechanisms of action; (ii) different pathogens can synthesize the same phytotoxin; and (iii) most phytotoxins are composed of simple carbohydrate and/or amino acid residues.

The paramount problem in most diseases still is to prove definitively that a particular phytotoxin does function in symptom production. As our knowledge increases it is becoming clear that the criteria given earlier for establishing a causal role for such substances are not sufficient for bacterial pathogens since: (i) the pathogen and the toxin may not exhibit the same host range, i.e., bacterial phytotoxins are not host specific, whereas many causal organisms are host specific; (ii) toxin concentration *in vitro* may not be correlated with virulence; (iii) toxigenicity is not always a requisite for pathogenicity; (iv) the rates of synthesis and degradation both *in vitro* and *in vivo* may vary depending upon the bacterial strain, environment and host; and (v) the phytotoxin may not act alone, but in concert with other pathogen- and host-produced substances in a dynamic, interrelated system.

Another aspect of the problem is that since many toxins can induce the same plant response, one has no guarantee that a compound isolated from culture is involved *in vivo* even if it can reproduce a portion of the disease syndrome. Therefore, as has already been stressed by earlier workers, one needs to isolate the phytotoxin, or alternatively one or more unique, related metabolites, from diseased tissue. However, merely demonstrating that such compounds are present in diseased tissue is not sufficient for it tells us nothing about how, or even whether, the phytotoxin is functioning.

The central problem is to determine the effective concentration of the phytotoxin in the diseased plant, that is to establish whether it is present at the appropriate time and location in amounts sufficient to account for the effect it is presumed to have on the host. Answering this question will undoubtedly be quite difficult. But we are now at a stage of understanding when it is both necessary and technically feasible.

In many instances the kinetics of the phytotoxin–plant interaction appear complex so that a number of possible considerations will have to be examined in a quantitative manner. For example, the phytotoxin may be intrinsically unstable and/or metabolized by the plant, the resulting products varying in phytotoxicity. Also, the toxin may irreversibly react with cell constituents. A further complication may be that the phytotoxin acts in combination with other components found in diseased tissue.

Understanding the action of phytotoxins has relevance not only for pathogenesis but for normal plant metabolism as well. Hopefully, it will also provide information that can be used in formulating new approaches to disease resistance.

References

Bajaj, Y. P. S., Spink, G. C., and Saettler, A. W. (1969). *Phytopathology* **59**, 1017 (abstr.).

Braun, A. C. (1955). *Phytopathology* **45**, 659–664.

Corey, R. R. and Starr, M. P. (1957). *J. Bact.* **74**, 137–140.

DeVay, J. E., Lukezic, F. L., Sinden, S. L., English, H., and Coplin, D. L. (1968). *Phytopathology* **58**, 95–101.

Dimond, A. E. and Waggoner, P. E. (1953). *Phytopathology* **43**, 229–235.

Erikson, D. and Montgomery, H. B. S. (1945). *Ann. appl. Biol.* **32**, 117–123.

Feder, W. A. and Ark, P. A. (1951). *Phytopathology* **41**, 804–808.

Freebairn, H. T. and Buddenhagen, I. W. (1964). *Nature* **202**, 313–14.

Gilpatrick, J. D. (1969). *Phytopathology* **59**, 973–978.

Gorin, P. A. J. and Spencer, J. F. T. (1961). *Can. J. Chem.* **39**, 2282–2289.

Grieve, B. J. (1939). *Ann. Bot.* **3**, 587–600.

Harris, H. A. (1940). *Phytopathology* **30**, 625–638.

Helgeson, J. P., and Leonard, N. J. (1966). *Proc. Nat. Acad. Sci. U.S.* **56**, 60–63.

Helgeson, J. P. (1968). *Science* **161**, 974–981.

Hodgson, R., Riker, A. J., and Peterson, W. H. (1945). *J. biol. Chem.* **158**, 89–100.

Hodgson, R., Peterson, W. H., and Riker, A. J. (1949). *Phytopathology* **39**, 47–62.

Hoitink, H. A. J. and Sinden, S. L. (1970). *Phytopathology* **60**, 1236–1237.

Holtzmann, O. V. and Thomas, W. D., Jr. (1953). *Phytopathology* **43**, 587 (abstr.).

Hopkins, D. L. and Durbin, R. D. (1971). *Can. J. Microbiol.* (In press).
Husain, A. and Kelman, A. (1958). *Phytopathology* **48**, 155–165.
Hutchinson, C. M. (1913). *Mem. Dep. Agr. India, Bacteriol. Ser.* **1**, 63–83.
Katunuma, N., Okada, M., and Nishii, Y. (1966). *In* "Advances in Enzyme Regulation" (G. Weber, Ed.), Vol. 4, pp. 317–335. Pergamon Press, New York.
Keen, N. T., Williams, P. H., and Upper, C. D. (1969). *Phytopathology* **59**, 703–704.
Kelman, A. (1969). *Conn. Agr. Exp. Sta. Bull. No.* **708**, 128–154.
Keen, N. T. and Williams, P. H. (1970). Unpublished work.
Kunz, R. (1952). *Phytopath. Z.* **20**, 89–112.
Lamar, C., Jr. and Sellinger, O. Z. (1965). *Biochem. Pharmacol.* **14**, 489–506.
Leach, J. G., Lilly, V. G., Wilson, H. A., and Purvis, M. R., Jr. (1957). *Phytopathology* **47**, 113–120.
Lesemann, D. and Rudolph, K. (1969). *Z. Pflanzenphysiol.* **62**, 108–115.
Lovrekovich, L. and Farkas, G. L. (1963). *Nature* **198**, 710.
Lovrekovich, L., Klement, Z., and Farkas, G. L. (1964). *Science* **145**, 165.
Lovrekovich, L., Lovrekovich, H., and Goodman, R. N. (1969). *Phytopathology* **59**, 1713–1716.
Lovrekovich, L., Lovrekovich, H., and Goodman, R. N. (1970). *Can. J. Bot.* **48**, 999–1000.
Orentas, D. G., Sloneker, J. H., and Jeanes, A. (1963). *Can. J. Microbiol.* **9**, 427–430.
Owens, L. D., Guggenheim, S., and Hilton, J. L. (1968). *Biochim. biophys. Acta* **158**, 219–225.
Owens, L. D. (1969). *Science* **165**, 18–25.
Patel, P. N. and Walker, J. C. (1963). *Phytopathology* **53**, 522–528.
Patil, S. S., Kolattukudy, P. E., and Dimond, A. E. (1969). *Phytopathology* **59**, 1043 (abstr.).
Patil, S. S. (1970a). *Phytopathology* **60**, 1307 (abstr.).
Patil, S. S. (1970b). Personal communication.
Phelps, R. H. and Sequeira, L. (1967). *Phytopathology* **57**, 1182–1190.
Puritch, G. S. and Barker, A. V. (1967). *Pl. Physiol.* **42**, 1229–1238.
Rai, P. V. and Strobel, G. A. (1969a). *Phytopathology* **59**, 47–52.
Rai, P. V. and Strobel, G. A. (1969b). *Phytopathology* **59**, 53–57.
Ries, S. M. and Strobel, G. A. (1970). *Phytopathology* **60**, 1309–1310 (abstr.).
Rudolph, K. (1969). *Naturwiss.* **56**, 569–570.
Schilperoort, R. A. (1969). "Investigations of Plant Tumors Crown Gall." Dissertation, University of Leiden, 170 pp.
Shear, G. M. and Ussery, H. D. (1940). *J. agr. Res.* **60**, 129–139.
Sinden, S. L. and DeVay, J. E. (1967). *Phytopathology* **57**, 102 (abstr.).
Sinden, S. L. and Durbin, R. D. (1968). *Nature* **219**, 379–380.
Sinden, S. L. and Durbin, R. D. (1970). *Phytopathology* **60**, 360–364.
Sloneker, J. H., Orentas, D. J., and Jeanes, A. (1964). *Can. J. Chem.* **42**, 1261–1269.
Smith, E. F., Brown, M. A., and Townsend, C. O. (1911). *U.S.D.A. Bur. Plant Ind. Bull.* **213**.

Spencer, J. F. T. and Gorin, P. A. J. (1960). *Can. J. Microbiol.* **7**, 185–188.
Steinberg, R. A. (1950). *Bull. Torrey bot. Club* **77**, 38–44.
Steinberg, R. A. (1951). *Pl. Physiol.* **26**, 807–811.
Steinberg, R. A. (1952). *Pl. Physiol.* **27**, 302–308.
Stewart, F. C. (1897). *N.Y. Agr. Exp. Sta. Bull. No.* 130.
Stewart, J. M. (1961). *J. Am. chem. Soc.* **83**, 435–439.
Strobel, G. A. and Hess, W. M. (1968). *Pl. Physiol.* **43**, 1673–1688.
Strobel, G. A. (1970). *J. biol. Chem.* **245**, 32–38.
Sutton, J. C. (1969). "The Relation of Extracellular Polysaccharide of Xanthomonas campestris to Xylem Plugging and Black Rot Lesion Development in Cabbage." Ph.D. Thesis, University of Wisconsin, 66 pp.
Sutton, J. C. and Williams, P. H. (1970). *Can. J. Bot.* **48**, 391–401.
Talboys, P. W. (1968). *In* "Water Deficits and Plant Growth" (T. T. Kozlowski, Ed.), Vol. 2, pp. 255–311. Academic Press, New York.
Taylor, P. A., Durbin, R. D., and Maxwell, D. P. (1970). *Phytopathology* **60**, 1317 (abstr.).
Thimann, K. V. and Sachs, T. (1966). *Am. J. Bot.* **53**, 731–739.
Thung, T. H. (1929). *Tijdschr. Plantenziekten* **35**, 263–269.
Vanderhoef, L. N. and Durbin, R. D. (1970). Unpublished work.
Visek, W. J. (1968). *J. Dairy Sci.* **51**, 286–295.
Wheeler, H., and Luke, H. H. (1963). *Ann. Rev. Microbiol.* **17**, 223–242.
Woolley, D. W., Schaffner, G., and Braun, A. C. (1952). *J. biol. Chem.* **198**, 807–813.
Woolley, D. W., Schaffner, and Braun, A. C. (1955). *J. biol. Chem.* **215**, 485–493.
Woolley, D. W. (1959). *In* "Plant Pathology Problems and Progress 1908–1958" (C. S. Holton, Ed.), pp. 130–134. University of Wisconsin Press, Madison, Wisconsin.

Discussion

Dimond: Would you accept the following additional points as part of your criteria or to be reckoned with in establishing these criteria?

1. That the amount of toxin in the diseased plant be demonstrably toxic? Mere presence of a toxin in infected tissues is not meaningful unless the concentration produces injury that corresponds to the amount of injury seen in naturally infected tissues.
2. That the concentration of toxin in naturally infected plants increases as severity of a portion of the symptoms increases and be a maximum when or soon before related symptoms are most severe. In other words, severity of a symptom should be related to concentration of toxin.
3. Toxic materials are often detoxified metabolically by plant tissues. This is true of introduced fusaric acid and probably of lycomarasmin. The true rate of synthesis of toxin is therefore not the same as the apparent rate. The apparent (net) rate is the difference between the true rate of synthesis and the sum of all processes that degrade the toxin. In dealing with this, one

must reckon with the fact that toxin is consumed when it reacts with host tissue to injure it, and the rate at which toxin is consumed in producing unit injury should be known as a characteristic of the toxin and measured.

Durbin: Yes.

Mirocha: Why should the addition of glutamine reverse or at least prevent the expression of chlorosis? If chlorosis was due to NH_4^+ accumulation, as you once suggested then the addition of glutamine would not tend to utilize the excess NH_4^+. It appears that glutamine is metabolized readily after addition, and that its nitrogen incorporated into the nitrogen pool.

It would be interesting to know whether the protection against chlorosis obtained by the addition of glutamine to a tobacco leaf together with the toxin, is associated with accumulation of NH_4^+ in the light.

Durbin: Presumably glutamine can protect the enzyme against the toxin. The experiment you suggest is a critical one. We are now in the process of testing the hypothesis that the toxin specifically affects glutamine synthetase.

Pringle: Stewart, Department of Experimental Pathology, National Institute of Mental Health, Bethesda, Maryland, U.S.A. has formulated a structure for tabtoxinine and has confirmed it by synthesis. It is an isomer of the one given by Durbin.

Rudolph: I have not found that chloroplasts degenerate in tissue made chlorotic by application of purified toxin. Small chlorotic haloes induced by low concentrations of toxin became green after about 20 days. This means that the effect of the toxin is reversible and so is any degeneration of chloroplasts that may occur. Yellow haloes caused by the *Pseudomonas phaseolicola* toxin are more pronounced at high light intensities.

Goodman: I want to point out that reports in the literature state chlorosis is reversible. Chlorosis induced by *P. tabaci* may be reversed or at least delayed by kinetin. The question that now arises is how long do chloroplasts photosynthesize after exposure to toxin? We have electron micrographic evidence which indicates that chloroplasts in seemingly poor condition (visually) still may photosynthesize because starch accumulates in them. One wonders just how long these chloroplasts remain photosynthetically competent.

Matta: In relation to point 5 of your criteria for the definition of toxin I should also consider the possibility that the reintroduction of a toxin reproduces disease symptoms only after deficiencies of some factors have been induced in the plant.

Ballio: I think that the definition of *vivotoxin* should include "... breakdown products, or undefined products related to the toxin".

3. Phytotoxins Produced by Fusaria

HEINZ KERN

Institut für Spezielle Botanik,
Eidg. Technische Hochschule,
Zürich 6, *Switzerland*

Numerous phytotoxins have been isolated from cultures of *Fusarium* spp. and their chemical and biological properties are known in more or less detail. Some of these toxins are discussed below.

Naphthazarin Toxins

Fusarium solani (Mart.) Sny. *et* Hans. f. sp. *pisi* (Jones) Sny. *et* Hans, causing a root and stem rot of peas, and other species of *Fusarium* from the group *Martiella* Wollenweber produce several phytotoxic red pigments with a common naphthazarin structure (Fig. 1; Kern and Naef-Roth, 1965, 1967). Six compounds have been isolated with different functional groups and with considerable differences in their biological activity (Table I). The isomeric compounds marticin and isomarticin are mainly phytotoxic; norjavanicin and novarubin are most toxic for fungi and bacteria and much less phytotoxic; fusarubin and javanicin show no activity against fungi and are intermediate in their toxicity for bacteria and higher plants.

The various fungal strains produce these compounds in very different amounts. A very pathogenic strain of *F. solani* f. sp. *pisi* forms more than 150 mg of marticins per litre after four days of shake culture whereas a slightly pathogenic strain forms considerable quantities of javanicin and fusarubin but only traces of marticins (Fig. 2).

Similar correlations between toxin formation *in vitro* and pathogenicity were found with related *Fusarium* strains and species and in the genus *Neocosmospora* (Kern and Naef-Roth, 1967). Marticins and other toxins of this group have also been extracted from roots, stems and leaves of diseased pea plants in quantities sufficient to cause serious damage to the tissues (22 μg/g in the roots, 34 μg/g fresh weight of the whole plant). No other toxins (neither fusaric acid nor phytolysin) have been found in cultures of these fungi.

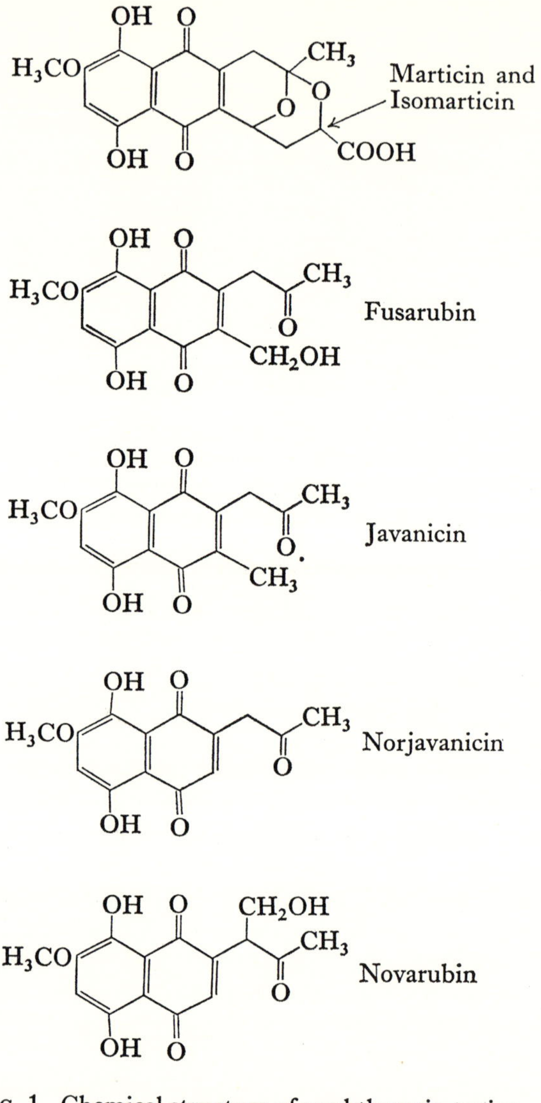

FIG. 1. Chemical structure of naphthazarin toxins.

Naphthoquinones interfere in various ways with the metabolism of micro-organisms and higher plants. They inhibit anaerobic decarboxylation of pyruvate through a reaction with coenzyme thiamine pyrophosphate (Kern *et al.*, 1970). The reactions may be partially overcome by an excess of coenzyme. Norjavanicin and novarubin, the most active inhibitors of

TABLE I. Biological activity of naphthazarin toxins

	Activity			Inhibition of decarboxylation	
Toxin	Phytotoxic μg/g*	Antibacterial μg/ml†	Antifungal μg/ml‡	Anaerobic μg/ml§	Oxidative μg/ml‖
Marticin, Isomarticin	8	7	> 500	248	750
Fusarubin	40	2	500	30	77
Javanicin	60	2	500	29	87
Norjavanicin	90	1	10	3	22
Novarubin	35	0·5	5	4	31

* Minimal dose per g fresh weight of tomato cuttings.
† Lowest inhibiting concentration per ml agar medium for *Bacillus subtilis*.
‡ LD_{50} for germination of spores of *Botrytis allii* Munn.
§ 50% inhibition of anaerobic decarboxylation of pyruvate.
‖ 50% inhibition of oxidative decarboxylation of α-ketoglutarate (data from Kern *et al.*, 1967, 1970)

germination of fungal spores and of bacterial growth, are most active in this reaction; much higher concentrations of the other compounds are required to obtain the same degree of enzyme inhibition (Table I). Oxidative decarboxylation of α-keto acids by suspensions of tomato mitochondria is inhibited in a similar way; the activity increases parallel to the toxicity for fungal spores and bacteria but the concentrations required are generally higher.

The toxicity of naphthazarin derivatives for fungal spores and the inhibition of seed germination and root growth of higher plants by them may be explained by inhibition of these enzyme reactions. The inhibition of bacterial growth and in particular the destruction of tissues of higher plants depend on other mechanisms of action. The toxicity of the toxins for bacteria and tissues of higher plants is reduced by certain metal ions (Cu, Al, Fe) and organic acids (glutamic acid, succinic acid, and others) (Kern and Naef-Roth, 1966). The degree of detoxification varies with the toxin and the compound added. Copper ions are most active and (in equimolar concentration) nullify the toxicity of the marticins to bacteria and tissues of higher plants. When pea plants are grown in nutrient solutions with increasing copper contents their sensitivity to these toxins and their susceptibility to the parasite decrease (Table II). The formation of marticins in pure culture is not influenced by increasing copper concentrations.

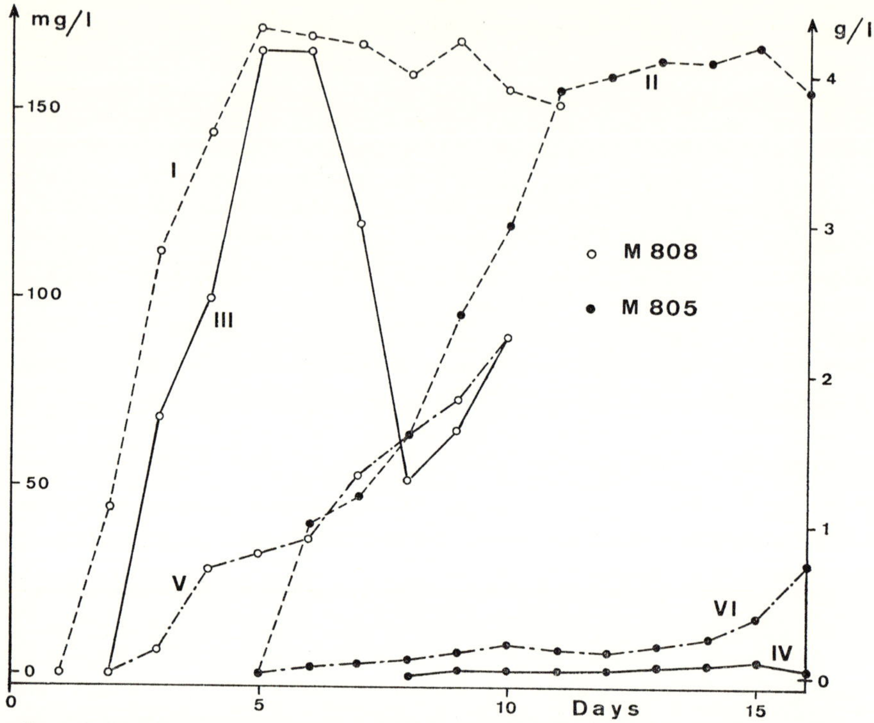

FIG. 2. Toxin production by a pathogenic (M 808) and a weakly pathogenic strain (M 805) of *Fusarium solani* f. sp. *pisi*. I, II mycelium (g/litre); III, IV marticins (mg/litre); V, VI fusarubin + javanicin (mg/litre). (Data from Kern and Naef-Roth, 1965).

Copper and other metals chelate the toxin molecules and probably interfere with their transport through vessels and protoplasmic membranes. The inhibition of the enzyme reactions above mentioned by naphthazarins is not influenced by copper or glutamic acid and the germination of fungal

TABLE II. Effect of copper ion concentration in the nutrient solution on sensitivity to toxin and susceptibility of peas to *Fusarium solani* f. sp. *pisi*

Cu^{++} μg/ml	0	0·32	0·63	1·25	2·5	5
Toxin sensitivity*	2·9	2·6	2·5	2·4	1·5	0·8
Disease index†	3·3	3·0	2·6	2·1	2·0	1·5

* Wilt index of pea cuttings treated with 50 μg isomarticin/g fresh weight.

† Disease index of pea plants; 0—no symptoms, 4—total necrosis (from Kern and Naef-Roth, 1966)

spores is only slightly affected. The main mechanism of action of the marticins in tissues of higher plants remains to be determined; inhibition of malate or citrate dehydration in the citric acid cycle is possible.

Fusaric Acid

Fusaric acid (5-n-butylpicolinic acid, Fig. 3; Gäumann, 1957, 1958) is produced by several *Fusarium* species mainly belonging to the group *Elegans* Wollenweber (*F. oxysporum* (Schl.) Sny. *et* Hans.) and causing wilt diseases in tomato, cotton, pea, banana and other plants. Some of these fungi also produce other toxic metabolites which may be closely related to fusaric acid, e.g. dehydrofusaric acid with a double bond at the end of the side chain, or entirely different with low or high molecular weight. This is one of the reasons why, despite numerous studies, the significance of fusaric acid in pathogenesis is not yet sufficiently understood.

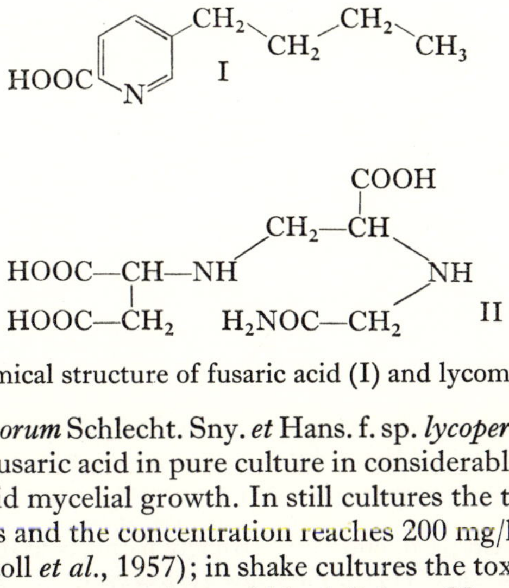

FIG. 3. Chemical structure of fusaric acid (I) and lycomarasmin (II).

Fusarium oxysporum Schlecht. Sny. *et* Hans. f. sp. *lycopersici* (Sacc.) Sny. *et* Hans. produces fusaric acid in pure culture in considerable amounts during the period of rapid mycelial growth. In still cultures the toxin is detectable after about 6 days and the concentration reaches 200 mg/litre or more after about 20 days (Stoll *et al.*, 1957); in shake cultures the toxin appears earlier and attains highest concentrations within 6–8 days. Nutritional conditions greatly affect toxin production. A pathogenic strain forms very little fusaric acid without zinc but large amounts are produced in media containing zinc ions; a weakly pathogenic strain forms considerable amounts of fusaric acid without zinc but much less with increasing concentrations of zinc ions (Fig. 4; Egli, 1969). The mechanism of the zinc effect is not yet known. Inoculation experiments with tomato plants confirm these results because the disease index caused by the pathogenic strain increases with the concentration of zinc ions in the nutrient solution. A similar relation has been found between iron nutrition, toxin formation and pathogenicity.

On disinfected stem pieces of "Tuckswood" tomatoes the pathogenic strain M 811c produced fusaric acid (about 100 μg/g fresh weight of plant material) whereas the non-pathogenic strain 5414 did not. *F. oxysporum* f. sp. *lini* (Bolley) Sny. *et* Hans. produced fusaric acid on tissue pieces (disinfected with propylene oxide) of a susceptible but not of resistant flax

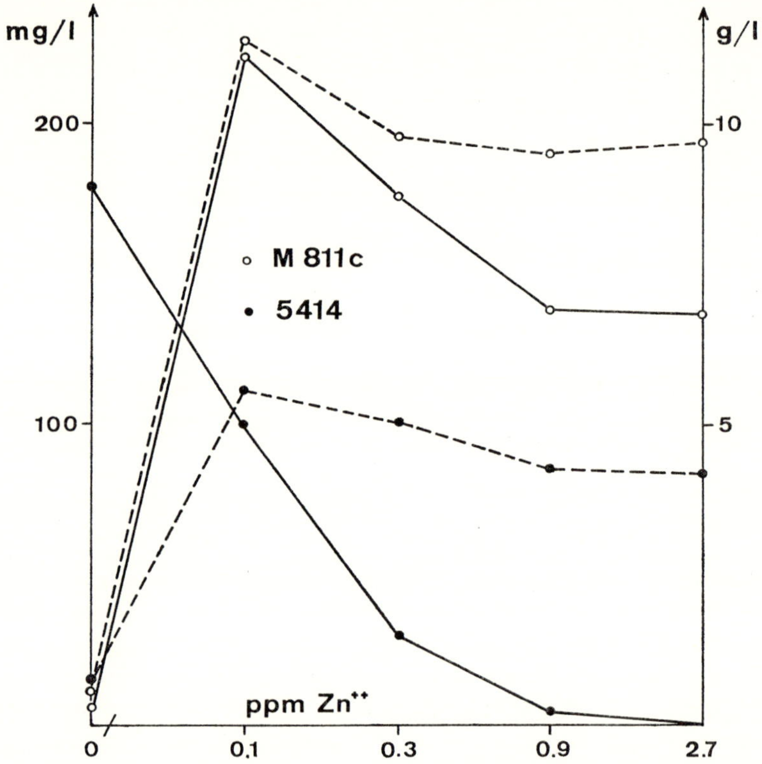

FIG. 4. Fusaric acid production by a strongly pathogenic strain (M 811c) and weakly pathogenic strain (5414) of *Fusarium oxysporum* f. sp. *lycopersici* depending on zinc nutrition. Twenty-five days of still culture on Richards' medium. Full lines; fusaric acid (mg/litre): broken lines; mycelial weight (g/litre). (Data from Egli, 1969).

varieties (Trione, 1960). On dead tissues of watermelon, a pathogenic strain of *F. oxysporum* f. sp. *niveum* (E. F. Sm.) Sny. *et* Hans. formed more than 150 μg/g fusaric acid whereas two weakly and 38 non-pathogenic strains formed less than 5 μg/g (Davis, 1967b).

Fusaric acid has been detected in various plants after inoculation with their wilt pathogens (Lakshminarayanan and Subramanian, 1955, in cotton;

Kalyanasundaram and Venkata Ram, 1956, in cotton, 17 μg/g after 2–3 weeks; Kern and Kluepfel, 1956, in tomato; Page, 1959, in banana rhizomes; Davis, 1967a, in flax, tomato and watermelon after 3, 6 and 8 days at concentrations of 3, 3 and 17 μg/g, respectively; Chandramohan and Mahadevan, 1968, in cotton, within 12 h after inoculation). A positive correlation between pathogenicity and the amount of fusaric acid formed in diseased plants was found for several strains of *F. oxysporum* f. sp. *niveum* (Davis, 1969). In our own experiments, a pathogenic strain of *F. oxysporum* f. sp. *pisi* (v. Hall) Sny. *et* Hans. Race 2 produced considerable amounts of fusaric acid *in vitro* (80 mg/litre) and *in vivo* (100–150 μg/g fresh plants), whereas a weakly pathogenic strain did not form fusaric acid *in vitro* or *in vivo*. Other strains, however, may behave differently and damage pea plants through other metabolites; in another pathogenic strain (*F. oxysporum* f. sp. *pisi*, Race 1) we did not detect any fusaric acid. Zinc ions have a different effect on these fungi compared to *F. oxysporum* f. sp. *lycopersici* by reducing fusaric acid production *in vitro*.

When fusaric acid is applied to tomato cuttings, about 10% of the toxin is broken down to smaller molecules and finally to carbon dioxide and water; $C^{14}O_2$ is given off by plants treated with fusaric acid labelled with C^{14} in the carboxyl group or in the pyridine ring (Sanwal, 1956; Kluepfel, 1957; Braun, 1960). Another part of fusaric acid is transformed into derivatives less toxic for the plant such as N-methylated fusaric acid amide; conjugation with glycine also is probable. The formation of methylated fusaric acid amide occurs to a greater extent in tomato cultivars resistant to *F. oxysporum* f. sp. *lycopersici* (25% in resistant Red Currant instead of 5–10% in susceptible Bonny Best). This difference, however, does not explain the basic difference in resistance of the cultivars which is essentially a resistance to growth and spreading of the parasite in the stem. Transformation of fusaric acid into another pyridine derivative which is more soluble in water and less phytotoxic occurs to a greater extent in pea plants which are less sensitive to fusaric acid than in tomato or in the very sensitive cotton plant (Jost, 1965). All transformations of fusaric acid known so far lead, therefore, to a detoxification and the toxicity appears to depend on the remaining fusaric acid.

Fusaric acid is toxic to higher plants and micro-organisms interfering with their metabolism in various ways. Before the appearance of visible symptoms, necrotic spots on the leaf blade, shrivelling and drying of leaves, shrinking and wilting of stem and petioles in tomato cuttings, the water permeability of protoplasts and the water balance of the whole plant are altered (Bachmann, 1956; Heinrich, 1962; Gäumann *et al.*, 1958), and the respiratory rate decreases (Naef-Roth, 1957). Fusaric acid inhibits polyphenol oxidase competitively in much lower concentrations than it affects several other enzymes (Bossi, 1959) and may also react with various phenol derivatives (Bär, 1963). In iron-deficient plants, fusaric acid chelates the remaining iron; the symptoms then caused are those of an increased iron deficiency and differ visibly from those that develop in plants with a balanced nutrition.

A comparison of fusaric acid with other pyridine derivatives shows that two parts of the molecule are essential for toxicity. Inhibition of fungal spore germination, yeast respiration, and polyphenol oxidase depends on the presence of a carboxyl group in α-position to the nitrogen. The aliphatic side chain in β-position determines primarily the effects on water permeability, the severity of damage increasing with the length of the side chain from methyl to butyl. Both types of toxicity lead to cell death and visible damage of the host tissues but the first appears the more important. 3-n-butyl pyridine which might be expected as a decarboxylation product of fusaric acid has a much greater effect on water permeability than does fusaric acid itself. Experimentally this very volatile compound has not been detected in plants treated with fusaric acid or in the atmosphere around them; as an intermediate product of metabolism it could be active in very small concentrations only and during a short time before further breakdown.

Other Low Molecular Weight Toxins

Less information is available about certain other toxins produced by various species of *Fusarium* (Dr. Naef-Roth, p. 49). Some species form several toxic metabolites belonging to various chemical groups but their significance in pathogenesis is difficult to evaluate at the present time. *F. oxysporum* f. sp. *niveum* produces phytonivein besides fusaric acid (Hiroe and Nishimura, 1956). *F. oxysporum* f. sp. *lycopersici* produces, besides fusaric acid, phytolycopersin (Hiroe and Matsuo, 1961) and lycomarasmin (Fig. 3; Gäumann *et al.*, 1950; Hardegger *et al.*, 1963). The latter compound has also been isolated from other *formae speciales* of *F. oxysporum* (f. sp. *vasinfectum* (Atk.) Sny. *et* Hans., f. sp. *melonis* Sny. *et* Hans.; Pouteau-Thouvenot and Barbier, 1965, 1966). On tomato plants, it causes necrotic spots at the tips of the leaf blade which are also found on plants inoculated with *F. oxysporum* f. sp. *lycopersici*. Lycomarasmin has not yet been detected in inoculated plants; in pure culture it appears within a few days. Lycomarasmin acts as a chelating agent; after application to tomato cuttings it forms a water soluble, unstable chelate with iron which is translocated into the leaves. During accumulation of the chelate in the leaves iron is set free and causes typical symptoms (Gäumann *et al.*, 1955). Iron increases the toxicity of lycomarasmin whereas copper which forms a stable chelate reduces toxicity; when lycomarasmin is replaced by synthetic compounds of the same chelating type such as ethylenediaminetetraacetic acid the same symptoms result. The toxicity of some structurally related fungal metabolites is correlated with their chelating properties. Lycomarasmic acid is a better chelating agent and more toxic than lycomarasmin (Gäumann and Naef-Roth, 1959) whereas anhydrolycomarasmic acid does not form chelates and is not toxic to tomato cuttings. These two compounds might occur in culture filtrates together with lycomarasmin; they are both produced by *Aspergillus oryzae* (Ahlb.) Cohn (Robert *et al.*, 1962) and lycomarasmic acid is also produced by *Colletotrichum gloeosporioides* Penz. (Ballio *et al.*, 1969).

The equimolar iron–lycomarasmin chelate causes a rapid increase in the respiration rate of tomato leaves (Naef-Roth, 1957); this might explain the increased respiration rate of inoculated plants in an early stage of pathogenesis before symptoms appear and respiration decreases.

Phytotoxic, High Molecular Weight Metabolites

Numerous studies have been carried out on the significance of enzymes and polysaccharides in *Fusarium* wilt diseases (e.g. Wood, 1967); many questions, however, remain open. Considerable attention has been given to the disintegration of plant tissues which may be caused not only by pectolytic enzymes but also by other metabolites. Phytolysin, a protein with the properties of an enzyme (Naef-Roth *et al.*, 1961) is produced by various *formae speciales* of *F. oxysporum* and several species from other genera; it does not occur in *F. solani*. Purified preparations disintegrate tissues in low concentration but exhibit very little pectolytic activity. Its mechanism of action and exact site of attack are not yet known; it damages protoplasmic membranes and in this way may liberate other phytotoxic enzymes. The activity of phytolysin is increased by cobalt but not by calcium ions. Dialysis removes a heat labile, cobalt-containing cofactor; activity may be restored by recombination. In a synthetic medium and on sterilized tomato tissues a pathogenic strain of *F. oxysporum* f. sp. *lycopersici* produced higher amounts of phytolysin than a weakly pathogenic strain; the same was found for strains of *F. oxysporum* f. sp. *pisi*. Phytolysin probably acts in the early phases of pathogenesis and facilitates growth of the parasite in the host tissues; it also increases the toxicity of fusaric acid. Phytolysin has been isolated in considerable quantities from apple tissues attacked by *Penicillium expansum* Link *et.* Thom. It is also produced by *Stereum purpureum* (Pers.) Fr. and *Corynebacterium betae* (Keyw. and How.) Dows. (causing silver leaf disease of stone fruit trees and red beets); after application to living plants the characteristic silvery symptoms appear on the leaf surface.

The work from our laboratory described here has been supported by the Swiss National Foundation for the Advancement of Scientific Research.

References

Bachmann, E. (1956). *Phytopath. Z.* **27**, 255–288.
Bär, H. (1963). *Phytopath. Z.* **48**, 149–177.
Ballio, A., Bottalico, A., Buonocore, V., Carilli, A., Di Vittorio, V., and Graniti, A. (1969). *Phytopath. medit.* **8**, 187–196.
Bossi, R. (1959). *Phytopath. Z.* **37**, 273–316.
Braun, R. (1960). *Phytopath. Z.* **39**, 197–241.
Chandramohan, D. and Mahadevan, A. (1968). *Experientia* **24**, 427.
Davis, D. (1967a). *Phytopathology* **57**, 339.
Davis, D. (1967b). *Phytopathology* **57**, 808.
Davis, D. (1969). *Phytopathology* **59**, 1391–1395.
Egli, T. A. (1969). *Phytopath. Z.* **66**, 223–252.
Gäumann, E. (1957). *Phytopath. Z.* **29**, 1–44.
Gäumann, E. (1958). *Phytopath. Z.* **32**, 359–398.
Gäumann, E., Naef-Roth, S., and Miescher, G. (1950). *Phytopath. Z.* **16,** 257–288.
Gäumann, E., Naef-Roth, S., and Kern, H. (1955). *Phytopath. Z.* **24**, 373–374.
Gäumann, E., Kern, H., Schüepp, H., and Obrist, W. (1958). *Phytopath. Z.* **32**, 225–244.
Gäumann, E. and Naef-Roth, S. (1959). *Phytopath. Z.* **34**, 426–431.
Hardegger, E., Liechti, P., Jackman, L. M., Boller, A., and Plattner, Pl. A. (1963). *Helv. Chim. Acta* **46**, 60–74.
Heinrich, G. (1962). *Protoplasma* **55**, 320–356.
Hiroe, I. and Matsuo, A. (1961). *Trans. Tottori Soc. agric. Sc.* **13**, 26–29.
Hiroe, I. and Nishimura, S. (1956). *Ann. phytopath. Soc. Japan* **20**, 161–216.
Jost, J. P. (1965). *Phytopath. Z.* **54**, 338–378.
Kalyanasundaram, R. and Venkata Ram, C. S. (1956). *J. Indian bot. Soc.* **35**, 7–10.
Kern, H. and Kluepfel, D. (1956). *Experientia* **12**, 181.
Kern, H. and Naef-Roth, S. (1965). *Phytopath. Z.* **53**, 45–64.
Kern, H. and Naef-Roth, S. (1966). *Phytopath. Z.* **57**, 289–297.
Kern, H. and Naef-Roth, S. (1967). *Phytopath. Z.* **60**, 316–324.
Kern, H., Naef-Roth, S. and Item, H. (1970). *Phytopath. Z.* **67**, 1–14.
Kluepfel, D. (1957). *Phytopath. Z.* **29**, 349–379.
Lakshminarayanan, K. and Subramanian, D. (1955). *Nature* **176**, 697–698.
Naef-Roth, S. (1957). *Z. PflKrankh. PflPath. PflSchutz* **64**, 421–426.
Naef-Roth, S., Gäumann, E., and Albersheim, P. (1961). *Phytopath. Z.* **40**, 283–302.
Page, O. T. (1959). *Phytopathology* **49**, 230.
Pouteau-Thouvenot, M. and Barbier, M. (1965). *C. r. hebd. Séanc. Acad. Sci., Paris* **260**, 5924–5927.
Pouteau-Thouvenot, M. and Barbier, M. (1966). *C. r. hebd. Séanc. Acad. Sci., Paris*, 262, 710–711.
Robert, M., Barbier, M., Lederer, E., Roux, L., Biemann, K., and Vetter, W. (1962). *Bull. Soc. chim. France* **1962**, 187–188.
Sanwal, B. D. (1956). *Phytopath. Z.* **25**, 333–384.
Stoll, Ch., Renz, J., and Gäumann, E. (1957). *Phytopath. Z.* **29**, 388–394.

Trione, E. J. (1960). *Phytopathology* **50**, 480–482.
Wood, R. K. S. (1967). "Physiological Plant Pathology." Blackwell, Oxford and Edinburgh, pp. 570.

Discussion

Durbin: Assays for enzymatic activity must be chosen with care because some "solubilizing" products of micro-organisms, e.g. LMP (lipomucopolysaccharide) of *P. lachrymans*, appear to be but are not enzymes. I mention this with a question as to the methods that have been used in assays of phytolysin.

Also, do resistant plants metabolize lycomarasmin differently from susceptible plants, as apparently happens with fusaric acid.

Kern: Crude phytolysin contains polygalacturonase, polymethylgalacturonase and pectin transeliminase; during purification the activities of these enzymes disappear. Viscosity measurements, thin layer chromatography, tissue disintegration was tested with tomato and cucumber tissues.

Metabolism of lycomarasmin by resistant and susceptible plants has not yet been studied.

Barbier: Detoxification of fusaric acid by N-methylation probably uses large amounts of methionine. It seems possible, therefore, that application of this amino acid could increase the resistance of the plant.

Wood: I should like to comment on the use of the word maceration. Most plant pathologists use it in the first place for cell separation along the line of the middle lamella and then, possibly, for the consequences of separation. Physiologists and others also use it for more drastic effects, usually brought about with strong acids, in which tissue is so disintegrated that individual cells have lost their identity. I suspect that Kern may be using maceration in a still other sense for disruption of cell membranes that may occur in the absence of cell separation. This ambiguity suggests that the sense in which the word is used should be made clear or be clear from the context.

Grossmann: It is difficult to imagine cell separation without pectolytic activity. Did you look for the pectolytic activity *in vivo* after application of phytolysin? It may be converted to a pectolytic enzyme by the plant tissue or, more probably, that it activates pectolytic enzymes of the host.

Kern: The activation or liberation of pectolytic enzymes produced by the host (e.g. by damaging plasmatic membranes of host cells) is quite possible. The word maceration is used for cell separation.

Page: We have noticed that fusaric acid isolated from diseased banana tissue was invariably chelated with iron and copper, that is, there were two fusaric acid chelates present in the tissues.

Introduction of fusaric acid-producing and non-producing isolates of *Fusarium oxysporum* and *F. solani* into tobacco plants did not result in disease although the fungi grew fully in the vascular tissues.

The fusaric acid-producing and non-producing strains behaved similarly before and after inoculation and subsequent re-isolation.

Can you explain the failure of symptoms to develop when fusaric acid producing isolates are inoculated into the tobacco plants?

Also, in connexion with the silvering symptom caused by phytolysin, have you investigated this phenomenon in the silver scurf disease of plants caused by *Spondylocladium atrovirens*?

Kern: Production of fusaric acid depends on nutritional and other environmental factors and might be reduced or blocked in tobacco plants; another explanation might be that fusaric acid is rapidly metabolized and inactivated.

Silver scurf disease of potato (*S. atrovirens*) has not been studied in this context.

Dimond: It is known that biosynthesis of fusaric acid is substrate-dependent. Can the results described by Page and earlier by Kern relating to fusaric acid production by strains of *F. pisi* not be correlated with pathogenicity? Will substrate specificity for biosyntheses and the possibility that this specificity differs between strains account for effects described by Page and Kern?

Kern: This explanation seems quite reasonable.

Graniti: Kern has stated that the ability of some phytotoxins to form complexes with metal ions has attracted considerable attention, because toxicity of these metabolites is changed by chelation. However, research along this line has been mostly done under laboratory conditions with purified preparations of toxins and with test plants or cuttings.

Field observations on the regression of olive anthracnose in Southern Italy make me think that these phenomena may be important under natural conditions.

After its appearance in Italy, the causal agent of this disease (*Gloeosporium olivarum* Alm. *Colletotrichum gloeosporioides* Penz.) caused heavy losses to olive orchards in the decade following the Second World War. Besides fruit, twigs and small branches of olive trees became infected and wilted. As chemical control became more generally used, severity of the disease was reduced. Obviously, this was achieved mainly because infections particularly of fruit was prevented. However, when copper fungicides were used, beneficial effects were also noticed in the trees whose branches and twigs were infected before application of sprays. Although no experimental evidence is yet available, I am inclined to think that detoxification by the copper of aspergillomarasmin B (lycomarasmic acid), which has been shown

(Ballio *et al.*, 1969, *Phytopath. medit.* **8**, 187–196) to be the toxic principle of *C. gloeosporioides*, had contributed to the recovery of these trees from wilt.

Main: The phenomenon of cross protection between *Fusarium* spp. *in vivo* is well documented. Kern indicated that in culture, weakly pathogenic isolates produce high levels of substances such as fusarubin and javanicin and low levels of fusaric acid and lycomarasmin. What is the spectrum of the antibiotic activity of these compounds? In cross protection, would the weakly pathogenic isolates grow sufficiently *in vivo* to produce these compounds and could they be antibiotic to the challenging pathogenic isolates?

Kern: These compounds exhibit weak to moderate antibiotic activity against fungi with the exception of lycomarasmin which is inactive. In cross protection experiments a weakly pathogenic isolate may be expected to grow sufficiently to exert some antibiotic activity on the challenging pathogenic isolate but the result will depend very much on the isolates used and on the conditions of the experiment; nutritional competition and other factors may also be important.

Matta: As I remember, Jost in Zürich made the interesting observation that the resistance of tomato cuttings to fusaric acid could be increased by pretreating them with low doses of the same toxin.

Mussell: There seems to be a correlation between marticin production and increased virulence of several of your *Fusarium* spp. Is the converse also true? Are the isolates which produce the materials with higher antibiotic activity such as novarubin better saprophytes?

Kern: Isolates of *F. solani* f. sp. *pisi* produce the antibiotic metabolites novarubin and norjavanicin in the rhizosphere and may inhibit other fungi (D. Meiler (1970), *Phytopath. Z.*, **68**, 289–322).

Wood: Cross protection is well known in diseases caused by *F. oxysporum*. Inoculation with an avirulent race induces changes that makes the host resistant to subsequent infection by a potentially virulent race. It would be interesting to know what role if any fusaric acid or other toxins play in cross protection. Here it may be relevant to refer to Keyworth's work with toxins (not, I believe, fusaric acid) produced by *F. oxysporum* f. sp. *lycopersici*. Cultivars resistant to the pathogen are much more sensitive to the toxin than are susceptible cultivars.

Mantle: Is the exponential production phase of fusaric acid *in vitro* coincident with the exponential growth phase?

Kern: Yes.

Scheffer: Kuo and Scheffer (*Phytopathology* (1964) **54**, 1041–1044) have shown that the biochemical effects of infection of tomato by *F. oxysporum* f. sp. *lycopersici* and the effects of fusaric acid treatment differ in kind. Infection results in increased respiration and increased synthesis and activities of mitochondrial enzymes. Fusaric acid causes drastic decreases in such activities. I have interpreted this to mean that fusaric acid does not have a major role in disease development. We do not rule out a possible minor and secondary role. I would like to hear any other interpretations of our results.

Kern: The increased respiration in infected plants may be ascribed to lycomarasmin after chelation with iron. In later stages of disease the respiration rate decreases significantly; this may be due to fusaric acid.

Durbin: There may not always be a good correlation between *in vitro* production of a toxin and virulence, particularly when one compares isolates of different virulence on different media and harvests at different times.

Kern: This is true and the relation between toxin production *in vitro* and virulence needs careful study in every single case.

Dimond: We know that oxygen tension in xylem vessels is low. Kern states that more fusaric acid is produced in shake than in quiet cultures. The limited supply of oxygen in vessels may be very relevant to the production of fusaric acid *in vivo*.

Kern: Fusaric acid appears earlier in shake culture but in still culture high amounts may also be produced. In both cases the composition of the nutrient solution has a great influence on toxin production.

Goodman: Might differences in results obtained by different workers with fusaric acid be explained by differences in cultivars of tomato used as test plants, e.g. Tuckswood against Bonny Best?

Kern: With plants grown and tested under the same conditions we did not find any significant differences in symptom expression between tomato races.

Kuć: I can relate the cross protection obtained with *Fusarium* spp. to our experience with *Colletotrichum lindemuthianum*. Non-pathogens of bean and incompatible races of *C. lindemuthianum* induced a resistant response in bean seedlings which includes biosynthesis of phenolics and accumulation and marked increased activity of phenylalanine deaminase. Inoculated plants became resistant to compatible races of *C. lindemuthianum*. Levels of enzyme and phenolics do not remain high during the period of protection. Thus it appears non-pathogens or incompatible races of the pathogen induce the expression of a resistance mechanism which after activation was not reversed or stopped by inoculation with a compatible race of the fungus.

4. Production and Bioassay of Phytotoxins

STEPHI NAEF-ROTH

Institut für Spezielle Botanik,
Eidg. Technische Hochschule,
Zürich 6, Switzerland

Biological Assay for Toxic Action

Injury Induced in Phanerogams

Host plant. Difficulties concerned with evaluation of phytotoxins arise from the fact that toxic action depends on a number of divergent factors. The toxicity of a substance is recognized by physiological effects which it produces in the host plant. The best assay is application to the living host at the site of parasitic attack. Thus, wildfire toxin causes the same injury in tobacco leaves as an infection with *Pseudomonas tabaci* (Wolf *et* Fost.) Stev. (Braun, 1955). Another example is the toxin of *Helminthosporium victoriae* Meehan *et* Murphy; extracts from culture filtrates applied to the rhizosphere of oats inhibit growth and development of roots of a susceptible variety in the same way as does the pathogen (Litzenberger, 1949). Another way to evaluate the action of a phytotoxin is to inject it into the stem of a living plant. If lycomarasmin, one of the toxins of wilt inducing species of *Fusarium*, is injected into the stem of a tomato plant, it injures the leaves severely, whereas the stem remains undamaged (Fig. 1). A solution of phytolysin, a toxic protein produced by various plant parasites (Kern and Naef-Roth, 1971a), injected into the stem of plum trees induces silvering similar to disease symptoms caused by *Stereum purpureum* (Pers. *et* Fr.) Fr. (Naef-Roth *et al.*, 1963). Fomannosin, produced by *Fomes annosus* (Fr.) Cooke, causes necrotic lesions around the area of application when injected into roots of pine seedlings (Bassett *et al.*, 1967). Aflatoxin B_1 which El-Khadem *et al.* (1966) found to be a vivotoxin of *Aspergillus flavus* Link *et* Fr., inhibits growth of peanut seedlings as is observed in diseased plants.

These techniques of direct application to the host plant can be used for routine work only in a few cases. Thus, the host specific toxins of *H. victoriae*, *H. carbonum* Ullstr. and *Periconia circinata* (Mang.) Sacc. can be isolated by means of their striking and specific growth inhibiting properties

(Scheffer and Pringle. 1967). Usually, however, and for practical considerations, a substitute assay must be chosen. Some of these are described below.

Plant cuttings. Toxins affecting higher plants can be applied to leaves or cuttings in solution taken up by the transpiration stream. Symptoms that develop can be used to compare the activity of culture filtrates, extracts and pure substances. Often they represent only a part of the syndrome.

Assays are carried out under controlled conditions of temperature, humidity and light. After uptake of the required dose of toxin the cuttings are transferred to water; 24–48 h later the symptoms are read using a scale from 0 to 4 (0 no damage, 4 heavy damage). The wilt index thus read is obtained as an average for 10–50 plants. Additionally water loss can be determined as a percentage of fresh weight; this usually shows a good correlation with the wilt indices (Table 3).

Symptoms differ with different toxins. Alternaric acid, a toxin from *Alternaria tenuis* Nees, produces black patches on potato leaflets and dark necrotic lesions on stem and petioles, whereas patulin, an antibiotic produced by several species of *Penicillium* and *Aspergillus*, affects only the veinlets of the leaf (Brian *et al.*, 1952). Fusicoccin, a glycoside isolated from *Fusicoccum amygdali* Del., induces marginal and intercostal necroses in leaves of almond and tomato (Graniti, 1964). After treatment with culmomarasmin, a toxin from *Fusarium culmorum* (W. G. Sin.) Sacc. (Kiss *et al.*, 1960), a tomato sprout stiffens and the leaves show extended, paper-like necroses (Fig. 2). Pea leaflets treated with marticin, one of the toxic naphthoquinones formed by *Fusarium* Section *Martiella* Wollenw. (*F. solani* (Mart.) Sny. *et* Hans.) (Kern and Naef-Roth, 1965), show greyish necrotic areas mainly at the bottom of the leaf. Lycomarasmin causes light brown necroses at the margin and the tips of tomato leaves (Kern, 1956), whereas the symptoms of fusaric acid injury, a second toxin from wilt inducing species of *Fusarium*, in tomato cuttings are yellow necroses near the veins as well as a collapsing of veins, petioles and stem (Gäumann *et al.*, 1952).

The way in which toxin solutions are applied to cuttings can be important (Table I). Thus, the toxic effect of the same dose of culmomarasmin or marticin on tomato plants decreases considerably with increasing length of uptake time. Fusaric acid injury is more severe after a short uptake time but it remains at the same level whether the same amount of toxin is applied during 3 or 7 h. With lycomarasmin the same dose causes more injury when applied over a longer period.

Another factor influencing the activity of toxins is the physiological state of the test plants. The toxicity of lycomarasmin depends on the season in which assays are carried out; the sensitivity of tomatoes grown in the greenhouse in winter is higher than in summer (Gäumann and Naef-Roth, 1953).

FIG. 1. Necroses induced in a tomato plant after injection of lycomarasmin.

FIG. 2. Necrosis of a tomato leaf after treatment with culmomarasmin.

TABLE I. Influence of toxin concentration on symptom development in tomato cuttings

Toxin	Dose (μg/g)	Conc. (μg/ml)	Time of uptake (min)	Wilt index
Culmomarasmin	1·5	12	80	3·2
		6	150	2·4
		3	280	2·0
		1·5	360	1·8
		0·8	600	1·2
Marticin	12	96	70	2·6
		48	150	2·0
		24	330	1·3
		12	450	0·9
		6	600	0·7
Fusaric acid	225	900	100	1·9
		450	160	1·0
		225	220	0·9
		115	300	0·8
		60	400	0·8
Lycomarasmin	225	1800	105	1·0
		900	180	1·2
		450	300	1·5
		225	480	2·3
		115	600	2·5

Besides seasonal variations the sensitivity of tomato tissues undergoes diurnal variations (Table II); the necrotic effect of fusaric acid is highest when plants are treated at noon. It also increases with the age of tissue (Table II); more mature leaves are more sensitive than younger ones. The light used for growing the test plants can be decisive. Thus, Ruffner (1970, unpublished) grew tomatoes on a standard medium (Zähner, 1954) under defined conditions of temperature and humidity and illumination by Sylvania Grolux S lamps (corresponding to the absorption spectrum of chlorophyll) or Philips TLF 55 (corresponding to daylight). The response of the plants to toxins was different. Fusaric acid and marticin induced heavier injury in tomatoes grown under Sylvania lamps than in those grown under Philips lamps (Table III). Water losses from plants in these experiments were well correlated to the wilt indices (correlation coefficients 0·65–0·79).

TABLE II. Influence of time of treatment and age of tissue on symptom development in tomato (225 μg fusaric acid per g tomato cutting) (after Jost, 1965)

Time of day	Wilt index	Age of leaf*	1-month-old plant Wilt index	2-month-old plant Wilt index
03.00	2·1	1	1·1	1·4
06.00	2·5	2	1·9	2·3
08.00	2·6	3	2·1	2·9
10.00	2·7	4	2·3	3·2
12.00 noon	3·0	5	2·3	3·1
15.00	1·8	6	2·4	3·0
19.00	1·6	7	2·5	3·2

* 1—leaf nearest apex.

The nutrition of test plants also has a considerable effect on sensitivity of tissues to the toxins of *F. oxysporum* f. sp. *lycopersici* (Sacc.) Sny. *et* Hans. Changes in the N/P ratio influences the susceptibility of tomatoes to lycomarasmin and its iron-chelate much less than to fusaric acid (Table IV). Nitrogen deficient plants show a high resistance to fusaric acid injury. Similar results are obtained with changes in the N/K ratio (Zähner, 1954). Optimum nitrogen levels in plants parallel maximum sensitivity of tissues to fusaric acid. Responses to the two other toxins is not significantly changed.

Similar results have been found in experiments on the influence of nutrition on disease resistance (Stoddard and Dimond, 1948). Plants grown at normal nutritional levels and infected with *F. oxysporum* f. sp. *lycopersici*

TABLE III. Influence of light on the sensitivity of tomato tissue to fusaric acid and marticin

Fusaric acid					Marticin				
Dose	Philips lights		Sylvania lights		Dose	Philips lights		Sylvania lights	
μg/g	W*	L†	W	L	μg/g	W	L	W	L
0	0	−10·2	0	−17·2	0	0	−10·2	0	−17·2
100	0·4	3·6	0·5	3·8	6	0·3	3·0	0·8	8·8
200	1·5	19·3	1·7	29·1	12	1·4	25·8	1·8	38·7
300	2·2	41·9	3·2	60·4	24	2·8	54·7	3·3	59·5

* Wilt index.

† Loss of water as % initial fresh weight (Ruffner, 1971)

TABLE IV. Influence of the N/P ratio in the growth medium for tomatoes on sensitivity to the toxins of *Fusarium oxysporum* f. sp. *lycopersici* (Zähner, 1954)

N/P		Fusaric acid (225 μg/g) Wilt index	Lycomarasmin (225 μg/g) Wilt index	Fe-Lycomarasmin (60 μg/g) Wilt index
1	66·5	0·4 ± 0·18	1·0 ± 0·20	1·0 ± 0·20
14·1	53·4	0·9 ± 0·28	1·8 ± 0·33	1·2 ± 0·07
27·2	40·3	3·5 ± 0·36	1·9 ± 0·25	1·4 ± 0·15
40·3	27·2	3·5 ± 0·40	1·8 ± 0·25	1·5 ± 0·16
53·4	14·1	3·6 ± 0·33	2·0 ± 0·16	1·9 ± 0·25
66·5	1	2·7 ± 0·58	2·3 ± 0·20	2·1 ± 0·23

developed severe symptoms whereas plants grown on a medium diluted 10 times showed no damage. Disease intensity increases with increasing amount of nitrogen in the nutrient solution (Walker and Foster, 1946).

The variation of tissue sensitivity to fusaric acid can be partly explained by differential degradation of fusaric acid which can be metabolized by

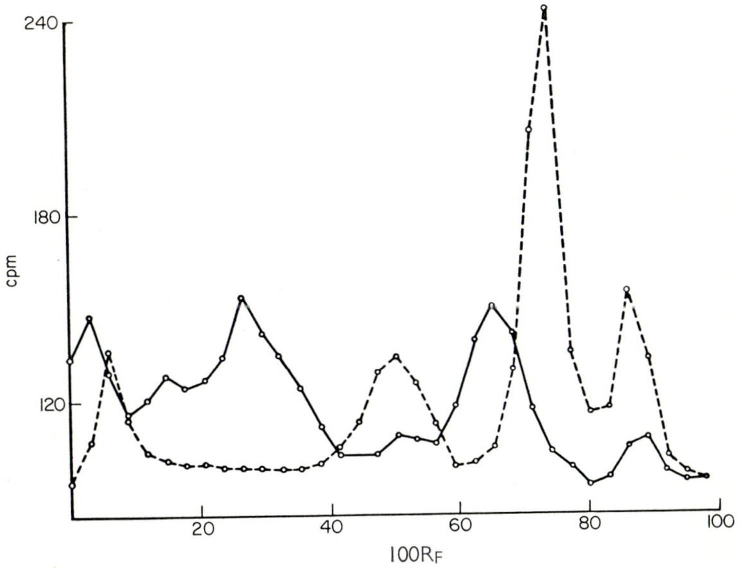

FIG. 3. Radiochromatograms of tomato extracts [cuttings treated with C^{14}-labelled fusaric acid in March (dotted line) and in October (full line); extraction 72 h after treatment (J. Harr and F. Ruffner, 1971)].

decarboxylation and methylation into a number of more water soluble substances. Application of C^{14}-labelled toxin to plants leads to a different pattern of metabolic products according to the type of plant used (Fig. 3). If tomato cuttings from plants grown under defined conditions in the greenhouse are treated in March or in October, the paper radiochromatograms of plant extracts differ markedly. A large peak of fusaric acid (100R_F75) in the extract of March tomatoes does not appear in the extract of October tomatoes, whereas in the October extract two additional metabolites appear with Rf 25 and 65 showing a rather high amount of radioactivity.

Tissue permeability. Alteration in cell permeability is one of the characteristic features of plant diseases. Thus, plants inoculated with the tomato wilt pathogen as well as plants treated with culture filtrates or with vascular fluids from infected plants show an increase of permeability (Collins and Scheffer, 1958; Thatcher, 1942; Gottlieb, 1944). A change in the level of inorganic ions in cotton plants after infection with *F. oxysporum* f. sp. *vasinfectum* is reported by Sadasivan and Kalyanasundaram (1956). The role of permeability changes in relation to the effects of phytotoxins has been demonstrated for fusaric acid on *Rhoeo* protoplasts (Bachmann, 1956; Gäumann, 1958), and on tomato (Gäumann and Löffler, 1957; Linskens, 1955). Cells of root hairs of oat and of hairs of tomato petioles loose the ability to plasmolyse after a short exposure to *Helminthosporium victoriae* toxin (Samaddar and Scheffer, 1968), or to fusaric acid (Zähner, 1954). Pronounced loss of

TABLE V. Material released by toxin treated tissues leached in distilled water

Tissue	Toxin	Electrical conductivity (mhos . 10^5)	Material released (* mg/100 cm^2 leaf surface, † mg/g leaf tissue) Na$^+$	K$^+$	Ca^{++}	Amino acids	
Tomato[1]	Fusaric acid	2·1	0·81	3·45	0·38	0·26	*
	Lycomarasmin	3·1	0·10	0·22	0·47	0·28	*
	Water control	1·1	0·02	0·01	0·05	0·09	*
Oat[2, 3]	*H. victoriae* toxin	5·7	0·20	1·15	0·13	0·47	†
	Water control	2·1	0·22	0·15	0·24	0·18	†
Tomato[4]	Fusicoccin A			6·0	3·0		†
	Water control			0·8	0·8		†

Data from: [1] Linskens (1955); [2] Wheeler and Black (1963); [3] Black and Wheeler (1966); [4] Ballio *et al.* (1968)

electrolytes, especially K, and amino acids (Table V) occur in oat tissue infected with *H. victoriae* or treated with its toxin (Wheeler and Black, 1963; Samaddar and Scheffer, 1968). Fusicoccin A causes an increased release of K, Ca and Mg from tomato leaf tissue (Ballio *et al.*, 1968).

Electrolyte loss can easily be measured by changes in electrical conductivity of water in which treated tissues are suspended and shaken for some hours (Table V). This assay is preferred to root inhibition tests for *H. victoriae* toxin as here a change can be induced by a toxin concentration 10 times lower than that required to inhibit root growth, in a shorter time and with more precision (Wheeler and Black, 1963).

Growth Inhibition

Techniques. The activity of non-specific phytotoxins (Table VIII) can be investigated in assays carried out with micro-organisms. Growth inhibition is the method mainly used to evaluate toxicity and to estimate the amount of toxin produced in culture or in diseased plants. Such methods are often highly sensitive and precise and they may allow identification of two or more toxic metabolites produced by the same pathogen on the basis of different effects on different organisms.

Inhibition of bacterial growth is determined by agar plate techniques such as are used with antibiotics (Kalyanasundaram, 1955; Kalyanasundaram and Venkata Ram, 1956). Similar techniques can be used with algae. Thus *Chlorella pyrenoidosa* Pring. has been used to determine the production of fomannosin in culture (Bassett *et al.*, 1967). A disadvantage of the agar plate assay is the difficulty in determining the effective concentration of a toxin in the medium.

The spore germination technique has also proved very satisfactory. Germination and growth of germ tubes of *Botrytis allii* Munn are very sensitive to the action of alternaric acid (Brian *et al.*, 1951). The same test was used to isolate two of the naphthoquinones produced by *Fusaria*, Section *Martiella*, novarubin and norjavanicin (Kern and Naef Roth, 1967). Lycomarasmin and fusaric acid can be separated by a germination test with *Ustilago maydis* (DC.) Cda. spores; germination is inhibited by fusaric acid but not by lycomarasmin (Kern, 1952).

Furthermore, the activity of non-specific phytotoxins can be evaluated by inhibition of growth of plant seedlings. Here stimulation at sub-inhibitory levels may also be recognized (Table VI). Under the influence of fusicoccin A growth of rice seedlings is affected in two ways; root growth is inhibited, whereas the growth of shoots is stimulated by low concentrations (Bottalico, 1969). The development of cress seedlings is inhibited by a toxin from *Neocosmospora africana* v. Arx (Kern *et al.*, 1971b). Under the influence of diacetoxyscirpenol isolated from *F. equiseti* (Cda.) Sacc. by Brian *et al.*,

TABLE VI. Inhibition or stimulation of growth of seedlings (mean length in mm after 4 days growth)

Fusicoccin A[1]			Diacetoxyscirpenol[2]		*Neocosmospora africana* Toxin[3]	
Ppm	Rice shoot	Rice root	Ppm	Cress root	Ppm	Cress root
0	4·9	11·3	0	26·8	0	25·3
1	6·6	10·0	0·0006	26·9	0·04	24·8
2·5	7·1	9·1	0·0025	26·7	0·16	25·0
5	7·8	7·9	0·01	30·3	0·75	22·3
10	7·8	4·8	0·04	31·5	1·5	18·5
20	7·7	3·7	0·16	32·2	3	16·1
50	6·8	2·5	0·625	31·2	6	15·3
100	5·6	1·6	2·5	30·0	12	14·0
250	3·8	0·6	10	13·4	50	10·5

[1] Bottalico (1969); [2] Brian *et al.* (1961); [3] Kern *et al.* (1971b)

(1961), root elongation of cress seeds is reduced at 10 ppm but is stimulated at higher dilutions.

Diacetoxyscirpenol furthermore inhibits the extension of pea sections thus counteracting the influence of auxins. In experiments in which the toxin was combined with indoleacetic acid at various concentrations the response of pea internode sections to the auxin was blocked (Table VII). A similar effect was found on elongation of pea sections induced by gibberellic acid (Brian *et al.*, 1961).

Interference by antimetabolites. All the techniques under this heading have the advantage that the test organisms are grown under defined conditions in

TABLE VII. Effect of diacetoxyscirpenol and indoleacetic acid (IAA) on extension of pea internode sections (mean increment in mm of 5 mm sections in 24 h at 15°C) (Brian *et al.*, 1961)

IAA μg/ml	Diacetoxyscirpenol (μg/ml) 0	0·008	0·04	0·2	1·0
0	1·8	1·8	1·4	1·2	0·8
5	3·2	3·0	2·8	1·8	1·1
10	4·1	3·6	2·9	2·0	1·2
20	3·6	3·8	3·1	2·2	1·2

TABLE VIII. Lowest effective concentration and host specificity of some phytotoxins

Toxin	Growth inhibition		Necrosis			
	Bacteria*	Fungi†	Plant root†	Plant shoot‡	Host specificity§	Reference
H. victoriae toxin	none	none	0·0002		+++	Scheffer *et al.*, 1963
victoxinine	none	none	75		++	Pringle *et al.*, 1960
H. carbonum toxin	none	none	0·5		+++	Pringle *et al.*, 1967
Carbtoxinine	none	none	25		++	Pringle *et al.*, 1967
Wildfire Toxin	none	none		0·05‖	+	Braun, 1955
Tentoxin	none	none		0·2‖	+	Saad *et al.*, 1970
Fusicoccin A	none	none	20	0·6	+	Graniti, 1964
Culmomarasmin	none	none	none	0·8	+	Kiss *et al.*, 1960
Fe-Lycomarasmic acid	none	none	none	10–30	+	Gäumann *et al.*, 1959
Fe-Lycomarasmin	none	none	none	15–40	+	Gäumann *et al.*, 1959
Lycomarasmic acid	none	none	none	50–70	+	Gäumann *et al.*, 1959
Lycomarasmin	none	none	none	50–200	+	Gäumann *et al.*, 1950
Alternaric acid	none	0·1	1	0·2	—	Brian *et al.*, 1952
Ascochitin	3	3	50	10	—	Oku *et al.*, 1963, 1966
Aflatoxins	1	25	25	1	—	Legator, 1969; El Khadem *et al.*, 1966
Diacetoxyscirpenol	none	100	10	10	—	Brian *et al.*, 1961
Enniatin A	60	450		35	—	Gäumann *et al.*, 1960
Enniatin B	500	120		20	—	Gäumann *et al.*, 1960
Baccatin A	60	65		7	—	Gäumann *et al.*, 1960

TABLE VIII. Lowest effective concentration and host specificity of some phytotoxins—*continued*

Toxin	Growth inhibition		Necrosis		Host specificity	Reference
	Bacteria*	Fungi†	Plant root†	Plant shoot‡		
Marticins	7	none	50	8	—	Kern *et al.*, 1967
Novarubin	0·5	5		35	—	Kern *et al.*, 1967
Fusarubin	2	none		40	—	Kern *et al.*, 1967
Javanicin	2	none		60	—	Kern *et al.*, 1967
Norjavanicin	1	10		90	—	Kern *et al.*, 1967
Fusaric acid	7	15	25	20–200	—	Gäumann, 1957
Dehydrofusaric acid	15	25		350	—	Stoll *et al.*, 1957

* Lowest inhibiting concentration; μg/ml medium.
† LD50 μg/ml medium.
‡ Minimal dose μg/g fresh weight.
§ Specific for susceptible host +++, susceptible and resistant host ++, phanerogams +, not specific —.
|| Chlorosis.

TABLE IX. Nullifying effect of peptone and glutamic acid on antibiotic action of naphthoquinones. Inhibition zone mm, plate diffusion test *Bacillus subtilis*, disks 6 mm

Test medium	Addition of glutamic acid*	Marticin 25	Novarubin 15	Fusarubin 15
		(μg/ml test medium)		
I Meat extract + peptone	0	7·0	9·7	11·5
II Glucose + tartrate + inorganic salts†	10×	8·8	17·3	11·0
	3·3×	9·5	19·5	12·7
	1×	10·0	21·0	14·7
	0	14·5	23·2	17·7

* Concentration of glutamic acid calculated on molecular weight of toxins.
† (Kern and Naef-Roth, 1967.)

a known medium. They can be carried out with small amounts of the toxic material. The results, however, may be affected by various substances, e.g. amino acids present in media or organisms. Thus, naphthoquinones are less toxic for bacteria in a medium containing peptone than in one containing glucose and tartrate. Glutamic acid acts as an antagonist (Table IX). Peptone present in the medium for germinating spores of *Cochliobolus miyabeanus* (Ito *et* Kurib.) Drechsl. decreases the antifungal activity of ascochitin a toxic metabolite from *Ascochyta pisi* Lib. (Oku and Nakanishi, 1966). Here, the interfering substances have been identified as glutamine, asparagine, alanine and arginine. Methionine neutralizes the inhibition of growth of *Chlorella vulgaris* Beij. caused by wildfire toxin but not the chlorosis induced in tobacco leaves (Braun, 1955). Rhizobitoxine, synthesized by *Rhizobium japonicum* (Kirch.) Buch. is detoxified by methionine and homocysteine; the development of *Salmonella typhimurium* (Loeffl.) Cast. *et* Chalm. in a nutrient medium is blocked by methionine deficiency caused by the toxin (Owens *et al.*, 1968).

Metal ions are another group of antimetabolites. Iron nullifies the inhibitory effect of ascochitin on germination of *Cochliobolus miyabeanus* spores and of mustard seed whereas other metals such as Mg, Cu, Co and Zn do not (Oku and Nakanishi, 1963). Iron and also Cu, Al, Co, Ba and Ni in very low concentrations reduce culmomarasmin injury to tomato. Iron detoxifies fusaric acid when tested with yeasts but not when bacteria, fungi and phanerogams are used (Bär, 1963). Chelation with Cu inactivates naphthoquinones whereas Al and Fe are less effective (Kern and Naef-Roth, 1966, 1967).

Cholesterol, cysteine and glutathione decrease injury caused by culmomarasmin in tomato. Sulphite affects the toxicity of *H. victoriae* toxin for oat seedlings considerably (Scheffer and Pringle, 1964) whereas it has little effect on the toxicity of *H. carbonum* toxin (Pringle and Scheffer, 1967).

Effect of Substrate on Toxin Formation

Culture Conditions

Phytotoxins are mostly formed by actively growing mycelia as has been shown for *Fusicoccum amygdali* (Graniti and De Leo, 1964), *H. victoriae* (Nishimura *et al.*, 1966), *Gibberella fujikuroi* (Saw.) Woll. (Stoll and Renz, 1957) and *Fusarium culmorum* (Table X).

TABLE X. Phytotoxicity of filtrates of cultures of *Fusarium culmorum*. Raulin medium, shake culture at 27°C, test with tomato cuttings, U unit per ml culture filtrate = dilution end point giving wilt index 1·5

	Age of culture (h)									
	45	55	62	70	80	87	103	120	144	164
Mycelium (mg/100 ml)	48	63	68	76	71	67	63	63	89	44
Activity (U/ml)	46	92	130	52	13	25	32	20	5	5

Conditions of aeration and temperature may greatly influence toxin production. *F. culmorum* produces no toxin in still culture but its main toxic metabolite, culmomarasmin, is formed readily in shake cultures at 27°C in 2–3 days. A second substance produced at 27°C after 4–5 days corresponds to the toxin found by Landolt (1952) in cultures shaken at 24°C. At the lower temperature of 24°C culmomarasmin does not appear. In contrast, maximum production of tentoxin by *Alternaria tenuis* occurs at 28°C and only in still cultures (Saad *et al.*, 1970).

Nutritional Factors

Another important factor for the formation of toxins by plant pathogens is optimum supply of essential nutrients.

The type of carbon source greatly affects the synthesis of toxin by *F. amygdali* (Graniti and De Leo, 1964). The yield of toxin is best with arabinose, glucose or sucrose. It decreases to less than a half with fructose and is very low with maltose and xylose. Formation of aflatoxins is favoured by sucrose; glucose, ribose, xylose and fructose give a medium yield only (Diener and Davis, 1969).

The kind and amount of the nitrogen source can also be important. Ammonium nitrogen favours toxin formation by *Fusarium scirpi* Lamb. *et* Faut. (Brian *et al.*, 1961) and by *Fusicoccum amygdali* (Graniti and De Leo, 1964). Ammonium sulphate and potassium nitrate are among the best inorganic nitrogen sources for production of aflatoxins. However, highest yields are produced in media containing corn steep liquor, casamino acids, yeast extract or peptone (Diener and Davis, 1969). For the pea pathogen, *Fusarium solani* f. sp. *pisi* (Jones) Sny. *et* Hans., synthesis of toxins depends mainly on the supply of ammonium. Although the pathogen grows well on a medium containing nitrate nitrogen, no toxins are formed. The standard Raulin medium used for production of naphthoquinones contains ammonium salts as the sole nitrogen source and the amount formed depends on the NH_4-content of the medium (Table XI). The best yield is obtained on the standard medium containing 425 mg N per litre. Toxin production falls to less than one third if the nitrogen content is increased to double or decreased to one half this concentration. Further decrease stops toxin production. Synthesis of the quinones can be stimulated by addition of very small amounts of glutamic or aspartic acid to a nitrogen deficient medium; these amino acids at 0·5–1 ppm nitrogen stimulate toxin production as much as does the addition of 50–100 ppm ammonium nitrogen as tartrate (Kern and Naef-Roth, 1971c). The two amino acids are among the main nitrogen compounds in roots and root exudates of pea and thus may play a part in pathogenesis.

Toxin synthesis may also be influenced by metal ions. Phytotoxicity of filtrates of cultures of *Fusicoccum amygdali* increases with traces of Cu or Mn and decreases with Co or with a high amount of Fe (Graniti and De Leo, 1964). *Fusarium solani*, growing in a medium with an optimum supply of nitrogen requires traces of Zn and Mg for production of pigments (Table XII). Zinc also in low concentrations stimulates the production of aflatoxins (Diener and Davis, 1969).

The influence of zinc on the physiology of *F. oxysporum* f. sp. *lycopersici* depends on various factors. In a medium without Zn two strains of differing pathogenicity showed very little growth, the non-pathogenic strain only producing fusaric acid. With increasing Zn content no toxin was formed by the non-pathogenic strain whereas the pathogenic strain synthesized large amounts of fusaric acid (Egli, 1969). In a medium containing 0·1–3 ppm Zn^{++} production of toxin by the two strains is related to their pathogenicity.

A similar situation has been found with *F. oxysporum* f. sp. *lini* (Bolley) Sny. *et* Hans. (Trione, 1960) and *F. oxysporum* f. sp. *niveum* (E. F. Sm.) Sny. *et* Hans. (Davis, 1969), both synthesizing fusaric acid in culture with an optimum supply of trace elements. The evidence suggests that selective

TABLE XI. Influence of the source of nitrogen on production of naphthoquinones by *Fusarium solani* f. sp. *pisi*. Raulin medium, shake culture at 27°C; 5 days (Kern and Naef-Roth (1971c))

Nitrogen		Mycelium	Naphthoquinones (mg/litre)			
Ppm	Source	(mg/100 ml)	Marticins	Novarubin	Fusarubin	Norjavanicin
850	Inorganic NH_4	476	36	18	18	2
425		481	130	45	65	6
320		475	84	32	60	10
213		466	22	8	30	5
106		446	2	0	2	0
43		86	0	0	0	0
213	Inorganic NH_4	466	22	8	30	5
+ 30	Glutamic acid	585	0	0	0	0
10		578	0	0	0	0
3·3		488	28	8	28	16
1·1		436	48	28	38	18
0·3		427	36	38	78	8
0·1		429	26	18	18	6
+ 30	Aspartic acid	580	0	0	0	0
10		585	2	8	0	0
3·3		545	2	4	10	2
1·1		527	22	22	38	4
0·3		454	34	24	42	4
0·1		474	28	12	28	2

TABLE XII. Influence of Zn^{++} and Mg^{++} on production of naphthoquinones by *Fusarium solani* f. sp. *pisi*. Raulin medium, shake culture 27°C, 5 days (Kern and Naef-Roth, 1971c)

Metal ions (ppm) Zn++	Mg++	Mycelium (mg/100 ml)	Naphthoquinones (mg/litre) Marticins	Novarubin	Fusarubin	Norjavanicin
0	79	560	0	1	1	0
0·1	79	535	2	4	4	2
0·4	79	456	28	18	56	6
1·1	79	455	114	40	56	3
11·4	79	440	130	48	84	4
11·4	26	442	112	56	72	8
11·4	9	375	20	28	56	8
11·4	3	334	4	9	17	1
11·4	0	18	0	0	0	0

pathogenicity of certain wilt inducing *Fusaria* may depend on the requirement of a special substrate for toxin synthesis.

Acknowledgement

Thanks are given to the Swiss National Foundation for Scientific Research for the support of our studies.

References

Bachmann, E. (1956). *Phytopath. Z.* **27**, 255–288.

Ballio, A., Graniti, A., Pocchiari, F., and Silano, V. (1968). *Life Sci.* **7**, II, 751–760.

Bar, H. (1963). *Phytopath. Z.* **48**, 149–177.

Bassett, C., Sherwood, R. T., Kepler, J. A., and Hamilton, P. B. (1967). *Phytopathology* **57**, 1046–1052.

Black, H. S. and Wheeler, H. (1966). *Am. J. Bot.* **53**, 1108–1112.

Bottalico, A. (1969). *Phytopath. medit.* **8**, 41–46.

Braun, A. C. (1955). *Phytopathology* **45**, 659–664.

Brian, P. W., Curtis, P. J., Hemming, H. G., Jefferys, E. G., Unwin, C. H., and Wright, J. M. (1951). *J. gen. Microbiol.* **5**, 619–632.

Brian, P. W., Elson, G. W., Hemming, H. G., and Wright, J. M. (1952). *Ann. appl. Biol.* **39**, 308–321.

Brian, P. W., Dawkins, A. W., Grove, J. F., Hemming, H. G., Lowe, D., and Norris, G. L. F. (1961). *J. exp. Bot.* **12**, 1–12.

Collins, R. P. and Scheffer, R. P. (1958). *Phytopathology* **48**, 349–355.

Davis, D. (1969). *Phytopathology* **59**, 1391–1395.
Diener, U. L. and Davis, N. D. (1969). *In* "Aflatoxins" (L. A. Goldblatt, Ed.), pp. 13–54. Academic Press, New York.
Egli, T. A. (1969). *Phytopath. Z.* **66**, 223–252.
El-Khadem, M., Menke, G. and Grossmann, F. (1966). *Naturwissenschaften* **20**, 532.
Gäumann, E. (1957). *Phytopathology* **47**, 342–357.
Gäumann, E. (1958). *Phytopathology* **48**, 670–686.
Gäumann, E. and Löffler, W. (1957). *Phytopath. Z.* **28**, 319–328.
Gäumann, E. and Naef-Roth, S. (1953). *Phytopath. Z.* **20**, 449–458.
Gäumann, E. and Naef-Roth, S. (1959). *Phytopath. Z.* **34**, 426–431.
Gäumann, E., Naef-Roth, S., and Kern, H. (1960). *Phytopath. Z.* **40**, 45–51.
Gäumann, E., Naef-Roth, S., and Kobel, F. (1952). *Phytopath. Z.* **20**, 1–38.
Gäumann, E., Naef-Roth, S., and Miescher, G. (1950). *Phytopath. Z.* **16**, 257–288.
Gottlieb, D. (1944). *Phytopathology* **34**, 41–59.
Graniti, A. (1964). *Phytopath. medit.* **3**, 125–128.
Graniti, A. and De Leo, P. (1964). *Phytopath. medit.* **3**, 109–118.
Harr, J. and Ruffner, F. (1971). Unpublished data.
Jost, J. P. (1965). *Phytopath. Z.* **54**, 338–378.
Kalyanasundaram, R. (1955). *J. Indian bot. Soc.* **34**, 43–46.
Kalyanasundaram, R. and Venkata Ram, C. S. (1956). *J. Indian bot. Soc.* **35**, 7–10.
Kern, H. (1952). *Phytopath. Z.* **19**, 351–382.
Kern, H. (1956). *Umschau* **18**, 548–551.
Kern, H. and Naef-Roth, S. (1965). *Phytopath. Z.* **53**, 45–64.
Kern, H. and Naef-Roth, S. (1966). *Phytopath. Z.* **57**, 289–297.
Kern, H. and Naef-Roth, S. (1967). *Phytopath. Z.* **60**, 316–324.
Kern, H. and Naef-Roth, S. (1971a). *Phytopath. Z.* in press.
Kern, H., Naef-Roth, S., and Défago, G. (1971b). *Phytopath. Z.* in press.
Kern, H. and Naef-Roth, S. (1971c) *Phytopath. Z.* in press.
Kiss, J., Naef-Roth, S., Hardegger, E., Boller, A., Lohse, F., Gäumann, E., and Plattner, Pl. A. (1960). *Helv. Chim. Acta* **43**, 2096–2101.
Landolt, E. (1952). *Phytopath. Z.* **19**, 126–128.
Legator, M. S. (1969). *In* "Aflatoxins" (L. A. Goldblatt, Ed.), pp. 107–149. Academic Press, New York.
Linskens, H. F. (1955). *Phytopath. Z.* **23**, 89–106.
Litzenberger, S. C. (1949). *Phytopathology* **39**, 300–318.
Naef-Roth, S., Kern, H. and Toth, A. (1963). *Phytopath. Z.* **48**, 232–239.
Nishimura, S., Scheffer, R. P. and Nelson, R. R. (1966). *Phytopathology* **56**, 53–57.
Oku, H. and Nakanishi, T. (1963). *Phytopathology* **53**, 1321–1325.
Oku, H. and Nakanishi, T. (1966). *Phytopath. Z.* **55**, 1–14.
Owens, L. D. and Wright, D. A. (1965). *Pl. Physiol.* **40**, 927–931.
Owens, L. D., Guggenheim, S., and Hilton. J. L. (1968). *Biochim. Biophys. Acta* **158**, 219–225.
Pringle, R. B. and Braun, A. C. (1960). *Phytopathology* **50**, 324–325.
Pringle, R. B. and Scheffer, R. P. (1967). *Phytopathology* **57**, 1169–1172.
Ruffner, F. (1971). Unpublished data.

Saad, S. M., Halloin, J. M., and Hagedorn, D. J. (1970). *Phytopathology* **60**, 415–418.
Sadasivan, T. S. and Kalyanasundaram, R. (1956). *Proc. Indian Acad. Sci.* **43**, 271–275.
Samaddar, K. R. and Scheffer, R. P. (1968). *Pl. Physiol.* **43**, 21–28.
Scheffer, R. P. and Pringle, R. B. (1963). *Phytopathology* **53**, 465–468.
Scheffer, R. P. and Pringle, R. B. (1964). *Phytopathology* **54**, 832–835.
Scheffer, R. P. and Pringle, R. B. (1967). *In* "The dynamic role of molecular constituents in plant-parasite interaction" (C. J. Mirocha and J. Uritani, Eds), pp. 217–234. Bruce Publishing Co., St. Paul, Minnesota.
Stoddard, E. M. and Dimond, A. E. (1948). *Phytopathology* **38**, 670–671.
Stoll, Ch. and Renz, J. (1957). *Phytopath. Z.* **29**, 380–387.
Thatcher, F. S. (1942). *Can. J. Res.*, C **20**, 283–311.
Trione, E. J. (1960). *Phytopathology* **50**, 480–482.
Walker, J. C. and Foster, R. E. (1946). *Am. J. Bot.* **33**, 259–264.
Wheeler, H. and Black, H. S. (1963). *Am. J. Bot.* **50**, 686–693.
Zähner, H. (1954). *Phytopath. Z.* **23**, 49–88.

Discussion

Paxton: I would like to add tentoxin produced by *Alternaria tenuis* to your list of toxins and refer you to the work done on it by Templeton and Halloin.

Naef-Roth: Thank you for mentioning tentoxin. My list is far from complete and was meant only to give some examples. But I shall now add tentoxin.

Durbin: How did you grow the plants which exhibited different responses at different times of the year?

Naef-Roth: Under defined, optimum conditions of humidity, temperature and soil; light only changed with the season.

Durbin: Do you know whether plants exhibit responses by themselves when grown in a growth room.

Naef-Roth: Not yet but we are now investigating this.

Graniti: The influence of growing conditions on the responses of plants has been described many times. For example, J.-P. Jost (*Phytopath. Z.* 1965, **54**, 338–378) demonstrated diurnal and seasonal variations in the sensitivity of tomato, pea and cotton plants to fusaric acid as assayed by the wilting test used at Zürich by Gäumann and his colleagues. Jost suggested that changes in the level of growth regulator substances in plants could account for some of these variations.

Page: We have been able to isolate lycomarasmin as a product of alkaline hydrolysis of walls of 6-day-old mycelium of *Fusarium oxysporum* f. *lycopersici*. The hydrolysate had the same absorption as lycomarasmin supplied by the late Professor E. Gäumann. Do you know whether this observation has been confirmed elsewhere?

Naef-Roth: Not so far as I know.

Meyer: What is the significance of toxins such as lycomarasmin that appear so late in cultures of the pathogen.

Kern: The toxic qualities of a substance formed in very old cultures would not be significant. Lycomarasmin, however, can be isolated also from young cultures, even after a few days of incubation.

Gottlieb: Has anyone added lycomarasmin to a young culture of the *Fusarium* and then determined whether or not the toxin has been incorporated into its cell wall?

Naef-Roth: Not to my knowledge.

Scheffer: We have used another assay of toxins, not yet mentioned. The host specific toxins, especially those of *Helminthosporium victoriae* and *Periconia circinata* cause rapid loss of electrolytes from tissues. Within the proper range of concentrations a good dosage/response is obtained. This assay was first suggested by Wheeler, and we now use it frequently. It can be completed in hours, whereas the growth assay requires days.

Naef-Roth: I did not mention permeability changes which will be discussed in other contributions, but the method certainly is most efficient and is now referred to in my paper.

Durbin: I should like to emphasize the importance of the environment in bioassays because by manipulating it, maximum responses can be obtained.

I should also like to point out that during purification, many substances that may interfere with the bioassay are eliminated. This makes it difficult to devise suitable controls for measurement of specific activities at successive stages in purification of a toxin.

Naef-Roth: The specific activity of a plant specific toxin in a convenient bioassay increases with purification. In the case of a non-specific toxin e.g., a growth inhibiting action against bacteria or fungi and the toxicity for plant cuttings must increase proportionally.

Goodman: Have you used suppression of protoplasmic streaming as an indicator of the action of a toxin?

Naef-Roth: Gäumann and co-workers investigated the effect of patulin and lycomarasmin on the protoplasmatic streaming of *Elodea canadensis*; patulin suppressed streaming abruptly whereas lycomarasmin had no effect.

Turner: How is the wilting index measured? It appears to be a very subjective and variable method.

Naef-Roth: Four leaf cuttings of standardized plants are treated with toxin solutions of different concentrations. After uptake of the required dose per g fresh weight they are transferred to tap water, 24–48 h after treatment the visible symptoms, mostly necroses in the leaves, are estimated for each leaf separately. The wilt index is an average of minimum of 10 plants, that is 40 leaves.

Kern: If assay conditions are standardized, the method is very useful and reliable as it helped to isolate several plant specific substances. Wilt index, as the term is used, means the reading of visible symptoms as a dosage response. The symptoms differ for each toxin. A small dose evokes limited necroses in the intercostal field, at tips, margin or veins. Only heavy doses produce wilting in the sense of the word, causing a collapse of petioles and stems.

Turner: If visible wilting is to be used as a measure of the effect of a phytotoxin on the water status of plant tissue, I suggest measurement of osmotic potential, water potential and hence turgor pressure as more objective. Do you know any experiments in which these parameters of water status were measured?

Naef-Roth: Another measure for injury in plants can be the determination of the pathological loss of water which gives a good correlation with the symptoms developed. A combination of the two methods gives best results.

Mussell: I find that objective judgement of wilt by a numerical (wilting) index is reproducible and reasonably accurate if two criteria are met.

First the investigator must not try to rate each observation too accurately; first impressions are best. Secondly individual leaves not entire cuttings or plants, must be rated. The reason for this is that at least with five leaf cotton cuttings, wilting of leaves caused by *Verticillium* toxins depends on age and position of the leaf on the plant or cutting.

Maraite: This morning you presented a slide showing vein yellowing on tomato leaves after treatment of cuttings with fusaric acid. I should like to know (1) whether there is yellowing between the veins; (2) a significant water loss during the first signs of yellowing, and (3) the further evaluation of symptoms, a general yellowing followed by necrosis or wilting.

Naef-Roth: (1) Yes, with small doses necroses between the veins appear 72 96 h after treatment; (2) Yes, depending upon the dose of toxin applied; (3) Most of the time there is a general necrosis, depending upon the doses and the toxin applied.

Kuć: Does yellowing occur in tomato plants before wilting and is the yellowing accompanied by a loss of water?

Naef-Roth: With fusaric acid yellowing and necroses in the leaves appear after treatment with small doses, and wilting, collapsing of stem and petioles appear after application of high doses. In both cases the macroscopically visible symptoms are accompanied by a loss of water.

5. Phytotoxins—An Exercise in the Chemistry of Biologically Active Natural Products

ALESSANDRO BALLIO

Laboratorio di Chimica delle Sostanze Naturali,
Istituto Chimico, Università di Napoli,
Italy

Biologically active natural products have attracted the attention of chemists for a very long time. Crude preparations of naturally occurring drugs, used therapeutically on a purely empirical basis, provided starting material in the nineteenth century for purification and study of the chemistry of their active components. Some of the first natural substances to be isolated and partly characterized were morphine from opium, quinine from *Cinchona* bark and atropine from *Atropa belladonna* L. During this period, research in physiological chemistry, from which the science of biochemistry later developed, resulted in the isolation of a series of metabolites from animal sources. Concurrently organic chemistry and its experimental techniques were developing rapidly so that, at the beginning of the twentieth century, conditions were ripe for extending the scope of research to include many more natural products. At first, chemists were mainly attracted by the problem of elucidating structures of new substances from animals or plants, paying little attention to their functions. This approach resulted in the discovery of a very large number of new compounds, and greatly stimulated the development of new synthetic methods. Indeed, chemical synthesis became important for confirming the correctness of structures deduced from analyses of degradation products, and for preparing new compounds modelled on natural prototypes which were often characterized by novel chemical features.

In the last twenty-five years there has been a formidable development in physical methods for use in investigations of chemical structure. Spectroscopic and optical techniques (ultraviolet, infrared, nuclear magnetic resonance and mass spectroscopy, optical rotatory dispersion and circular dicroism) are now such that a suitably trained chemist can apply them empirically by using commercially available instruments. Information obtained by the judicious use of these methods is so instructive that the ingenuity previously required to exploit the older, conventional methods of organic chemistry for degradation and synthesis is rapidly becoming less important. This is particularly true for another physical method, X-ray diffraction which at present is one of the most powerful available but is also the most difficult, except for experts (Stout and Jensen, 1968; Woolfson, 1970). Nevertheless, its importance for chemists must not be underestimated because it is the only physical method giving a direct representation of molecular structure. It is, therefore, the best technique whenever a natural product, or one of its derivatives, can be prepared as a single crystal. The

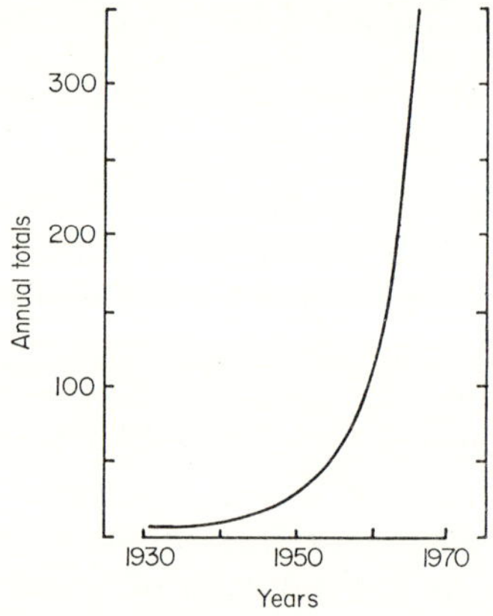

FIG. 1. Numbers of Organic Structures Determined by X-Ray Diffraction during the period 1931–66 (Drawn from data of Dr. Olga Kennard, Cambridge University, provided through the courtesy of Prof. A. Vaciago, Rome University).

ready availability of automated, single-crystal, X-ray diffractometers operated on line to a high-power computer has contributed greatly to the rapid expansion of X-ray analysis. Figure 1 illustrates the increase in use of this technique for determination of structures of natural and synthetic organic compounds in the period 1931–66. A number of phytotoxins, all terpenoid substances each with several chiral centres, have been investigated by X-ray analysis and are listed in Table I.

The molecular complexity of compounds amenable to analysis by X-ray diffraction also continuously increases, as demonstrated by the outstanding work on globular proteins pioneered by Perutz (1964) and Kendrew (1961), and more recently extended to include enzymes (Perutz, 1969).

The rapid development of "automatic" techniques for elucidation of structure has induced some frustration in a few organic chemists who feel that their aptitude for ingeniously dissecting a novel molecule to unveil the arrangement of its atoms is becoming superfluous. But this is only partly true. Firstly, some compounds are not yet amenable to X-ray analysis. Secondly, the conventional methods of organic chemistry are still of great value when the biosynthetic pathways involving natural products are investigated because here it is necessary to prepare appropriate degradation products in order to establish sites of incorporation of specifically labelled precursors. Thirdly, even when X-ray analysis is the primary analytical method the crystallographer may still depend on help from the chemist. Thus, heavy atom derivatives used for X-ray analysis must be chosen so that their relation to the original substance is understood if the structure of this compound is to be derived unambiguously. Although synthesis as a means of confirming structure is becoming less important it is being used increasingly to produce potential intermediates for experiments in biosynthesis and to attempt new reactions along stereochemical and non-stereochemical lines.

One also remembers those prerequisites to elucidation of structure, isolation and purification of the active principle. For natural products these can present the chemist with formidable problems particularly when the substances are present in organisms only in very low concentrations. With biologically active substances, such as phytotoxins, successful extraction of a particular compound, or class of compounds, usually requires the development of a suitable bioassay; this is when the activities of the chemist integrate with those of the biologist.

The methods used for extraction and fractionation procedures depend on the solubility and stability of the active principles, two properties which must be determined in small scale preliminary tests. Clearly, different extraction methods are used for metabolites that accumulate in the culture medium and for those that remain inside cells.

A few examples, all phytotoxins, will illustrate typical procedures. The culture filtrate of *Ascochyta fabae* Speg., the fungus causing brown spot disease of broad bean, is phytotoxic and is an antibiotic (Oku and Nakanishi, 1963). With germination of rape seed as a bioassay, it was found that a toxin could be extracted from an acidified culture filtrate with chloroform. The residue after evaporation of the solvent was repeatedly extracted with *n*-hexane which on evaporation left a solid which crystallized from ethanol, benzene or acetone as the yellow, active compound ascochitine (I). The solubility properties on which this easy purification procedure is based are readily accounted for by the structure of the compound which was determined from simple degradation reactions (Iwai and Mishima, 1965), and later confirmed by total synthesis of the racemate (Galbraith and Whalley, 1966).

A similar procedure, supplemented by chromatography with alumina, was used to isolate zinniol (II), a phytotoxin produced by *Alternaria zinniae* Pape in culture (White and Starratt, 1967; Starratt, 1968). This fungus

TABLE I. Structures of phytotoxins determined by X-ray diffraction

Compound		Source	Heavy atom derivative used for structure determination	References
Ophiobolin A	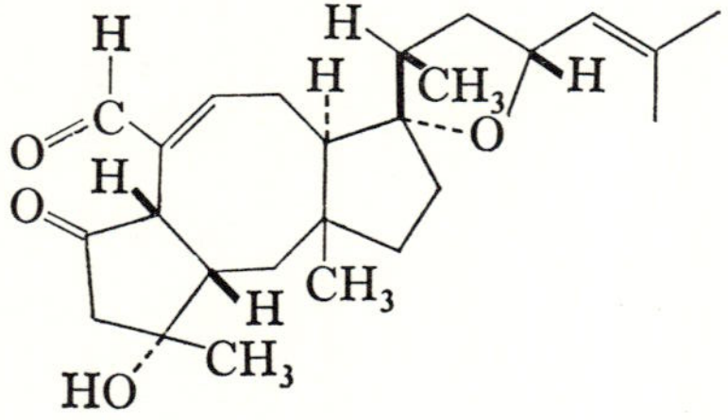	*Cochliobolus miyabeanus* (Ito et Kurib.) Drechs. ex Dast. (Conidial stage: *Helminthosporium oryzae* Breda de Haan)	Methoxybromide	Nozoe *et al.*, 1965
Trichodermin*	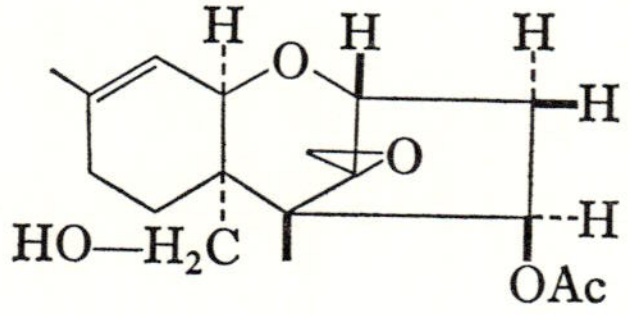	*Trichoderma viride* Pers. ex Fr.	*p*-Bromobenzoate of trichodermol†	Abrahamsson and Nilsson, 1964, 1966
Siccanin	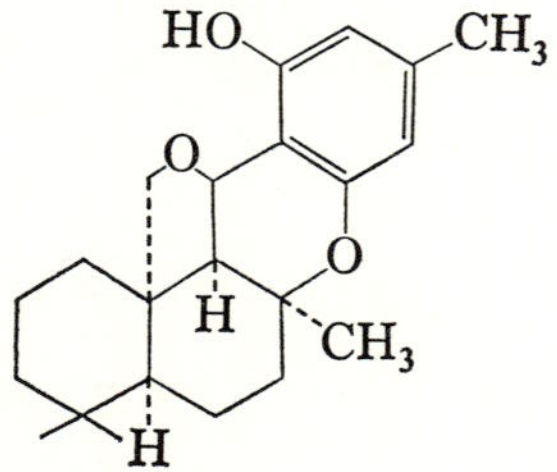	*Helminthosporium siccans* Drechsler	*p*-Bromobenzene-sulphonate	Hirai *et al.*, 1967

Fomannosin	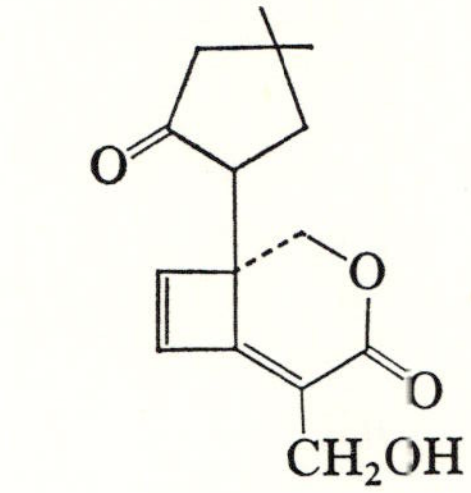	*Fomes annosus* (Fr.) Cooke	*p*-Bromobenzoyl-urethane of dihydrofomannosin‡	Kepler *et al.*, 1967; McPhail and Sim, 1968
Fusicoccin	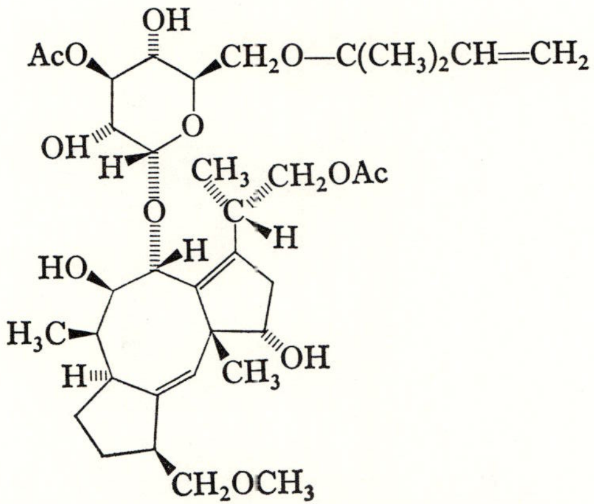	*Fusicoccum amygdali* Del.	*p*-Iodobenzene-sulphonate Mercuribromide of fusicoccin desacetylaglycone	Ballio *et al.*, 1968a Hough *et al.*, 1968

* Trichodermin is an antifungal antibiotic closely related to the fusarial phytotoxin diacetoxyscirpenol (Flury *et al.*, 1965; Dawkins *et al.*, 1965)

† Trichodermol is the deacetylderivative of trichodermin

‡ Dihydrofomannosin is the derivative obtained on hydrogenation of the disubstituted double bond of fomannosin

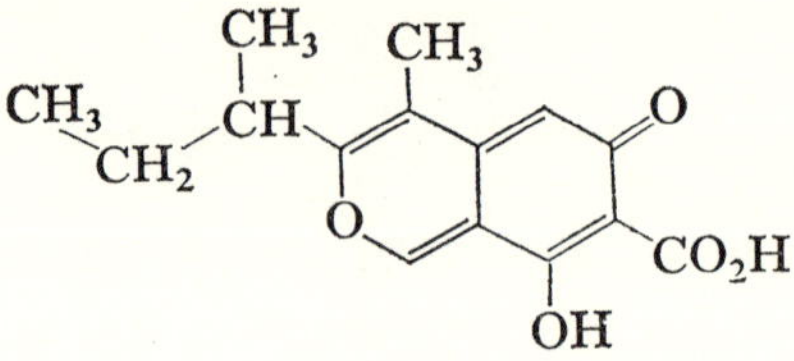

(I)

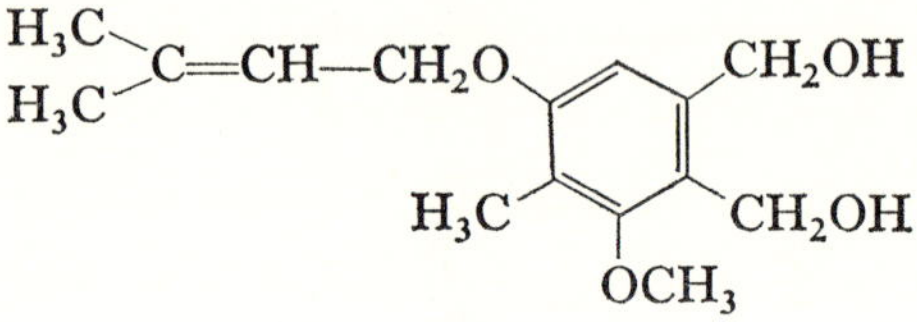

(II)

causes leaf and stem blight of zinnia, sunflower, and marigold (McDonald and Martens, 1963).

Other pathogenic species of *Alternaria* often contain alternariol (III) and its monomethylether (IV) (Raistrick *et al.*, 1953; Thomas, 1961b; Freeman, 1965; Starratt and White, 1968; Tirokata *et al.*, 1969). Only IV is phytotoxic. Both compounds are intracellular and have been isolated by fractional extraction of mycelia with various solvents.

The above examples refer to substances extractable with solvents. Difficulties arise when the active principle has partition coefficients much in favour of water. Such is the case of 1-amino-2-nitrocyclopentane-carboxylic acid (V) (Burrows *et al.*, 1965; Burrows and Turner, 1966),

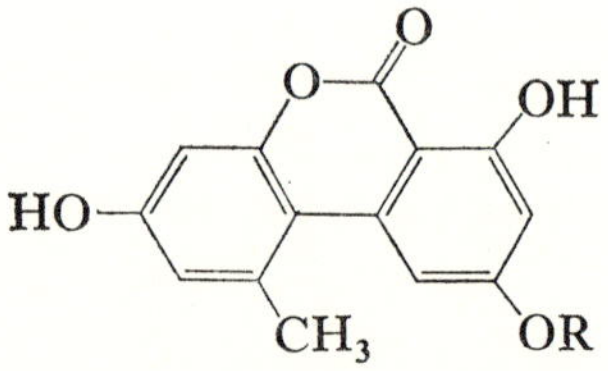

III: R = H; IV: R = CH_3

(III)

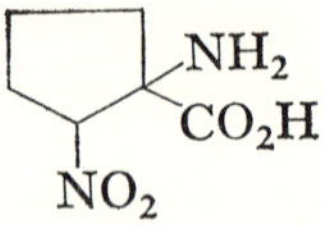

(V)

a compound with plant growth-regulating properties (Brian *et al.*, 1965), and which induces the syndrome of yellow strapleaf disease of chrysanthemum (Woltz and Littrell, 1968). This metabolite is secreted in culture by *Aspergillus wentii* Wehmer, which interestingly also synthesizes another aliphatic nitro-compound, nitropropionic acid, readily purified by chromatography on ion-exchange resins (Burrows and Turner, 1966). A similar procedure has been used to extract and purify the phytotoxin in cultures of *Colletotrichum gloeosporoides* Penz. (Ballio *et al.*, 1969), which causes olive anthracnose. It was found to be identical with aspergillomarasmine B, or lycomarasmic acid (VI), produced by *A. flavus-oryzae* (Robert *et al.*, 1962; Haenni *et al.*, 1965).

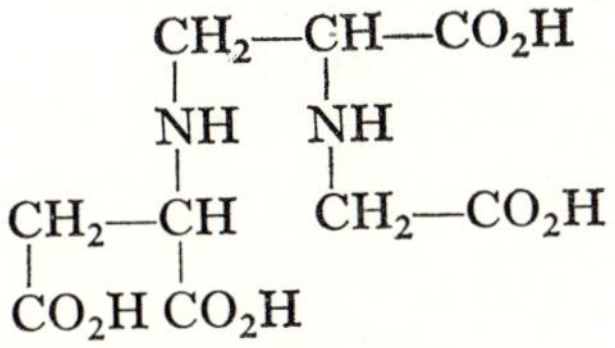

(VI)

Besides ion exchange chromatography, other chromatographic techniques have been used for the isolation of phytotoxins not extractable with solvents. The most recent is gel filtration which separates molecules of different size by passing them through a column containing an insoluble hydrophilic polymer behaving as a molecular sieve (Determann, 1968). Solutes above a certain molecular weight are excluded, whereas solutes of lower molecular weight can diffuse into the polymer and, therefore, pass through the column more slowly. Gel filtration has been largely used for purification of the halo blight phytotoxin from *Pseudomonas phaseolicola* (Burkh.) Dow. (Rudolph, 1969), of the wilting glycopeptides produced by *Corynebacterium sepedonicum* (Spieck. et Kotth.) Dows. (Strobel, 1967) and *C. michiganense* (E. Sm.) Jens. (Rai and Strobel, 1969), to isolate a phytotoxic oligosaccharide produced by *Didymella applanata* (Niessl) Sacc. (Salemink *et al.*, 1966), a pathogen of raspberries, and a very high molecular weight phytotoxic glycoprotein present in filtrates from cultures of *Ceratocystis ulmi* (Buisman) Moreau (Salemink *et al.*, 1965; Rebel and Salemink, 1968), the well-known pathogen of Dutch elm disease. Modified gel filtration polymers, possessing additionally ion exchange properties, have been widely used to purify proteins and will probably also prove useful in the study of some high molecular weight phytotoxins.

Another separation and purification technique which can be very satisfactory, particularly for isolation of labile compounds in complex mixtures, is countercurrent distribution (Craig, 1962). It repeatedly partitions between two immiscible or partially miscible solvents, so that with a suitable system, substances with only slightly different partition coefficients are cleanly separated. The whole operation can be carried out quickly in the commercially available, automatic Craig apparatus. Countercurrent distribution has been used to purify the host-specific *Periconia circinata* (Mangin) Sacc. toxins A and B (Pringle and Scheffer, 1967a), and the toxins of *Helminthosporium carbonum* Ullstrup (Pringle and Scheffer, 1967b).

This brief survey of methods used for the isolation of phytotoxins needs to be supplemented with the following. It is frequently found that the active metabolites are produced only in very small amounts. It then becomes the task of the microbiologist to increase fermentation yields, either by improving methods of culture, or by isolating better strains through mutation and selection. Once higher yields have been attained the chemist often finds quicker, simpler, and more efficient procedures for extraction and purification. An example is given by the isolation of fusicoccin (VII) (Ballio *et al.*,

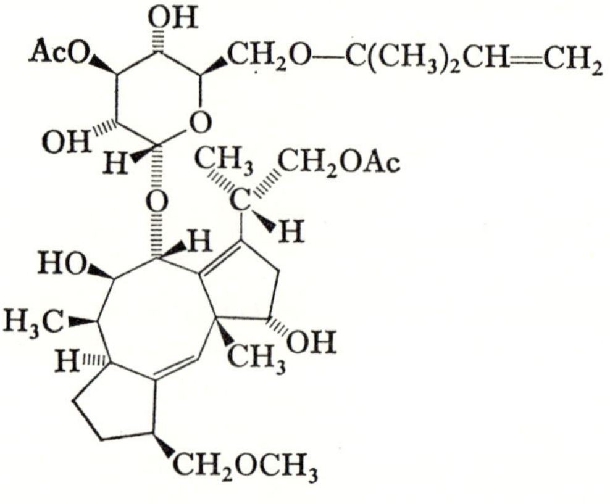

(VII)

1968a; Barrow *et al.*, 1968) from *Fusicoccum amygdali* Del., the pathogen of almond and peach canker. With the original strain the isolation procedure required solvent extraction of the culture filtrate, followed by a long and tedious chromatographic fractionation on a Florisil column (Ballio *et al.*, 1964). With high-yielding strains extracts are treated with charcoal from which the active principle is eluted with acetone. Evaporation of the acetone followed by partition between water and chloroform, gives a chloroform solution which, after evaporation and simple crystallization of the residue, gives pure fusicoccin (Ballio *et al.*, 1968b).

The role of the chemist does not end with isolation and determination of the structure of the active compound because knowledge of its chemistry will almost certainly be essential in the investigation of its biological role. The final structure is only one aspect of a complicated and fascinating sequence, but it must be determined as precisely and quickly as possible, so that it can be unambiguously used to suggest the nature of the reactions in which the compound may be involved *in vivo*. I should like to mention here a few phytotoxins for which the structures proposed earlier had to be revised later after further investigations. These are lycomarasmin (VIII) (Woolley, 1948; Brenner *et al.*, 1958; Hardegger *et al.*, 1963), the enniatins (IX, X) (Plattner and Nager, 1948a, b; Shemyakin *et al.*, 1962; Plattner *et al.*, 1963; Quitt *et al.*, 1963; Shemyakin *et al.*, 1963a, b, c, d; Shemyakin *et al.*, 1966), patulin (XI) (Engel *et al.*, 1949a, b; Woodward and Singh, 1950), and the wildfire toxin (Woolley *et al.*, 1955; Sinden and Durbin, 1968, 1970).

Present knowledge of the ways in which phytotoxins damage higher plants is the subject of several other lectures in this Institute so I shall illustrate only the sort of contribution which can be made in this field by close cooperation between chemists and biologists.

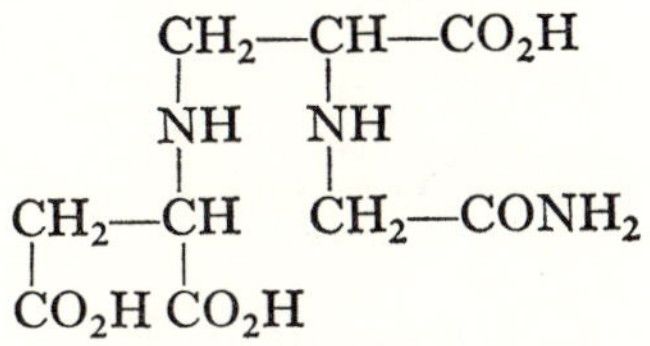

(VIII)

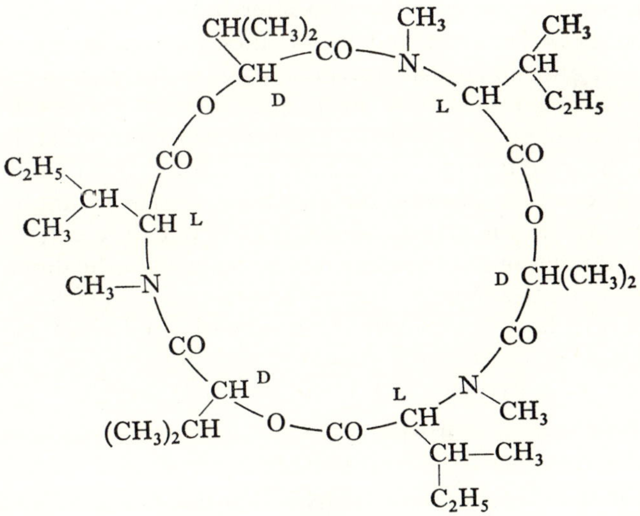

(IX)

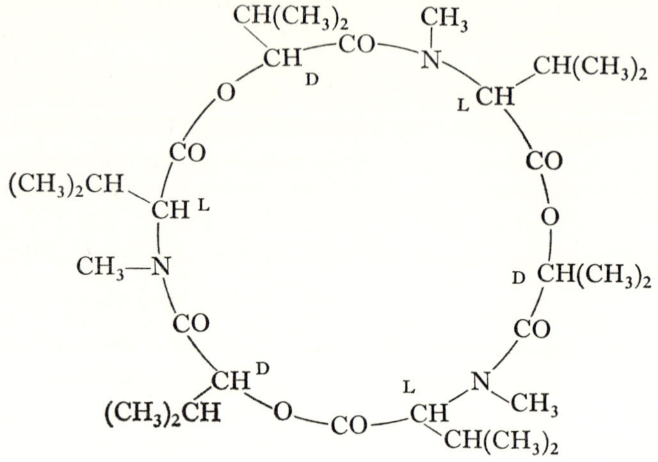

(X)

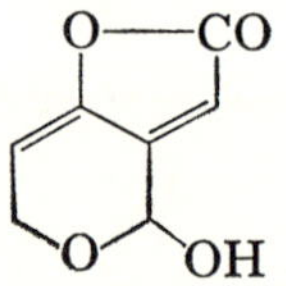

(XI)

In evaluating the role of a phytotoxin in disease it is, of course, vital to establish whether the substance is synthesized *in vivo*. This requires selective isolation of toxins present in the host after but not before infection, and comparisons between them and metabolites formed by the parasite in culture. Here again, and usually to a greater degree, extraction and fractionation of what are usually very small quantities of active substances in diseased plants can present difficulties. It may also happen that the products isolated are different from those produced in culture. It is then necessary to establish and compare the structures of the materials synthesized *in vivo* and *in vitro*. When the structures are related but different, comparative biosynthetic studies might reveal metabolic reasons for the differences and, more particularly, whether the fate of the substances *in vivo* is affected by enzymes of the host. Furthermore, structural similarities may suggest relationships between chemical structure and biological activity. This is where the chemist, by modifying the original phytotoxin, or synthesizing

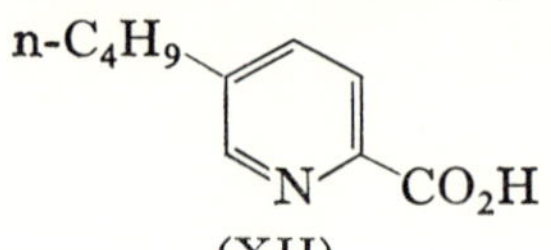

(XII)

TABLE II. Structure, occurrence and biological properties of some microbial metabolites related to diaporthin

Compound		Source	Biological activity	References
Diaporthin	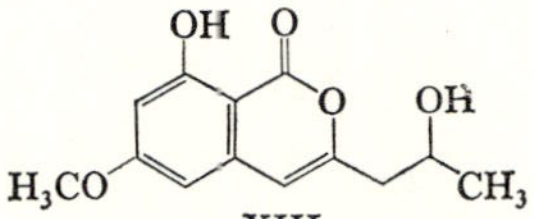 XIII	*Endothia parasitica* (Murr.) P. et H. And.	Phytotoxin	Boller *et al.*, (1957); Hardegger *et al.*, (1966)
Mellein (Ochracin) Mellein antipode	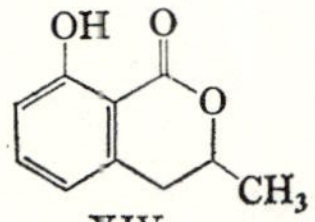 XIV	*Aspergillus melleus* Yukawa *A. ochraceus* Wilhelm Unidentified fungus		Nishikawa (1933a, b); Burton (1950) Yabuta and Sumiki (1933, 1934) Patterson *et al.*, (1966)
5-Methylmellein	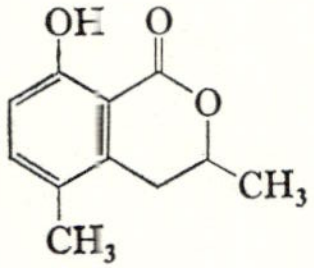 XV	*Fusicoccum amygdali* Del. (some strains)	Inhibits germination of spores of some fungi; not phytotoxic	Ballio *et al.*, (1966)
6-Hydroxymellein	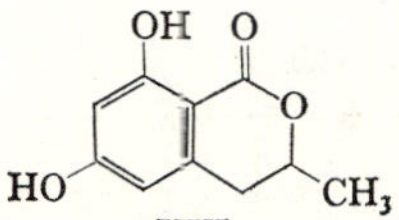 XVI	*A. terreus* Thom (mutant strain)		Curtis *et al.*, (1966)

TABLE II—*continued*

Compound		Source	Biological activity	References
6-Methoxymellein	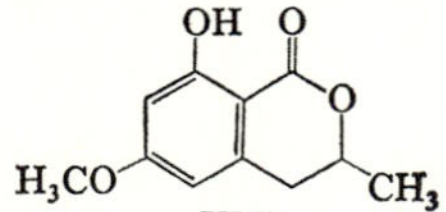 XVII	Carrots held in cold storage or inoculated with fungi *Sporormia bipartis* Cain *S. affinis* Sacc., Bomm. et Rouss.	Phytoalexin	Sondheimer (1957, 1961); Condon and Kuć (1960, 1962); Hampton (1962) Aue *et al.*, (1966) McGahren and Mitscher (1968)
8-Hydroxy-3-methyl-*iso*-coumarin	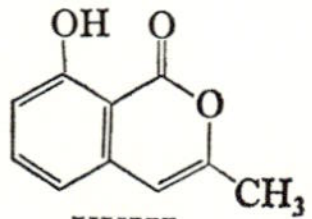 XVIII	*Marasmius ramealis* (Bull.) Fr.		Bendz (1959)
6,8-dihydroxy-3-methyl-*iso*coumarin	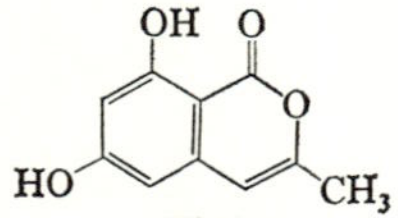 XIX	*Ceratocystis fimbriata* Ell. et Halst.		Curtis (1968)
8-Hydroxy-6-methoxy-3-methyl-*iso*coumarin	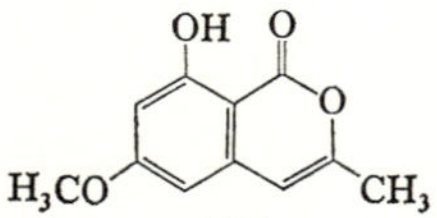 XX	*C. fimbriata*		Curtis (1968); Stoessl (1969)
Reticulol	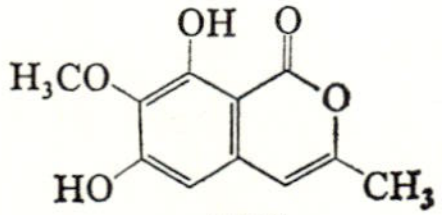 XXI	*Streptomyces rubrireticuli* (Waks.) Waks. et Henr.		Mitscher *et al.*, (1964)

Sclerin	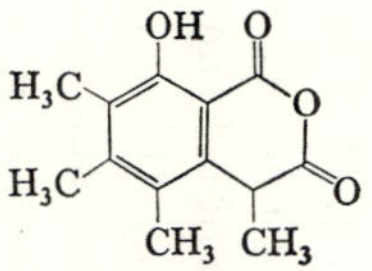 XXII	*Sclerotinia sclerotiorum* (Lib.) deBary	Plant growth promoter	Satomura and Sato (1965); Kubota *et al.*, (1966); Tokoroyama *et al.*, (1968); Sassa *et al.*, (1968)
Sclerotinin A	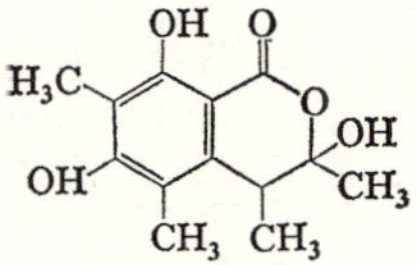 XXIII	*S. sclerotiorum*	Plant growth promoter	Sassa *et al.*, (1968)
Sclerotinin B	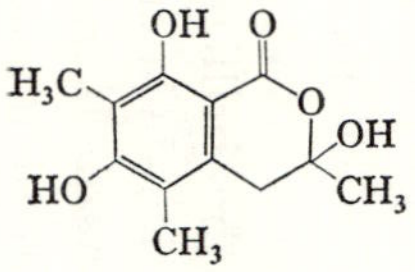 XXIV	*S. sclerotiorum*	Plant growth promoter	Sassa *et al.*, (1968)
3,4-dihydro-6,8-dihydroxy-7-carboxy-3,4,5-trimethyl-*iso*coumarin	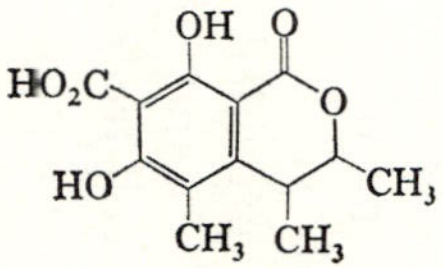 XXV	*A. terreus* Thom (mutant strain)		Hassall and Jones (1962)

analogues and model compounds, can make substantial contributions. Examples are lycomarasmic acid (VI) (Barbier *et al.*, 1963; Bogdanovsky and Barbier, 1964), fusaric acid (XII) (Tamari and Kaji, 1952; Gäumann *et al.*, 1958), and fusicoccin (VII) (Ballio *et al.*, 1971). In this connexion I think that it would be of interest to see whether natural compounds structurally related to known phytotoxins are also phytotoxic.

To illustrate the above I shall consider a series of natural isocoumarins and closely related products (Table II). All are strikingly similar to diaporthin (XIII), a phytotoxin produced by *Endothia parasitica* (Murr.) P. *et* H. And. (Boller *et al.*, 1957; Hardegger *et al.*, 1966), the fungus responsible for chestnut blight. Most are metabolites of parasitic fungi, and have some kind of biological activity. Some are antibiotics, one is a phytoalexin, and sclerin as well as sclerotinins A and B promote growth of plants. However, none except methylmellein, has yet been tested for phytotoxicity.

Finally, I should like to comment on the role played by the chemist in the study of biosynthesis of natural products, a topic dealt with in another lecture (Canonica, pp. 157–172). In considering biosynthesis at least three different types of approach must be considered, (i) isolation and determination of the structure of minor metabolites chemically related to the main product, (ii) use of labelled precursors of the metabolite and determination of the distribution of the labels in the metabolite by suitable degradations, and (iii) study of enzymes involved in reactions leading to synthesis of the metabolite. The first approach leads only to the isolation of products related to the final metabolite. Their role in the biosynthetic pathway may be established by the second and third approaches. I shall not comment further on the second which is the one most commonly adopted. It is the third that probably yields the most convincing and conclusive results. A good example is provided by the brilliant analysis of Gatenbeck and Hermondsson (1965), and Sjöland and Gatenbeck (1966) of the synthesis of alternariol (III) by cell-free extracts of *Alternaria tenuis* Nees. They obtained decisive evidence for a pathway involving malonyl and acetyl-coenzyme A, through a mechanism postulated earlier on the basis of results obtained with radioactive precursors (Thomas, 1959, 1961a). One can predict that in the near future biosynthesis will be studied more and more frequently with molecules labelled with the stable isotopes such as ^{2}H, ^{13}C and ^{15}N. The position of these labels can be determined directly by mass spectrometry whenever fragmentation occurring after ionization can be rationalized. This approach was used extensively by Lederer's group (1969) to elucidate the mechanism of biological methylations in a series of natural products. In the case of ^{13}C, the position of the tracer may often be located by nuclear magnetic resonance, again without chemical degradation. This procedure was used to study biosynthesis of fusaric acid (XII) (Desaty *et al.*, 1968), and it is expected to become increasingly popular as adequate equipment and appropriate precursors become available.

The importance of detailed knowledge of pathways leading to biosynthesis of phytotoxins cannot be underestimated. Once they are known it is possible in principle to find compounds which by selectively inhibiting specific steps of biosynthesis nullify the harmful effect of the parasite.

Let me conclude by expressing my firm conviction that many problems involved in the understanding of host–parasite interactions will be solved only by close collaboration between chemists and plant pathologists in ways that during the past twenty years or so have already led to the solution of problems of comparable complexity in other fields.

References

Abrahamsson, S. and Nilsson, B. (1964). *Proc. chem. Soc.* **188**.

Abrahamsson, S. and Nilsson, B. (1966). *Acta chem. Scand.* **20**, 1044–1052.

Aue, R., Mauli, R., and Sigg, R. P. (1966). *Experientia* **22**, 575.

Ballio, A., Chain, E. B., De Leo, P., Erlanger, B. F., Mauri, M., and Tonolo, A. (1964). *Nature* **203**, 297.

Ballio, A., Barcellona, S., and Santurbano, B. (1966). *Tetrahedron Lett.* 3723–3726.

Ballio, A., Brufani, M., Casinovi, C. G., Cerrini, S., Fedeli, W., Pellicciari, R., Santurbano, B., and Vaciago, A. (1968a). *Experientia* **24**, 631–635.

Ballio, A., Carilli, A., Santurbano, B., and Tuttobello, L. (1968b). *Ann. Ist. Sup. Sanità* **4**, 317–332.

Ballio, A., Bottalico, A., Buonocore, V., Carilli, A., Di Vittorio, V., and Graniti, A. (1969). *Phytopath. medit.* **8**, 187–196.

Ballio, A., Bottalico, A., Framondino, M., Graniti, A., and Randazzo, G. (1971). *Phytopath. medit.* **10**, 26-32.

Barbier, M., Bogdanovsky, D., Vetter, W., and Lederer, E. (1963). *Liebigs Ann. Chem.* **668**, 132–138.

Barrow, K. D., Barton, D. H. R., Chain, E. B., Ohnsorge, U. F. W., and Thomas, R. (1968). *Chem. Comm.* 1198–1200.

Bendz, G. (1959). *Ark. Kemi* **14**, 511–518.

Bogdanovsky, D. and Barbier, M. (1964). *Bull. Soc. chim. Fr.* 2778–2782.

Boller, A., Gäumann, E., Hardegger, E., Kugler, F., Naef-Roth, S., and Rosner, M. (1957). *Helv. chim. Acta* **40**, 875–880.

Brenner, M., Tamm, R., and Quitt, P. (1958). *Helv. chim. Acta* **41**,763–773.

Brian, P. W., Elson, G. W., Hemming, H. G., and Radley, M. (1965). *Nature* **207**, 998–999.

Burrows, F. T., Mills, S. D., and Turner, W. B. (1965). *Chem. Comm.* 75.

Burrows, F. T. and Turner, W. B. (1966). *J. chem. Soc.* (C), 255–260.

Burton, H. S. (1950). *Nature, Lond.* **165**, 274–275.

Condon, P. and Kuć, J. (1960). *Phytopathology* **50**, 267–270.

Condon, P. and Kuć, J. (1962). *Phytopathology* **52**, 182–183.

Craig, L. C. (1962). *In* "Comprehensive Biochemistry" (M. Florkin and E. H. Stotz, Eds), Vol. 4, pp. 1–31. Elsevier, Amsterdam, London and New York.
Curtis, R. F. (1968). *Experientia* **24**, 1187–1188.
Curtis, R. F., Harries, P. C., Hassall, C. H., Levi, J. D. and Phillips, D. M. (1966). *J. chem. Soc.* (C), 168–174.
Dawkins, A. W., Grove, J. F., and Tidd, B. K. (1965). *Chem. Comm.*, 27–28.
Desaty, D., McInnes, A. G., Smith, D. G., and Vining, L. C. (1968). *Can. J. Biochem.* **46**, 1293–1300.
Determann, H. (1968). "Gel Chromatography." Springer-Verlag, New York.
Engel, B. G., Brzeski, W., and Plattner, Pl. A. (1949a). *Helv. chim. Acta* **32**, 1166–1175.
Engel, B. G., Brzeski, W., and Plattner, Pl. A. (1949b). *Helv. chim. Acta* **32**, 1752–1758.
Flury, E., Mauli, R., and Sigg, H. P. (1965). *Chem. Comm.*, 26–27.
Freeman, G. G. (1965). *Phytochem.* **5**, 719–725.
Galbraith, M. N. and Whalley, W. B. (1960). *Chem. Comm.*, 620.
Gatenbeck, S. and Hermondsson, S. (1965). *Acta chem. scand.* **19**, 65–71.
Gäumann, E., Kern, H., Schüepp, H., and Obrist, W. (1958). *Phytopath. Z.* **32**, 225–244.
Haenni, A. L., Robert, M., Vetter, W., Roux, L., Barbier, M., and Lederer, E. (1965). *Helv. chim. Acta* **48**, 729–750.
Hampton, R. E. (1962). *Phytopathology* **52**, 413–415.
Hardegger, E., Liechti, P., Jackman, L. M., Boller, A., and Plattner, Pl. A. (1963). *Helv. chim. Acta* **46**, 60–74.
Hardegger, E., Rieder, W., Walser, A., and Kugler, F. (1966). *Helv. chim. Acta* **49**, 1283–1290.
Hassall, C. H. and Jones, D. W. (1962). *J. chem. Soc.*, 4189–4191.
Hirai, K., Nozoe, S., Tsuda, K., Iitaka, Y., Ishibashi, K., and Shirasaka, M. (1967). *Tetrahedron Lett.*, 2177–2179.
Hough, E., Hursthowe, M. B., Neidle, S., and Rogers, D. (1968). Chem Comm., 1197–1198.
Iwai, I. and Mishima, H. (1965). *Chemy. Ind.*, 186–187.
Kendrew, J. C. (1961). *Sci. Am.* **205**, (6), 96–110.
Kepler, J. A., Wall, M. E., Mason, J. E., Bassett, C., McPhail, A. T., and Sim, G. A. (1967). *J. Am. chem. Soc.* **89**, 1260–1261.
Kubota, T., Tokoroyama, T., Kamikawa, T., and Satomura, Y. (1966). *Tetrahedron Lett.*, 5205–5210.
Lederer, E. (1969). *Quart. Rev.* **23**, 453–481.
McDonald, W. C. and Martens, J. W. (1963). *Phytopathology* **53**, 93–96.
McGahren, W. J. and Mitscher, L. A. (1968). *J. org. Chem.* **33**, 1577–1580.
McPhail, A. T. and Sim, G. A. (1968). *J. chem. Soc.* (B), 1104–1109.
Mitscher, L. A., Andres, W. W., and McCrae, W. (1964). *Experientia* **20**, 258–259.
Nishikawa, E. (1933a). *J. Agric. chem. Soc. Japan* **9**, 772–774 (*Chem. Abstr.* **28**, 2751 (1934)).
Nishikawa, E. (1933b). *J. Agric. chem. Soc. Japan* **9**, 1059–1063 (*Chem. Abstr.* **28**, 1073 (1934)).

Nozoe, S., Morisaki, M., Tsuda, K., Iitaka, Y., Takahashi, N., Tamura, S., Ishibashi, K., and Shirasaka, M. (1965). *J. Am. chem. Soc.* **87**, 4968–4970.
Oku, H. and Nakanishi, T. (1963). *Phytopathology* **53**, 1321–1325.
Patterson, E. L., Andres, W. W., and Bohonos, N. (1966). *Experientia* **22**, 209–210.
Perutz, M. F. (1964). *Sci. Am.* **211** (5), 64–76.
Perutz, M. F. (1969). *Eur. J. Biochem.* **8**, 455–466.
Plattner, Pl. A. and Nager, U. (1948a). *Helv. chim. Acta* **31**, 665–671.
Plattner, Pl. A. and Nager, U. (1948b). *Helv. chim. Acta* **31**, 2129–2203.
Plattner, Pl. A., Vogler, K., Studer, R. O., Quitt, P., and Keller-Schierlein, W. (1963). *Helv. chim. Acta* **46**, 927–935.
Pringle, R. B. and Scheffer, R. P. (1967a). *Phytopathology* **57**, 530–532.
Pringle, R. B. and Scheffer, R. P. (1967b). *Phytopathology* **57**, 1169–1172.
Quitt, P., Studer, R. O., and Vogler, K. (1963). *Helv. chim. Acta* **46**, 1715–1720.
Rai, P. V. and Strobel, G. A. (1969). *Phytopathology* **59**, 47–52.
Raistrick, H., Stickings, C. E., and Thomas, R. (1953). *Biochem. J.* **55**, 421–433.
Rebel, H. and Salemink, C. A. (1968). 5, *Symp. Chromat. Soc. Belge Sci. Pharm*[e], 309–313.
Robert, M., Barbier, M., Lederer, E., Roux, L., Biemann, K., and Vetter, W. (1962). *Bull. Soc. chim. Fr.*, 187–188.
Rudolph, K. (1969). *Naturwissenschaften* **56**, 569–570.
Salemink, C. A., Rebel, H., Kerling, L. C. P., and Tchernoff, V. (1965). *Science* **149**, 202–203.
Salemink, C. A., Schuring, F., Kerling, L. C. P., and Schippers, B. (1966). *Experientia* **22**, 248.
Sassa, T., Aoki, H., Namiki, M., and Munakata, K. (1968). *Agr. biol. Chem. Japan* **32**, 1432–1439.
Satomura, Y. and Sato, A. (1965). *Agr. biol. Chem. Japan* **29**, 337–344. (*Chem. Abstr.* **63**, 2319).
Shemyakin, M. M., Ovchinnikov, Yu. A., Kiryushkin, A. A., and Ivanov, V. T. (1962). *Tetrahedron Lett.*, 301–305.
Shemyakin, M. M., Ovchinnikov, Yu. A., Ivanov, V. T., and Kiryushkin, A. A. (1963a). *Tetrahedron* **19**, 581–591.
Shemyakin, M. M., Ovchinnikov, Yu. A., Kiryushkin, A. A., and Ivanov, V. T. (1963b). *Tetrahedron Lett.*, 885–890.
Shemyakin, M. M., Ovchinnikov, Yu. A., Kiryushkin, A. A., and Ivanov, V. T. (1963c). *Chem. Abstr.* **59**, 7645a.
Shemyakin, M. M., Ovchinnikov, Yu. A., Kiryushkin, A. A., and Ivanov, V. T. (1963d). *Chem. Abstr.* **59**, 10234a.
Shemyakin, M. M., Ovchinnikov, Yu. A., Kiryushkin, A. A., and Ivanov, V. T. (1966). *Chem. Abstr.* **64**, 2157g.
Sinden, S. L. and Durbin, R. D. (1968). *Phytopathology* **58**, 1067–1068.
Sinden, S. L. and Durbin, R. D. (1970). *Phytopathology* **60**, 360–364.
Sjöland, S. and Gatenbeck, S. (1966). *Acta chem. scand.* **20**, 1053–1059.
Sondheimer, E. (1957). *J. Am. chem. Soc.* **79**, 5036–5039.
Sondheimer, E. (1961). *Phytopathology* **51**, 71–72.
Starratt, A. N. (1968). *Can. J. Chem.*, 767–770.

Starratt, A. N. and White, G. A. (1968). *Phytochem.* **7**, 1883–1884.
Stoessl, A. (1969). *Biochem. biophys. Res. Comm.* **35**, 186–192.
Stout, G. H. and Jensen, L. H. (1968). "X-Ray Structure Determination." The Macmillan Company. Macmillan Limited, London.
Strobel, G. A. (1967). *Plant Physiol.* **42**, 1433–1441.
Tamari, K. and Kaji, J. (1952). *J. Agric. chem. Soc. Japan* **26**, 298–303. (*Chem. Abstr.* **48**, 9483).
Thomas, R. (1959). *Proc. chem. Soc.*, 88.
Thomas, R. (1961a). *Biochem. J.* **78**, 748–758.
Thomas, R. (1961b). *Biochem. J.* **80**, 234–240.
Tirokata, H., Ohkawa, M., Sassa, T., Yamada, T., Ohkawa, H., Tanaka, H., and Aoki, H. (1969). *Nippon Shokubutsu Byori Gakkaiho* **35**, 62–66. (*Chem. Abstr.* **71**, 19620).
Tokoroyama, T., Kamikawa, T., and Kubota, T. (1968). *Tetrahedron* **24**, 2345–2355.
White, G. A. and Starratt, A. N. (1967). *Can. J. Bot.* **45**, 2087–2090.
Woltz, S. S. and Littrell, R. H. (1968). *Phytopathology* **58**, 1476–1480.
Woodward, R. B. and Singh, G. (1950). *J. Am. chem. Soc.* **72**, 1428.
Woolfson, M. M. (1970). "An Introduction to X-Ray Crystallography." University Press, Cambridge.
Woolley, D. W. (1948). *J. biol. Chem.* **176**, 1291–1298.
Woolley, D. W., Schaffner, G., and Braun, A. C. (1955). *J. Biol. Chem.* **215**, 485–493.
Yabuta, T. and Sumiki, Y. (1933). *J. Agr. chem. Soc. Japan* **9**, 1264–1275. (*Chem. Abstr.* **28**, 2350 (1934)).
Yabuta, T. and Sumiki, Y. (1934). *J. Agr. chem. Soc. Japan* **10**, 703–714. (*Chem. Abstr.* **29**, 160 (1935)).

Discussion

Stoessl: There is reason to believe that the dihydroisocoumarin from carrots may be a fungal product and not a carrot phytoalexin. This is not yet certain but we should bear it in mind.

Salemink: In making comparisons between infra-red and nuclear magnetic resonance spectroscopy it must be emphasized that the latter is still under development.

Kuć: I know of no evidence to support Stoessl's statement that 3-methyl-6-methoxy-8-hydroxy-3,4-dihydroisocoumarin which accumulates in carrots following inoculation is of fungal origin and not synthesized by the carrot. Though *Ceratocystis fimbriata* does produce trace quantities of 3-methyl-6-methoxy-8-hydroxy-isocoumarin, it would be dangerous to assume that this serves as the precursor for the dihydroisocoumarin which accumulates in carrot after inoculations with non-pathogens.

Surface sterilized whole carrots stored at low temperatures accumulate the dihydroisocoumarin. Cutting intact, surface sterilized carrots under aseptic conditions result in the accumulation of the dihydroisocoumarin, chlorogenic acid and various *p*-coumaryl derivatives. Cut carrot slices incorporate C^{14}-acetate and malonate into dihydroisocoumarin. The dihydroisocoumarin is not detected in parsnips or sweet potato root or Irish potato tubers inoculated with *C. fimbriata.* Some 15 fungi induce accumulation of the dihydroisocoumarin in carrot slices. Accumulation of the compound is also evident after treatment of slices with $HgCl_2$, $CuCl_2$ and a number of organic compounds. Ethylene induces accumulation in carrot slices held under aseptic conditions at levels equal to or greater than that following inoculation with *C. fimbriata.* Frozen and thawed slices, or autoclaved slices inoculated with *C. fimbriata* do not accumulate the dihydroisocoumarin. The trace quantity of isocoumarin produced by *C. fimbriata* is unlikely to explain the high concentration of dihydroisocoumarin found following inoculation. These facts lead me to stick to my conclusion that the accumulation of 3-methyl-6-methoxy-8-hydroxy-3,4-dihydroisocoumarin is due to synthesis by the carrot root.

Stoessl: In reply to Kuć I should like to state that: (1) other isocoumarins induced in carrots should not be drawn into the argument until they have been isolated and characterized; (2) the fact that *C. fimbriata* does not induce isocoumarins in parsnips etc., is not evidence in either direction because the critical importance of media for the production of compounds is well known; (3) increases in dihydroisocoumarin concentrations in carrot, after treatment with $CuCl_2$, $HgCl_2$, ethylene etc., are much smaller than the levels produced by *C. fimbriata*; (4) evidence against the view that the dihydroisocoumarin is a carrot metabolite is as follows: (a) a number of isocoumarins are known from fungi, but only one biogenetically typical isocoumarin is as yet reported from plants other than carrot; (b) the dihydroisocoumarin of the carrot has been isolated from two *Sporormia* spp. in pure culture; (c) the compound, by structural analysis is acetate-malonate derived; the fact that it is a simple phenolic compound favours a fungal over a plant origin; (d) it is difficult to accept as a coincidence the production of isocoumarin by *C. fimbriata* and production of its dihydro derivative independently by the carrot particularly when the latter occurs in high concentrations only after inoculation with *C. fimbriata.*

Barbier: I am attracted to the idea of a systematic search for phytotoxins in natural products. Many biologically active compounds have chelating properties and I just wonder if this is not a key to their activity. We have tested certain industrial chelating agents for phytotoxicity and have observed some astonishing effects, which could interest biologists.

Durbin: Several students at Wisconsin have shown in a class experiment that carrot disks inoculated with *P. solanacearum* produced large amounts of dihydroisocoumarin, but that this substance was not produced when the bacterium was grown in a carrot extract medium.

Goodman: Can you state which heavy atoms can be used to prepare specific organic compounds for study by the crystallographer ? For example would acidic, hydroxyl or amino groups suggest particular derivative atoms ?

Ballio: Acids can be transformed into salts (for instance rubidium salts in the case of penicillins), or esterified with halogen and/or sulphur containing alcohols or phenols. Hydroxy and amino groups can be acylated with halogen and/or sulphur containing compounds, as shown in the case of terpenoid phytotoxins in the first slide I have projected; amines can often be examined as crystalline hydrobromides, hydroiodides, sulphates, etc.

Turner: Potassium movement has been implicated in turgor changes resulting in stomatal movement and nyctinastic closing in *Albizzia julibrissin.* Certain antibiotics such as valinomycin and the cyclic polyethers manufactured by Du Pont have been shown to move potassium through membranes in animals, particularly those of blood cells. Does Ballio know whether these compounds move more potassium through plant membranes ?

Ballio: I do not know whether valinomycin and similar compounds have ever been tested on plant membranes.

6. The Chemistry of Some Amino-acid Derived Phytotoxins

MICHEL BARBIER

Institut de Chimie des Substances Naturelles,
91-*Gif-sur-Yvette, France*

Lycomarasmine

Chemical Structure

Lycomarasmine was isolated from liquid cultures of *Fusarium oxysporum* Schl. f. sp. *lycopersici* (Sacc.) Sny. *et* Hans. Sacc. by Clauson-Kaas *et al.* in 1944. The name of Gäumann is indeed bound to the story of this phytotoxin (Gäumann *et al.*, 1950, 1955, 1956; Clauson-Kaas *et al.*, 1944). The wilting caused by this toxin is enhanced by ferric ions. A second compound was given the name "Substance I" (the I standing for "Inactive"). The formula $C_9H_{15}O_7N_3$ has been established for lycomarasmine and $C_9H_{12}O_7N_2$ for "Substance I". These formulae have been confirmed by different methods including measurement of the rate of diffusion of the sodium salt and X-ray analysis of the copper complex. (For reviews on the chemistry of lycomarasmine see Hardegger *et al.*, 1963a; Liechti, 1958; Boller, 1951). The phytotoxin is insoluble in water and the more common organic solvents; bases are usually added to increase solubility. However, in alkaline media, lycomarasmine is transformed into Substance I with liberation of one mole ammonia. Acid hydrolysis is difficult; one mole of glycine, one of aspartic acid and one of pyruvic acid have been obtained in this way. No acetic acid is produced after Kuhn–Roth oxidation and lycomarasmine is not reduced catalytically. Glycinimid was isolated after treatment with $KMnO_4$ and there is no free amino group. From the result of a Hofmann degradation, Woolley (1948) concluded that lycomarasmine is a derivative of asparagine because no more aspartic acid was found after hydrolysis. Boller (1951) repeated this reaction but found that the conditions were so drastic that even aspartic acid was destroyed. From Substance I the methyl ester picrolonate was prepared as a crystallized derivative. Analysis showed the presence of three ester functions and one basic amino group. On the basis of these results a series of structures has been successively proposed by Plattner and Clauson-Kaas (1945a, b), Woolley (1948), Boller

Successive formulae proposed for lycomarasmine and anhydro-lycomarasmic acid

Lycomarasmine	Corresponding anhydro-lycomarasmic acid

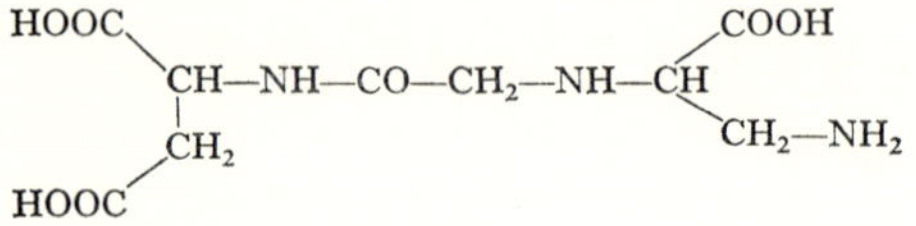

I

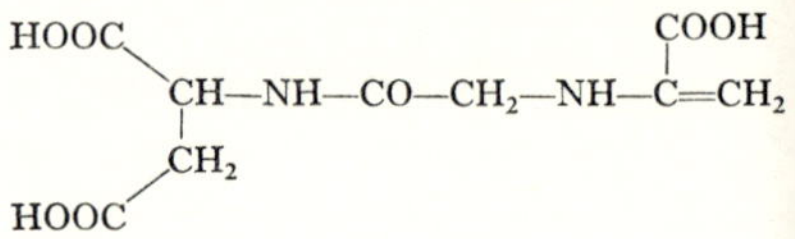

II

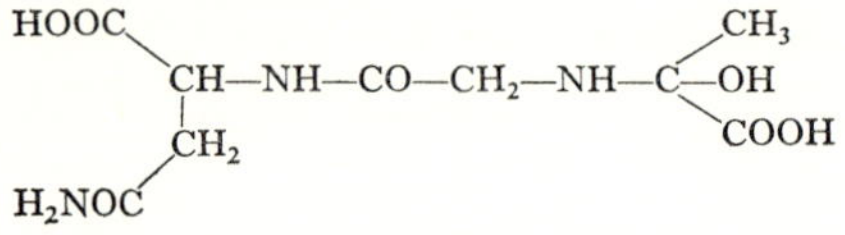

III

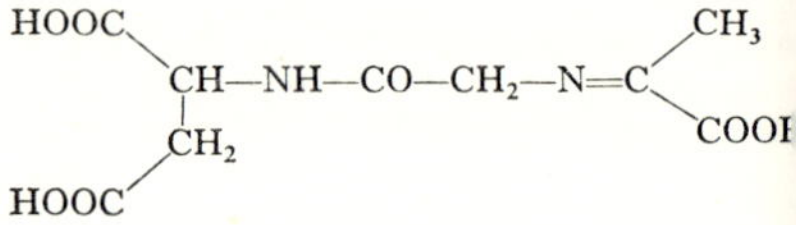

IV

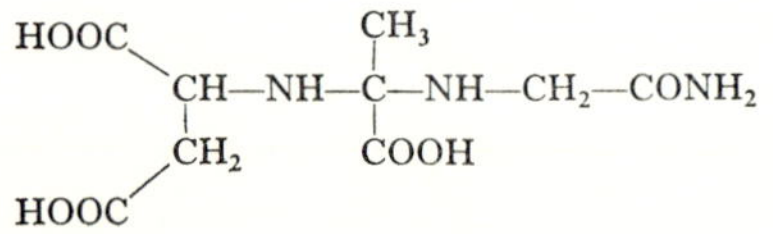

V

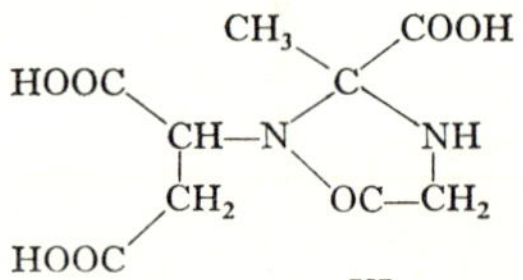

VI

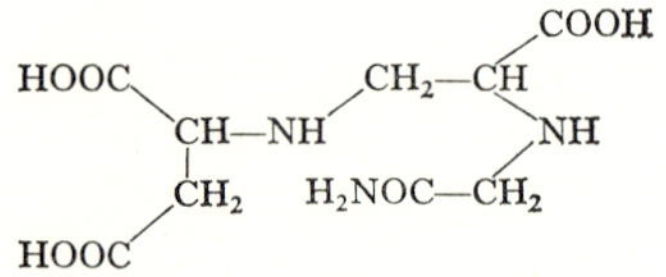

VII

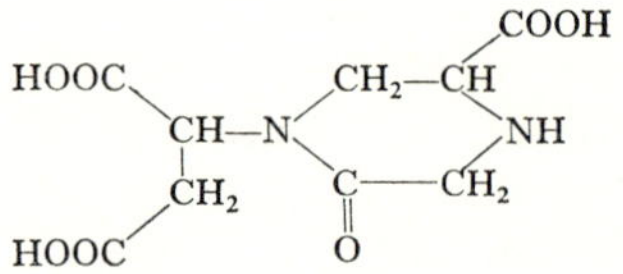

VIII

(1951) and Hardegger *et al.* (1963a) (I to VIII). Woolley (1948) has synthesized III supposedly the correct structure for lycomarasmine. The substance is active in the wilting test, but is very soluble in water in contrast with the natural product. No more details were given about the synthetic compound and its transformation into the anhydro derivative (Substance I). Boller (1951) was unable to repeat Woolley's synthesis of III.

The different products formed by acid hydrolysis can be understood by a mechanism of β-elimination which will be dealt with in detail later.

The correct structures VII and VIII for lycomarasmine and Substance I were demonstrated by comparative NMR study of VIII with aspartic acid, N-acetyl aspartic acid and N-(β-carboxy-β-aminoethyl)aspartic acid; (Hardegger *et al.*, 1963a). Finally synthesis of VIII was accomplished by Hardegger *et al.* (1963b).

Synthesis of Anhydro-lycomarasmic Acid (*Substance I*)

Several unsuccessful attempts to synthesize lycomarasmic acid (or

Synthesis of anhydro-lycomarasmic acid

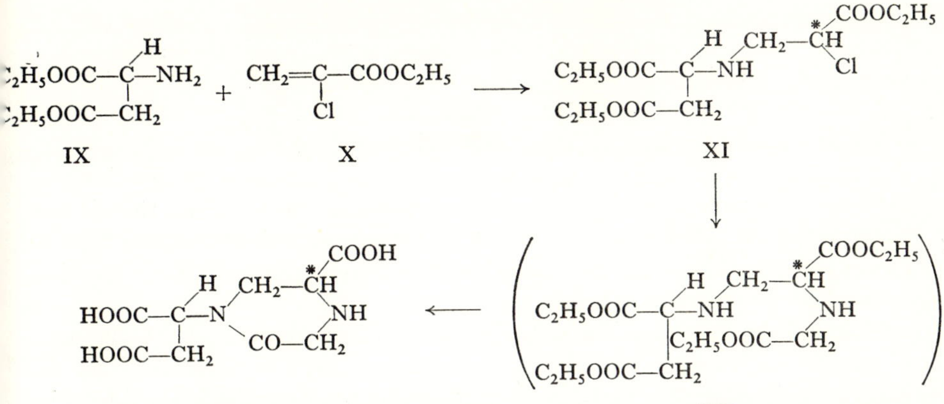

IX X XI XIII XII

aspergillomarasmine B) were made by Liechti (1958). The synthesis of the anhydro compound VIII was performed by the Hardegger team (1963b); the reaction sequences are summarized by formulae IX–XIII.

The synthetic product XIII was identical in all respects except stereochemistry to the anhydro derivative obtained from lycomarasmine.

By its curious structure, in which amino acids are bound through C—N—C bonds, lycomarasmine is a member of a new family of natural compounds. The aspergillomarasmines are other constituents of this family, as well as the imino-acids octopine, lysopine and saccharopine (see for example Lederer, 1963). The chelating ability of lycomarasmine with Fe^{+++} ions and that of the aspergillomarasmines, is related to the presence of two NH groups in the 1,2-position (cf. EDTA).

Aspergillomarasmines A and B (XIV and XV)

Chemical Structure

In 1956, research workers from Société d'Etude et d'Applications Biologiques isolated three crystalline compounds from cultures of *Aspergillus flavus-oryzae*. Later, two were found to be biologically active in the wilting test on tomato leaves and then called aspergillomarasmines A and B. The third, biologically inactive, is the lactam of B and was called anhydro-aspergillomarasmine B. The activity of compounds A and B is enhanced by ferric ions and their similarity to lycomarasmine was established mainly after elementary analysis and the results of acid hydrolysis (Robert *et al.*, 1962; Haenni *et al.*, 1962, 1965).

The chemical structure of aspergillomarasmine B (XV) was determined before that of aspergillomarasmine A (XIV). It was shown to be identical with lycomarasmic acid. However, the name aspergillomarasmine B was retained because of its natural occurrence in *A. flavus-oryzae* group (not in *F. oxysporum* f. sp. *lycopersici*). The formation of aspartic acid and glycine after acid hydrolysis of aspergillomarasmine B, and of aspartic acid and 2,3-diamino-propionic acid from aspergillomarasmine A are explained by a β-elimination mechanism described below.

The amino acids are liberated as soon as the substances are dissolved in HCl but complete hydrolysis takes many hours at 100°C. From this, it has been supposed that hydrolysis of the intact compound occurs rapidly but much more slowly when the anhydro derivatives are used; these are also formed in the presence of acids.

Aspergillomarasmines A and B give tetramethyl esters, $C_{13}H_{22}O_8N_2$ and $C_{14}H_{25}O_8N_3$ respectively. Mass spectrometry was used extensively for all determinations of structure and interpretation of spectra was considerably helped by comparison with those of synthetic analogues.

Hydrolysis of aspergillomarasmine B (XV)

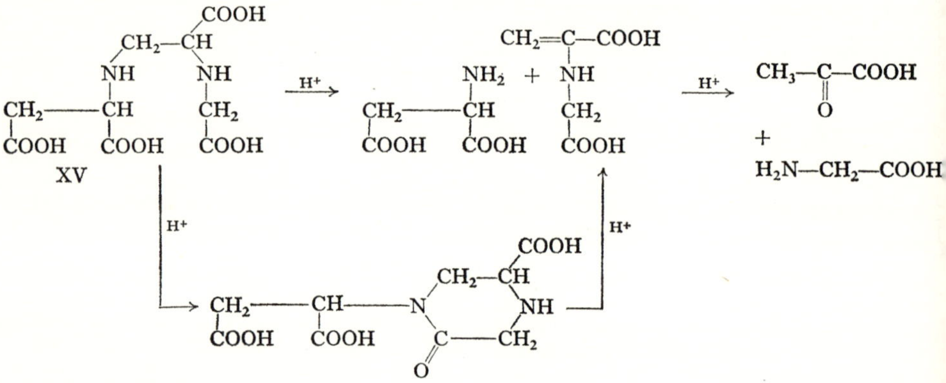

Aspergillomarasmine A was found in young cultures whereas aspergillomarasmine B was found in older cultures together with the anhydro-aspergillomarasmine B. This suggested the hypothesis that aspergillomarasmine A may be a precursor of B. Van Slyke oxidation of aspergillomarasmine A showed one free amino group; acid hydrolysis repeated after treatment with HNO_2 produced aspartic acid, glycine and isoserine (in place of 2,3-diamino-propionic acid). The finding of isoserine and not serine is in agreement with the proposed structure. Another proof of the structure was the isolation of α-alanine after catalytic hydrogenolysis of the benzyl ester (corresponding to the three carbon atoms of the central part of the molecule).

Stereochemistry. The aspartic acid from the hydrolysis of the aspergillomarasmines is L(S). Unfortunately, the 2,3-diamino-propionic acid and α-alanine which would have given the solution for the other asymmetric centers are racemized. Attempts at enzymatic oxidation of the free amino group in aspergillomarasmine A with the D-amino acid oxidases from sheep kidneys and from *Trigonopsis variabilis* Schach., as well as with the L-amino-acid oxidase from *Crotalus adamanteus* Beauv. have been unsuccessful. However, success was obtained with the D-amino-acid oxidase from *A. ustus* (Bain.) Thom *et* Church. Twice the calculated amount of oxygen was used. The configuration D(R) was thus tentatively proposed for the two

Hydrolysis of aspergillomarasmine A (XIV)

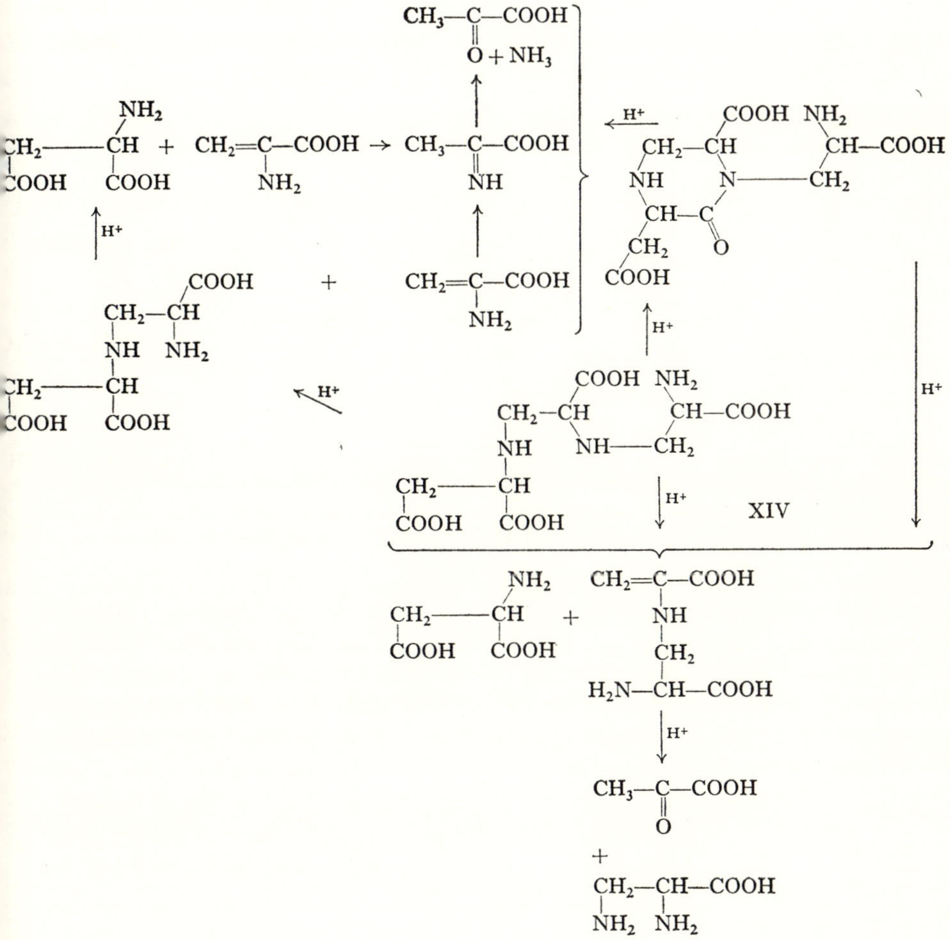

Proposed mechanism for the enzymatic oxidation of aspergillomarasmine A

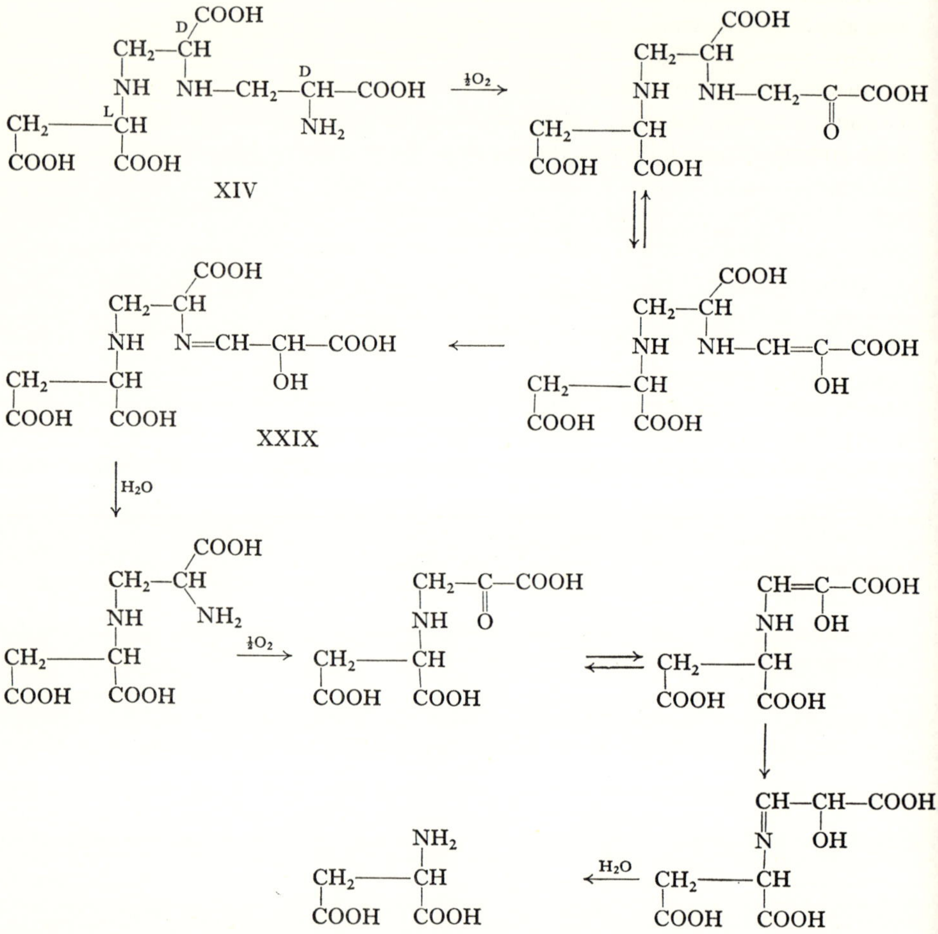

asymmetric centres due to 2,3-diamino-propionic acid in aspergillomarasmine A. The uptake of a second oxygen was explained by enolization of the keto acid after the first oxidation (see later); such a mechanism would also explain the formation of aspartic acid through the enzymatic oxidation (Haenni *et al.*, 1965).

Aspergillomarasmine B has been isolated from *Colletotrichum gloeosporioides* Penz. (*Gloeosporium olivarum* Alm.) which attacks olive trees, the first time that this toxin has been found in a plant pathogen (Ballio *et al.*, 1969).

Aspergillomarasmine A has been recently found in a strain of *C. gloeosporioides* a pathogen of willows (Bousquet *et al.*, 1971): this phytotoxin has also been isolated from *Fusarium oxysporum* Schl. f. sp. *melonis* (Leach *et* Curr.) Sny. *et* Hans. cultures, the alternative production of aspergillomarasmine A or lycomarasmine depending on the culture conditions (Trouvelot *et al.*, 1971). This is a new argument in favour of the hypothesis that considers aspergillomarasmine A as a precursor of lycomarasmine.

Synthesis of Aspergillomarasmine B (racemic)

Because of its insolubility in the usual organic solvents and its instability, a special pathway was used for synthesis of aspergillomarasmine B (XV). The reaction between a maleic monobenzyl ester and an amine is known to give a N-substituted benzyl β-aspartate which can be later transformed into the N-substituted aspartic acid by hydrogenolysis. This was first applied to ethylene-diamine and 2,3-diamino propionic acid to give the compounds XVI, XVII and XVIII which were found active in the wilting bioassay on tomato leaves.

For synthesis of aspergillomarasmine B (racemic), the maleic condensation was applied to the amino acid XIX (Barbier *et al.*, 1963; Bogdanovsky and Barbier, 1964, 1965; Bogdanovsky, 1968). The synthetic aspergillomarasmine B was identical with the natural product except for stereochemistry.

XVI

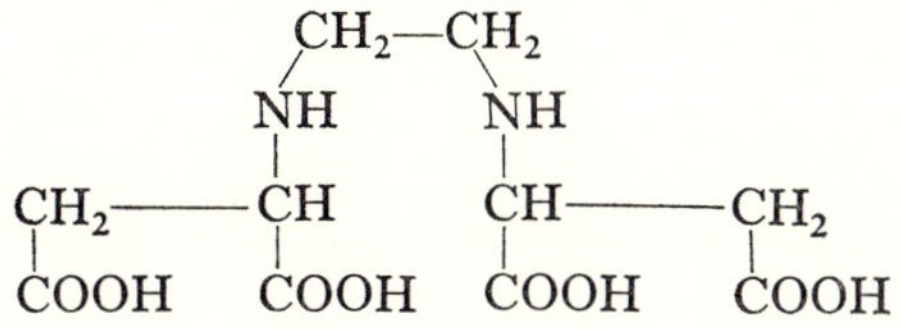

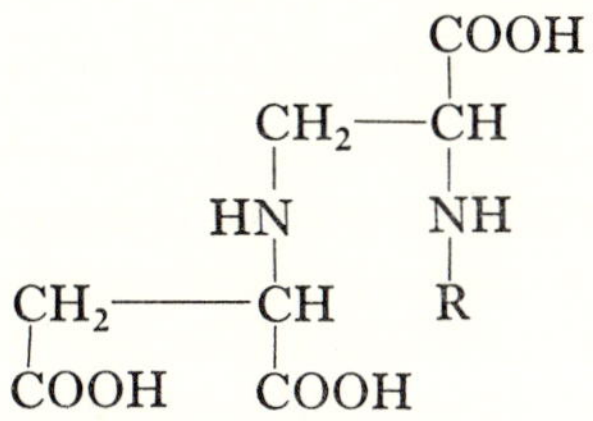

XVII R = H

XVIII R =

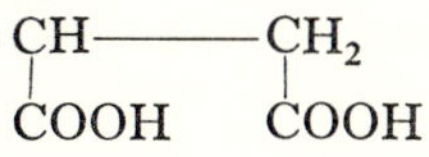

Synthesis of aspergillomarasmine B

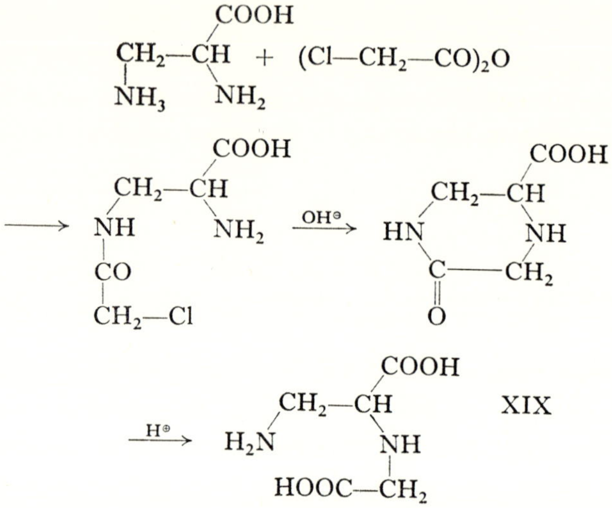

XIX

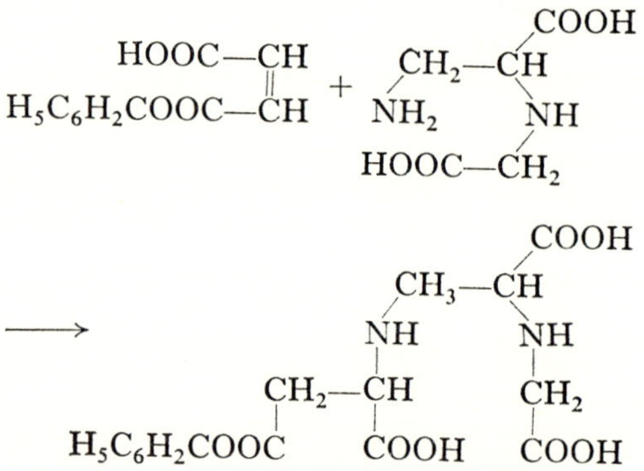

XIX

Fusaric Acid (XX)

Fusaric acid was isolated from species of *Fusarium* in 1934 by Yabuta *et al.* and the formula XX was proposed (Yabuta *et al.*, 1940). The chemical structure was confirmed some 20 years later by Plattner *et al.* (1954). With CaO the acid is decarboxylated to the 3-n-butylpyridine identical to the synthetic product. Fusaric acid has been synthesized by different methods (Plattner *et al.*, 1954; Hardegger and Nikles, 1956; Biland *et al.*, 1960; Schreiber and Adam, 1960). The most interesting synthesis consists in hydrogenating dehydrofusaric acid (also a natural compound from *Fusarium* spp.). Dehydrofusaric acid has been prepared with a 50% yield by the

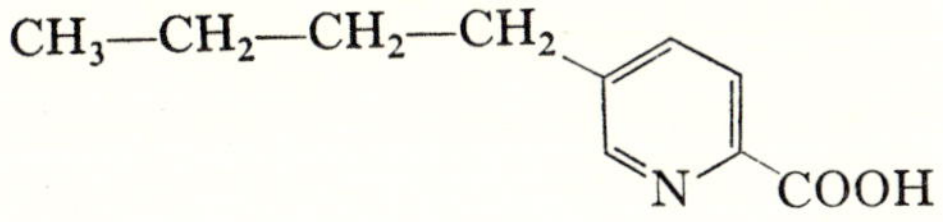

XX

reaction of allylchloride with 5-methyl picolinic acid (Steiner *et al.*, 1963).

Two possible pathways were considered for the biosynthesis of fusaric acid: I) polyacetate condensation as shown in XXI; 2) polyacetate +

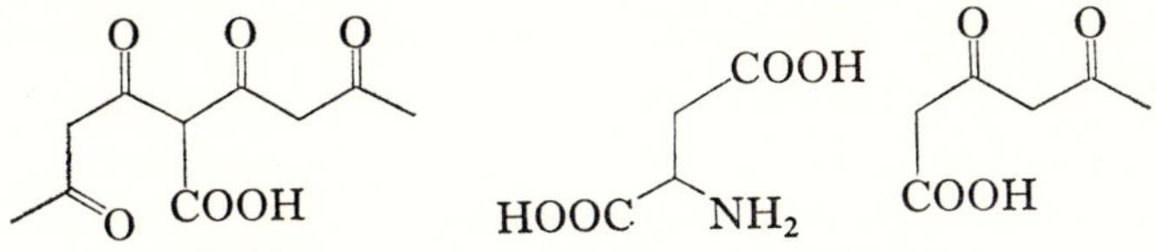

XXI XXII

aspartate condensation as in XXII. Pathway 2) (XXII) has been demonstrated by Hill *et al.* (1966) using 1,4-$C^{14}N^{15}$-aspartic acid, 1-C^{14}-acetic acid and 2-C^{14} acetic acid as precursors.

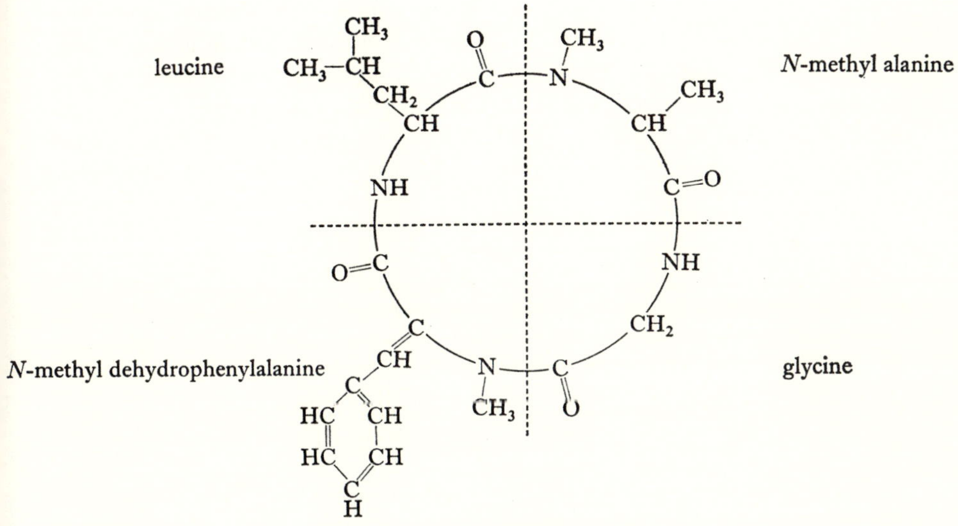

XXIII

Tentoxin

Alternaria tenuis Nees. causes chlorosis in cotyledon of cotton seedlings. Many *Alternaria* species but not all produce tentoxin which induces chlorosis in test plants (Fulton *et al.*, 1960, 1965; Templeton *et al.*, 1965, 1967; Grable *et al.*, 1966). Tentoxin has the molecular formula $C_{24}H_{32}N_4O_4$, shows a UV maximum at 285 nm; from ORD, IR and NMR spectra it is a peptide. Mass spectrometry indicates a mass of 414. Acid hydrolysis has given leucine, glycine and N-methylalanine. If the toxin is catalytically hydrogenated before hydrolysis, these amino acids are still obtained but two additional ones are present, N-methylphenylalanine and N-methylcyclohexylalanine. The peptide thus contains glycine, leucine, N-methylalanine, N-methyldehydrophenylalanine. Their sequence has been established by mass spectrometry and as no free amino or carboxyl groups are found, it must be a cyclic peptide corresponding to structure XXIII (Templeton *et al.*, 1967).

Rhizobitoxine

Rhizobitoxine has been isolated from *Rhizobium japonicum* (Kirch.) Buch. and produces necrosis on the leaves of soybean. It is a basic, sulphur-containing amino-acid, the precise structure of which is not yet known to my knowledge. Desulphurization leads to a non-toxic amino-acid derivative called "Y" that also occurs naturally both in soybean plants and in *R. japonicum* cultures. Substance Y has been tentatively identified as an ether of homoserine as shown in XXIV where R is a three carbon atoms moiety with one amino and one hydroxyl group. The toxin could consist of two molecules of Y joined through a thio-ether linkage. Growth of *Salmonella typhimurium* (Loeffl.) Cast. *et* Chalm. is inhibited by rhizobitoxine and this activity is reversed by L-methionine or homocysteine. An interesting

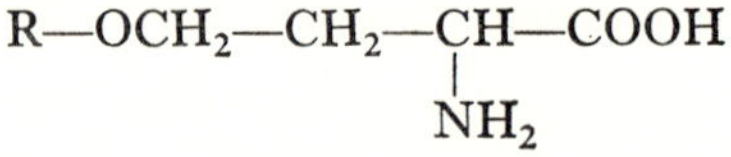

XXIV

XXV

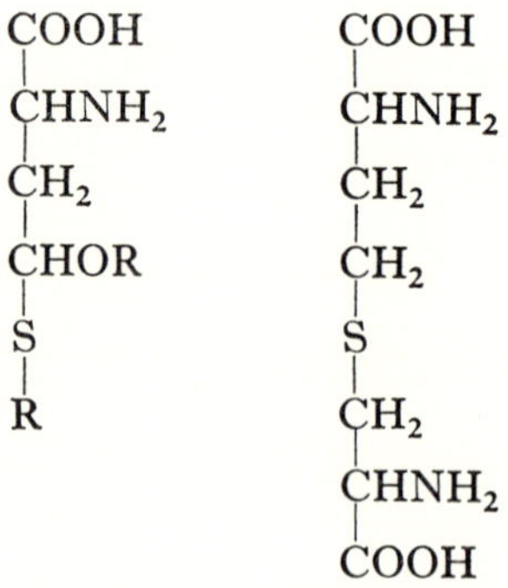

rhizobitoxine cystathionine

hypothesis suggests that rhizobitoxine inhibits enzymatic cleavage of cystathionine to homocysteine (inhibition of cystathionase) (Owens and Wright, 1965; Owens and Thomson, 1967; Owens *et al.*, 1968; Owens, 1969).

A possible partial structure of rhizobitoxine illustrating its relation to the four-carbon moiety of cystathionine is shown in XXV (Owens, 1969).

References

Ballio, A., Bottalico, A., Buonocore, V., Carilli, A., Di Vittorio, V., and Graniti, A. (1969). *Phytopath. medit.* **8**, 187–196.

Barbier, M., Bogdanovsky, D., Vetter, W., and Lederer, E. (1963). *Annalen* **668**, 132–138.

Biland, H., Lohse, F., and Hardegger, E. (1960). *Helv. chim. Acta* **43**, 1436–1440.

Bogdanovsky, D. (1968). Thesis (Ingenieur-Docteur, Paris).

Bogdanovsky, D. and Barbier, M. (1964). *Bull. Soc. chim. France* 2778–2782.

Bogdanovsky, D. and Barbier, M. (1965). *Bull. Soc. chim. France* 1832–1834.

Boller, A. (1951). Thesis (ETH, Zürich).

Bousquet, J. F., Vegh, I., Pouteau-Thouvenot, M., and Barbier, M., *Ann. Phytopath.*, 1971, in press.

Clauson-Kaas, N., Plattner, P. A., and Gäumann, E. (1944). *Ber. schweiz. Bot. Ges.* **54**, 523.

Fulton, N. D., Bollenbacher, K., and Moore, B. J. (1960). *Phytopathology* **50**, 575.

Fulton, N. D., Bollenbacher, K., and Templeton, G. E. (1965). *Phytopathology* **55**, 49–51.

Grable, C. I., Templeton, G. E., and Meyer, W. L. (1966). *Phytopathology* **56**, 897.

Haenni, A. L., Barbier, M., and Lederer, E. (1962). *C. r. Acad. Sc. Paris* **255**, 1476–1478.

Haenni, A. L., Robert, M., Vetter, W., Roux, L., Barbier, M., and Lederer, E. (1965). *Helv. chim. Acta* **48**, 729–750.

Hardegger, E. and Nikles, E. (1956). *Helv. chim. Acta* **39**, 505–513.

Hardegger, E., Liechti, P., Jackman, L. M., Boller, A., and Plattner, P. A. (1963a). *Helv. chim. Acta* **46**, 60–74.

Hardegger, E., Seres, J., Andreatta, R., Szabo, F., Zandowska-Jasinska, W., Romeo, A., Rostetter, C., and Kindler, H. (1963b). *Helv. chim. Acta* **46**, 1065–1067.

Hill, R. D., Unrau, A. M., and Canvin, D. T. (1966). *Can. J. Chem.* **44**, 2077–2082.

Gäumann, E., Naef-Roth, S., and Miescher, G. (1950). *Phytopathology* **16**, 257.

Gäumann, E., Naef-Roth, S., and Kern, H. (1955). *Phytopathology* **24**, 373.

Gäumann, E. and Naef-Roth, S. (1956). *Phytopathology* **25**, 418.

Lederer, E. (1963). *Pure appl. Chem.* **7**, 247–268.

Liechti, P. (1958). Thesis (ETH, Zürich).

Owens, L. D. (1969). *Science* **165**, 18–25.
Owens, L. D. and Wright, D. A. (1965). *Pl. Physiol.* **40**, 927.
Owens, L. D. and Thomson, J. F. (1967). *Bact. Proc.* **19**.
Owens, L. D., Guggenheim, S., and Hilton, J. L. (1968). *Biophys. Biochim. Acta* **158**, 219–225.
Plattner, P. A. and Clauson-Kaas, N. (1945a). *Helv. chim. Acta* **28**, 188–195.
Plattner, P. A. and Clauson-Kaas, N. (1945b). *Experientia* **1**, 195–196.
Plattner, P. A., Keller, W., and Boller, A. (1954). *Helv. chim. Acta* **37**, 1379–1392.
Robert, M., Barbier, M., Lederer, E., Roux, L., Biemann, K., and Vetter, W. (1962). *Bull. Soc. chim. France* 187–188.
Schreiber, K. and Adam, G. (1960). *Ber.* **93**, 1848–1851.
Steiner, K., Graf, U., and Hardegger, E. (1963). *Helv. chim. Acta* **46**, 690–694.
Templeton, G. E., Grable, C. I., Fulton, N. D., and Bollenbacher, K. (1965). *Phytopathology* **55**, 1079.
Templeton, G. E., Meyer, W. L., Grable, C. I., and Sigel, C. W. (1967). *Phytopathology* **57**, 833.
Trouvelot, A., Camporota, P., Barbier, M., and Pouteau-Thouvenot, M., (1971). *C. r. Acad. Sc. Paris*, 272, Ser.D, 754–756.
Woolley, D. W. (1948). *J. biol. Chem.* **176**, 1291–1298.
Yabuta, T., Kambe,K., and Hayasi, T. (1934). *J. agr. Chem. Soc. Jap.* **10**, 1059.
Yabuta, T., Kambe, K., and Hayasi, T. (1940). *J. Imp. agr. Expt. Sta. Jap.* **3**, 365.

Discussion

Paxton: Since you have mentioned the rapid conversion of aspergillomarasmine B to its anhydro derivative, I should also like to mention that the ethyl ester of glutamic acid readily cyclizes under basic conditions. This is important because amino acids are often extracted from plants under acidic conditions in ethanol. This would form some of the ethyl ester and subsequent elution from columns with NH_4OH leads to cyclization and loss of ability to react with ninhydrin.

Ballio: I wonder whether aspergillomarasmine A is an inhibitor of catalase, as are some other acids. In such a case the keto-acid found by oxidation might take up more oxygen, being oxidatively decarboxylated by H_2O_2; this would explain the high consumption of oxygen.

Barbier: This is an interesting idea. However, one has also to explain the formation of free aspartic acid in the enzymatic oxidation of aspergillomarasmine A.

Strobel: What is known about the biosynthesis of aspergillomarasmine?

Barbier: To my knowledge, nothing is known about the biosynthesis of the aspergillomarasmines. One can imagine a process starting from pyruvic acid or an enol pyruvate such as shown in a and b, but I am not sure that such reactions would be biologically possible. Aspergillomarasmine A can

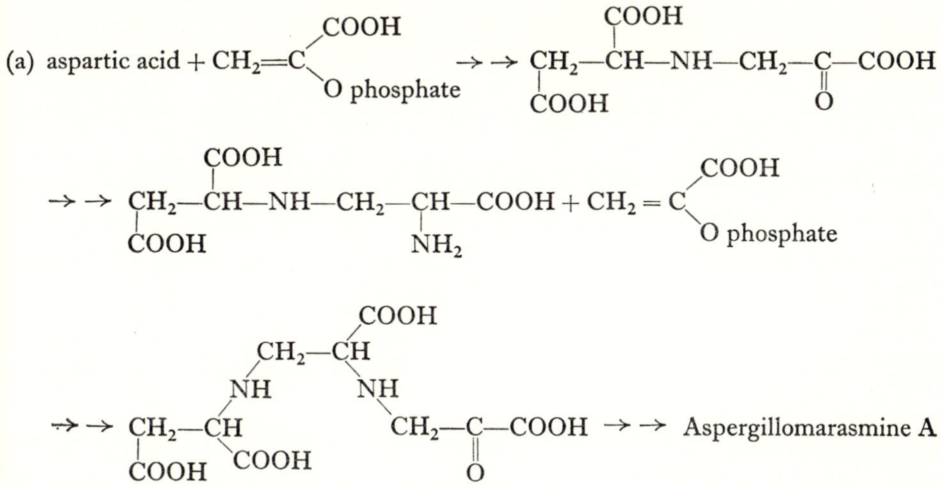

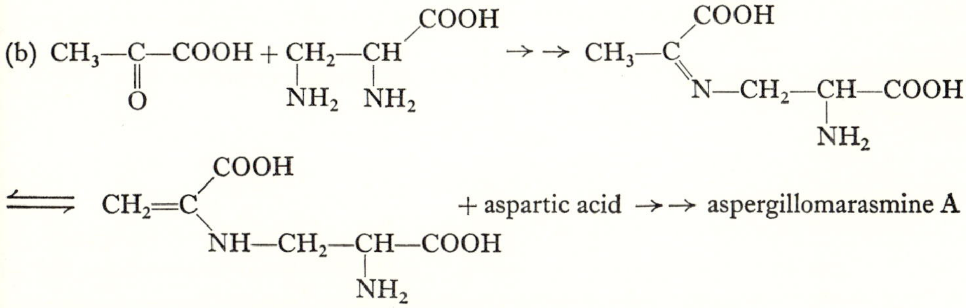

be a precursor of the aspergillomarasmine B, after oxidation of the free amino group and decarboxylation as follows:

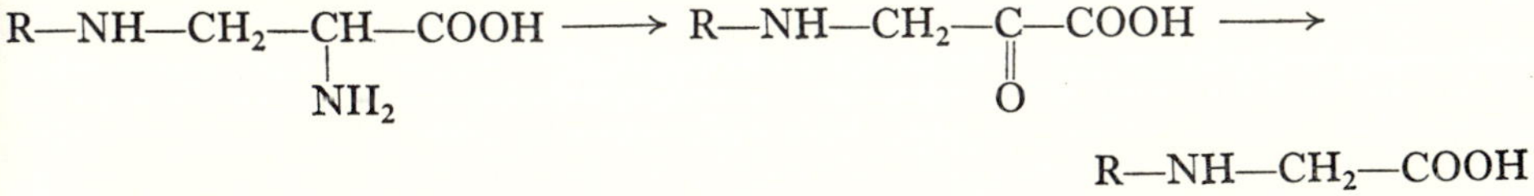

Mirocha: My question is of practical significance to the plant pathologist. Is it possible to detect the aspergillomarasmines which you have described when present in small concentrations, i.e. physiological amounts by TLC, GLC or other chromatography? Have aspergillomarasmines any mammalian toxicity?

Barbier: I suggest precipitation of a calcium salt, followed by the paper electrophoresis as we have described.

The aspergillomarasmines do not appear to be very toxic to mammals; aspergillomarasmine B has an anti-inflammatory effect on skin lesions produced by injection of carragenine (Mrs. Robert's thesis).

7. Chemistry of the Terpenoid Phytotoxins

CARLO G. CASINOVI

Laboratori di Chimica Biologica,
Istituto Superiore di Sanità,
Rome, Italy

A number of phytotoxic compounds have structures which are clearly of terpenoid origin. It is the aim of this paper to survey the chemistry of these substances with the main emphasis on the facts which allowed their structures to be elucidated.

To save space, non-essential considerations and experimental observations will generally be omitted.

Because of their chemical derivation the substances to be dealt with will be classified as sesquiterpenoid, diterpenoid and sesterterpenoid phytotoxins.

Sesquiterpenoid Phytotoxins

Fomannosin (I)

Although not obtained in crystalline form fomannosin (I), a metabolite of *Fomes annosus* (Fr.) Cooke a parasite of *Pinus taeda* L. was shown by Bassett *et al.* (1967) to have a $C_{15}H_{18}O_4$ crude formula by elemental analysis, nuclear magnetic resonance data (NMR) and high resolution mass spectrometry of its dihydroderivative (II); under forcing conditions, a tetrahydroderivative was formed (Fig. 1). A second paper (Kepler *et al.*, 1967) reported that it contained the functional groups shown in Fig. 2.

One of the oxygens in fomannosin (I) must be OH because of its infra-red (IR) peak at 3450 cm^{-1}. Another oxygen belongs to a cyclopentanone moiety, and is responsible for the peak which appears at 1750 cm^{-1} and is practically unchanged in the dihydro-(II) and in tetrahydroderivative (III). Accordingly the UV maxima of dihydrofomannosin-2,4-dinitrophenylhydrazone (IV) (Fig. 1) occur at 360, 274 and 232 nm. The remaining oxygens must be accounted for by a conjugated δ-lactone in order to explain the second carbonyl oxygen which is unreactive towards 2,4-dinitrophenylhydrazine, and whose peak appears at 1715 cm^{-1} in (I), and at 1720 cm^{-1} in

Fomannosin (I)
$C_{15}H_{18}O_4$

↓

Dihydrofomannosin (II)
$C_{15}H_{20}O_4$

↙ ↓ ↘

Tetrahydrofomannosin $C_{15}H_{22}O_4$ (III)

Dihydrofomannosin-2,4-DNPH (IV) $C_{21}H_{24}N_4O_7$

Dihydrofomannosin acetate (V) $C_{17}H_{22}O_5$

FIG. 1. Principal reactions of fomannosin.

(a) —OH: $\nu_{max}^{CS_2}$ 3450 (I)

(b) 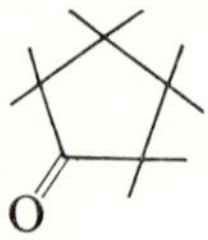 : $\nu_{max}^{CS_2}$ 1750 (I); 1745 (II); 1740 (III); 1725 (IV)
λ_{max}^{EtOH} 360, 274, 232 (IV)

(c) 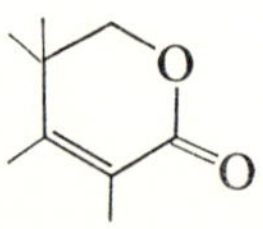 $\nu_{max}^{CS_2}$ 1715 (I); 1720 (II); 1740 (III); no bis-2,4DNPH (II)
: λ_{max}^{EtOH} 261 (I), 205-217 (II); <205 (III)

$$\text{II} \underset{H^+}{\overset{OH^-}{\rightleftarrows}} \text{Acid} \underset{H^+}{\overset{CH_2N_2}{\rightleftarrows}} \text{Ester}$$

FIG. 2. Chemical functions of fomannosin.

(II). This conjugation disappears in (III) which absorbs at 1740 cm^{-1}. Correspondingly, the UV absorption maximum shifts from 261 nm in (I) to 205–217 nm in (II), and disappears in (III). Alkaline hydrolysis transforms (I) into an acid, giving an ester with CH_2N_2; this is reversed with acid.

With the four oxygens accounted for, further information about the environment of the δ-lactone moiety can be deduced. The changes in UV absorption following the two successive hydrogenation stages certainly indicate that the lactone is α,β,γ,δ unsaturated. Furthermore, the γ,δ double bond must be a disubstituted ethylene, because the I → II transformation is characterized in NMR spectra by the disappearance of a typical AB system ($\delta_A = 6{\cdot}91, \delta_B = 6{\cdot}66, J_{AB} = 2{\cdot}5$ cps). The low J value suggests a cyclobutene ring. In contrast, the α,β double bond is fully substituted, because apart from the system already mentioned, no other signals in the region characteristic of the olefinic protons are shown by (I). If the β carbon atom were substituted by the cyclobutene ring, the substituent on the α carbon would be indicated by the 2-proton singlet which in (I) lies at 4·36 δ and in (II) lies at 4·28 δ and is characteristically shifted downfield following acetylation (4·78 δ in *V*): it is undoubtedly a primary alcoholic function.

The correctness of these deductions (Fig. 3) was eventually demonstrated by X-ray analysis of the *p*-bromobenzoylurethane of (II) (McPhail and Sim, 1968). The structure and relative stereochemistry of (I) and (II) must then be those shown in Fig. 4, in which three isoprene units are discernible and which fully explain their NMR spectra.

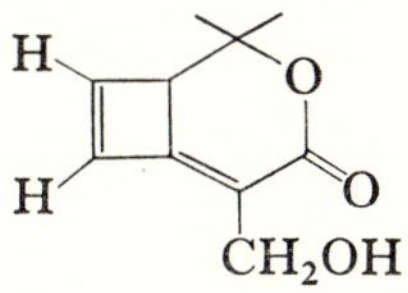

λ_{max}^{EtOH} 261 (I); 205-217 (II); <205 (III)

2 H, 6·91 and 6·66δ; $J_{AB} = 2{\cdot}5$ cps (I), disappearing in (II)

2 H, 4·36 (I), 4·28 (II), 4·78δ (V)

FIG. 3. The skeleton of fomannosin.

A discussion of the biogenesis of fomannosin would be of great interest but beyond the scope of this paper.

Helminthosporal (*VI*)

This toxin, produced by *Helminthosporium sativum* Pamm., King *et* Bakk. has the basic formula $C_{15}H_{22}O_2$, both oxygens of which could be accounted for as two aldehyde groups on chemical and spectroscopic considerations (IR, UV, NMR). NMR data showed that one of the two CHO is α,β-unsaturated (ν_{max} 1685 cm^{-1}), the ethylenic linkage being completely substituted (no signal for olefinic protons in the NMR spectrum), one of the substituents being a —CH_3 (3 proton singlet at 2·09 δ). In the absence of more multiple bonds, the compound must be bicyclic (de Mayo *et al.*, 1961). Demonstration of the structure of (VI) started from the

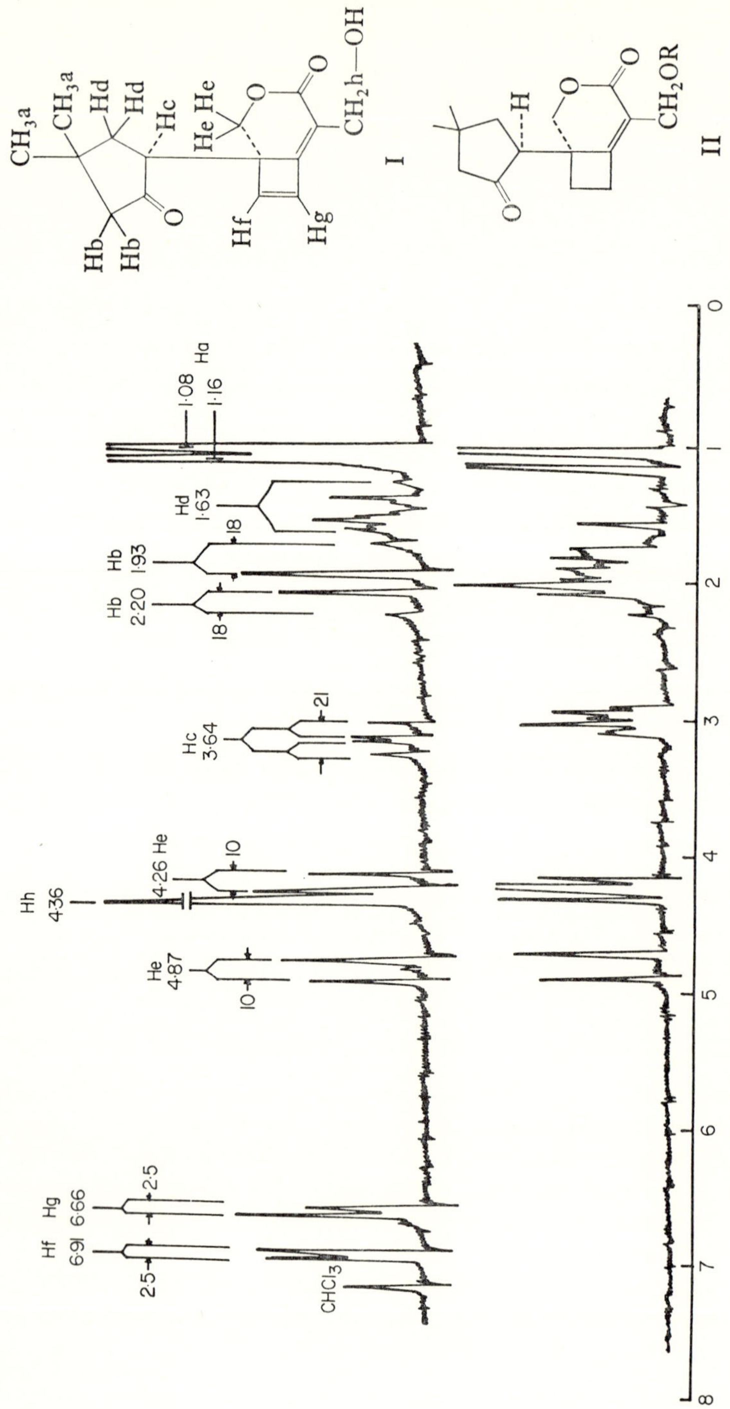

FIG. 4. Structures and NMR spectra of fomannosin (I) and dihydrofomannosin (II).

corresponding saturated monoacid obtained from (VI) by Ag_2O oxidation (de Mayo *et al.*, 1962a). Reduction with borohydride gave a compound with an elemental analysis and IR absorption that suggested the formation of a γ-lactone ring. However, such a ring does not arise from the reaction of the carboxyl with the primary alcohol expected from the reduction of the CHO; the lactone NMR spectrum has in fact no signals in the region expected for an esterified alcoholic function but instead has a pattern

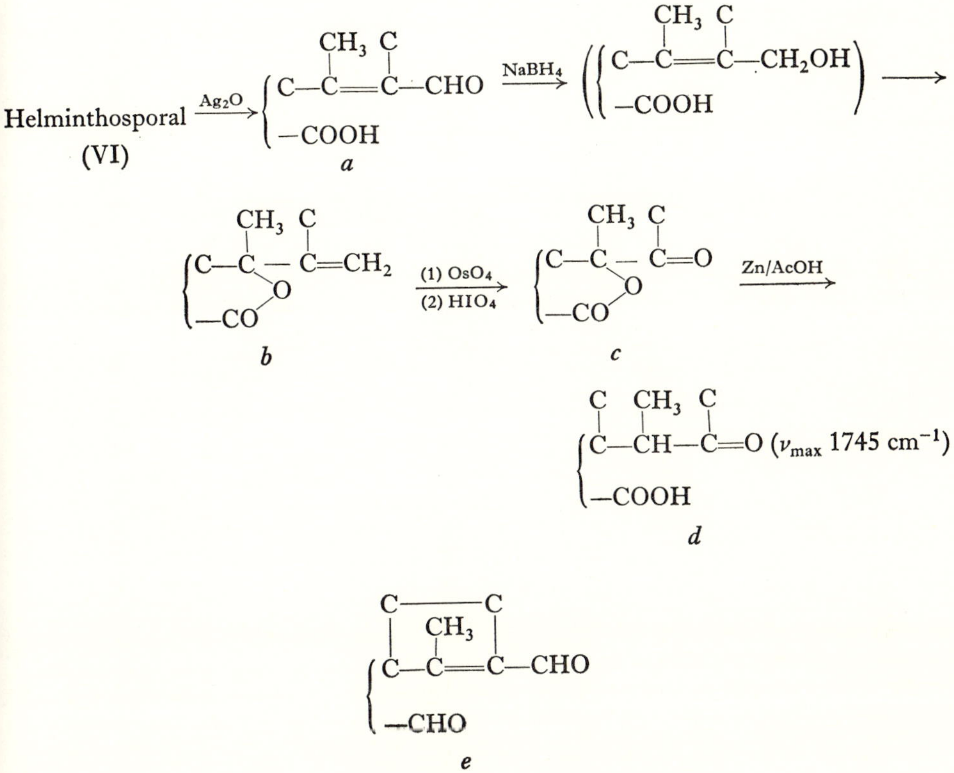

FIG. 5. Partial structure of helminthosporal.

typical for a > $C{=}CH_2$ group, and the CH_3 above mentioned has moved to 1·4 δ. Such a lactone treated with ozone gave CH_2O, therefore it must have arisen from an allylic rearrangement. The most plausible explanation of lactone formation is to assume the presence of the moiety *a* in (VI) and *b* in the lactone (Fig. 5).

OsO_4 and successive HIO_4 oxidation of the lactone, followed by elimination of the alcoholic function by Zn/acetic acid reduction (indicative of its

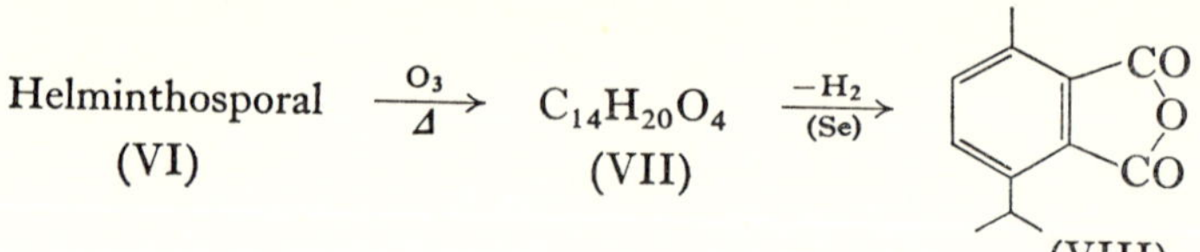

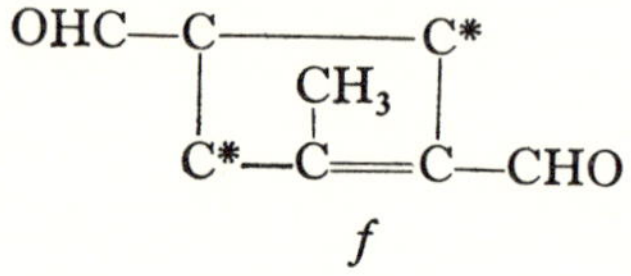

FIG. 6. The skeleton of helminthosporal.

α relationship to the ketone) gave a 5-membered ring carbonyl (ν_{max} 1745 cm^{-1}). This means that a ring of this size occurs in helminthosporal for which we can write the partial structure *e* shown in Fig. 5. That the second ring must be six-membered, follows from the isolation of 3-isopropyl-6-methyl-phthalic anhydride (VIII) following the reactions given in Fig. 6.

As a consequence, bearing in mind that one of the carbonyls in this phthalic anhydride must derive from the unconjugated CHO of helminthosporal and the other from the α-ketoaldehyde resulting on ozonolysis, partial structure *e* (Fig. 5) becomes *f* (Fig. 6).

Because two carbons are lost in the step from (VII) to (VIII) a six-membered ring can be closed only between the two carbons marked *, leading to either (IX) or (X) for helminthosporal (Fig. 7).

Because the NMR spectra of (VII) and several other derivatives have a singlet for the methyl on the six-membered ring, structure (IX), which requires a doublet, could be excluded and (X) assigned to helminthosporal. But this assumes that there is no rearrangement during dehydrogenation. It was demonstrated that this was so by studying the complete structure of (VII). On formula (X) for helminthosporal, (VII) should contain a methyl

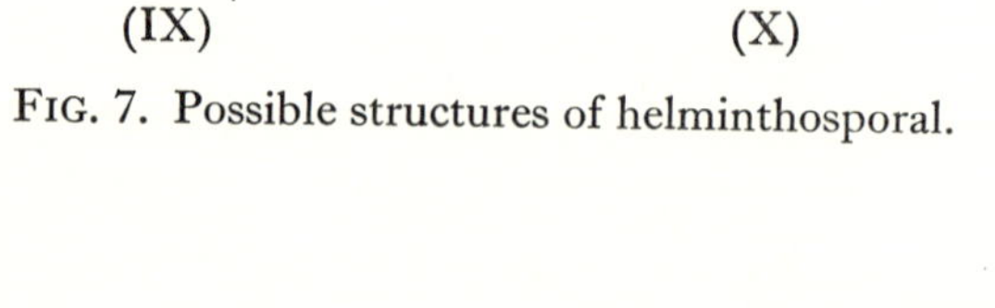

(IX) (X)

FIG. 7. Possible structures of helminthosporal.

ketone group, but its IR spectrum did not show this. However, hydrolysis gave a diacid $C_{14}H_{22}O_5$ (from which a diester (XI) to be described later, can be formed), with such a peak. This probably means that (VII) has the structure shown in Fig. 8 where the keto group is formulated as a dilactone and for which there are several examples in the literature.

The NMR spectrum of (VII) provides evidence for this formula by exhibiting the characteristic two doublet (6 protons) pattern for the isopropyl, 2 singlets (3 protons each, 1·15 and 1·60 δ respectively) for the two tertiary methyls, a doublet (1 proton, 2·4 δ, $J = 3{\cdot}6$ cps) and a multiplet (1 proton, 3·11 δ) which may well be attributed to the Ha and Hb respectively in formula (VII). These spectral data exclude derivation of (VII) from structure (IX) and rule out any possibility of rearrangement during Se dehydrogenation.

In structure (X) the stereochemistry at the two bridgeheads is imposed

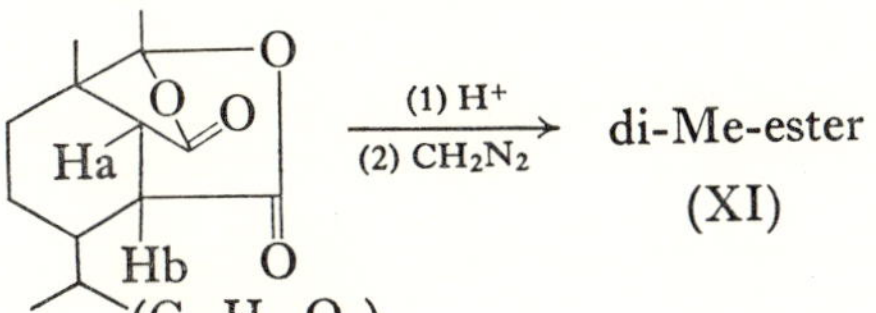

$(C_{14}H_{20}O_4)$

(VII)

(XI)

FIG. 8. The nature of the six-membered ring of helminthosporal.

by geometric considerations. This means that only the relative configuration of the carbons bearing the saturated CHO and the isopropyl remains to be demonstrated. The saturated CHO must be *cis* with respect to the ethylenic bridge, to explain the γ-lactone, as formed by allylic rearrangement following reduction by borohydride. Slightly more sophisticated considerations are needed for the asymmetric centre carrying the isopropyl. The diester (XI) can now be formulated either as (XIa) or (XIb) (Fig. 9), but, should (XIb) be correct, its conformation must be reversed as in XIc where we would expect for H_b a large coupling constant, because of its axial–axial relationship to H_c. However, this does not occur, and the diester must therefore have the stereochemistry of (XIa) in which the 1,3-diaxial relationship of the $COOCH_3$ to the —$COCH_3$ explains its resistance to carbonyl functionalization and its easy bridging condensations (e.g., upon treatment with *t*-BuOK). Hence, helminthosporal must be (VI) or its mirror image (de Mayo *et al.*, 1963); its synthesis from (—)-carvomenthone (Corey and Nozoe, 1963) also shows that (VI) represents its absolute structure.

Biosynthesis of helminthosporal utilizes mevalonate (de Mayo *et al.*, 1962b). Its immediate precursors are almost certainly helminthosporol

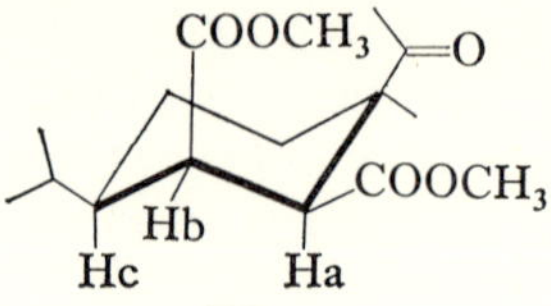

XIa

J_{Hb-Hc}: ~3 cps

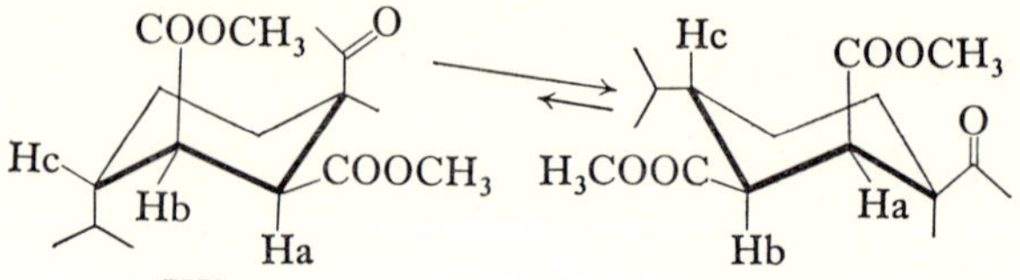

XIb (less stable conformer) XIc (more stable conformer)

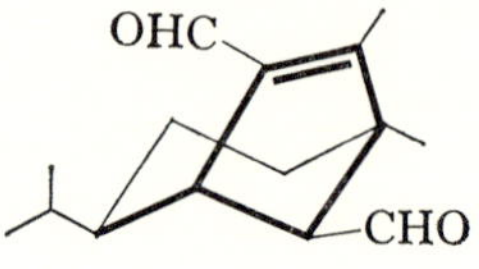

VI
Helminthosporal

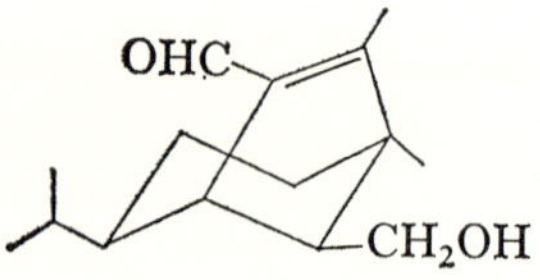

XII
Helminthosporol

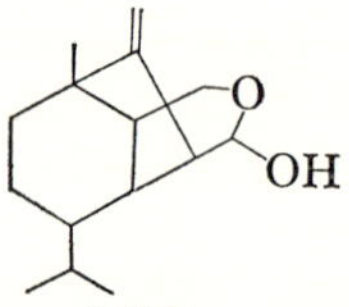

XIII
Prehelminthosporol

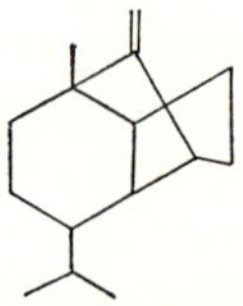

XIV
Sativene

FIG. 9. Stereochemistry of helminthosporal and its biosynthetic precursors.

(XII) and prehelminthosporol (XIII) (Spencer *et al.*, 1965). A more remote possibility is sativene (XIV) (de Mayo and Williams, 1965). All are metabolites of *H. sativum* (Fig. 9).

Diacetoxyscirpenol (XV)

This toxin, and some of its derivatives, are produced by many microorganisms, typically by *Fusarium roseum* (Link) Sny. *et* Hans. var. *gibbosum* (Woll.) Mess. *et* Cass. (=*F. equiseti* (Cda.) Sacc.) and *F. scirpi* (Lamb. *et* Fautr.). It was isolated in 1960, and shown to have the composition $C_{19}H_{26}O_7$ (Brian *et al.*, 1961). On the evidence given below, it contains the functional groups shown in Fig. 10.

Because the acetate does not show OH absorption in the IR and diacetoxyscirpenol contains no alkoxyls, the two oxygens remaining must be ethereal. Also, because diacetoxyscirpenol is neutral to permanganate and has only

(a) —OH: ν_{max}^{nujol} 3·415, 3·350 cm^{-1}; formation of an acetate $C_{21}H_{28}O_8$

(b) 2 —O—CO—CH_3: ν_{max}^{nujol} 1·739 cm^{-1}; 2 mols AcOH formed upon saponification

(c) 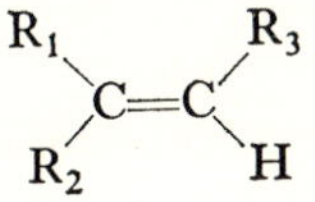

: ν_{max}^{nujol} 1·681, 820 cm^{-1}

FIG. 10. Chemical functions in diacetoxyscirpenol.

end-absorption in the UV, it can be considered to be a diacetate of a tetracyclic alcohol, scirpentriol.

The interesting biological properties of this toxin stimulated further research, and, in 1965, two groups independently and simultaneously reported on its structure (Dawkins *et al.*, 1965; Flury *et al.*, 1965). An important clue to the structure was given by NMR spectroscopy which revealed that scirpentriol had several features in common with verrucarol (XVI) (Fig. 11) (Tamm and Gutzwiller, 1962; Gutzwiller and Tamm, 1963; Gutzwiller *et al.*, 1964). This is the sesquiterpene moiety present in verrucarin A (Harri *et al.*, 1962; Tamm and Gutzwiller, 1962; McPhail and Sim, 1966), the structure of which had been proved by relating it chemically to trichodermin, an antifungal antibiotic for which formula (XVII) (Godtfredsen and Vangedal, 1964, 1965) was substantiated by X-ray analysis (Abrahamsson and Nilsson, 1964) (Fig. 11).

Diacetoxyscirpenol shows a tertiary methyl (0·82 δ), an olefinic proton (5·54 δ) and an AB pattern (3·07 and 2·78 δ, $J_{AB} = 4$ cps) which suggests an oxyrane ring of the kind present in verrucarol. A series of chemical transformations established a firm chemical correlation between diacetoxyscirpenol and verrucarol. Thus, diacetoxyscirpenol methane- or *p*-toluenesulphonate, treated with $LiAlH_4$, suffers elimination of an OH with concomitant opening of the oxirane ring, giving a triol, which can be acetylated to give the product similarly obtainable from verrucarol. This

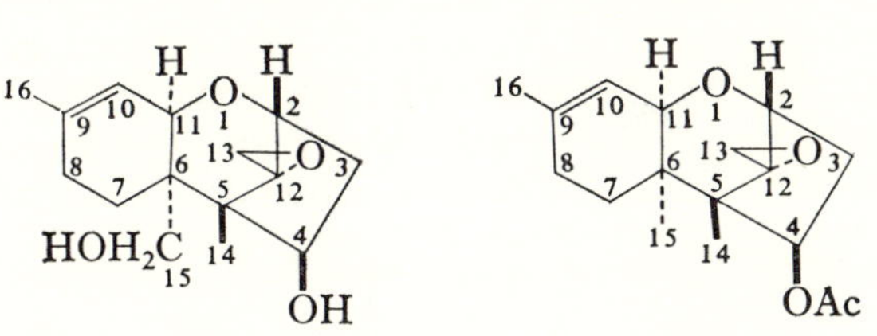

XVI XVII

FIG. 11. Structures of verrucarol and trichodermin.

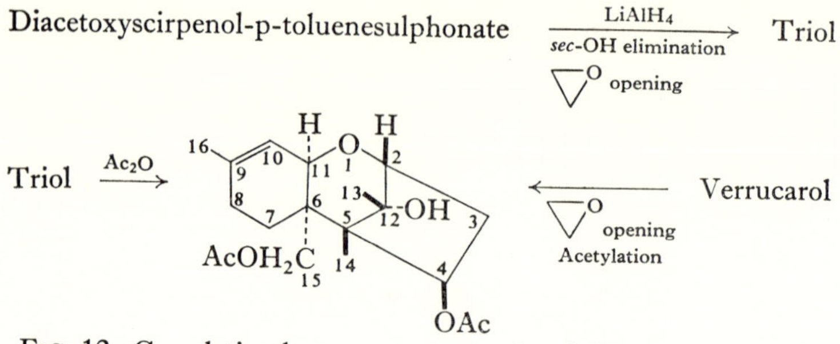

FIG. 12. Correlation between verrucarol and diacetoxyscirpenol.

means that scirpenetriol is a monohydroxyderivative of verrucarol (Fig. 12).

The location and the stereochemistry of this alcoholic function was deduced from inspection of the NMR spectrum of its triacetate, in which the H at C-4 appears at 5·76 δ as a doublet ($J = 3$ cps) and the H at C-3 at 5·2δ as a quadruplet ($J = 3$ and 5 cps). Therefore, the OH under consideration must be on C-3, and because scirpentriol is resistant to cleavage by periodic acid and lead tetraacetate (although cleavage readily occurs with manganese dioxide) (Dawkins and Grove, 1970), the α-glycol system must be *trans*. The absolute stereochemistry (XV) of diacetoxyscirpenol is thus established to be as shown in Fig. 13.

Two other phytotoxins closely related to diacetoxyscirpenol have been found, compounds $C_{19}H_{24}O_9$, m.p. 135–6°, and $C_{21}H_{28}O_{10}$, m.p. 185–8° (Brian *et al.*, 1961). NMR spectra of the first gave good evidence for the structure (XVIII) (Dawkins *et al.*, 1965). This was confirmed later by Grove (1969, 1970a) in the way shown in Fig. 14.

The second compound, $C_{21}H_{28}O_{10}$, was related to the first by sodium borohydride reduction of (XVIII) giving the corresponding triol (XIX) (Fig. 15) in which the stereochemistry of the new hydroxyl at 8 must be α (attack by the reagent from the less crowded β side of the molecule), the diacetate (XX) of which is identical to a monoacetate of the $C_{21}H_{28}O_{10}$ toxin, which as a consequence has structure (XXI), as shown by a comparison of NMR spectra of (XXI) and (XX) (Grove, 1969, 1970b).

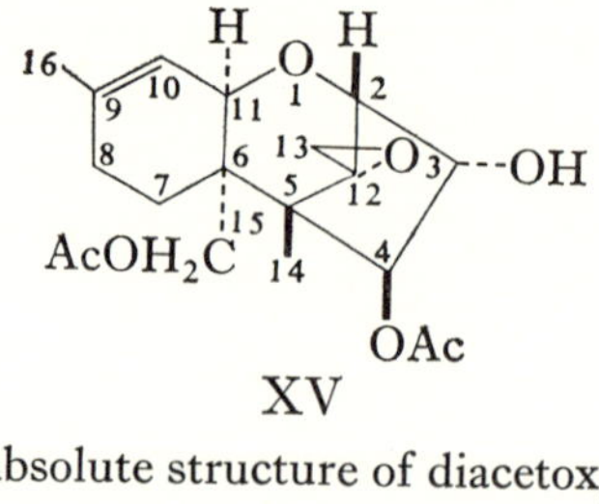

XV

FIG. 13. Absolute structure of diacetoxyscirpenol.

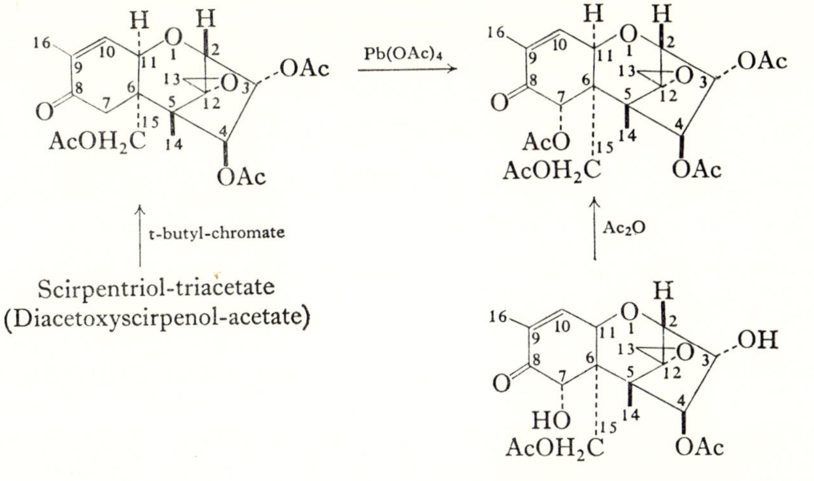

XVIII ($C_{19}H_{24}O_9$ toxin)

FIG. 14. Structure of the $C_{19}H_{24}O_9$ phytotoxin.

Another closely related sesquiterpenoid phytotoxin is the antifungal antibiotic trichothecin (Grove and Mortimer, 1969). Its structure but not its absolute stereochemistry is shown by (XXI) (Godtfredsen and Vangedal, 1965) on the basis that hydrolysis gives isocrotonic acid and an alcohol,

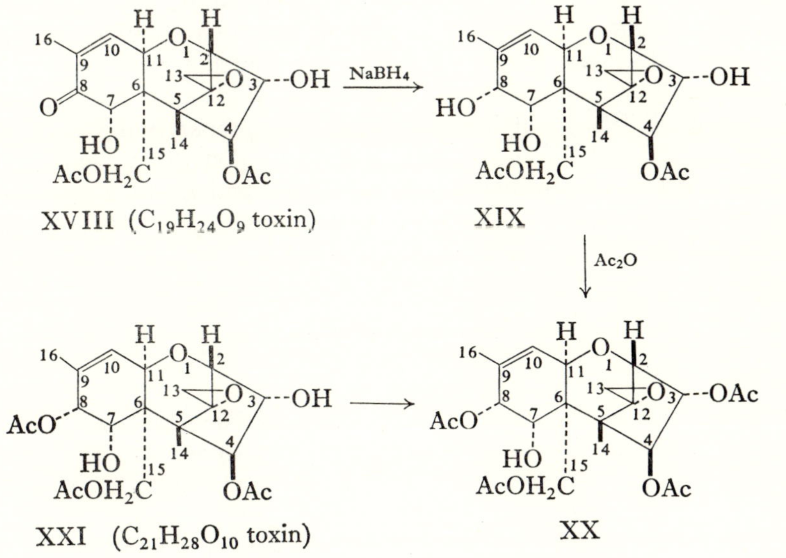

FIG. 15. Structure of the $C_{21}H_{28}O_{10}$ phytotoxin.

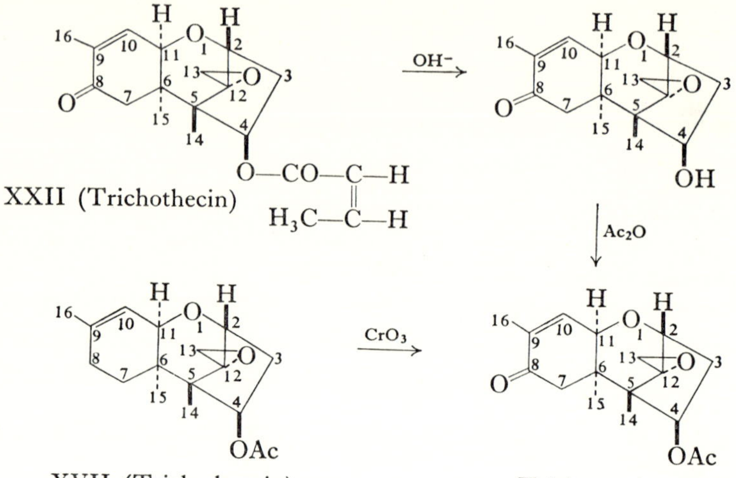

XXII (Trichothecin)

XVII (Trichodermin)

Trichotecolone

FIG. 16. Structure of trichothecin.

trichothecolone, the acetate of which had already been isolated after CrO_3 oxidation of trichodermin (XVII) (Fig. 16).

It is interesting to note how chemical structure and biological properties (toxicity to plants and animal cells, antibiotic and insecticidal properties) are related in this series of compounds, all containing the 12,13-epoxy-trichothecane nucleus, the most important of which are given in Fig. 17 (Grove and Mortimer, 1969). All can be considered metabolites, with the possible exception of nivalenol (Tatsuno *et al.*, 1968), and fusarenone (Tsunoda *et al.*, 1968), which may be artefacts of preparation (Grove, 1969).

Experiments with labelled compounds show that the nucleus of these substances derives from mevalonic acid (Jones and Lowe, 1960; Hanson and Achilladelis, 1967) through 6,7-*trans*-farnesylpyrophosphate (Achilladelis *et al.*, 1970).

Diterpenoid Phytotoxins

Fusicoccin (XXIII)

The structure of this phytotoxin produced by *Fusicoccum amygdali* Del. (Graniti, 1962, 1964), has been elucidated in our laboratories (Ballio *et al.*, 1964, 1968) and confirmed by Imperial College workers (Barrow *et al.*, 1968a, b; Hough *et al.*, 1968). Elemental analysis, NMR and mass spectrometry give the formula $C_{36}H_{56}O_{12}$. Its main structural features are based on the evidence shown in Fig. 18.

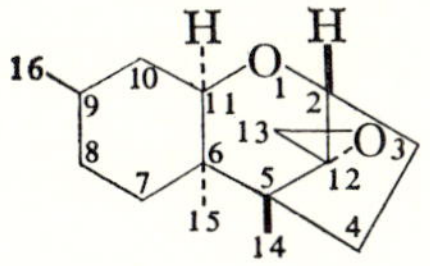

12.13-epoxytrichothecan

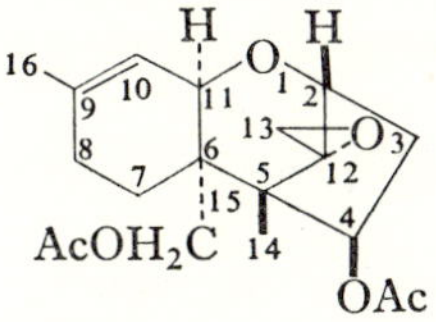

Diacetylverrucarol

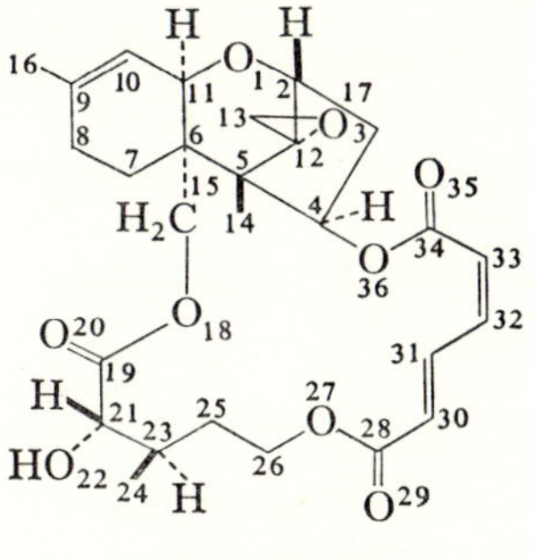

Verrucarin

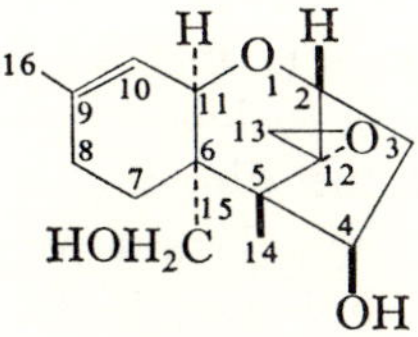

Roridin C
(Trichodermol)

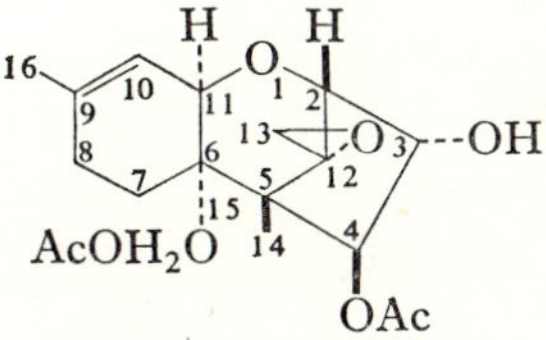

Diacetoxyscirpenol (XV)

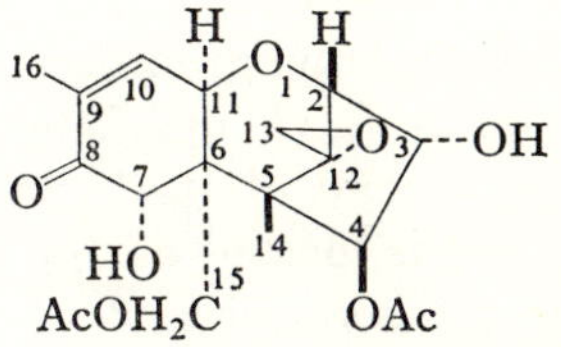

Phytotoxin $C_{19}H_{24}O_9$ (XVIII)

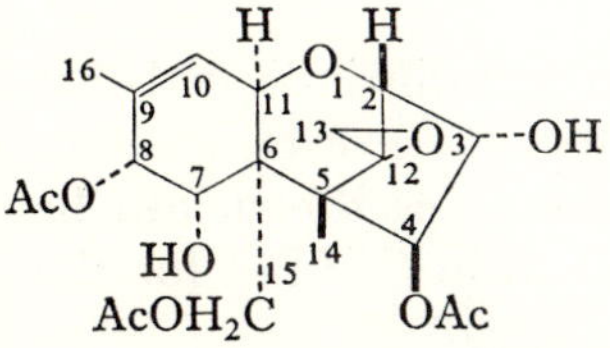

Phytotoxin $C_{21}H_{28}O_{10}$ (XXI)

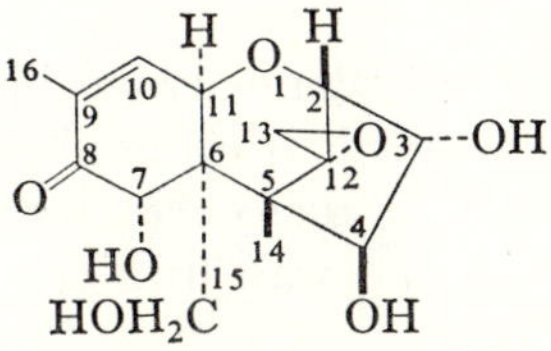

Nivalenol

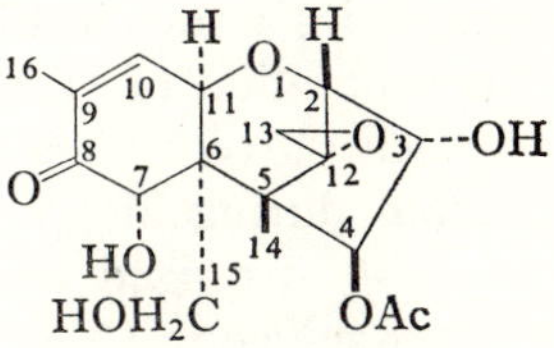

Fusarenone

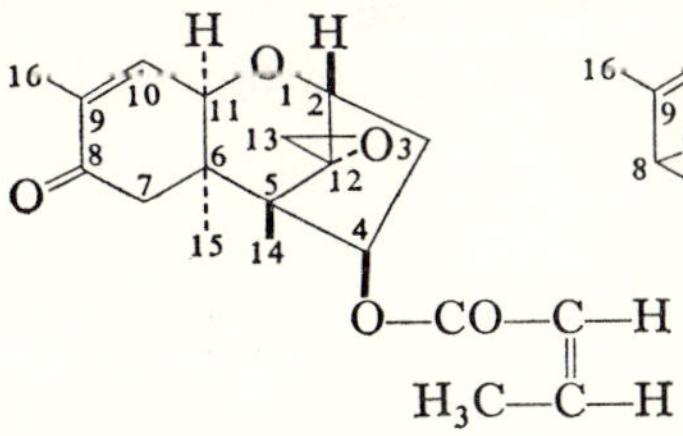

Trichothecin (XXII)

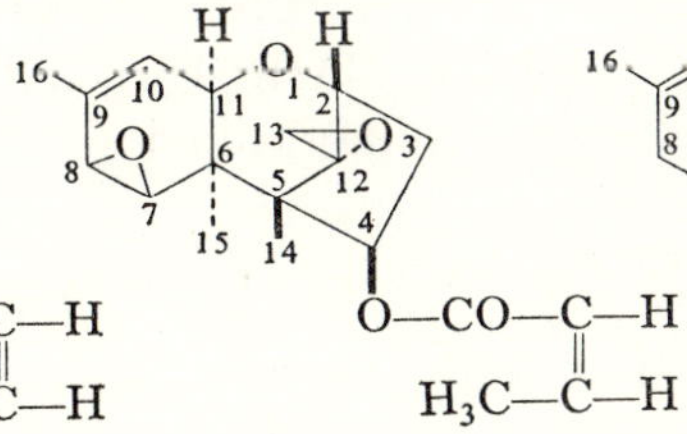

Crotocin

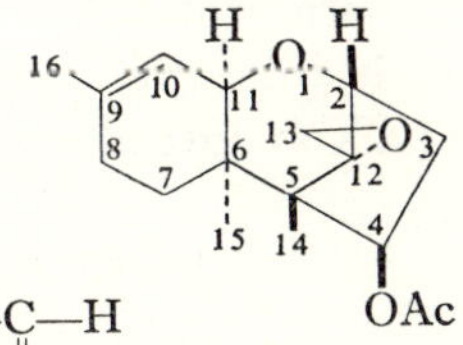

Trichodermin (XVII)

FIG. 17. Principal 12,13-epoxytrichothecane derivatives.

$CH_2{=}CH{-}C\lessgtr$	typical ABX pattern in NMR
$-CH{=}C<$	1H triplet ($J\sim 1{\cdot}5$ cps), $\delta\sim 5{\cdot}3$
2X $>CH{-}CH_3$	$2\times$ 3H doublets ($J\sim 6{\cdot}5$ cps), δ 0·82 and 1·10
3X $-\!\!\!>C{-}CH_3$	$3\times$ 3H singlets, δ 1·20–1·24

FIG. 18. Chemical groups in fusicoccin.

Both low and high resolution mass spectrometry gave additional information which led us to deduce the elemental formula of the aglycone and the presence of a monoacetylglucosyl-residue, and some of the interrelations between the functional groups shown in Fig. 19. Hydrolysis removes the dimethylallyl group easily with the formation of *t*-pentenol (2-methylbut-3-en-2-ol). Comparison of the mass spectra of the acetate of fusicoccin with the acetate of its hydrogenolysis product shows that the dimethylallyl group is located on the glucose moiety. Consequently, the aglycone must be the monoacetate of a $C_{21}H_{34}O_5$ alcohol. This alcohol was, in fact, produced by acidic hydrolysis of fusicoccin and comparison of its NMR spectrum with that of its acetate, in various solvents and the extensive use of nuclear magnetic double resonance (NMDR) indicated the presence of the moieties shown in Fig. 20.

Incorporation of labelled mevalonolactone in fusicoccin indicated that fusicoccin aglycone has a diterpenic tricyclic skeleton. If no rearrangement occurred during biosynthesis of fusicoccin the sequences shown can be reasonably arranged only in one way to give structure (XXIV) for the deacetyl-aglycone (Fig. 21).

The same ring system occurs in ophiobolins, to be discussed later. Additional evidence for the cyclopentene ring and its environment came from oxidation of the acetonide of the deacetyl-aglycone with dimethylsulphoxide in acetic anhydride. NMR, UV, and IR data of the resulting ketone (XXV) agreed with this interpretation.

Aglycone = $C_{23}H_{36}O_6$ (HRMS)

Monoacetylglycosyl— (HRMS)

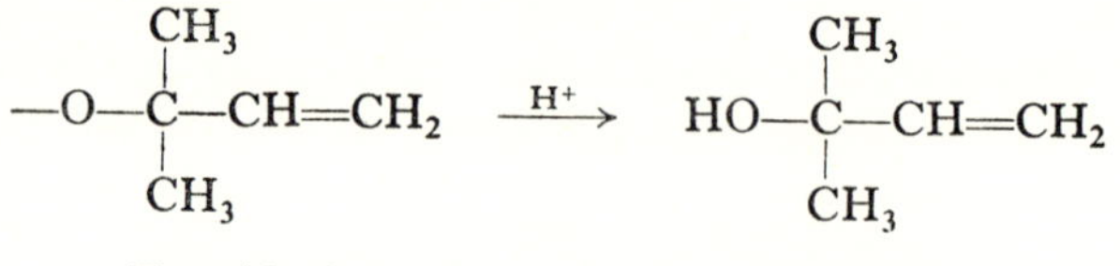

FIG. 19. Structural features of fusicoccin.

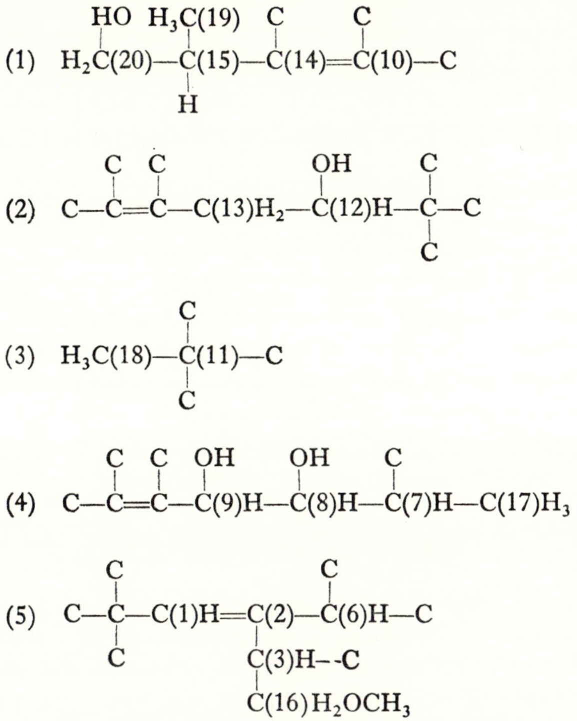

FIG. 20. Structural features of fusicoccin aglycone.

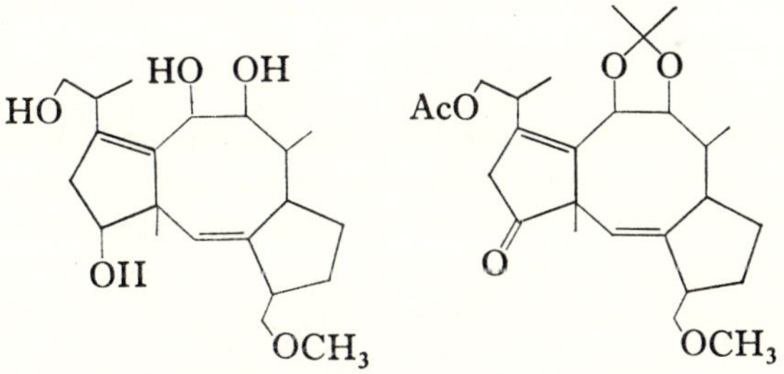

XXIV XXV

FIG. 21. Substantiation of fusicoccin aglycone structure.

The complete structure of fusicoccin was established by X-ray analysis of the *p*-iodobenzenesulphonyl-derivative of the toxin the NMR spectrum of which showed that the secondary alcoholic function in the five membered ring had reacted. This showed that the substituted glucose moiety was on either C_8 or C_9. Because glucose is the natural D-antipode, (XXIII) must represent the absolute structure of fusicoccin (Fig. 22).

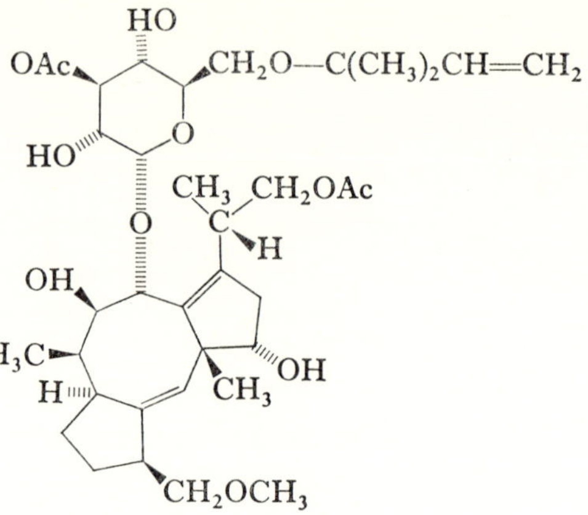

FIG. 22. Absolute structure of fusicoccin.

Sesterterpenoid Phytotoxins

Ophiobolin A (XXVII)

Formerly known as cochliobolin (Orsenigo, 1957), ophiobolin A is a phytotoxin from *Helminthosporium oryzae* v. Breda de Haan, with the formula $C_{25}H_{36}O_4$. The essential evidence which enabled Canonica *et al.* (1966a) to establish its structure is summarized in the following way. Catalytic hydrogenation shows that ophiobolin A has two double bonds and hence four rings, since all oxygenated functions on the molecule are accounted for in the way shown in Fig. 23.

The relationship between the two carbonyls was proved as follows: $LiAlH_4$ reduction of ophiobolin A gave two epimeric triols, one of which after oxidation by CrO_3 or MnO_2 gave an α,β-unsaturated γ-lactone; UV, IR, NMR and mass spectral data demonstrated that the reactions could be explained in the way shown in Fig. 24.

Considering first the oxide ring, hydrogenation of the same triol resulted in hydrogenolytic reduction of the allylic primary alcohol to a CH_3, in cleavage of the oxide ring and in reduction of the adjacent double bond. The easy loss of a C_8H_{17} fragment in the mass spectrum of this hydrogenolysis product suggests that it has such a saturated side chain which obviously originates from the opening of the oxide ring. Thus it was concluded that the ring must be at least five-membered.

The environment of the single tertiary methyl which NMR data indicates is present in ophiobolin A was deduced from the isolation of an acid with the structure shown in Fig. 25 after oxidation of ophiobolin A with HNO_3. Failure to isolate an acid with a longer chain suggests that there is a methine

(a) 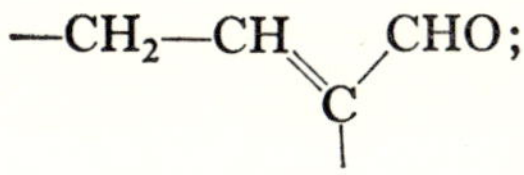

λ_{max}^{MeOH} 236; 1H singlet (9·23δ) and 1H triplet (7·21δ)

(b) 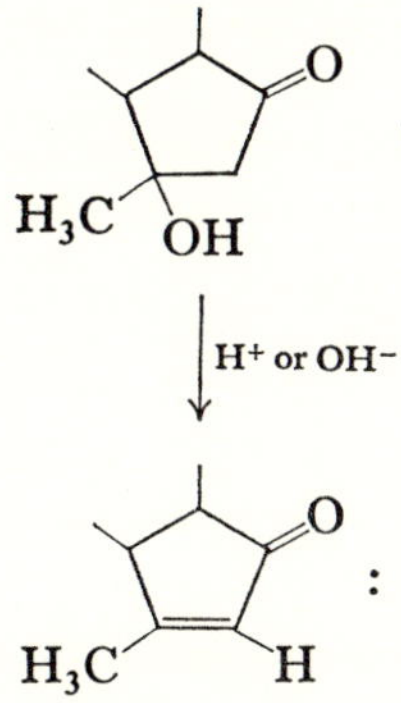

ν_{max}^{nujol} 1742 cm^{-1}; 3H singlet (1·37δ) and 2H, AB system, 2·65δ

: λ_{max} = 232; ν_{max}^{nujol} = 1695; 3H broad singlet (2·09δ) and 1H singlet (6·11δ)

(c) 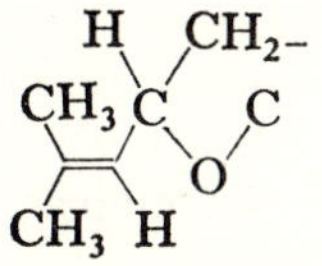

: 6H broad singlet (1·78δ), 1H broad doublet (5·18δ) and 1H multiplet (4·45δ)

All couplings proved by NMDR

FIG. 23. Chemical groups in ophiobolin A.

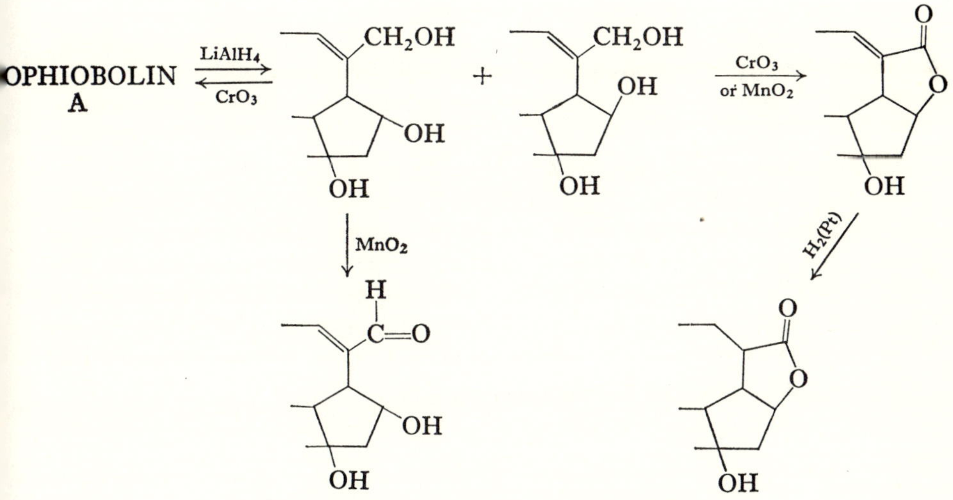

FIG. 24. Partial structure of ophiobolin A.

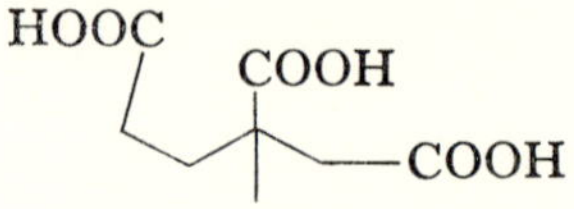

FIG. 25. Deduction of the environment of the tertiary $-CH_3$ of ophiobolin A.

carbon β to the tertiary methyl. Because, apart from the oxide bridge, the ophiobolin A skeleton is tricyclic (one ring of which contains the ketone), the methine carbon and the adjacent quaternary carbon must be the bridgeheads of the remaining two rings so that ophiobolin A must be either (XXIV) or (XXV) (Fig. 26).

The decision in favour of (XXV) was made on the basis of the appearance of ion (XXVI) in the mass spectrum.

Nozoe *et al.* (1965) confirmed the correctness of this formula by X-ray analysis which demonstrated that the structure of ophiobolin A is as shown in Fig. 27.

Ophiobolin A was the first representative of the new family of C_{25} terpenes (sesterterpenes) to be discovered; several others are now known (Tsuda *et al.*, 1967; Iitaka *et al.*, 1968; Nozoe *et al.*, 1968; Tschesche *et al.*, 1968). This has stimulated much work on its biosynthesis (Canonica *et al.*, 1966b, 1967a, b, 1968).

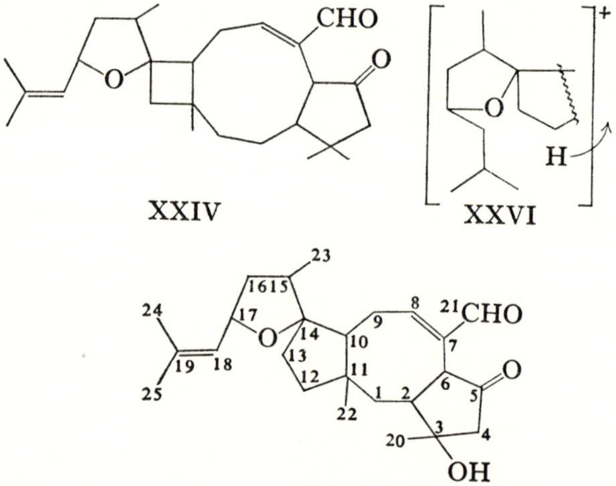

FIG. 26. Possible structures of ophiobolin A.

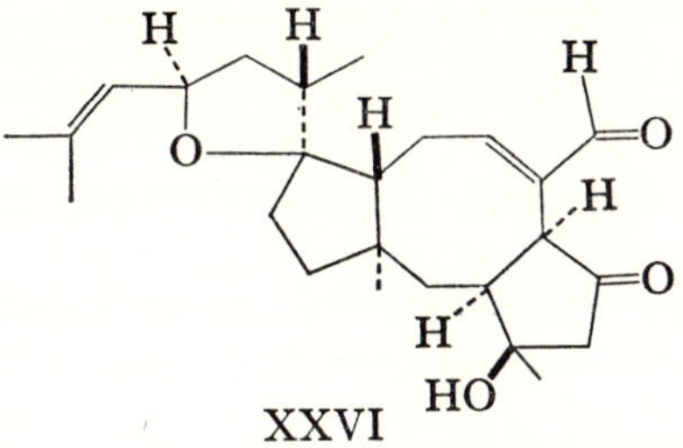

XXVI

FIG. 27. Absolute structure of ophiobolin A.

References

Achilladelis, B., Adams, P. M., and Hanson, J. R. (1970). *Chem. Communs* 511.

Abrahamsson, S. and Nilsson, B. (1964). *Proc. chem. Soc.* 188.

Ballio, A., Chain, E. B., De Leo, P., Erlanger, B. F., Mauri, M., and Tonolo, A. (1964). *Nature* **203**, 297.

Ballio, A., Brufani, M., Casinovi, C. G., Cerrini, S., Fedeli, W., Pellicciari, R., Santurbano, B., and Vaciago, A. (1968). *Experientia* **24**, 631–635.

Barrow, K. D., Barton, D. H. R., Chain, E. B., Conlay, C., Smale, T. V., Thomas R., and Waight, E. S. (1968a). *Chem. Communs* 1195–1197.

Barrow, K. D., Barton, D. H. R., Chain, E. B., Ohnsorge, U. F. W., and Thomas, R. (1968b). *Chem. Communs* 1198–1200.

Bassett, C., Sherwood, R. T., Kepler, J. A., and Hamilton, P. B. (1967). *Phytopathology* **57**, 1046–1052.

Brian, P. W., Dawkins, A. W., Grove, J. F., Hemming, H. G., Lowe, D., and Norris, G. L. F. (1961). *J. exp. Bot.* **12**, 1–12.

Canonica, L., Fiecchi, A., Galli Kienle, M., and Scala, A. (1966a). *Tetrahedron Lett.* 1211–1218.

Canonica, L., Fiecchi, A., Galli Kienle, M., Ranzi, B. M., and Scala, A. (1966b). *Tetrahedron Lett.* 3035 3039.

Canonica, L., Fiecchi, A., Galli Kienle, M., Ranzi, B. M., Salvatori, T., and Pella, E. (1967a). *Tetrahedron Lett.* 3371–3376.

Canonica, L., Fiecchi, A., Galli Kienle, M., Ranzi, B. M., and Scala, A. (1967b). *Tetrahedron Lett.* 4657–4659.

Canonica, L., Fiecchi, A., Galli Kienle, M., Ranzi, B. M., and Scala, A. (1968). *Tetrahedron Lett.* 275–279.

Corey, E. J. and Nozoe, S. (1963). *J. Am. chem. Soc.* **85**, 3527–3528.

Dawkins, A. W., Grove, J. F., and Tidd, B. K. (1965). *Chem. Communs* 27-28.

Dawkins, A. W. and Grove, J. F. (1970). *J. chem. Soc.* (C), 369–375.

Flury, E., Mauli, R., and Sigg, H. P. (1965). *Chem. Communs* 26–27.

Godtfredsen, W. O. and Vangedal, S. (1964). *Proc. chem. Soc.* 188–189.

Godtfredsen, W. O. and Vangedal, S. (1965). *Acta Chem. Scand.* **19**, 1088–1102.

Graniti, A. (1962). *Phytopath. medit.* **1**, 182–185.
Graniti, A. (1964). *Phytopath. medit.* **3**, 125–128.
Grove, J. F. (1969). *Chem. Communs* 1266–1267.
Grove, J. F. and Mortimer, P. H. (1969). *Biochem. Pharmacol.* **18**, 1473–1478.
Grove, J. F. (1970a). *J. chem. Soc.* (C), 376–378.
Grove, J. F. (1970b). *J. chem. Soc.* (C), 378–379.
Gutzwiller, J. and Tamm, Ch. (1963). *Helv. chim. Acta* **46**, 1786–1790.
Gutzwiller, J., Mauli, R., Sigg, H. P., and Tamm, Ch. (1964). *Helv. chim. Acta* **47**, 2234–2262.
Hanson, T. R. and Achilladelis, B. (1967). *Chemy Ind.* 1643–1644.
Härri, E., Löffler, W., Sigg, H. P., Stähelin, H., Stoll, Ch., Tamm, Ch., and Weisinger, D. (1962). *Helv. chim. Acta* **45**, 839–853.
Hough, E., Hursthowe, M. B., Neidle, S., and Rogers, D. (1968). *Chem. Communs* 1197–1198.
Iitaka, Y., Watanabe, I., Harrison, I. T., and Harrison, S. (1968). *J. Am. chem. Soc.* **90**, 1092–1093.
Jones, E. R. H. and Lowe, G. (1960). *J. chem. Soc.* 3959–3962.
Kepler, J. A., Wall, M. E., Mason, J. E., Bassett, C., McPhail, A. T., and Sim, G. A. (1967). *J. Am. chem. Soc.* **89**, 1260–1261.
McPhail, A. T. and Sim, G. A. (1966). *J. chem. Soc.* (C), 1394–1406.
McPhail, A. T. and Sim, G. A. (1968). *J. chem. Soc.* (B), 1104–1109.
de Mayo, P., Spencer, E. Y., and White, R. W. (1961). *Can. J. Chem.* **39**, 1608–1612.
de Mayo, P., Spencer, E. Y., and White, R. W. (1962a). *J. Am. chem. Soc.* **84**, 494–495.
de Mayo, P., Robinson, J. R., Spencer, E. Y., and White, R. W. (1962b). *Experientia* **18**, 359–360.
de Mayo, P., Spencer, E. Y., and White, R. W. (1963). *Can. J. Chem.* **41**, 2996–3004.
de Mayo, P. and Williams, R. E. (1965). *J. Am. Chem. Soc.* **87**, 3275.
Nozoe, S., Morisaki, M., Tsuda, K., Iitaka, Y., Takahashi, N., Tamura, S., Ishibashi, K., and Shirasaka, M. (1965). *J. Am. chem. Soc.* **87**, 4968–4970.
Nozoe, S., Morisaki, M., Fukushima, K., and Okuda, S. (1968). *Tetrahedron Lett.* 4457–4458.
Orsenigo, M. (1957). *Phytopath. Z.* **29**, 189.
Spencer, E. Y., Ludwig, R. A., de Mayo, P., White, R. W., and Williams, R. E. (1965). Advances in Chem. **53**, 106–111. Academic Press, London & New York.
Tamm, Ch. and Gutzwiller, J. (1962). *Helv. chim. Acta* **45**, 1726–1731.
Tatsuno, T., Saito, M., Enomoto, M., and Tsunoda, H. (1968). *Chem. Pharm. Bull.* **16**, 2519–2520.
Tschesche, R., Brock, F. X., and Duphorn, I. (1968). *Tetrahedron Lett.* 2905–2909.
Tsuda, K., Nozoe, S., Morisaki, M., Hirai, K., Itai, A., Okuda, S., Canonica, L., Fiecchi, A., Galli Kienle, M., and Scala, A. (1967). *Tetrahedron Lett.* 3369–3370.

Discussion

Barbier: Does the observed biological activity of the compounds with an extracyclic epoxide function, depend on this function? Such compounds are known to have alkylating properties *in vivo* and thus have potent anti-cancer activities. I wonder therefore if it would be of interest to test the anticancer properties?

Casinovi: Grove and Mortimer (1969) have studied the cytotoxicity of a number of these compounds and shown that it is associated with the epoxy group although other chemical features may also be important. Certainly it would be desirable to assay them as potential anticancer compounds.

Wood: I should like to refer to the work of Professor Brian and his colleagues which showed that *Fusarium equiseti* which readily produces the highly phytotoxic diacetoxyscirpenol in culture is no more than a weak pathogen.

Mirocha: Which of the compounds of the trichothecene series shown in your slides have been shown to be phytotoxic or have been isolated from material naturally infected with the fungi concerned? I ask this question because many of the trichothecene compounds although extremely toxic to animals (causing haemorrhages) and thought to be mycotoxins, have never been isolated from contaminated food-stuffs.

As Wood indicated, *F. scirpi* which produces at least one of the scirpenols is weakly pathogenic. However, scirpenol compounds are produced by *F. roseum "graminearum"*, *F. tricinctum*, *Trichothecium* sp., some of which are pathogenic.

I believe that more significance can be attached to the chemistry of the spiroepoxytrichothecenes which has been presented when we consider that they are suspected of being mycotoxins and some are strongly being considered as the possible cause of alimentary toxic aleukia, described by Russian scientists a number of years ago.

Casinovi: Phytotoxicity has been reported for diacetoxyscirpenol and its cometabolites, (XVIII) and (XXI), and for trichothecin; there is no evidence, however, that the other members of the series are not phytotoxic.

8. The Chemistry of Polysaccharide and Glycopeptide Phytotoxins

J. J. CH. LOUSBERG AND C. A. SALEMINK

Organic Chemistry Laboratory,
State University, Utrecht,
The Netherlands

The concept that plant pathogens cause diseases by producing toxic substances dates back for about a century. Evidence for the general validity of this concept was only recently obtained. Research in this field is difficult because most plant diseases show symptoms which are the result of a complex of several factors. Some pathogens, in addition to producing one or more toxins, may also secrete enzymes which degrade cell walls and cause tissue disorganization. Pathogens may also destroy plant hormones causing abnormal growth of the host and they may physically block the water-conducting vessels of their hosts by prolific growth or by the production of viscous polysaccharides.

It is known that polysaccharides secreted by certain phytopathogenic bacteria cause wilting of the host plant by physical obstruction of water-conducting vessels. There are, however, some polysaccharide toxins which possess a polypeptide moiety. This increases the possibility that these toxins act not only by physical obstruction.

Goodman (1959, 1960, and 1963) isolated and partly characterized colletotin, a non-dialysable toxin, produced by the pathogen *Colletotrichum fuscum* Laub.

Digitalis lanata Ehrh. and *D. purpurea* L. plants inoculated with a suspension of spores of *C. fuscum* wilt six days later, and the plants are dead 8–9 days after inoculation. These symptoms strongly suggest that a toxin, or toxins are produced *in vivo*.

A toxin obtained from filtrates from cultures of *C. fuscum* caused similar symptoms to develop in *Digitalis* plants.

The toxin has polysaccharide and peptide characteristics. It is non-dialysable, soluble in water, and insoluble in organic solvents. Acid or alkali treatment at room temperature or boiling in water for one hour, did not alter its toxicity.

After hydrolysis of the toxin with 6 N HCl for 1 h at 100° C the following monosaccharides were detected: D-glucose, D-galactose, and D-mannose. Chromatograms of the hydrolysate sprayed with ninhydrin reagent showed several unidentified amino acids. Mild hydrolysis with 0·5 N nitric acid yielded the same sugars.

After both hydrolyses toxicity was destroyed. The toxin appears to act by disintegrating leaf mesophyll and palisade tissue.

The next toxins to be discussed are those produced by the fungus *Verticillum albo-atrum* Reinke *et* Berth., which cause wilting and necrosis in alfalfa (*Medicago sativa* L.) and tomato plants. These symptoms appear when alfalfa seedlings are placed in culture filtrates (Kiesig and Häller-Kiesig, 1957; Michael and Carr, 1966).

From culture filtrates two fractions have been isolated, one consisting of high molecular weight substances and the second of a mixture of low molecular weight substances. The first fraction (I) which was studied by Stoddart and Carr (1966) possesses most of the phytotoxicity and contains proteins and polysaccharides. The second fraction (II) is far less toxic and is a mixture of peptides and sugars.

The two fractions were separated from the culture filtrate by gel filtration on Sephadex G25 with de-ionized water as eluent. The polysaccharides in I were detected with the anthrone reaction of Shields and Burnett (1960). Micro Kjeldahl analysis confirmed the presence of proteins. After freeze-drying of I the non-hygroscopic, colourless fibrous material which remained caused rapid wilting of the leaves of alfalfa seedlings at a concentration of 250 μg/litre or higher. Gel filtration of this material on Sephadex G100 yielded two fractions, one containing a protein, mol.wt *c.* 100,000, and the other containing a polysaccharide, mol.wt *c.* 5000–10,000 (MW determination by osmosis and paper electrophoresis).

Both components showed similar phytotoxicity, but wilting appeared sooner with the polysaccharide fraction. Heating of fraction I, containing protein and polysaccharide, denatured the protein. The same result was obtained by adding a solution of $ZnSO_4.7H_2O$ (10 mg/litre). 40% of the phytotoxicity was lost after this treatment.

Talboys (1958) suggested that part of the pathogenic activity of the fungus could be attributed to the formation of a cellulase. The protein part of the high molecular weight fraction had a weak but reproducible cellulase activity.

Acid hydrolysis of the protein with 6 N HCl at 105° C for 24 h gave the following amino acids: L-lysine, L-aspartic acid, L-alanine, L-valine, L-leucine, and glycine. Hydrolysis of the polysaccharide with 0·1 N HCl for one hour at 90° C yielded fructose and glucose in the ratio 4:1.

The low molecular weight fraction was a mixture of four components,

A, B, C, and D. B showed little and C showed no wilting activity. The molecular weights of A, B, C, and D were 700 or more, 350, 200, and 100 or less. Sucrose was principally located in B. The carbohydrate portion of A consisted of oligosaccharides of four units or more. B and C contained ninhydrin-positive compounds; at least 6 amino acids were found in the hydrolysates. Fraction D consisted predominantly of mineral salts. Detailed experiments revealed that the toxicity of A and B was due to the carbohydrate. The authors, however, do not exclude the possibility that the wilting had been caused by osmotic activity of the sugars.

Kordosova *et al.* (1969a, b) isolated mixtures of phytotoxic polysaccharides from water extracts of the fruit bodies of the wood-destroying fungi *Fomes fomentarius* (L. *ex* Fr.) Kickx, *F. igniarius* (L. *ex* Fr.) Kickx and *Polyporus squamosus* Huds. *ex* Fr. The mixtures consisted of a glucose and a mannofucogalactan (or fucomannogalactan).

The release of polysaccharides from species of phytopathogenic bacteria into their external environment, *in vivo* and *in vitro*, has been linked to their virulence. This fact and because many phytopathogenic bacteria seem to produce these substances justifies extensive discussion of their nature.

These organic polymers are commonly referred to as capsules when they occur as a well defined layer around the bacterium. The term slime or slime layer is used when the polymers are more extensive and diffuse.

In some bacteria, the polysaccharides are internal and may be released on autolysis. If slime production can be linked to pathogenesis the factors which control it may be considered as controlling virulence, at least in part.

This was established by Corey and Starr (1957) for bean plants inoculated with *Xanthomonas phaseoli* (E. Sm.) Dows. Four colony types of the pathogens, rough, smooth, semi-mucoid, and mucoid, produced, in the order listed, increasing amounts of polysaccharides in culture and also larger colonies on agar and more lesions on bean leaves. It was found that the relative viscosity of culture filtrates was an excellent index of polysaccharide production and wilt-inducing capacity.

Lesley and Hochster (1959) reported on the composition of this extracellular polysaccharide. The product, shown to be homogeneous by moving boundary electrophoresis in the analytical centrifuge and by gradient precipitation with ethanol, has a molecular weight of *c.* 19×10^6 and hydrolysis gave the monosaccharides D-glucose, D-mannose, and D-glucuronic acid, in the proportions 1:1:1.

Other bacterial pathogens shown by Sloneker and Alleane Jeanes (1962), and Gorin and Spencer (1963) to produce polysaccharides are *Xanthomonas hyacinthi* (Wakk.) Dows., *X. translucens* (Jones *et al.*) Dows., *X. maculifolii-gardeniae* (Ark) Elr. *et* Braun, *X. campestris* (Pamm.) Dows., and *X. oryzae* (Uy. *et* Ish.) Dows. A number of *Corynebacterium* spp. also do so.

But except for *X. phaseoli* (see above) and *C. sepedonicum* (Spieck. *et* Kott.) Skapt. *et* Burkh. the relations between polysaccharide production and symptom production were not studied.

The mode of action of polysaccharides in wilt induction may certainly be attributed to blockage of water flow in the vascular system of plants. Toxicity is a function of molecular size and cannot be assigned to any specific functional group of the molecule.

It has to be emphasized that the above effects are non-specific. They can be induced in almost any plant by a variety of organic polymers of synthetic and bacterial origin.

Spencer and Gorin (1961) and Gorin and Spencer (1961) have described the wilt pathogens, *C. insidiosum* (McCull.) Jens., parasitizing alfalfa plants, and *C. sepedonicum*, parasitizing potato plants, and have studied the chemical compositions of polysaccharide slime secreted by these and other species of *Corynebacterium*, including most of the more important phytopathogens. They found that *C. insidiosum* and *C. sepedonicum* produce a polysaccharide which contains galactose, glucose and fucose. Fucose is an unusual sugar and is generally not found in terrestrial plants. The same hexoses and also mannose were detected in slime produced by *C. michiganense* (E. Sm.) Jens. *C. fascians* (Tilf.) Dows. produced a polysaccharide similar to that of *C. michiganense* except that it did not contain fucose.

The extracellular slime of *C. flaccumfaciens* (Hedges) Dows. contained in addition to glucose, mannose and fucose, L-rhamnose and aldobiuronic acid.

The polysaccharides of *C. tritici* (Hutch.) Burkh. and *C. rathayi* (E. Sm.) Dows. were similar in that both contained glucose, mannose and aldobiuronic acid. However, the former also contained rhamnose, whereas the latter contained an unidentified hexose.

The extracellular polysaccharide isolated from cultures of *C. insidiosum*, induced severe wilting of succulent cuttings of alfalfa within 10–15 min, at concentrations as low as 20 μg/ml.

As mentioned above it contains D-glucose, D-galactose, and L-fucose in the ratio 2:3:4, with a small amount of an unknown acid (one acid grouping for every six sugar units).

Much of this information was obtained from partial hydrolysis of the polysaccharide and fractionation of the products by charcoal and cellulose chromatography.

So far two neutral oligosaccharides from *C. insidiosum* have been obtained, 4-O-β-D-glucopyranosyl-L-fucose and 4-O-β-D-glucopyranosyl-(3-O-D-galactopyranosyl)-L-fucose. Two other sugars contained acidic functions, not uronic in nature, D-galactose, substituted in position 4 with an unknown acid, and 4-O-β-D-glucopyranosyl-(3-O-D-galactopyranosyl)-L-fucose, substituted with the unknown acid at position 4 of the galactopyranosyl moiety. It appears that these oligosaccharides represent the main structure of the polysaccharides.

Strobel (1967) described some properties of a phytotoxic polysaccharide produced by *C. sepedonicum*. The molecular weight of the compound with the empirical formula $C_{48}H_{96}O_{48}N$ is about 21,450. A hydrolysate contains D-glucose, D-mannose, L-fucose, and 2-keto-3-gluconic acid and a small percentage of unknown sugars. The following amino acids were detected: alanine, aspartic acid, arginine (trace), glutamic acid, glycine, histidine, isoleucine (trace), leucine, lysine, methionine, phenylalanine (trace), serine, threonine, and tyrosine (trace). The purified bacterial glycopeptide has its iso-electric point at about pH 2·1. A high titre antiserum to it was prepared.

Strobel (1970) also isolated a glycopeptide from potato plants infected by *C. sepedonicum* and showed that it was the same as the one from the bacterium with respect to mobility by paper and acrylamide gel electrophoresis, optical rotation, infrared spectrum, X-ray diffraction pattern, molecular weight as judged by elution volumes from Sephadex G200 columns, serological activity and analytical data for amino acids, sugars and sugar derivatives.

The glycopeptide was present in infected plants in quantities sufficient to give symptoms in biological assays. It can thus be considered a vivotoxin.

Rai and Strobel (1969) have also studied the phytotoxicity of cultures of *C. michiganense* from which they isolated an acid labile, heat resistant, water soluble exotoxin. Three phytotoxic fractions were obtained from it by gel filtration on Sephadex G200 with estimated molecular weights of (I) 200,000, (II) 130,000, and (III) 35,000. Chromatographic analyses showed that fractions I and III, after hydrolysis gave alanine, glycine, lysine, methionine, serine, threonine, and an unknown compound. A hydrolysate of fraction II contained alanine, glycine, lysine, serine and an unknown product. The monosaccharides of fractions I and II were galactose, glucose, mannose and two unknown compounds, whereas fraction III contained fucose, galactose, glucose, mannose, and two unknown reducing compounds.

Recently Rudolph (1969) has described the isolation of a phytotoxic polysaccharide from *Pseudomonas phaseolicola* (Burkh.) Dowson. This bacterium causes halo blight in *Phaseolus vulgaris* L. characterized by inhibition of growth and synthesis of chlorophyll. At the same time ornithine accumulates. This toxin is not specific, affecting many green plants and also chlorophyll-free callus cultures.

McIntire *et al.* (1942), and Hodgson *et al.* (1945, 1947, and 1949) isolated from cultures of *Agrobacterium tumefaciens* (Sm. *et* Town.) Conn a polysaccharide with a molecular weight of 3600 ± 200. Hydrolysis of the polysaccharide gave only D-glucose, as indicated by specific rotation, iodometric titration and a yield of glucobenzimidazole. Feder and Ark (1951) reported the isolation of a phytotoxic endotoxin from four strains of *A. tumefaciens* which was not the same as the one obtained by Hodgson *et al.* (1947) from the filtrate of growing cultures of the bacterium because this had a low negative optical rotation whereas the endotoxin had a positive optical rotation.

The endotoxin, prepared by the Boivin trichloroacetic acid technique from dried bacterial cells, was a polysaccharide–polypeptide–lipid complex, non-dialysable, soluble in organic solvents, insoluble in water, highly opalescent and pale yellow in colour. Acetic acid hydrolysis split the endotoxin into its polysaccharide and polypeptide–lipid components (Mesrobeanu, 1936). The polypeptide–lipid fraction was rapidly dialysable, soluble in organic solvents, but insoluble in water. The polysaccharide, hydrolysed by refluxing for one hour in 1 N HCl, gave glucose units (*c.* 98%) and a small amount of other residues, probably mainly galacturonic acid.

Mesrobeanu (1936) also isolated endotoxins able to wilt plants from two non-wilt-inducing phytopathogenic bacteria, *Erwinia carotovora* (Jones) Holl. and *X. phaseoli.*

Harris (1940) also described an extracellular factor with wilting properties from cultures of *E. tracheiphila* (E. Sm.) Holl. and *X. stewartii* (E. Sm.) Dows. but the endotoxin was not isolated.

Another endotoxin was obtained from *P. mors-prunorum* Worm. by Erikson and Montgomery (1945). The substance caused lesions in the bark of plum trees similar to those caused by the pathogen. Filtrates of older cultures (5–7 weeks) were generally more active than filtrates of younger cultures (2 weeks), suggesting that the toxin is a product of autolysis. Acetic acid extracts of dried bacterial cells gave a protein–polysaccharide complex which induced lesions in the bark similar to cankers caused by the pathogen. After heating an acetic acid extract or extracting dried cells with trichloroacetic acid—both protein denaturating procedures—relatively inactive substances were obtained, indicating that the toxin is either a protein or a polysaccharide complex thereof.

A phytotoxin of carbohydrate character has been isolated from *Didymella applanata* (Niessl) Sacc. cultures by Salemink *et al.* (1966), and Schuring (1968). This caused the same symptoms in raspberry shoots as does the fungus.

The phytotoxin was isolated from liquid cultures of *D. applanata* by absorption on charcoal (Norit R II), and eluting with pyridine–water (1:1). The eluate was concentrated *in vacuo*, taken up in water and concentrated again. This procedure was repeated several times to eliminate the pyridine and the residue was submitted to gel filtration (G25 "coarse") with water as eluent. The active fractions were combined and the solution was concentrated before the next separation by elution chromatography over Norit R II with a water–alcohol gradient. The active fractions from this column were combined, the solution concentrated again and then submitted to partition chromatography over cellulose; 0·5–2 mg phytotoxin per litre culture filtrate could be isolated by these procedures.

The colourless product was still active on young raspberry shoots in solutions of 1 mg/litre, is neutral, contains no nitrogen, sulphur or phosphorus, and leaves no residue after combustion. It has a molecular weight 1682 ± 150, and is an oligosaccharide with no (or very little) reducing power. One of the monosaccharides of the toxin is D-glucose. The ultraviolet spectrum, however, indicates other groups besides carbohydrates.

Monosaccharide derivatives obtained after methylation (Hakamori, 1964) followed by hydrolysis and trimethylsilylation (Sweeley *et al.*, 1963, Wood *et al.*, 1965) were isolated by gas chromatography and examined by mass spectrometry. Structures I, II, III, and IV were found.

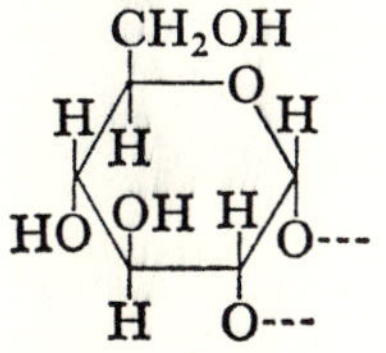

I

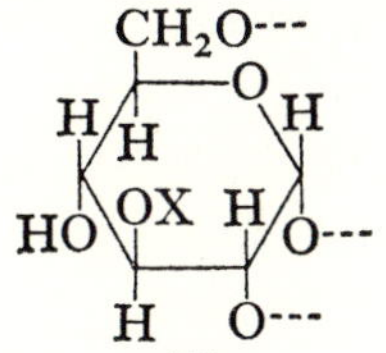

II

$X = CH_2COC_6H_4C_7H_{15}$

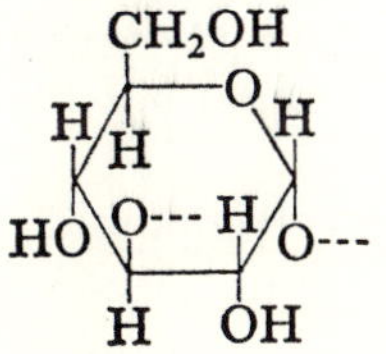

III

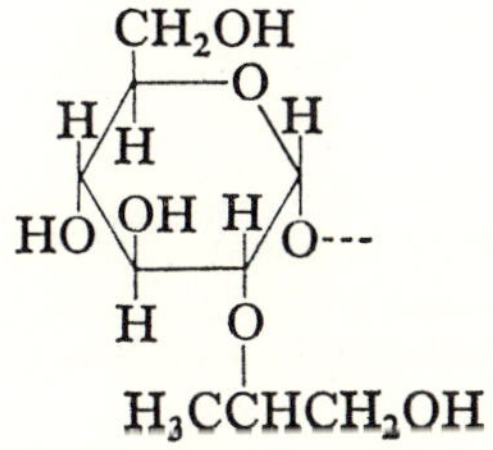

IV

Phytotoxins are also known to be produced by the fungus *Ceratocystis ulmi* (Buisman) C. Moreau the cause of Dutch elm disease which has been and still is important in The Netherlands, North America and elsewhere. The pathogen is carried to elms by several kinds of bark beetles.

Zentmyer (1942) suggested that a phytotoxin was present in the filtrate of cultures of *C. ulmi* and this has been confirmed by Dimond (1947), Feldman *et al.* (1949), Kerling (1955), and Beckman (1956).

According to Zentmyer and Horsfall (1942) this toxin which is thermostable consists of more than one component. Dimond (1947) characterized one substance as a polysaccharide. Feldman *et al.* (1949) and Lafayette and Howard (1951), however considered that the polysaccharide is not the phytotoxin. Beckman (1956) showed that a pectinase and cellulase are present in culture filtrates.

Salemink *et al.* (1965), and Rebel (1969) reported the isolation of phytotoxic material from liquid cultures of *C. ulmi*. One component of the material, at that time thought to be pure, induced disease symptoms similar to those produced by the fungus in elm shoots and trees.

The toxic material could be isolated from three week old cultures by filtration, concentration of the filtrates, precipitation of a polysaccharide from the concentrate with ethanol, dialysis of the residue, followed by freeze-drying. The non-dialysable, electrophoretically neutral substance contained all the toxicity of the culture filtrate. The material is stable in air, resists heating in water, and to a certain extent, also heating in dilute acid. Prolonged treatment with dilute sodium hydroxide at room temperature hardly alters its phytotoxicity, nor did incubation with several proteolytic enzymes. Treatment with sodium periodate inactivated the toxin.

When Sepharose 4 B became available, the crude toxin was resolved into fractions after many vain efforts. Two main fractions, each of molecular weight 25,000–30,000, toxic and heterogeneous, were obtained. Purification of these compounds is expected in the near future.

All or most of the separated components have carbohydrate and peptide properties; the carbohydrate is more than 90% of the total.

The capacity of a neutralized culture filtrate to destroy its phytotoxicity is probably due to certain enzymes it contains. Salemink *et al.* (1965) did not find pectinase or cellulase activities in the crude toxin (*cf.* Beckman, 1956).

At present, the chemistry of the polysaccharide and glycopeptide phytotoxins leaves much to be desired. More detailed structural analyses and the application of modern and powerful techniques should eventually lead in combined efforts with plant pathologists to a better understanding of the complex problem of the mode of action in biological systems of these intriguing types of natural products.

References

Beckman, C. H. (1956). *Phytopathology* **46**, 605.
Corey, R. R. and Starr, M. P. (1957). *J. Bact.* **74**, 137.
Dimond, A. E. (1947). *Phytopathology* **37**, 7.
Erikson, D. and Montgomery, H. B. S. (1945). *Ann. appl. Biol.* **34**, 117.
Feder, W. A. and Ark, P. A. (1951). *Phytopathology* **41**, 804.
Feldman, A. W., Caroselli, N. E., and Howard, F. L. (1949). *Phytopathology* **39**, 6.
Goodman, R. N. (1959). *Phytopath. Z.* **37**, 187.

Goodman, R. N. (1960). *Phytopathology* **50**, 325.
Goodman, R. N. (1963). *Phytopathology* **52**, 1273.
Gorin, P. A. J. and Spencer, J. F. T. (1961). *Can. J. Chem.* **39**, 2274.
Gorin, P. A. J. and Spencer, J. F. T. (1963). *Can. J. Chem.* **41**, 2357.
Hakamori, S. (1964). *J. Biochem.* (*Tokyo*) **55**, 205.
Harris, H. A. (1940). *Phytopathology* **30**, 625.
Hodgson, R., Riker, A. J., and Peterson, W. H. (1945). *J. biol. Chem.* **158**, 89.
Hodgson, R., Riker, A. J., and Peterson, W. H. (1947). *Phytopathology* **37**, 301.
Hodgson, R., Peterson, W. H., and Riker, A. J. (1949). *Phytopathology* **39**, 47.
Kerling, L. C. P. (1955). *Acta bot. Neerl.* **4**, 398.
Kiesig, R. and Häller-Kiesig, R. (1957). *Phytopath. Z.* **31**, 185.
Kordosova, A., Kabor, K., Rosik, J., and Kubala, J. (1969a). *Chem. Zvesti* **23**, 454.
Kordosova, A., Kabor, K., Rosik, J., and Kubala, J. (1969b). *Chem. Zvesti* **23**, 462.
Lafayette, F. and Howard, F. L. (1951). *Phytopathology* **41**, 82.
Lesley, S. M. and Hochster, R. M. (1959). *Can. J. Biochem. Physiol.* **37**, 513.
McIntire, F. C., Peterson, W. H., and Riker, A. J. (1942). *J. biol. Chem.* **143**, 491.
Mesrobeanu, L. (1936). Thesis, University of Paris.
Michael, S. H. and Carr, A. J. H. (1966). *Trans. Br. mycol. Soc.* **49**, 133; compare also Pegg, G. F. (1965). *Nature* **208**, 1228.
Rai, P. V. and Strobel, G. A. (1969). *Phytopathology* **59**, 47.
Rebel, H. (1969). Thesis, State University of Utrecht.
Rudolph, K. (1969). *Nature* **56**, 569.
Salemink, C. A., Rebel, H., Kerling, L. C. P., and Tchernoff, V. (1965). *Science* **149**, 202.
Salemink, C. A., Schuring, F., Kerling, L. C. P., and Schippers, B. (1966). *Experientia* **22**, 248.
Schuring, F. (1968). Thesis, State University of Utrecht.
Shields, R. and Burnett, W. (1960). *Anal. Chem.* **32**, 885.
Sloneker, J. H. and Alleane Jeanes (1962). *Can. J. Chem.* **40**, 2066.
Spencer, J. F. T. and Gorin, P. A. J. (1961). *Can. J. Microbiol.* **7**, 185.
Stoddart, J. L. and Carr, A. J. H. (1966). *Ann. appl. Biol.* **58**, 81.
Strobel, G. A. (1967). *Pl. Physiol.* **42**, 1433.
Strobel, G. A. (1970). *J. biol. Chem.* **245**, 32.
Sweeley, C. C., Bentley, R., Makita, M., and Wells, W. W. (1963). *J. Am. Oil Chem. Soc.* **85**, 2497.
Talboys, P. W. (1958). *Trans. Br. mycol. Soc.* **41**, 242.
Wood, R. D., Raju, P. K., and Reiser, E. (1965). *J. Am. Oil Chem. Soc.* **42**, 161.
Zentmyer, G. A. (1942). *Phytopathology* **32**, 20.
Zentmyer, G. A. and Horsfall, J. S. (1942). *Science* **95**, 512.

Discussion

Mirocha: Could you comment from the chemist's point of view, on the problems of isolating polymeric phytotoxins where hydrolysis may occur during isolation e.g. during concentration. Data now accumulating indicate that many polymers in which we are interested are glycopeptides, which are readily hydrolysed.

Salemink: Isolation and purification of uncharacterized biological active products from natural sources should always be carried out under the mildest effective conditions. It would be ideal to know beforehand what type of compound is involved. When one becomes aware in a later stage of the isolation procedure of the type of compound one is dealing with, it might be necessary to repeat the isolation e.g. with pH-conditioned buffers, according to the properties of the object to be extracted or to the environmental conditions, followed by subsequent concentration and comparison of the two differently isolated toxic substances. This provides an important control. Moreover, when a bioassay is available, alterations of the compound, e.g. by hydrolysis or concentration, will generally result immediately in changes of specific toxicity symptoms.

Dimond: What is known about the relation between amino acid and sugar residues in the structure of glycopeptides ? Are amino acids attached in peptide bonds or are they bonded randomly over a polysaccharide skeleton ?

Salemink: Structural determinations are not advanced enough to allow a firm answer to your question. Generally for glycopeptides evidence has been obtained for both types of bonds, the peptide bond and randomly association bonding types.

Goodman: Is it possible that the accumulation of soluble proline in plants infected with *Ceratocystis ulmi* reflects on interference with its incorporation into wall protein as hydroxyproline ?

Elgersma: Singh and Smalley suggested that the high level of proline in diseased elms might indicate the water stress which occurred in infected trees. The high level of proline in healthy resistant elms might be related to resistance against water stress.

Wood: Is the polysaccharide produced by *C. ulmi* at all specific in its action ? In discussions about substances such as these one cannot repeat this type of question too often because usually the active principles produced *in vitro* by pathogens that are very specific in their parasitism are quite non-specific in their action on plants. This is not to discount the importance of studying these active principles and how they affect plants but sooner or later we have to return to the problem as to what permits or does not permit establishment of a pathogen in a plant in events that may well happen before the active principles are produced in amounts sufficient to affect the host.

Elgersma: According to Rebel (1969) the polysaccharide induces wilting only in the elm sprouts in contrast to the glycopeptide which causes primarily necrosis and wilting. The toxin does not seem to be specific because resistant elm trees and tomato plants react to it.

Dimond: I suspect that the different responses of plants to polysaccharides and to glycopeptides depend more on differences in their molecular weights than on specific chemical groupings.

Let me revert to the question raised by Wood. Immunological responses are characteristic of polymers in which different units occur in a sequence. Specificity can arise from either the sequence of amino acids in peptide linkage or from saccharides in unique sequence. This was the intent of my original question. What Wood has asked is a most important question.

Pringle: I endorse the words of caution of the speaker about the difficulties of isolating and characterizing toxins and also Wood's point about the importance of specificity in pathology. The quality of host-specific toxicity is a very subtle and evanescent property of a molecule and it is relatively difficult to isolate host-specific toxins with this property intact. On the other hand, a number of compounds with non-specific toxicity can easily be isolated from culture filtrates, or can be produced from host-specific toxins by very mild treatment.

Kuć: What is the linkage between the aromatic group and the carbohydrate portion of the molecule?

Salemink: The linkage is an ether-type bond between the C-3 hydroxyl of glucose and a methylene, which in itself is bond to a carbonyl, attached to a substituted aromatic ring. The linkage is stable under acid-hydrolysis conditions which release the monosaccharides. Evidence for the aromatic group was obtained from UV and MS data.

Strobel: Work in our laboratory has demonstrated the glycopeptide nature of toxins from *Corynebacterium michiganense*, *C. insidiosum* and *C. sepedonicum*. Structural studies on the toxin of *C. sepedonicum* have shown that the glucose residues in the molecule are linked α 1 $\rightarrow$ 6, and mannose residues are primarily linked 1 $\rightarrow$ 2 and 1 $\rightarrow$ 3. The peptide occurs as a chain attached to the sugar portion of the molecule. Perhaps this answers some of the questions of Wood and Dimond.

9. Chemistry of Host-Specific Phytotoxins*

ROSS B. PRINGLE

Cell Biology Research Institute,
Research Branch,
Canada Department of Agriculture,
Ottawa, Ontario, Canada

Introduction

Host-specific toxins are becoming more important in infectious plant diseases as highly-bred crops are being planted more widely. An increasing number of newly developed cultivars of cereals have been found to be highly susceptible to certain strains of fungi that do not cause noticeable diseases in non-selected populations of the crop. In each case that has been thoroughly investigated, it has been found that this abnormal susceptibility of the host was correlated with the production by the pathogen of a substance that is highly toxic to the susceptible plant and which is much less toxic to resistant plants. In these diseases, the toxic substances are true chemical determinants of pathogenicity. For convenience they are generally referred to as host-specific toxins. For general reviews of host-specific toxins see Pringle and Scheffer (1964) and Scheffer and Pringle (1967).

These host-specific toxins, after being isolated and purified, reproduce all the visible and all the known biochemical symptoms of the associated diseases. They are very useful substances for studying disease, because many complexities of the host–parasite interaction are avoided. Three examples of host-specific toxins have been studied most intensively with the aim of establishing the chemical constitution of these unusual substances and of correlating these structures with their unique biological properties. These are the host-specific toxin of *Helminthosporium victoriae* Meehan *et* Murphy, the host-specific toxin of *H. carbonum* Ullstrup and a number of different host-specific toxins produced by *Periconia circinata* (Mangin) Sacc. The biology of host-specific toxins is described by Scheffer elsewhere in this book (pp. 251–269).

* Contribution No. 690 Cell Biology Research Institute, Research Branch, Canada Department of Agriculture, Ottawa, Ontario, Canada.

Although the concept of the role of toxins in disease is very old, little definitive work was done on plant toxins until relatively recently. The study of the chemistry of host-specific toxins is an outgrowth of more general studies begun in the 1940's on the specific toxic effects of antimetabolites (Woolley, 1952). Woolley had discovered that chemical analogues of animal growth factors or vitamins, when fed to animals, produced diseases similar to those caused by a deficiency of the corresponding vitamin and moreover that these diseases could be cured by administering the appropriate vitamin.

This early work on the chemical synthesis and biological testing of antimetabolites was done during the early years of antibiotic research, so a good deal of attention was paid to metabolic products of fungi. At that time there was great interest in developing a rational approach to chemotherapy and most research on antimetabolites was directed towards this end. It was only with the pioneering work of Gäumann, Plattner and associates (Plattner and Clauson-Kaas, 1945) on lycomarasmin that the idea arose that phytotoxins might be naturally occurring antimetabolites. It occurred to Woolley (1948) that lycomarasmin might be a peptide of serine, glycine and aspartic acid and that it might be a metabolic antagonist of the streptococcus growth factor strepogenin which at that time was thought to be serylglycylglutamic acid. Woolley was able to demonstrate that strepogenin preparations would prevent the wilting of excised tomato leaves dipped in lycomarasmin solutions.

Although it is now realized that the wilting syndrome is a complicated phenomenon (Dimond, pp. 289–306) and that the chemistry of lycomarasmin is also more complex (Barbier, pp. 91–102), these experiments were valuable in directing attention to the possibility of reversing the action of plant toxins. Braun (1950) was able to demonstrate that an enzymatic digest of casein, which is a rich source of strepogenin, would reverse the inhibition of the growth of *Chlorella vulgaris* Beij. caused by the sterile culture filtrate of *Pseudomonas tabaci* (Wolf *et* Forst.) Stev.

When the individual constituents of enzymatic casein hydrolysates were tested separately, it was found that methionine was active in reversing the toxicity to *C. vulgaris*. To examine the hypothesis that wildfire toxin was a naturally occurring analogue of methionine, the toxin was isolated and purified (Woolley *et al.*, 1952). The chemical structure of this toxin has only recently been elucidated completely (Stewart, 1970).

During the course of isolation of wildfire toxin it became apparent that the disease was more complicated than had been expected. Wildfire toxin was non-specific in its action being highly toxic to all living organisms that were tested. This made it difficult to determine its importance in the etiology of one particular disease.

By this time, the early 1950's, the chemistry of phytotoxins had become a study in its own right. Among those known at that time, the most important was the toxin of *H. victoriae* because of its very special role in the notorious epiphytotic Victoria blight of oats (Meehan and Murphy, 1947). Since the role of a toxin in this disease was so clear-cut and prominent, work was started on its isolation and characterization.

This *Helminthosporium* blight of Victoria oats was an important landmark in the evolution of theoretical pathology, of great interest, not only for plant disease, but also for animal and human disease. Although the concept of the importance of toxins in disease was very ancient, no "model" diseases had hitherto been uncovered. The toxin of *H. victoriae* is only harmful to those plants which bear the gene for susceptibility to the pathogen. This susceptibility of the host had been accidentally emphasized by breeding because susceptibility to *H. victoriae* is closely linked genetically to hypersensitive resistance to crown rust caused by *Puccinia coronata* Cda.

It is now evident why clear-cut examples of toxins in disease had been hard to find. The highly susceptible individuals would be quickly eliminated under natural conditions. To obtain material for a study of the mechanism of toxin action it is necessary to breed plants for susceptibility to a particular disease. This may be done directly or, as for the examples used in this chapter, accidentally while seeking to emphasize other characters.

Helminthosporium victoriae Toxin

Although the presence of this remarkable substance, which we now call a host-specific toxin, was first recognized in the early 1940's (Meehan and Murphy, 1947; Litzenberger, 1949), it was not until more than ten years later that the toxin from *H. victoriae* was first isolated (Pringle and Braun, 1957).

As a guide in following the isolation of the toxin, a bioassay depending on inhibition of growth of roots of susceptible seedlings was developed.

Selected high yielding strains of the fungus were used for toxin production (Scheffer *et al.*, 1964). Fries' modified solution was found to be the best culture medium (Luke and Wheeler, 1955). This medium is buffered to about pH 5·4, and remains acid until cultures begin to lyse. Toxin production is very poor in aerated cultures but good in stationary cultures grown at 21–25°C.

The isolation of the host-specific toxin of *H. victoriae* proved to be exceedingly difficult because upon purification the toxin became extremely unstable (Pringle and Braun, 1957; Scheffer and Pringle, 1963). Steps in the isolation procedure involved concentrating the toxin by removing water, salts and other inert material, while keeping the toxin below pH 4 and in solution. Final purification was achieved by chromatography, first on alumina, then on potato starch.

The toxin was active at a dilution of 2×10^{-4} μg per ml when freshly prepared. Samples of material retained some activity for up to 3 years when kept *in vacuo* at –20°C. The preparation lost activity quickly when stored at room temperature or in solution, even if kept frozen.

Very little chemical work was done on the intact toxin because it was very unstable. Almost all the information on its constitution has been obtained from a study of degradation products (Pringle and Braun, 1958). The chemical constitution of one part, victoxinine, has been given most attention because it represents a new class of chemical compound. The following is largely a description of the production and study of this substance.

When an aqueous solution of the purified toxin was adjusted to pH 8·4 by saturating with sodium bicarbonate, the toxin began to break down. The solution gradually became cloudy and in the course of a few hours an opaque, white mixture resulted.

A freshly prepared, biologically active solution of the toxin did not react with ninhydrin but after treatment with sodium bicarbonate the mixture gave a strongly positive ninhydrin reaction. When the mixture was chromatographed on paper in propanol/acetic acid/water (200:3:100), two ninhydrin positive spots appeared with R_F values 0·67 and 0·84. When these substances were eluted separately from the chromatogram, hydrolysed with hydrochloric acid and rechromatographed, the 0·67 spot changed to 5 ninhydrin-reacting spots but the 0·84 spot remained as a single spot.

The ninhydrin reacting material of higher R_F could be extracted from the sodium bicarbonate-treated toxin solution with diethyl ether. After evaporation of the ether extract and solution of the residue in N-hydrochloric acid, the compound was precipitated with a saturated aqueous solution of mercuric chloride. The mercury complex was washed with N-hydrochloric acid and decomposed with hydrogen sulphide. The filtrate from this treatment was saturated with sodium bicarbonate and extracted with diethyl ether. When a dry ether solution of the compound was treated with ether saturated with hydrogen chloride, the material crystallized in colourless needles as the hydrochloride. After recrystallization from ethanol/ether the hydrochloride melted at 172–73°C.

Found: C: 68·10%, H: 9·95%, N: 4·58%, Cl: 11·75%. Calc. for $C_{17}H_{30}NOCl$ C: 68·08%, H: 10·08%, N: 4·67%, Cl: 11·83%.

There was no sulphur in the molecule so the empirical formula of the hydrochloride was $C_{17}H_{30}NOCl$. This was confirmed by crystallizing and analysing the Reineckate salt. In 1968 the molecular formula was confirmed by mass spectrometry. The hydrochloride showed only end-absorption in the ultraviolet region of the spectrum; thus it was not aromatic. It was found to be an optically active molecule $[\alpha]_D^{25} = -78°$ ($c = 3{\cdot}2$ in 95% ethanol). The compound was one that had not been previously described. The trivial name "victoxinine" was given following the precedent in naming the alkali degradation product, tabtoxinine. The classical suffix -ine was used to designate the major constituent of an active principle (Woolley *et al.*, 1955).

From the molecular formula of victoxinine, $C_{17}H_{29}NO$ (free base), it can be calculated that the compound has a combination of 4 rings and/or double bonds. When hydrogenated with a palladium-charcoal catalyst, victoxinine took up one mole of hydrogen. Two atoms of bromine were added when victoxinine was brominated in carbon tetrachloride solution. The carbon–carbon double bond stretch absorption found at 1660 cm^{-1} in the infrared spectrum of victoxinine free base was not present in the dihydro derivative.

The nature of the single oxygen atom in victoxinine was also established by elimination. It was not in the form of a hydroxyl because there was no absorption at the —OH stretching frequency in its infrared spectrum. Moreover, the compound was not acetylated by acetic anhydride and pyridine. This confirmed the absence of hydroxyl and also indicated that the nitrogen was not in a readily acetylatable form. Neither was the oxygen in the form of carbonyl because there was no C=O stretch absorption in its infrared spectrum. By elimination, the oxygen must be in the ether linkage. A Zeisel determination yielded no methyl or ethyl iodide so the ether was not a methoxyl or ethoxyl group. Taking all the available data into consideration, it would appear that the oxygen atom is a member of one of the rings in victoxinine. Because victoxinine contains one double bond it must, by difference, contain three rings in the absence of a carbonyl group, i.e. it has a tricyclic structure.

The nitrogen in victoxinine appears to be unusual. Both methyl and isopropyl absorptions were detected in the infrared absorption spectrum of the free base. Considering that victoxinine contains three rings, this suggested that the compound might have a sesquiterpenoid structure. Whether it is diterpenoid or sesquiterpenoid will depend on whether all of the carbons or only 15 of the 17 are part of a mevalonate-derived system. There are very few examples of sesquiterpenoid compounds containing nitrogen. Victoxinine was sufficiently basic to form a hydrochloride salt but was an unusually weak base with a pK_a of about 6·2. The free base was insoluble in water but the hydrochloride salt was quite soluble. It was such a weak base that it was displaced from a solution of its hydrochloride salt by sodium bicarbonate solution. When the double bond was saturated by hydrogenation a much stronger base resulted which was not displaced from its hydrochloride by bicarbonate. This indicated that the double bond was near enough to the nitrogen atom to influence its electronic structure. Victoxinine free base exhibited secondary amine absorption in its infrared spectrum and also gave a positive dithiocarbonate test for secondary amine. On the other hand, it also gave a positive iodoplatinate test showing that it has tertiary amine properties. The most satisfactory assignment for the nitrogen atom, in the light of the available data, is in a ring; possibly the same ring as the oxygen atom.

A tentative assignment for the location of the double bond in victoxinine can be made from a consideration of its chemical properties and infrared spectrum. When victoxinine free base was fused with benzoyl peroxide, formaldehyde was produced. This was detected by the colour it gave with chromotropic acid. The formaldehyde could originate from a methoxyl group, an N-methyl group or an exocyclic methylene group. The possibility of a methoxyl group was eliminated by the finding of a negative Zeisel determination. An analytical $>N-CH_3$ determination yielded less than half a mole of methyl iodide. Thus, again by elimination, the double bond was assigned to an exocyclic methylene group. This was confirmed by the finding of a distinct peak of absorption at 3090 cm^{-1} in the infrared spectrum of the free base.

When victoxinine free base dissolved in carbon tetrachloride was treated with bromine dissolved in carbon tetrachloride, a typical addition reaction first took place. However, on evaporation of the solvent, a hydrobromide salt of a monobromo derivative was formed. This facile elimination of hydrogen bromide would be likely with an exocyclic double bonded methylene. The finding of less than half a mole of methyl iodide in the N-methyl determination could be due to thermal rearrangement on treatment with hydriodic acid.

The assignment of methyl, isopropyl and exocyclic methylene groups in victoxinine accounts for five of the 17 carbon atoms in the molecule. If it is assumed that the compound is a sesquiterpene, a further 10 carbon atoms may be assigned to two fused alicyclic rings (two 6-membered, or one 5- and one 7-membered) leaving only two carbon atoms unaccounted for. Since the compound contains three rings, the remaining two carbons must be in the third ring along with the nitrogen and oxygen atoms.

It is possible to draw a number of positional isomeric structures for victoxinine. However, since the structure is new, chemical degradation studies will be required to establish the final structure. This will require much more material than was available in the studies that have been described. Alternatively, it may be possible to solve the problem with less material using high resolution nuclear magnetic resonance spectrometry.

Although spin-spin decoupling experiments at 100 MHz could not resolve overlapping signals in the alicyclic area, it is possible that this could be accomplished using a modern 220 MHz apparatus.

Toxins of *Periconia circinata*

Except for the modern spectrometric evidence, most of the information presented on the structure of *H. victoriae* toxin had been obtained by 1958 (Pringle and Braun, 1958). It was evident at that time, that before investing in the effort necessary to obtain more information on this toxin, it would be wise to determine whether this was an isolated phenomenon or whether host-specific toxins had a wider role in the etiology of plant diseases (Scheffer and Pringle, 1961). As the result of a deliberate search, the host-specific toxin produced by *Periconia circinata*, the causal agent of the milo disease of grain sorghum, was discovered. The biological aspects of the *P. circinata* problem are described by Scheffer in pp. 258–9. The situation was found to be qualitatively the same as the oat disease discussed above. It was quickly discovered that *P. circinata* toxin was of the same general nature as *H. victoriae* toxin in that it was a low molecular weight polypeptide-like substance. It was also found that it was much easier to produce and was more stable than *H. victoriae* toxin so for the next few years it received major attention (Scheffer and Pringle, 1961; Pringle and Scheffer, 1963, 1966, 1967).

This toxin was unlike the *H. victoriae* toxin in that it was not inactivated by ion-exchange resins. This proved to be a very efficient way of purifying the toxin. Columns of loosely cross-linked cation exchanger (Dowex 50W-X2) were used. The purified toxin was a colourless amorphous powder which gave results in bioassays at concentrations 1·0–0·1 μg/ml.

The culture filtrates of *P. circinata* first examined in 1961 contained only a single host-specific toxin. This was demonstrated by paper chromatography, ion-exchange column chromatography and countercurrent distribution. During the next year a very small amount of a second host-specific toxin was found in certain batches of culture filtrates. The following year other batches were found to contain relatively large amounts of this second toxin which was distinctly different from the one first identified. Careful examination of a large number of batches of culture filtrate produced since that time has revealed other closely related host-specific toxins. Those that were still specifically toxic only to plants susceptible to the pathogen could be separated by the ion-exchange column chromatography system described above or by 1000 transfer counter-current distribution using secondary butanol and 0·01 N HCl. They have been named *P. circinata* toxin A, *P. circinata* toxin B, *etc.*, in the order in which they were discovered.

Periconia circinata toxin A is the only one of these toxins that has been studied chemically and, because of difficulties in obtaining sufficient material, the studies are incomplete (Pringle and Scheffer, 1966). It was obtained in crystalline form from butanol-ether extracts of material purified by chromatography on Norite-Celite and recrystallized from absolute ethanol. The recrystallized material was very unstable in the dry state and no elemental analyses were obtained. Unlike the *H. victoriae* toxin, this toxin gave a strong ninhydrin reaction, indicating the presence of free amino groups. The compound was resistant to complete acid hydrolysis, but when this was carried out at 125°C, four common amino acids, aspartic, glutamic, alanine and serine were identified. No other ninhydrin-positive products were found in the hydrolysate. A quantitative amino acid analysis accounted for 80% of the molecule as 6 mol alanine, 4 mol aspartic acid, and 2 mol each of glutamic acid and serine.

An estimation of the molecular weights of the *P. circinata* toxins was made by chromatographing various toxin preparations in aqueous solutions on columns of Sephadex G-25. All the toxins entered this gel so all must have a molecular weight of less than 2000, the limit of exclusion for Sephadex G-25.

Production and characterization of the *P. circinata* toxins has been postponed pending the solution of the problem of controlled production of a homogeneous product. The cause of the variation in the various toxins produced in culture is unknown. One possibility is that the fungus isolate was a mixture of strains. Isolates were derived from single conidiophores because spores can seldom be germinated so the nuclear condition of the heterocaryotic fungus is unknown. Another possibility is that the variation can be traced to differences in the amino acid composition of the medium. *P. circinata* requires a yeast extract supplement to the basic medium for toxin production and different sources of yeast extract were used.

Helminthosporium carbonum Toxin

The host-specific toxin of *H. carbonum* was discovered about the time it became apparent that further chemical study of *P. circinata* toxins should be postponed (Scheffer and Ullstrup, 1965). Since this toxin appeared to be free of the production problems associated with the ones discussed above, it was convenient to turn attention to a study of its chemical properties (Pringle and Scheffer, 1967, 1970; Pringle, 1970).

H. carbonum toxin was found to be another example of a toxin that was only toxic to the plant susceptible to the pathogen and not to resistant plants. There were, however, certain interesting qualitative differences between the action of this toxin and the ones that had been studied earlier, so a study of its chemical properties would be useful in determining the range of structures connected with this phenomenon. The biological aspects of the *H. carbonum* problem are covered in pp. 259–261.

A seedling assay patterned after those described above proved to be very effective. It was possible to use hybrid seed where the toxin-susceptible and toxin-resistant seedlings differed only in the gene controlling resistance to *H. carbonum*. Thus, results were quite uniform. It was also possible to measure 50% inhibition of growth of roots of susceptible seedlings. The action of this toxin was again different in that very low concentrations were found to stimulate root growth (Kuo *et al.*, 1970).

A simple qualitative bioassay was also developed for this toxin taking advantage of its relative stability (Pringle and Scheffer, 1967). A measured volume of the solution to be tested (e.g., from countercurrent distribution tubes) was placed on a 7 cm circle of filter paper. Solvents were removed by placing the paper in a current of air for 3 h. The paper disk was placed on the bottom of a petri dish; seeds and 10 ml of White's solution were added. Toxin was located on paper chromatograms in a similar way after cutting the paper into segments.

High toxin-yielding isolates, developed in breeding experiments (Scheffer *et al.*, 1967), were used for toxin production following the conditions used for producing the *P. circinata* toxins. Again yeast extract was needed and only shallow cultures were productive. The culture filtrate was concentrated and partially deproteinized by methods developed for use with *P. circinata* toxin. In this case, crude concentrates were relatively stable and could be shipped between laboratories without refrigeration.

H. carbonum toxin was found to be much less polar than the other host-specific toxins. This allowed a relatively simple isolation procedure to be developed (Pringle and Scheffer, 1967).

The crude concentrated toxin filtrate was shaken with successive 100 ml portions of chloroform. The first chloroform extractions caused additional protein denaturation, producing emulsions. These were broken by centrifugation and the phases separated. The aqueous phase was treated in this way until no more colour was extracted into chloroform and deproteinization was complete. Five extractions were needed usually to remove the host-specific toxin from the aqueous phase completely.

The residue from evaporation of the chloroform extracts was washed several times with absolute diethyl ether and purified by countercurrent distribution in 100 mg batches using a solvent system consisting of *n*-butanol:glacial acetic acid:water (4:1:5, v/v). The maximum of weight, toxin activity, and colour with bromcresol green occurred in tube 70 of a 100-transfer distribution, corresponding to a partition coefficient of 1·2.

The toxin could be crystallized from an ethanol-ether mixture. However, material from which solvents had been replaced by water prior to lyophilization was found to be more stable in the dry state. In spite of the fact that *H. carbonum* toxin assays are relatively more precise, it is still difficult to determine the absolute activity of highly purified preparations, because these are very unstable and may be inactivated during the assay procedure. Best values were less than one μg/ml.

Because *H. carbonum* toxin was more easily produced, its chemical constitution has been more extensively studied than has been possible with other host-specific toxins (Pringle, 1970). Chemical studies are still relatively difficult because of the great instability of highly purified preparations. Pure crystalline toxin was found to be so unstable that it lost biological activity even when dry, under vacuum, at –20°C.

A number of preparations have been analysed for carbon, hydrogen and nitrogen, so that for the first time a conventional empirical formula can be written for a complete host-specific toxin molecule.

Found (Crystallized)	C = 55·76%; H = 7·44%; N = 12·43%
(Crystallized)	C = 55·96%; H = 7·43%;
(Lyophilized)	C = 56·52%; H = 7·19%; N = 12·32%
Calculated for $C_{32}H_{50}N_6O_{10}$	C = 56·61%; H = 7·44%; N = 12·38%

The formula $C_{32}H_{50}N_6O_{10}$ can only be considered an approximation for an unstable molecule of this size. A more precise formula will evolve as degradation work establishes the formulae of smaller pieces. The empirical formula can also be considered the molecular formula because the molecular weight produced by this formula is consistent with molecular sieving experiments. Ten milligrams of lyophilized toxin, purified by countercurrent distribution, and dissolved in 0·25 ml of 10% (v/v) acetic acid was applied to a 1·2- × 23 cm column of Sephadex G-10 that had been equilibrated with 10% acetic acid. One-half ml fractions of eluate were collected. One drop of effluent from every second tube was evaporated on a 7 cm circle of filter paper and was bioassayed as described previously. The void volume of the column, as determined by Blue Dextran 2000, was 16·5 ml. The toxin first emerged in the fourth tube (18·5 ml) following the void volume. The elution volume of the column was calculated to be 22·5 ml. The toxin appears to have a molecule weight less than 700, which is the exclusion limit of this gel. The calculated molecular weight for $C_{32}H_{50}N_6O_{10}$ is 679.

H. carbonum was found to be very weakly basic to bromcresol green but did not react with ninhydrin by the Stein and Moore procedure or on paper chromatograms. The infrared absorption spectrum of freshly lyophilized toxin, purified by countercurrent distribution is shown in Fig. 1. This

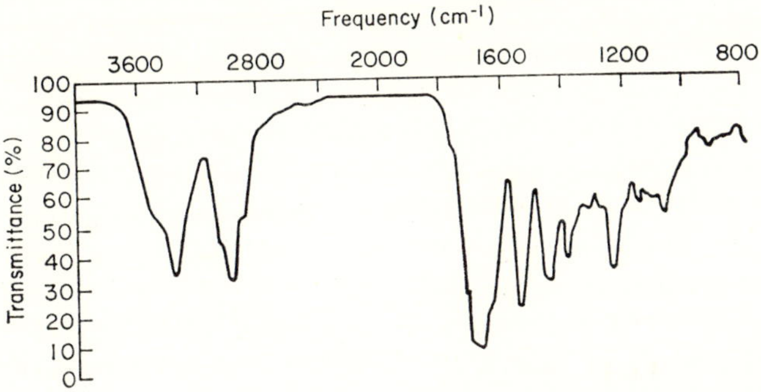

FIG. 1. Infrared absorption spectrum of *Helminthosporium carbonum* toxin.

spectrum was recorded from a film deposited on NaCl windows from chloroform solution.

The principal absorption maxima have been assigned as follows: 3500 (shoulder), free secondary amine; 3320, bonded secondary amine; 2950, methyl; 1710, unsubstituted amide; 1650–80, 1525 substituted amide; 1440, 1375, methyl; 1230, 1045, ester.

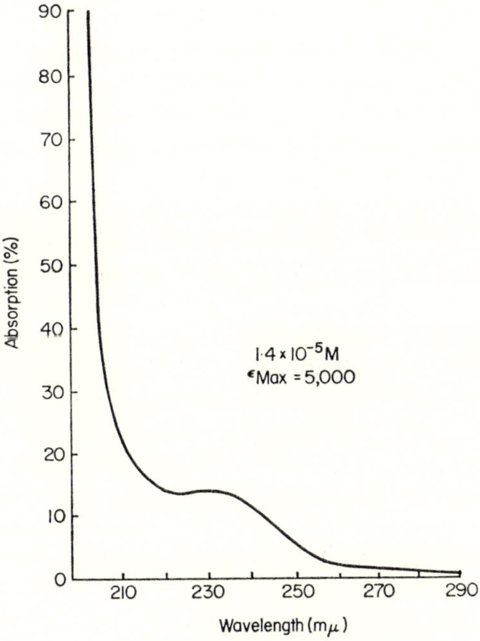

FIG. 2. Ultraviolet absorption spectrum of *Helminthosporium carbonum* toxin.

The principal structural feature revealed by infrared spectrometry was that of a substituted polyamide. Because the presence of ester absorption was unexpected, a number of crude preparations that had not been exposed to acetic acid, were checked. Each showed strong absorption at 1230 and at 1045 cm^{-1}.

The ultraviolet absorption spectrum of freshly lyophilized toxin dissolved in freshly distilled absolute ethanol is shown in Fig. 2. The molar absorptivity at 230 mm of $1{\cdot}4 \times 10^{-5}$ molar solution (assuming a molecular weight of 679) was approximately 5000. The absorption of light in the ultraviolet region of the spectrum is unusual for a peptide. The position of the maximum and its relatively high intensity suggested a double bond in the vicinity of a carbonyl group, possibly an α,β-unsaturated amide.

A complete acid hydrolysis of *H. carbonum* toxin yielded five different ninhydrin reacting products. Two of these were identified as alanine and proline by comparison with standard solutions on paper chromatography. Proline was confirmed by the yellow colour it gave with ninhydrin and by the characteristic blue colour with isatin. The other three spots have not been identified. They appeared at locations different from amino acids or other common ninhydrin-reacting substances on two dimensional paper chromatography and in the profile from a Beckman automatic amino acid analyser. Each of the three unknowns yielded considerably less colour than alanine or proline with ninhydrin. However, because all ninhydrin reactions took place at room temperature in the dark, the colour must be due to amino groups on aliphatic carbons. None of these appear to be methylamino groups since none of the compounds reacted on paper with *p*-nitrobenzoyl chloride.

With the finding of five different amino groups and unsubstituted amide absorption in the infrared region of the spectrum, the six nitrogen atoms of the toxin were assigned to amide groups, probably peptide and probably all different.

H. carbonum like all the other host-specific toxins was found to be unstable to alkali. When freshly purified toxin in aqueous solution was treated with saturated aqueous sodium bicarbonate solution the mixture turned milky immediately, but no precipitate formed. A product was extracted from the reaction mixture with chloroform and converted to the hydrochloride which was recrystallized and analysed.

Found:	C = 55·94%; H = 7·08%; N = 10·37%; Cl^- = 4·98%
Calculated for:	
$C_{32}H_{50}N_6O_{10}$ HCl	C = 53·73%; H = 7·20%; N = 11·75%; Cl^- = 4·96%
$C_{32}H_{48}N_5O_{10}$ HCl	C = 54·96%; H = 7·08%; N = 10·02%; Cl^- = 5·07%

The analysis indicated that treatment with bicarbonate had split off the elements of ammonia. Examination of the infrared spectrum of the crystalline hydrochloride revealed essentially a similar pattern except for the loss of absorption at 1710 cm^{-1}. This is consistent with the loss of nitrogen from a labile unsubstituted amide. It does not appear that simple hydrolysis of an amide to form a carboxylic acid is sufficient to explain this change because the oxygen content of the molecule did not change.

The value of 10 oxygen atoms was calculated by difference from the carbon, hydrogen and nitrogen analyses of the toxin. The assignment of six of these oxygen atoms to amide groups left four neutral oxygen atoms to be accounted for. The finding of ester absorption in the infrared region of the spectrum suggested that these extra oxygen atoms were present in two ester linkages. The most plausible assignment seemed to be acetates of primary alcohols. The methyl absorption at 1440 and at 1375 cm^{-1} were consistent with acetate.

Because the toxin did not react with ninhydrin nor with 2,4-dinitrofluorobenzene, it must be in the form of a ring or the terminal amino group must be acylated in some way. It if were a linear peptide with the terminal amino group acylated, the molecule as a whole would be acidic. Because it is basic, the cyclic configuration seems more probable. The molecule is predominantly non-polar in character since it may be extracted from acidic or basic aqueous solutions by chloroform. This property is also consistent with a cyclic peptide.

From a consideration of the valency requirements of the elements involved, it was calculated from the molecular formula that the toxin contains 11 rings and/or double bonds. Ten of these are accounted by six amide double bonds, two ester double bonds, one α,β-unsaturated double bond, and one proline ring. The one remaining ring is evidently the macro ring of the cyclic molecule.

If *H. carbonum* toxin were a simple cyclic peptide, it would be more stable than it appears. It was found to be unusually unstable to acids. Treatment with concentrated hydrochloric acid at room temperature produced a large number of peptides and amino acids rather than a single linear peptide expected from the opening of a cyclic peptide. On treatment with 0·1 N HCl or with glacial acetic acid made 0·1 N with respect to hydrochloric acid, alanine was more quickly released than the remaining components from the molecule.

The loss of biological activity of the toxin may be due to polymerization since insoluble material is formed as the material loses activity. This implies one or more chemically reactive sites in the molecule. An α,β-unsaturated double bond would be such a reactive site.

Relation Between Chemical Structure and Biological Activity

It is evident that host-specific toxins are complex chemical substances and that much more work must be done on them. Nevertheless, a certain amount of basic knowledge of their chemistry has been obtained and the difficult exercise of attempting to relate chemical structure with biological activity must be undertaken.

In each case investigated so far, specific toxicity has been found to be a subtle and evanescent quality of the compounds. It is relatively difficult to isolate any of these substances in pure form and still retain this property of specific toxicity. In the case of two host-specific toxins, those from *H. victoriae* and *H. carbonum*, very mild treatment will destroy specificity but at the same time will produce artefacts that are still relatively toxic in a non-specific manner.

Victoxinine, a breakdown product of the *H. victoriae* toxin was found to be toxic at a dilution of 1×10^{-4} M (Pringle and Braun, 1960; Scheffer and Pringle, 1963). Solutions of this concentration inhibited growth of oats with no selectivity between cultivars, completely at pH 7·0 and pH 8·0, almost completely at pH 6·0 and 50% at pH 5·0. This indicates that victoxinine, a base, is much more toxic in the non-ionized form at high pH than in the ionized form at low pH. This finding and the observation that ethanol also increased the toxicity of victoxinine indicates that the toxin is acting on non-polar sites in cell membranes. The intact *H. victoriae* toxin molecule is less polar than its degradative products since it does not react with ninhydrin. This indicates that there is no nitrogen in the form of ionizable primary amino groups in the toxin molecule.

Bicarbonate-inactivated *H. carbonum* toxin was also found to have lost most of its specific toxicity (Pringle, 1970). It retained about the same non-specific toxicity for corn that victoxinine had shown for oats (1×10^{-4} M). It has been suggested as a working hypothesis to explain the loss of specific toxicity, that specificity and toxicity are properties of different parts of host-specific toxin molecules (Braun and Pringle, 1959). Some support of this hypothesis was given by the finding that alkali-inactivated *H. victoriae* toxin would counteract the effects of the complete toxin (Scheffer and Pringle, 1964). Competition between the peptide fraction of the toxin molecule and the complete toxin molecule for receptor sites is one suggested explanation of the results.

Work with the *P. circinata* toxins has demonstrated that host-specific toxins, shown to have different chemical properties, still possess the same specific biological activity. This would be the case if similar segments in different molecules were responsible for specific toxicity.

H. carbonum toxin has been found to resemble some of the peptide antibiotics in being a cyclic molecule containing unsaturated amino acid residues. *H. victoriae* toxin has a biological action on membrane permeability that resembles in some ways the action of certain peptide antibiotics. It is possible that future work will discover other links between these classes of microbial peptides.

References

Braun, A. C. (1950). *Proc. Nat. Acad. Sci.* **36**, 423–427.

Braun, A. C. and Pringle, R. B. (1959). *In* "Plant Pathology Problems and Progress 1908–1959" (C. S. Holton, *et al.*, Eds), pp. 88–99. University of Wisconsin Press, Madison, Wis.

Kuo, M. S., Yoder, O. C., and Scheffer, R. P. (1970). *Phytopathology* **60**, 365–368.

Litzenberger, S. C. (1949). *Phytopathology* **39**, 300–318.

Luke, H. H. and Wheeler, H. E. (1955). *Phytopathology* **45**, 435–458.

Meehan, F. and Murphy, H. C. (1947). *Science* **106**, 370–371.

Plattner, P. A. and Clauson-Kaas, N. (1945). *Helv. chim. Acta* **28**, 188–195.

Pringle, R. B. (1970). *Pl. Physiol.* **46**, 45–49.

Pringle, R. B. and Braun, A. C. (1957). *Phytopathology* **47**, 369–371.

Pringle, R. B. and Braun, A. C. (1958). *Nature* **181**, 1205–1206.

Pringle, R. B. and Braun, A. C. (1960). *Phytopathology* **50**, 324–325.

Pringle, R. B. and Scheffer, R. P. (1963). *Phytopathology* **53**, 785–787.

Pringle, R. B. and Scheffer, R. P. (1964). *A. Rev. Phytopath.* **2**, 133–156.

Pringle, R. B. and Scheffer, R. P. (1966). *Phytopathology* **56**, 1149–1151.

Pringle, R. B. and Scheffer, R. P. (1967). *Phytopathology* **57**, 530–532.

Pringle, R. B. and Scheffer, R. P. (1967). *Phytopathology* **57**, 1169—1172.

Pringle, R. B. and Scheffer, R. P. (1970). *Phytopathology* **60**, 565–566.

Scheffer, R. P. and Pringle, R. B. (1961). *Nature* **191**, 912–913.

Scheffer, R. P. and Pringle, R. B. (1963). *Phytopathology* **53**, 465–468.

Scheffer, R. P. and Pringle, R. B. (1964). *Phytopathology* **54**, 832–835.

Scheffer, R. P. and Pringle, R. B. (1967). *In* "The dynamic role of molecular constituents in plant-parasitic interaction" (C. J. Mirocha, and I. Uritani, Eds), pp. 217–236. Bruce Publishing Co., St. Paul, Minn.

Scheffer, R. P., Nelson, R. R., and Pringle, R. B. (1964). *Phytopathology* **54**, 602–603.

Scheffer, R. P., Nelson, R. R., and Ullstrup, A. J. (1967). *Phytopathology* **57**, 1288–1291.

Scheffer, R. P. and Ullstrup, A. J. (1965). *Phytopathology* **55**, 1037–1038.

Stewart, W. (1970). Private communication.

Woolley, D. W. (1948). *J. biol. Chem.* **176**, 1291–1298.

Woolley, D. W. (1952). "A study of antimetabolites". John Wiley and Sons, New York.

Woolley, D. W., Pringle, R. B., and Braun, A. C. (1952). *J. biol. Chem.* **197**, 409–417.

Woolley, D. W., Schaffner, G., and Braun, A. C. (1955). *J. biol. Chem.* **215**, 485–493.

Discussion

Barbier: Did I understand well that there are at least two main components in the phytotoxin of *Helminthosporium victoriae*?

Pringle: When purified *H. victoriae* toxin is treated with sodium bicarbonate solution at room temperature, two ninhydrin-reacting products are formed.

Barbier: As a consequence of my preceding question, I would like to know whether victoxinine was isolated from a pure phytotoxin or from a mixture?

Pringle: Victoxinine was first isolated from toxin purified by chromatography on a column of starch. It was subsequently isolated directly from culture filtrates of low toxin-producing strains of the fungus.

Barbier: Could you give the main features of the NMR spectra of victoxinine?

Pringle: The NMR spectrum of victoxinine is being studied by Dr. Boris Weinstein, University of Washington and Dr. Lois J. Durham, Stanford University. They have evidence for the presence of an isopropyl group, a tertiary methyl group and two separate ring protons.

Lukezic: If the host-specific toxins are peptides would they fit into Kuć's hypothesis that antibody–antigen reactions are involved in resistance?

Pringle: Amino acids are found among the products of total acid hydrolysis in each case. It has not yet been shown that true peptide linkages occur in host-specific toxins.

The biological mechanism of action is discussed in detail in Dr. Scheffer's lecture. However, it should be remembered that susceptibility to *H. victoriae* and to its toxin is inherited as a dominant character. In this case, resistance is simply lack of this character.

Ballio: I was very surprised by the amount of work done on host-specific phytotoxins. Just their high specificity makes them most interesting and it would be desirable to speed up elucidation of their structures. Some of these toxins are probably obtained only in small amounts through long and tedious isolation and purification procedures, and represent, therefore, typical examples of substances best investigated by non-classical organic chemistry methods. I think that many of the chemists present here, who are more lucky than Dr. Pringle in having costly NMR and MS equipment, are willing to help in solving structural problems in the field of host-specific toxins chemistry that are still unsolved.

Pringle: I would like to thank Professor Ballio for his kind offer of assistance involving costly NMR and MS equipment. The progress that has been made to date in producing and isolating host-specific toxins would not have been possible without the extensive international collaboration that we have enjoyed. I think it would be very rewarding to all concerned if this collaboration could be extended.

Durbin: Will any of the degradation products of host-specific toxins protect susceptible plants against subsequent treatment with the active toxins?

Pringle: Yes. These experiments are discussed in Dr. Scheffer's lecture and specifically answered in the next comment.

Scheffer: The breakdown products of *H. victoriae* toxin will give partial protection against toxin. Victoxinine gives no protection. By elimination, the peptide product appears to be responsible for the protection.

Lukezic: Could cross-protection by breakdown products of the toxin be considered similar to the use of attenuated toxin in animal diseases?

Scheffer: The simplest explanation of cross protection is the breakdown products compete with toxin for receptor sites. That is our working hypothesis. More complex explanations will be considered if this one is ruled out later.

Paxton: When plants are treated with the *H. victoriae* peptide then inoculated with the toxin, does the concentration of the peptide determine the amount of protection?

This would indicate whether sites for binding the toxin are blocked by the peptide, and whether site blockage is responsible for protection.

Scheffer: We do not yet have the data which would allow us to determine whether or not the protection is competitive.

Strobel: With reference to linkages between amino acids in host specific toxins how else besides a peptide linkage could you image that 3–6 or more amino acids could be coupled together to make the toxin? Next, have any of the host specific toxins been subjected to pronase digestion in attempt to confirm the nature of the amino acid linkages?

Pringle: There are many examples of complex nitrogen-containing natural products that yield amino acids on complete acid hydrolysis, e.g. alkaloids and other heterocyclic products, as well as the more common esters and amides. Since the complete quantitative analyses of the products of acid hydrolysis are not yet known, and since compounds other than amino acids have been found in some host-specific toxins, it would be premature to assume that the amino acids are present exclusively in peptide linkages. The fact that host-specific toxins are much more unstable than known peptides, also calls for caution in pronouncing them to be peptides.

Pronase or other enzymatic digestion of host-specific toxins has not been studied. This could be a study, in its own right, in connection with biological mechanism of action. However, as a method for determining structure, there are more precise physical-organic methods such as NMR which require much less material.

10. The Biosynthesis of Phytotoxins

LUIGI CANONICA

Istituto di Chimica Organica della Facoltà di Scienze, Università degli Studi di Milano, Italy

The literature on phytopathogenic micro-organisms shows that they produce a great number of phytotoxins with completely different structural features and biogenetic origin. Thus some phytotoxins are related to polysaccharides, whereas others contain peptide chains in which various amino acids (aspartic and glutamic acids, glycine, valine and isoleucine) are linked by normal peptide bonds. Moreover a great number of fungi produce anthraquinone derivatives with an acetogenin-like origin whereas others produce terpenoid phytotoxins. Up to now very little experimental information has been obtained about the biosynthetic pathways leading to these different substances and most of the biogenetic hypotheses are based only on the examination of their structures.

Further developments can now be expected in the field of the biosynthesis of phytotoxins. Research on those produced by phytopathogenic micro-organisms will be facilitated by the fact that a great number of these organisms can grow and produce phytotoxins in synthetic media and because small changes in cultural conditions may give rise to the accumulation of different intermediates in biosynthesis. Furthermore the incorporation of intermediates of biosynthesis by this kind of micro-organism is usually very high compared with that obtained with higher plants. In addition, because of their short life cycle, it is inexpensive to grow the micro-organisms in an atmosphere enriched with ^{18}O. All these facts favour detailed study of the biosynthesis of these compounds.

In contrast it is difficult to isolate and purify the enzyme complexes responsible for the different steps of biosynthesis; the oxidases likely to be important in the formation of some phytotoxins are particularly difficult in this respect.

In recent years the method of incorporating labelled precursors for studying a biosynthetic process has been introduced. However here it must be remembered that the incorporation of a radioactive precursor into the final product of a biosynthesis is not proof that it is a true precursor. Labelled glucose, for instance, is very easily incorporated into ergosterol by yeast cells but it is not a true intermediate; in fact its incorporation is preceded by degradation to acetic acid.

The first investigation in the field of phytotoxin biosynthesis was performed on helminthosporal (I) (Fig. 1). This substance, which shows a terpene origin, was isolated in 1961 by de Mayo *et al.* (1961) from cultures of *Helminthosporium sativum* Pamm. King *et* Bakke. The structure and the stereochemistry of this molecule was elucidated by the same author (de Mayo *et al.*, 1963). Later, Tamura *et al.* (1963) isolated helminthosporol (II) from cultures of the same organism. In 1965 de Mayo *et al.* demonstrated that both these substances are artefacts formed from the true phytotoxins, prehelminthosporal (III) and prehelminthosporol (IV), during the extraction procedure.

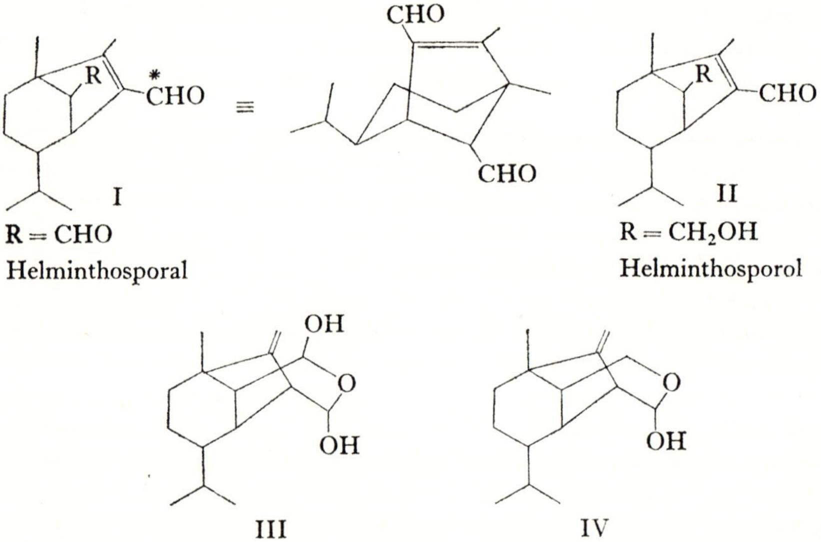

I — R = CHO, Helminthosporal

II — R = CH_2OH, Helminthosporol

III

IV

FIG. 1

Biosynthetic experiments have shown that about 5% of 2-^{14}C mevalonic acid is incorporated by *H. sativum* into helminthosporal (de Mayo *et al.*, 1962). Moreover it was possible to demonstrate that 38% of the total radioactivity is located at carbon atom 13. (*)

From these findings it was easy to hypothesize a sesquiterpene origin for helminthosporal. The probable biosynthetic pathway is represented in Fig. 2. The cyclization of farnesol leads to the cation (V) in which a ten membered ring is present; a 1,3 hydrogen shift occurs with the formation of the bicyclic cation (VI) which gives rise to the hydrocarbon (VII) through the mechanism shown in the figure. No experimental evidence has been reported as yet to support this cyclization hypothesis. The mechanism of the oxidative cleavage of the C_{13}–C_{14} bond is also unknown. The presence of helminthosporol in the cultures does not clarify this point because it is still obscure whether this compound is oxidized by the micro-organism to helminthosporal or whether the opposite reaction takes place.

De Mayo and Williams (1965) were able to isolate from cultures of *H. sativum* the hydrocarbon sativene (VII). In this case this fact allows sativene to be considered as a true intermediate in the biosynthesis of helminthosporal though more generally such a conclusion would not be valid. Neither the

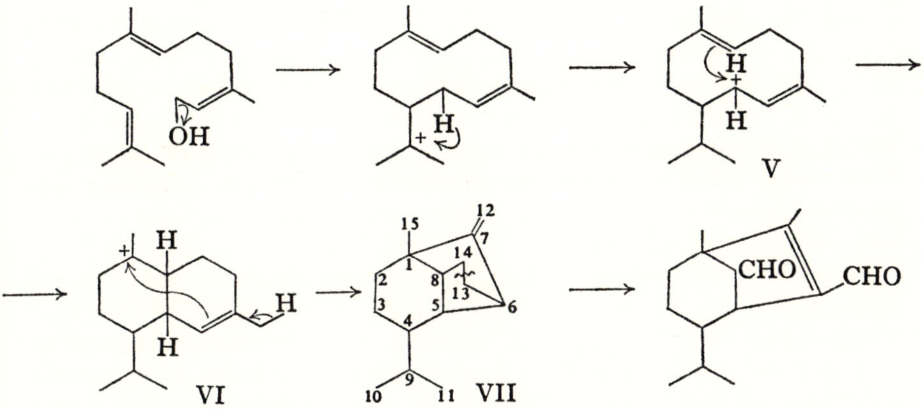

FIG. 2

incorporation of a compound into a biological system nor its isolation from a tissue is sufficient evidence to consider it a true intermediate because in both instances the compound could be the product of a secondary biogenetic pathway.

To establish the true intermediary role of a compound, the method of the "metabolic trap" has been used in some cases. The method is based on the following considerations. In a general biological sequence (Fig. 3) the radioactive precursor A is incorporated into the final product C probably through the intermediate compound B. After supplying the biological system with labelled A and a large amount of unlabelled B we can distinguish two cases. If B is a true intermediate a decrease in the incorporation yield of A into C will occur and labelled B will be recovered from the incubation mixture. If B is not a true intermediate B unlabelled will be recovered. Up to now the method of the "metabolic trap" has been used only with partially purified enzymes but never with whole cells.

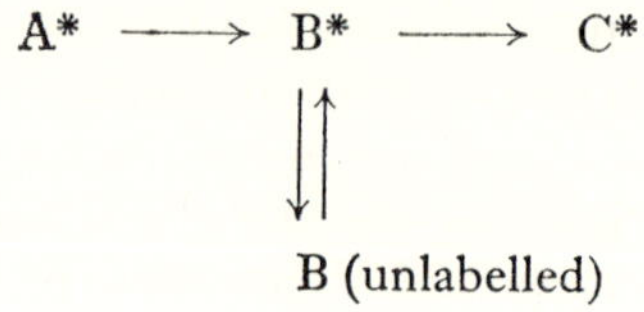

FIG. 3

In the field of phytotoxins we have made detailed studies on the biosynthesis of ophiobolins (Canonica *et al.*, 1966b, 1967a, b, 1968). The structures of the ophiobolins which have been isolated from different strains of *Helminthosporia* are shown in Fig. 4. All can be related to the hypothetical hydrocarbon ophiobolan (XII). Ophiobolin A (VIII) was first isolated from *Cochliobolus miyabeanus* (Ito *et* Kurib.) Drechsl. *ex* Dast. (Orsenigo, 1957), and its structure was contemporaneously elucidated by Nozoe *et al.* (1965) with X-ray analysis and by us with chemical methods (Canonica *et al.*, 1966a). The chemistry of ophiobolins is particularly difficult because of their sophisticated structures and unusual reactivity. No correlation was possible between ophiobolins and other natural compounds owing to their new and rather peculiar skeleton.

From the same organism which produces ophiobolin A, ophiobolin B (IX) can also be obtained in quite good yield whereas ophiobolin C (X) is usually present only in very small amounts. We were able to demonstrate that ophiobolin B is the biological precursor of ophiobolin A; in fact the incubation of radioactive ophiobolin B with *Helminthosporium oryzae* Breda de Haan gave rise to radioactive ophiobolin A. Later, Japanese workers established that the same organism transforms labelled ophiobolin C into ophiobolin A.

The structure of ophiobolins is particularly interesting from a biosynthetic point of view because the skeleton of these substances is derived from the head to tail linkage of five isoprene units. In terpene classification the discovery of this kind of substance filled the gap between diterpenes with four isoprene units, and triterpenes with six isoprene units. Arigoni

Gascardic acid

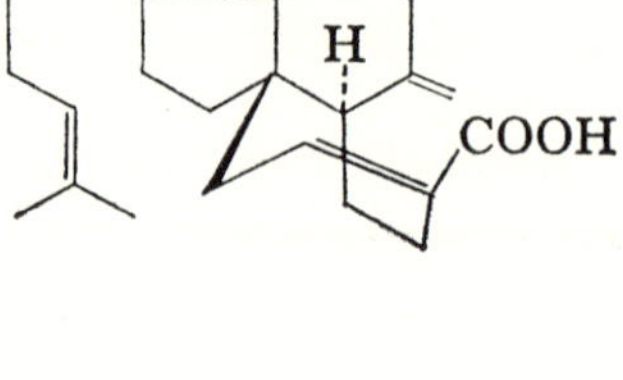

(1965) who determined the structure of the first compound of this class, gascardic acid, isolated from *Gascardia madagascariensis*, named them sesterterpenes.

To demonstrate the isoprene origin of ophiobolins, 2-^{14}C mevalonic acid was incubated with *H. oryzae* (Canonica *et al.*, 1966b). Under suitable cultural conditions the conversion values of mevalonic acid into ophiobolin

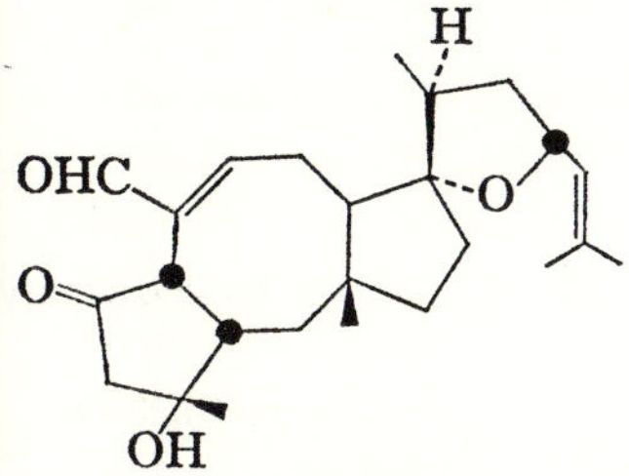

Ophiobolin A
VIII

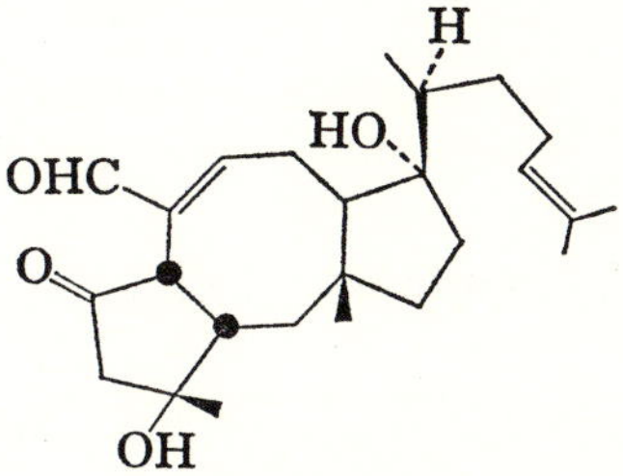

Ophiobolin B
IX

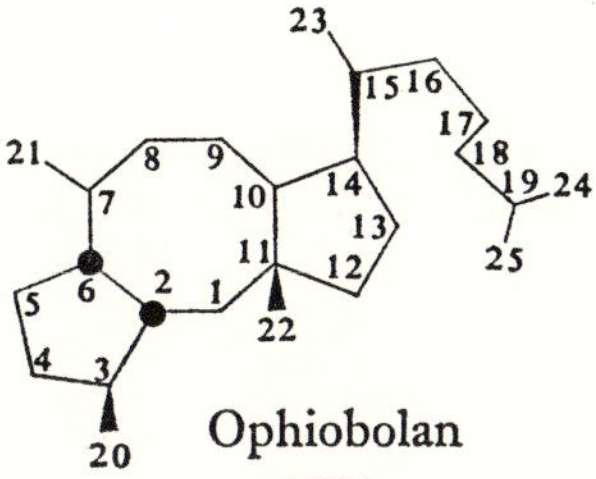

Ophiobolan
XII

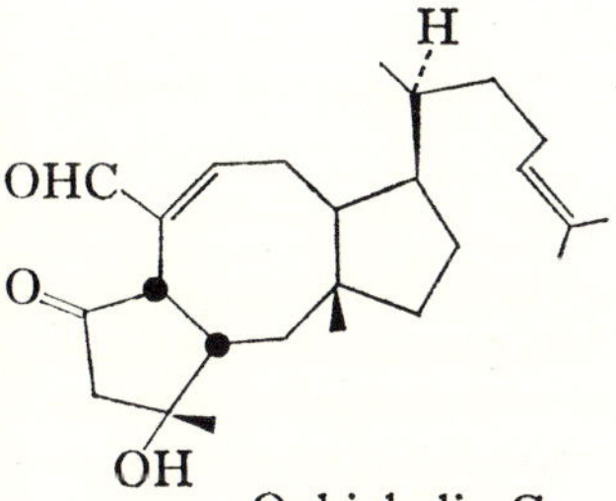

Ophiobolin C
X

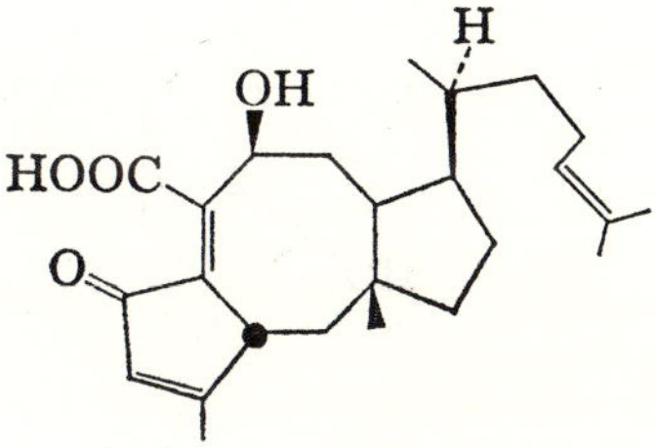

Ophiobolin D
XI

FIG. 4

A were in the range 3–10%. The chemical degradation of labelled ophiobolins obtained from these cultures was performed according to the scheme reported in Fig. 5. The determination of the molar radioactivity of the degradation products allowed us to localize the radioactive atoms in the molecule. Acetone (XIII) obtained after ozonolysis or mild chromic oxidation of ophiobolin A retained 20·4% of the molar radioactivity of ophiobolin A indicating that five mevalonic units had been incorporated into the molecule and that one carbon atom of the isopropylidene is derived from carbon 2 of mevalonic acid. The absence of radioactivity in the acetic acid (XIV) obtained by Kuhn-Roth oxidation from carbon atoms 3, 11, 15, 20, 22, and 23 forced us to exclude their origin from carbon 2 of mevalonic acid. Catalytic hydrogenation of ophiobolin A followed by nitric oxidation yielded the lactone (XV) which contained two ^{14}C atoms. Further oxidation of this lactone gave rise to isovaleric acid containing only one radioactive carbon atom. This result and those obtained from chromic and Kuhn-Roth

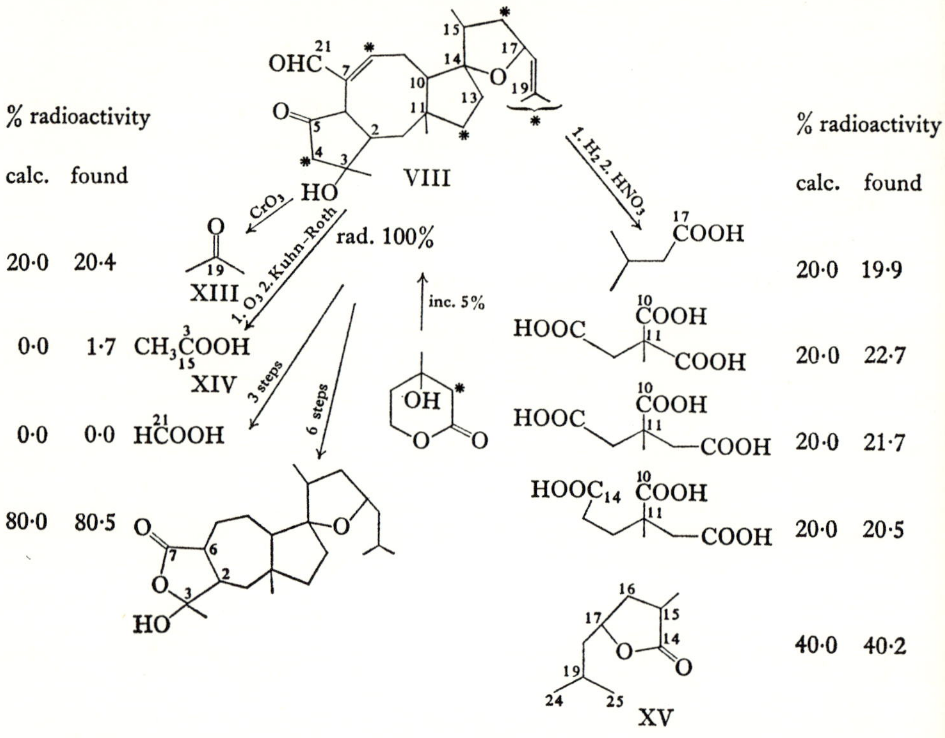

FIG. 5

oxidation located the radioactivity of compound (XV) at positions 16 and 24 (or 25).

The determination of the molar radioactivity of all the other degradation products indicated in Fig. 5 allowed us to localize the five radioactive centres as shown in Fig. 6.

Further information on the biosynthesis of ophiobolins was obtained by growing a culture of *H. oryzae* in an atmosphere enriched with ^{18}O (Canonica *et al.*, 1967a). Ophiobolin A isolated from this culture, and its derivatives (represented in Fig. 7) were checked for ^{18}O enrichment. From the results

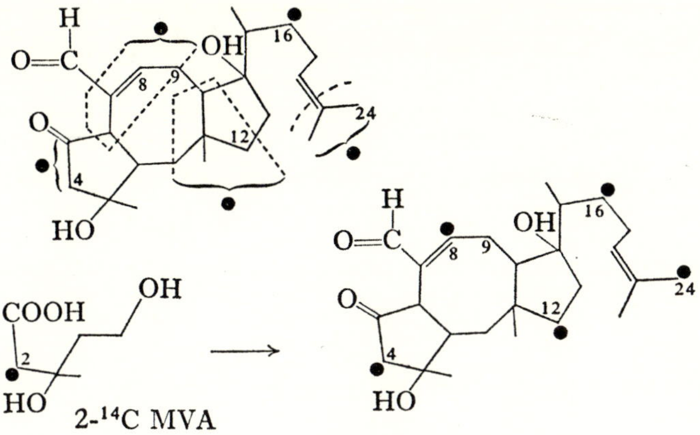
H
O=C
OH
16
8
9
O
12
24
4
HO
COOH
OH
2
HO
2-¹⁴C MVA

Fig. 6

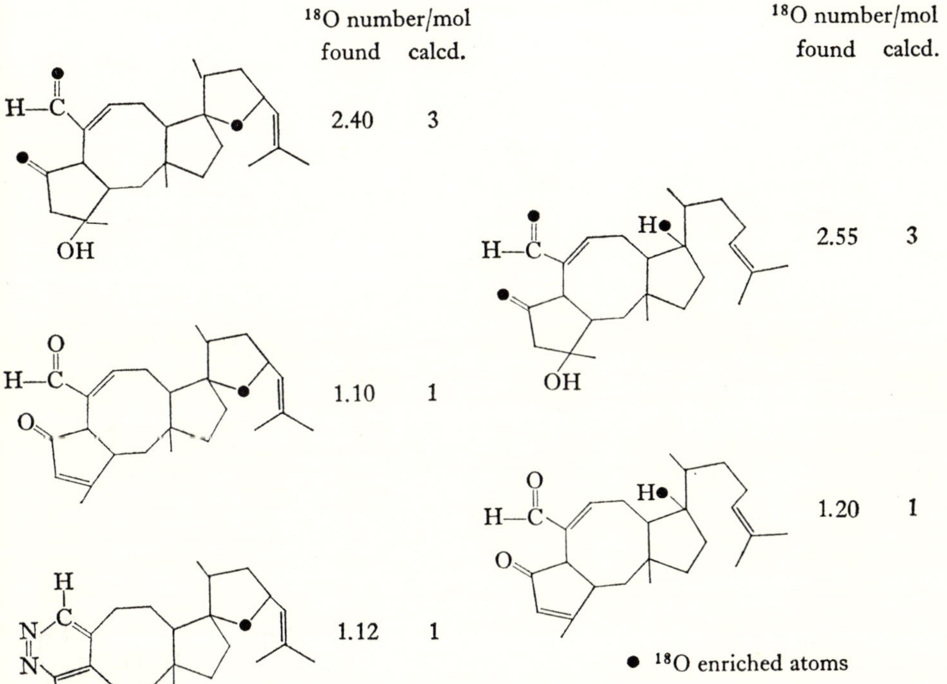
¹⁸O number/mol
found calcd.
H—C
OH
2.40 3
1.10 1
H
N
N
1.12 1
¹⁸O number/mol
found calcd.
H
2.55 3
1.20 1
● ¹⁸O enriched atoms

Fig. 7

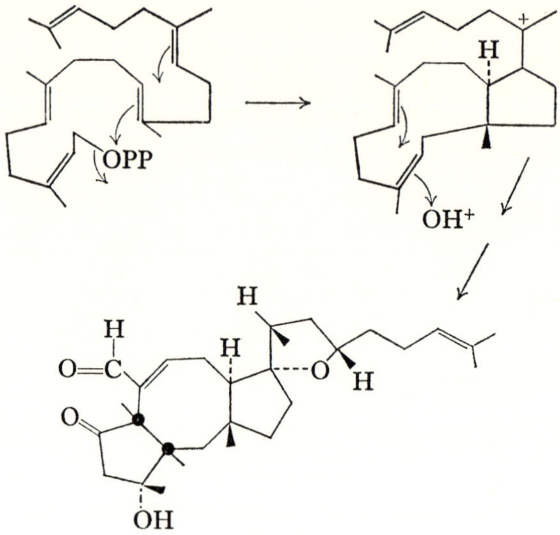
OPP
H
OH+
H
H
O=C
H
O
O
H
OH

FIG. 8

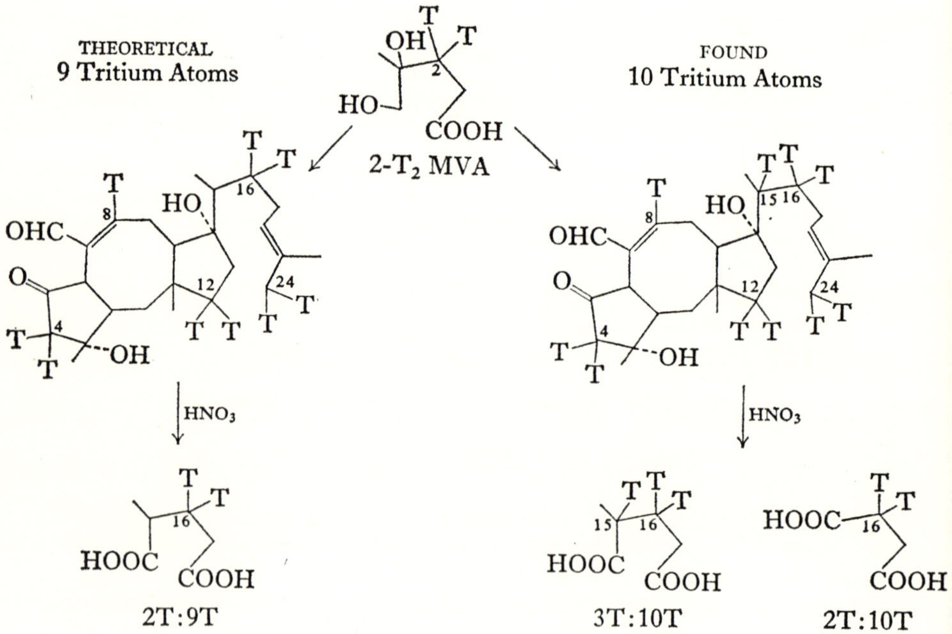
THEORETICAL
9 Tritium Atoms
FOUND
10 Tritium Atoms
2-T2 MVA
HO
OH
COOH
OHC
HNO3
HOOC
COOH
2T:9T
3T:10T
2T:10T

FIG. 9

shown in the figure we concluded that both the hydroxy group at position 14 of ophiobolin B and the ethereal oxygen of ophiobolin A derive from the atmosphere while the hydroxy group at position 3 derives from the water.

These results were not in agreement with the biosynthetic mechanism reported by Nozoe *et al.* (1965) (Fig. 8). However, they were not sufficient to allow us to postulate a different mechanism.

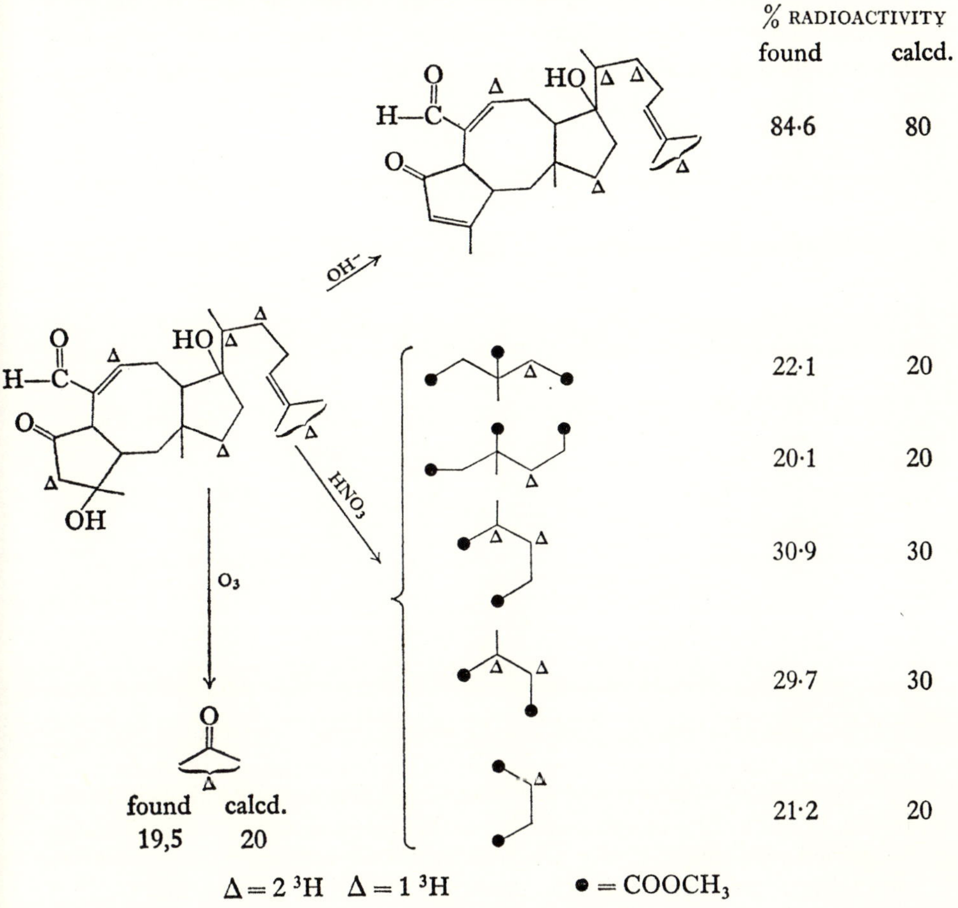

Fig. 10

More details were obtained from the analysis of ophiobolins biosynthesized from 2-$^3H_2$-mevalonic acid (Canonica *et al.*, 1967a). These compounds were supposed to contain 9 tritium atoms as indicated in Fig. 9. Surprisingly both α-methyl glutaric acid obtained by nitric acid oxidation from carbon atoms 14, 15, 16, 17, 18, and 23 of ophiobolin B and α-methyl succinic acid deriving from the same carbon atoms (except 18) contained three instead of two tritium atoms per molecule. These results allowed us to localize one tritium atom at position 15, position 23 being ruled out by the results of the Kuhn-Roth oxidation reported above. On the other hand all the other tritium atoms of the molecule were localized as indicated in Fig. 10. Therefore the tritium atom found at position 15 of ophiobolin B had to be the one eliminated in the formation of the double bond at position 7,8. As a consequence, the cyclization mechanism leading to ophiobolins must involve the migration of one hydrogen atom from position 8 to position 15.

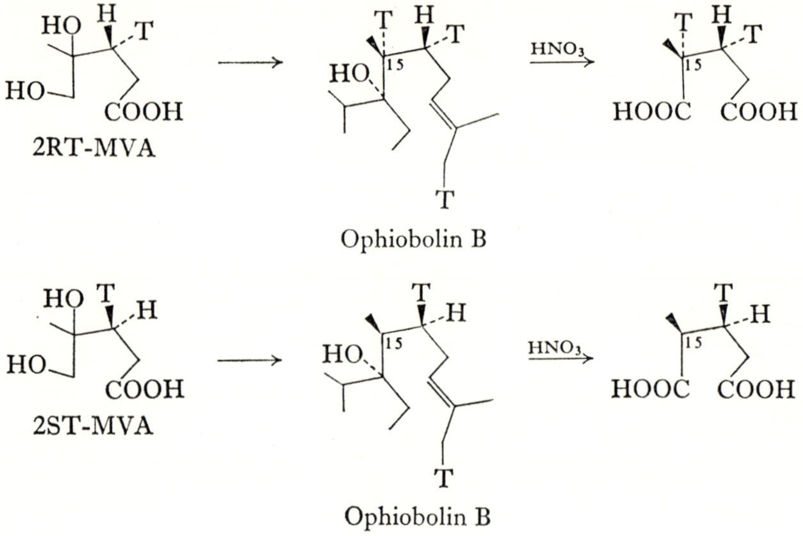

FIG. 11

It was now of interest to elucidate the stereochemical aspect of this unusual migration. To do this ophiobolin B was biosynthesized both from 2R [2-$^3H_1$]-and 2S[2-$^3H_1$]-mevalonic acids (Canonica *et al.*, 1967b). As indicated in Fig. 11, in the first case α-methylglutaric acid obtained by nitric oxidation contained two tritium atoms whereas in the second one it contained only one tritium atom and this was demonstrated to be at position 16. This stereospecific 1,5 hydrogen shift combined with the above reported results allowed us to hypothesize the cyclization mechanism shown in Fig. 12. The hypothetical acyclic precursor is here represented with the terminal double bond in the *cis*-configuration. To establish whether the true acyclic precursor is as indicated in the figure or whether it has an all-*trans*-configuration, ophiobolins were biosynthesized both from 4R [2-^{14}C,4-$^3H_1$]- and 4S [2^{14}C,4$^3H_1$]-mevalonic acids (10). According to Popják and Cornforth

(1966) in the formation of a *trans* double bond the 4 proR hydrogens of mevalonic acid are retained whereas the 4 proS hydrogens are lost. As indicated in Fig. 13 no loss of tritium was observed in the formation of ophiobolin C from 4R tritiated mevalonic acid. As a result the acyclic precursor of ophiobolins must have all the double bonds in the *trans* configuration. This is taken into account in the mechanism shown in Fig. 14. All-*trans* geranylfarnesylpyrophosphate (XVI) may either isomerize to geranylnerylpyrophosphate (XVII) or cyclize directly yielding the hypothetical cation (XIX). Assuming that the last hypothesis is valid, inversion

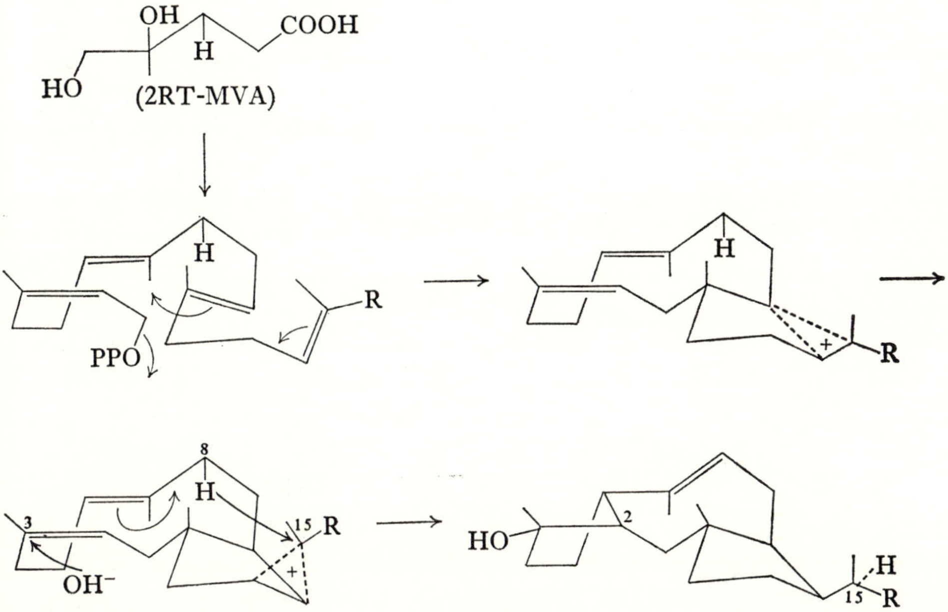

FIG. 12

at position 3 may occur in the tricyclic cation (XXI) with the formation of compound (XVIII). The same tricyclic alcohol would arise from the cyclization of the pyrophosphate (XVII).

The next biosynthetic steps leading to ophiobolin C from the cyclized intermediate (XVIII) are reported in Fig. 15. The isomerization of the double bond in position 7,8 must occur as in pathway A. In fact, as reported in Fig. 13, no exchange of the hydrogen at position 6 occurs in the formation of ophiobolin B.

Up to now no experimental evidence has been reported on the sequence by which the oxygenated functions are introduced at positions 5, 14, and 21 of ophiobolins. A number of different pathways could be suggested but there are no criteria for establishing which one is operative. Because this is a very general problem it is likely that it will be solved in the not too distant future.

The biosynthetic pathway leading to ophiobolins proposed by us has been confirmed by much indirect evidence. Very recently geranylfarnesol

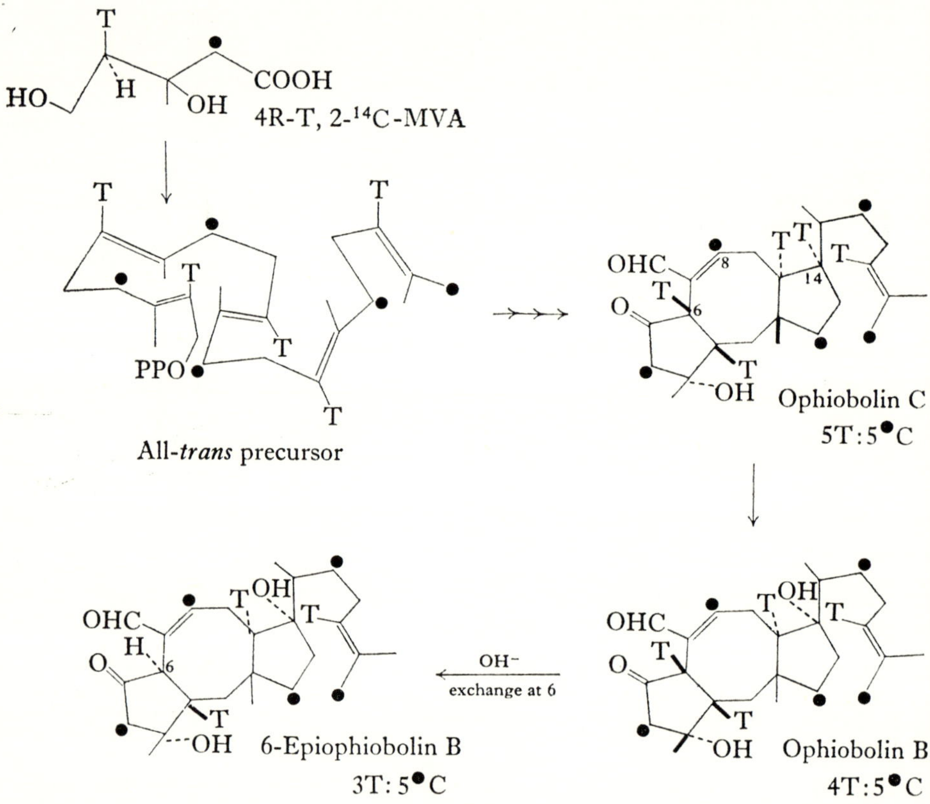

FIG. 13

was isolated from *Ceroplastes albolineatus* Ckll., a wax-producing insect (Rios and Perez, 1969). Furthermore Nozoe *et al.* (1968b) isolated from cultures of *Colchliobolus heterostrophus* (Drechsl.) Drechsl. both geranylnerolidol, which could arise from geranylfarnesol by allylic transposition, and a tricyclic compound with structural features corresponding to those of compound (XVIII). The same compound was isolated by Nozoe and Morisaki (1969) by incubating mevalonic acid with *H. oryzae* homogenates.

Compounds with structures similar to that of ophiobolins are produced by some species of insects. The constitution of both ceroplastol (XXII)

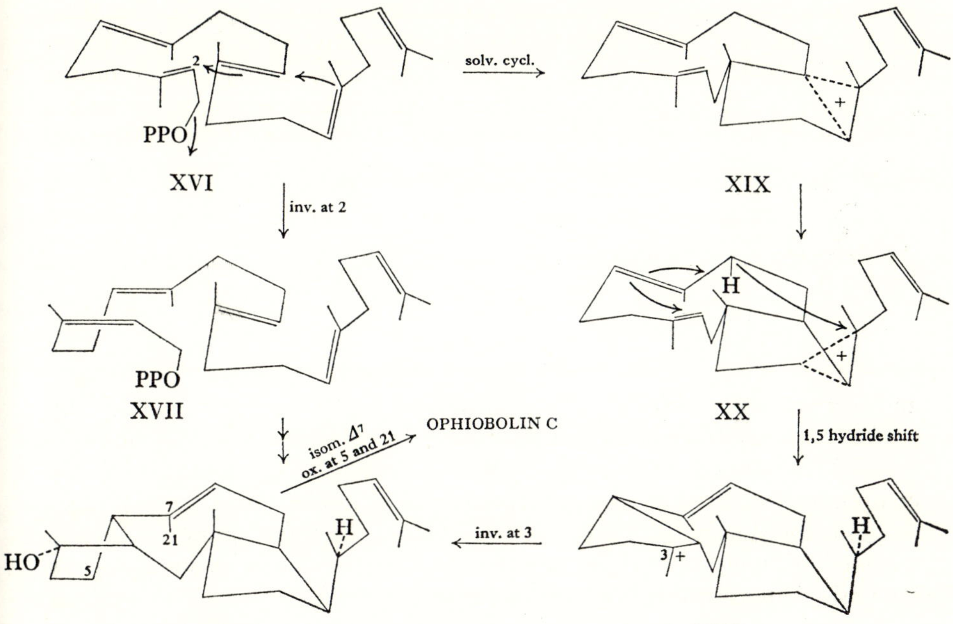
PPO
XVI
solv. cycl.
XIX
inv. at 2
PPO
XVII
H
XX
isom. Δ7
ox. at 5 and 21
OPHIOBOLIN C
1,5 hydride shift
HO
7
21
5
H
inv. at 3
3
H
XVIII
XXI

FIG. 14

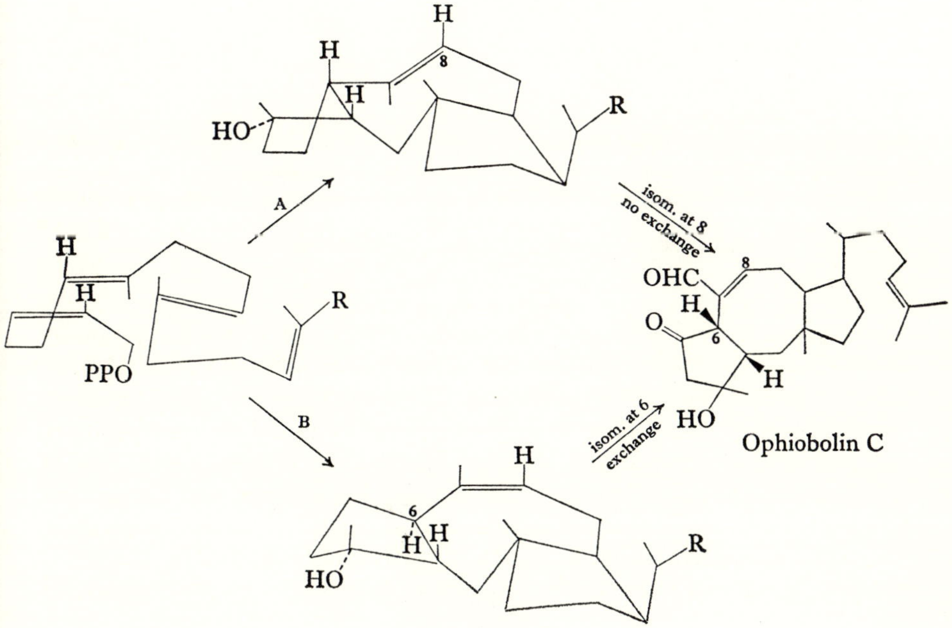
H
H
8
H
HO
R
A
isom. at 8
no exchange
H
H
R
PPO
OHC
8
H
O
6
H
HO
B
isom. at 6
exchange
Ophiobolin C
H
6
H
H
R
HO

FIG. 15

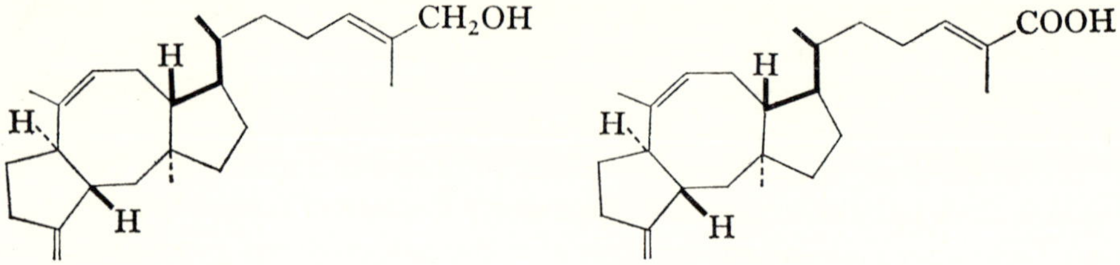

Ceroplastol
XXII

Ceroplasteric acid
XXIII

FIG. 16

and ceroplasteric acid (XXIII), produced by *Ceroplastes albolineatus* (Itaka *et al.*, 1968) are shown in Fig. 16.

It is interesting to note that from different strains of Helminthosporia the compounds reported in Fig. 17 (Nozoe *et al.*, 1968a; Hirai *et al.*, 1967; Arigoni, 1968) have been isolated. It is evident that these phytotoxins have a mixed biosynthetic origin; apparently a farnesol unit is linked to an aromatic system probably of acetic derivation.

There is, at present, no experimental evidence on the biosynthesis of these phytotoxins. However, mixed biosynthesis has been demonstrated

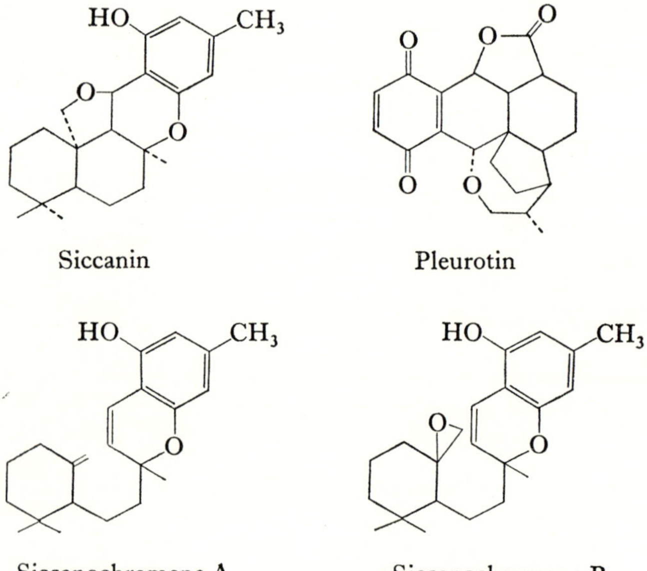

Siccanin

Pleurotin

Siccanochromene A

Siccanochromene B

FIG. 17

for important classes of natural products such as the ubiquinones and K vitamins. Furthermore, very recently we isolated a new compound containing a quinone moiety and from preliminary experiments we established that only 15 out of 30 carbon atoms had a terpene derivation.

From the same micro-organism which produces this quinone we isolated ophiobolins and also ergosterol. Apparently the fate of farnesol biosynthesized from the organism may involve either elongation to yield geranylfarnesol or condensation on an aromatic molecule. A third possibility would be head to head dimerization to squalene which is transformed into ergosterol.

These studies on the biosynthesis of certain phytotoxins have shown that many steps are involved in their formation. But little or nothing is yet known about the relationships between the chemical structure and the phytotoxic activity of these compounds. These relationships and the search for inhibitors of the biosynthesis of phytotoxins present an attractive field for future research.

References

Arigoni, D. (1965). Plenary Lecture at Autumn Meeting of the Chemical Society, Nottingham.

Arigoni, D. (1968). 5th International Symposium on the Chemistry of Natural Products, London, 8–13 July 1968.

Canonica, L., Fiecchi, A., Galli Kienle, M., and Scala, A. (1966a). *Tetrahedron Lett.* 1121.

Canonica, L., Fiecchi, A., Galli Kienle, M., Ranzi, B. M., and Scala, A. (1966b). *Tetrahedron Lett.* 3035.

Canonica, L., Fiecchi, A., Galli Kienle, M., Ranzi, B. M., Scala, A., Salvatori, T., and Pella, E. (1967a). *Tetrahedron Lett.* 3371.

Canonica, L., Fiecchi, A., Galli Kienle, M., Ranzi, B. M., and Scala, A. (1967b). *Tetrahedron Lett.* 4657.

L. Canonica, Fiecchi, A., Galli Kienle, M., Ranzi, B. M., and Scala, A. (1968). *Tetrahedron Lett.* 275.

Hirai, K., Nozoe, S., Tsuda, K., Itaka, Y., Ishibashi, K., and Shirasaka, M. (1967). *Tetrahedron Lett.* 23.

Itaka, Y., Watanabe, I., Harrison, I. T., and Harrison, S. (1968). *J. Am. chem. Soc.* **90**, 1092.

de Mayo, P., Spencer, E. Y., and White, R. W. (1961). *Can. J. Chem.* **39**, 1608.

de Mayo, P., Spencer, E. Y., and White, R. W. (1963). *Can. J. Chem.* **41**, 2996.

de Mayo, P., Robinson, J. R., Spencer, E. Y., and White, R. W. (1962). *Experientia* **18**, 359.

de Mayo, P. and Williams, R. E. (1965). *J. Am. chem. Soc.* **87**, 3275.

de Mayo, P., Williams, R. E., and Spencer, E. Y. (1965). *Can. J. Chem.* **43**, 1357.

Nozoe, S. and Morisaki, M. (1969). *Chem. Commun.* 1319.

Nozoe, S., Morisaki, M., Tsuda, K., Takahashi, N., Tamura, S., Ishibashi, K., and Shirasaka, M. (1965). *J. Am. chem. Soc.* **87**, 4968.

Nozoe, S., Suzuki, K. T., and Okuda, S. (1968a). *Tetrahedron Lett.* 3643.
Nozoe, S., Morisaki, M., Fukushima, K., and Okuda, S. (1968b). *Tetrahedron Lett.* 4457.
Orsenigo, M. (1957). *Phytopathol. Z.* **29**, 189.
Popják, G. and Cornforth, J. W. (1966). *Biochem. J.* **101**, 553.
Rios, T. and Perez, C. S. (1969). *Chem. Commun.* 214.
Tamura, S., Sakurai, A., Kainuma, K., and Takai, M. (1963). *Agr. biol. Chem., Tokyo* **27**, 738.

Discussion

Barbier: Is there any chance of finding side-chain substituted (methyl for instance) ophiobolins?

Canonica: This possibility cannot be excluded. The biosynthesis of isoprenoids is a highly specific process which apparently excludes the incorporation of carbon atoms units different from the one deriving from the decarboxylation and dehydration of mevalonic acid. On the other hand, as in the case of the side chain of sterols C-alkylation could occur after the cyclization. A similar process cannot be ruled out also in the case of ophiobolins.

Dimond: One is impressed with the similarity in the skeletal structures of ophiobolin and fusicoccin. Perhaps they are biosynthesized by similar routes. Ballio has pointed out that ophiobolin is found inside hyphae. But in contrast fusicoccin would seem to be transported in diseased plants and to be translocated when introduced into plants.

Canonica: The problem of the biosynthesis of fusicoccin is still open. We have no experimental evidence for it, it would well be a degradation product of ophiobolins but we cannot exclude that fusicoccin is derived from the cyclization of an open-chain intermediate with four isoprene units. Biogenetic schemes are often so unpredictable that we are not allowed to favour one hypothesis or the other.

Dimond: The different substituents on the skeleton that give to fusicoccin its solubility are a part of how it can pass out of mycelium and be transported in the plant. When these substituents are added during biosynthesis and when the aglycone becomes glycoside are also important parts of the picture. One suspects that ophiobolin would not escape from mycelium, but that it could be transported physically if it did.

Canonica: Up to now we have no basis on which to answer this question. The toxicity of these products is very high and we have to study the problem of the transport of very small amounts of toxins in the plant. Feeding the plant with labelled ophiobolins could give interesting but not conclusive informations. As a matter of fact finding radioactivity in the tissues could be attributed to the presence of inactive metabolites of ophiobolins which are produced in the fungus or in other tissues.

Goodman: Is it possible to conceive that phosphatidase produced by either the pathogen or the host plant (as a consequence of infection) would influence phospholipid in the membrane to an extent that would improve translocatability of ophiobolin ?

Canonica: It is possible but up to now we have no evidence on this argument, either.

Kuć: Has helminthosporal been isolated from infected plants at some distance from the fungus ?

Canonica: The analytical methods available at this time can probably only detect amounts of helminthosporal smaller than the toxic ones. On the other hand, to my knowledge no systematic research has been performed yet on this point.

Ward: Helminthosporal has not been isolated from infected tissue, and the evidence that it does occur there depends on the demonstration that browning of root tissue occurs when infection originates from inoculated amended soil. This is unsatisfactory because it is uncertain whether the helminthosporal is merely produced outside the plant and causes browning on contact with the roots.

11. Mechanism of Action of Antibiotics as Models for Studies on Phytotoxins

DAVID GOTTLIEB

Department of Plant Pathology,
University of Illinois,
Urbana, Illinois 61801, U.S.A.

Introduction

The deleterious effects of many toxins can best be understood if their mechanisms of action are first studied on a cellular basis. From such underlying alterations flow all other effects that together make up the syndromes of plant diseases. Relatively little work has been done on the effects of fungal or bacterial toxins on cells in higher plants and on their relationship to pathogenesis. The most intensive investigations of the cellular basis of toxic action have been done on the mechanism of action of antibiotics. These compounds affect the metabolic functions of microbes and many are also toxic to higher plants. Results of studies point to the cellular pathways that are vital and sensitive so that they are readily interrupted, thereby causing cellular death. These microbial systems can be used as models for similar research on the effects of toxin on higher plants. But to emphasize the role of toxins in disrupting basic metabolic pathways is not to deny that the ultimate injury and death of plants can be caused by morphological changes in tissues due to a pathogen as for example loss of roots or vessel plugging and the subsequent wilting of a host plant. Similarly, necrosis of leaves or the rupture of epidermal layers, as in heavy infections by rust fungi, can result in decreased photosynthesis. Such relatively non-specific effects would also include the production by a pathogen of a series of hydrolytic enzymes, such as the pectic enzymes which cause soft rots, by destroying the spatial continuity of cells and thus their tissue functions.

Site of Action

To determine the site in the cell where the lethal action of an antibiotic takes place is a complex process. In early studies, investigators sought a single site to explain mode of action of an antibiotic; however, it was often found that antibiotics could cause multiple effects, particularly in whole cells. These results should probably not have been surprising because the processes which occur in a cell are not independent of one another. One could predict, for example, that a compound which inhibited respiration would also decrease all energy-requiring cellular activities. Materials that alter the permeability of the cell membrane have indirect secondary effects, such as decreased rate (inhibition) of synthesis of cellular constituents because of the lack of exogenous nutrients and the loss of essential metabolites. Cycloheximide, an inhibitor of protein synthesis, for example, has been shown to decrease ribosome formation in mouse liver, not because it inhibits RNA synthesis, but probably because it inhibits synthesis of ribosomal proteins required for the formation of complete ribosomes (Trakatellis and Montjar, 1968). This antibiotic also inhibits light-induced carotenoid biosynthesis in *Neurospora crassa* Shear *et* Dodge, yet it is the light-induced synthesis of proteins required for synthesis of the carotenoid pigments that is disturbed (Harding and Mitchell, 1968; Raw *et al.*, 1968). Nystatin, an antibiotic whose primary mode of action appears to be the alteration of cell membranes, also inhibits the synthesis of an extracellular protease by *Aspergillus niger* van Tiegh. (Tomonaga, 1968). It has been suggested that some, as yet unknown, process in the hyphal tip of the fungus requires amino acids from an endogenous pool for induction of the protease and that nystatin depletes this pool by causing the amino acids to leak through the cell membrane. Lamb *et al.* (1968) reported that chloramphenicol and other antibiotics which inhibit protein synthesis in bacteria, but not cytoplasmic protein synthesis in eukaryotic cells, inhibited respiration in *Saccharomyces cerevisiae* Hansen. Inhibition was due to the absence of cytochromes a, a_3, b, and c in the yeast mitochondria. Since these cytochromes are thought to be synthesized in the yeast cytoplasm, the authors suggested that loss of respiratory activity was caused by an inhibition of synthesis of mitochondrial structural protein so that the cytochromes had no sites on which to be bound to the mitochondria.

The dramatic increase over the past few years in our understanding of the chemical events which occur within cells has provided tools to study mechanism of antibiotic action in a more precise fashion. Even with our current knowledge, however, it is often impossible to assign a single primary site of action to an antibiotic or to state with confidence the precise lethal effect of the antibiotic on the cell. The primary site of action of antimycin A, for example, would appear to be in the mitochondrial electron transport system, but certain bacteria are apparently inhibited by antimycin A by entirely different mechanisms. Another example of multiple activities involves the antibacterial compound streptolydigin. Siddhikol *et al.*

(1969) reported that this antibiotic is an inhibitor of the RNA polymerase from *Escherichia coli* (Mig.) Cast. *et* Chalm. Other workers found that this antibiotic, at concentrations about ten times higher than the concentration required to inhibit RNA synthesis, decreased respiration by uncoupling oxidative phosphorylation in rat liver mitochondria (Reusser, 1969). Since these experiments were done in separate *in vitro* systems, one must conclude that antimycin A and streptolydigin each have at least two distinct primary sites of action. Once again, one should not be surprised at these results for it is conceivable that two entirely different biological processes might contain components essential for their activities which have sufficiently similar structures or structural environments as to be capable of binding the same antibiotic. In such cases both processes might be inhibited even though they are functionally quite different. Another possible explanation for multiple sites of inhibition is that an antibiotic may have a sufficiently complex structure so that certain of its functional groups are involved in the inhibition of one process while other functional groups might be involved in the inhibitory action of a second completely unrelated process.

Specificity of Toxins

Toxins vary greatly in the width of their spectra of inhibition. Some inhibit the growth of a wide range of species whereas others inhibit only one or at most a few species. Though specificity is usually thought to be an innate property of the antibiotic-organism relationship, it is not absolute, but is often dependent on the concentration of antibiotic that is present. Nevertheless, the question to be asked is still the same. What metabolic mechanisms are responsible for the ability of one cell to withstand the presence of an antibiotic while other microbes are susceptible to it?

The permeability of a cell to an antibiotic is obviously of extreme importance for the sensitivity of an organism to a toxic compound. Thus the specificity of streptomycin to only two genera of fungi, among all those that have been tested, has been attributed to the ease with which this antibiotic enters their cells. The data of Vörös (1965) indicate that *Pythium* and *Phytophthora* species of the *Oomycetes* removed streptomycin from a medium two to eight times faster than did members of the *Mucorales*. Such phenomena may also be responsible for the resistance of some mutants that arise from sensitive cultures. *Pyricularia oryzae* Cav., which is usually very sensitive to Blasticidin S, gave rise to a mutant that was resistant to the antibiotic. This antibiotic inhibits the protein synthesizing system of the fungus. Yet, when cell-free protein synthesizing systems of the parent culture and the mutant were studied, there were no differences in their sensitivity to Blasticidin S which probably could not permeate the membrane of the mutant resistant strain but could in the parent strain.

In other cases, the metabolic mechanism that is inhibited in a sensitive species might be refractory in a resistant species. For example, *Corticium sasakii* (Shirai) Mats. is resistant to Blasticidin S but *Pyricularia oryzae* is sensitive. In this case, the cell-free protein synthesizing system of *C. sasakii* was not inhibited whereas that of *P. oryzae* was inhibited by the compound, indicating an innate resistance of the protein synthesizing system to Blasticidin S.

Another factor in resistance to toxins is the absence of a given metabolic system in resistant organisms and its presence in sensitive ones. Fungi usually conduct their aerobic respiration through a well organized mitochondrial system whereas most bacteria do not have this system. Antimycin A inhibits the terminal electron transport system of mitochondria and thus blocks growth and respiration of fungi but not of bacteria. It is obvious that antimycin A would inhibit only those micro-organisms with a mitochondrial terminal respiratory system. Inhibition of fungi but not bacteria by filipin is also in the same category. The sterols in fungal membranes are essential for some of their permeability characteristics; but bacteria do not have nor do they require sterols for their growth. Filipin reacts with the sterols in fungi, destroying the functional integrity of their membranes and causing cell death. Since bacterial membranes do not have or require sterols, filipin cannot prevent bacterial growth. The specificity of polyoxin to fungi and of penicillin to some bacteria can also be understood in this manner. Bacteria have mucopeptides in their cell walls whereas fungal walls contain chitin or cellulose but not mucopeptides (Endo and Misato, 1969; Gottlieb and Shaw, 1967). Penicillin inhibits only the synthesis of mucopeptides and thus only the growth of Gram positive bacteria. Polyoxin, on the other hand, inhibits the incorporation of glucosamine into chitin, preventing cell wall formation and growth in fungi with cell walls containing chitin. It cannot inhibit bacteria since they need no chitin for their cell walls (Endo and Misato, 1969).

A fourth general mechanism responsible for the resistance of some species is their ability to alter enzymatically an otherwise toxic compound. For example, many staphylococci that are resistant to penicillin produce penicillinase which hydrolyses the antibiotic and inactivates it (Schnitzer and Grunberg, 1957). Such mechanisms have also been found in *E. coli*. In this case, the inactivation was initiated by a "chloramphenicol reductase" which reduced the nitro group in chloramphenicol.

Finally, one must allow for the role of alternate pathways to account for the resistance of some microbes even though there are no well documented studies to support this concept. Yeast, for example, can respire and derive energy from either an aerobic or anaerobic pathway. The aerobic pathway could be blocked by antimycin A yet the yeast might still derive some energy from its anaerobic pathway and growth would occur, though less efficiently. The role of alternative pathways is also indicated by the fact that sometimes an antibiotic inhibits respiration incompletely, suggesting that another respiratory system which is not sensitive to the antibiotic can carry on the process.

Respiration

Respiration is the *sine qua non* for life but unfortunately can be prevented by a wide variety of toxic materials. This process is important both in the production of energy for all synthetic activities of organism and in the formation of various intermediates for general cell syntheses. The hexose monophosphate shunt could be a source for the ribose in nucleic acid. The Embden-Meyerhof-Parnas pathway (EMP) could give the acetate for the glyoxylic acid pathway. The Krebs cycle is a source of aspartic and glutamic acid precursors, and many others. Yet there are no microbially produced toxins that have been unequivocally demonstrated to prevent cellular growth by acting on any of the steps for the degradation of hexoses by the EMP or hexose monophosphate shunt mechanisms. This situation might be ascribed to the general versatility of the cell in using one or another of these systems or modifications of them in the degradation of sugars. Interference with the citric acid cycle as a prime site for toxic action is also unusual.

In contrast to the relative scarcity of metabolites that are toxic to the preceding pathways are the number that inhibit the terminal electron transport system. This mitochondrial system seems very similar in higher plants, animals, and fungi and current concepts of it are summarized in Fig. 1. Different antibiotics interfere at different stages in the transfer of electrons down this pathway so that energy is not generated at or beyond the stage affected. Oxygen consumption is inhibited because oxygen is the terminal electron acceptor and is not utilized if the flow of electrons is prevented. Antimycin A has received the greatest attention as an inhibitor of electron transport. It interferes with both succinic and NADH oxidases and also with the respective cytochrome c reductases. Coenzyme Q oxidase is also inhibited, but reduced cytochrome oxidase is not affected by the antibiotic. When heart mitochondria were fractionated into four components, antimycin A inhibited complex III between cytochrome b and c_1 (Fig. 1). Pyrrolnitrin and flavensomycin, on the other hand, inhibited electron transfer even before the antimycin-sensitive site, somewhere between NADH or succinate and CoQ, preventing the reduction of enzyme CoQ (Gottlieb and Inoue, 1967; Inoue and Gottlieb, 1967; Tripathi and

Gottlieb, 1969). Flavensomycin, in addition, inhibited the transfer of electrons in enzyme systems that have nucleotide electron transfer intermediates or donors.

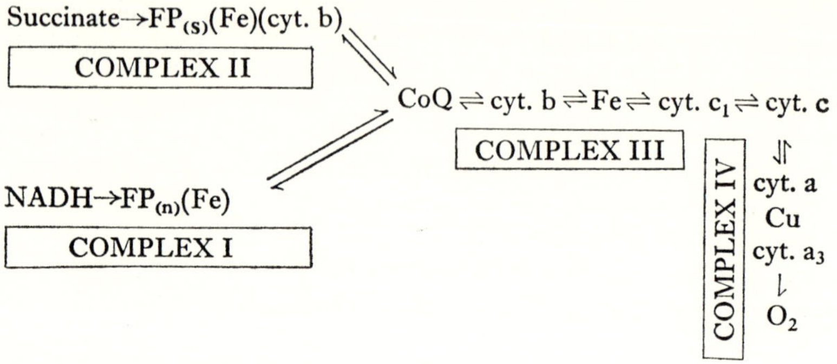

FIG. 1. Terminal electron transport system with its four distinct complexes and loosely bound cytochrome c.

Energy Transfer

The energy transfer mechanisms of the mitochondria also present an extremely sensitive system with which microbial metabolites interfere. This mechanism involves the formation of adenosine triphosphate (ATP) from adenosine diphosphate as well as a direct energy transfer for other work functions of the mitochondria, such as in ion transfer and mitochondrial swelling. Current concepts postulate the involvement of a number of hypothetical and presumably the same intermediates ($A \sim I$, $X \sim I$, and $X \sim P$) that successively transfer energy from their origin in the different steps of the terminal electron transport system to the formation of a high energy phosphate bond. The phosphorylated intermediate can be used for its energy directly in the work functions or to form ATP. It is from ATP that energy is derived for most cell synthetic functions. The oligomycin antibiotics inhibit ATP-driven processes and presumably act between $X \sim P$ and ATP, whereas dinitrophenol which uncouples oxidative phosphorylation acts at a site prior to that intermediate (Lardy *et al.*, 1958; Lardy and McMurray, 1959; Lardy *et al.*, 1965; Shaw, 1967).

Protein Synthesis

Protein synthesis can be blocked by a large number of microbial toxins, indicating its extreme vulnerability to toxins that might be produced by a pathogen in the host plant. The different steps in the synthesis are so specific and essential that inhibiting any one of them prevents protein formation or alters it, thereby inhibiting cellular activity. Five steps can be isolated in the process and are potential sites of attack. The first is the activation of the amino acids with ATP in the presence of a specific synthetase for each amino acid to form an aminoacyl-AMP-enzyme conjugate with the release of inorganic phosphorus. The second stage involves the same synthetase but in this event it catalyses the movement of the aminoacyl portion to transfer RNA (tRNA), releasing AMP. The third step is the formation of a complex between aminoacyl-tRNA, messenger RNA (mRNA) and ribosome. In the fourth, the different aminoacyl-tRNA molecules are transferred to the ribosome, in a sequence that is dictated by the code in the mRNA. Finally, the growth of the chain of amino acids is terminated as determined by the code. The cytoplasmic ribosomes in fungi (80S) consist of two subunits; the smaller unit (40S) seems to be attached to the mRNA and the larger unit (60S) receives the amino acids to form the peptide chain. The mitochondria of fungi and chloroplasts of green plants have another and apparently independent protein synthesizing system in which the ribosomes resemble those in bacteria rather than those of the cytoplasm of the cells.

The first two steps in protein synthesis seem relatively insensitive to agents that inhibit protein synthesis. Blasticidin S is an exception for it has been reported to prevent completely the activation step and only slightly affects the formation of the aminoacyl-tRNA. But even in this instance, total inhibition of activation required higher concentrations of antibiotic than were needed to prevent the transfer reaction from the aminoacyl-tRNA to protein (Huang *et al.*, 1964a; 1964b). Cycloheximide is still somewhat of a puzzle; the best data indicate that protein inhibition could take place because it inhibits polysome breakdown (Wettstein *et al.*, 1964). The polysomes are assembled when the ribosomes make their initial attachment to mRNA and then move along the message. Cycloheximide could prevent the movement of the ribosomes from that point of attachment thereby preventing the turnover of ribosomes into polysomes (Siegel and Sisler, 1964a, b; 1965). A large number of materials will prevent the transfer reaction. Gougerotin acts as a non-functional analogue which is structurally similar to aminoacyl-tRNA and probably therefore competitively competes with it at an enzyme site. It could inhibit movement along mRNA by blocking the release of free ribosomes from the complex (Casjens and Morris, 1965; Clark and Chang, 1965). Chloramphenicol attaches to the 50s subunit of the ribosome and there inhibits peptide bond formation (Gottlieb and Shaw, 1967; Nathans, 1964; Vazquez, 1964). Puromycin inhibits protein synthesis because it is structurally related to tRNA and acts as a functional inhibitor of tRNA. It takes the place of tRNA in the formation of the normal peptide chain. The inclusion of puromycin causes the premature release of the growing chain thus forming small peptides containing puromycin at their termini (Nathans, 1967). In contrast, tenuazonic acid inhibits the release of newly formed proteins from the ribosomes and thus inhibits further synthesis (Shiguera, 1967).

Nucleic Acids

All cellular activity depends on efficient functioning of the information code that is contained in and passed on by deoxyribonucleic acid (DNA) to messenger ribonucleic acid (mRNA). Blocking the replication of DNA or the synthesis of RNA soon causes protein synthesis to cease and with it all other metabolic functions. For the synthesis of either of these ribonucleic acids a DNA template is necessary as well as soluble polymerases that are specific for each synthesis. DNA consists of two complementary strains which must separate during the synthesis of additional DNA and a new strand is synthesized complementary to one of the original strands. A similar separation of strands occurs during the synthesis of mRNA and the new mRNA is synthesized so that its base components are complementary to those in the parent DNA strand.

Many cellular toxins act to stabilize the double stranded structure of DNA, preventing separation of the strands and thus inhibiting DNA synthesis; among them are the anthracyclines, mitomycins, porphyromycins and others (DiMarco, 1967; Szybalski and Iyer, 1967). Toxic metabolites also inhibit only DNA by binding to primer DNA. The best known interference with the synthesis of RNA is that of actinomycin (Reich, 1966). This metabolite from various actinomycetes binds to DNA so strongly that it prevents separation of the two strands and thus the formation of more RNA. It is the 2-amino purines in the helical DNA that bind this antibiotic. Even in situations where RNA synthesis is primarily inhibited, as with chromomycin, olivomycin and mithramycin, the reduction of the DNA dependent RNA synthesis is caused by complex formation with DNA. Very different from the action of these DNA binding toxins is the action of sarkomycin, for this antibiotic prevents the action of the soluble enzyme, DNA polymerase.

The synthesis of RNA can also be inhibited independently of an effect on DNA. Lomofungin (Gottlieb and Nicolas, 1969), phytoactin (Lynch and Sisler, 1967) and griseofulvin (El-Nakeeb and Lampen, 1964; El-Nakeeb *et al.*, 1965; Huber and Gottlieb, 1968) are in this group but the exact mechanisms of their inhibition have not been established.

Structural Antagonists

Many microbial toxins are structurally closely related to normal essential intermediates of cells. Such compounds produced by pathogens could interfere with the host plant by taking the place of the normally produced compound in its metabolic systems. Toxic compounds of this type could even be incorporated into DNA, for example, to give a faulty code, or might just bind to an enzyme site more strongly than the natural compound and prevent a normal reaction from taking place. Psicofuranine, for example, is a nucleoside resembling adenosine; it prevents bacterial growth by inhibiting the amination of xanthosine-5′-phosphate to guanosine-5′-phosphate. The interference is due to the binding of psicofuranine to the xanthosine-5′-phosphate aminase so that the normal amination to the guanosine analogue does not occur (Hanka, 1960). Adding guanosine-containing compounds allows subsequent reactions which require guanosine to be carried on by the cell and growth again proceeds normally. Other nucleosidic antibiotics such as tubercidin are also toxic. These are phosphorylated in the cell and are incorporated into cell nucleotides and nucleic acids and thus might form analogues that are not physiologically active (Acs and Reich, 1967).

Membrane Permeability

The organization of a barrier for the separation of cytoplasm from the environment is essential to life. It is important for the selective movement of materials in and out of cells in proper balance and control. The functioning of such a cell membrane can be affected in two ways. The integrity of the membrane itself can be altered by toxins so that the selective permeability of the cell membrane is lost; materials then do not concentrate in the cell and essential metabolites leak out. The second type of action is different, for in this the toxic metabolite might attack or combine with the active functional elements of the membrane and prevent the movement of materials across the membrane. The polyene type of antibiotics such as filipin, nystatin, pimaricin and many others, are examples of metabolic toxins that seem to destroy the integrity of the membrane. Such compounds act only on sterol-containing membranes and prevent fungal growth by allowing metabolites to leak out of the cell (Gottlieb *et al.*, 1961; Gottlieb *et al.*, 1960). Even organisms that are insensitive can be made sensitive by allowing them to incorporate sterols (Feingold, 1965; Weber and Kinsky, 1965). The polyenes form complexes with sterols in aqueous environments such as would occur in the cell. A deficiency of sterols makes the membranes more fragile to pressure and temperature. In artificial membrane systems exposure to polyenes also results in leakage of metabolites from the cells. Resultant structural changes are also evident in the pitting that occurs in both artificial and red blood cell membranes (Kinsky *et al.*, 1966; Lampen *et al.*, 1959). The addition of exogenous sterols of the types predominating in organisms, such as cholesterol, ergosterol, stigmasterol, and sitosterols, offsets the deleterious effects of these toxins.

Another group of compounds acts primarily on the specific transport of ions and water across membranes, especially the mitochondrial membranes. The transport of ions and water is energy dependent and is associated with oxidative phosphorylation. A number of these compounds are peptides and cyclic peptides. Enniatin A and B from *Fusarium oxysporum* Schl., and valinomycin from *Streptomyces fulvissimus* (Jens.) Waks. *et* Henr. are of this type (Shemyakin *et al.*, 1967; Shemyakin *et al.*, 1965). They uncouple oxidative phosphorylation and stimulate an energy dependent uptake of potassium in which the energy is derived from some high energy intermediate normally involved in ATP synthesis (Hunter and Schwartz, 1967; McMurray and Begg, 1959). This type of compound is thought to act as a carrier for alkali metal cations into the lipid-rich mitochondrial membrane. The cations are then transferred to an ion pump that moves them into the mitochondria at the expense of energy.

Permeability changes in host plants have also been associated with pathogenesis in which the fungal pathogen produces a toxic metabolite. *Helminthosporium victoriae* Meehan *et* Murphy, for example, produces on susceptible oat cultivars a lesion that is caused by the action of the metabolite victorin. This toxin causes a loss of electrolytes from the host cells; there is also a decreased uptake of phosphate by tissues exposed to victorin (Wheeler and Black, 1963; Wheeler and Hanchey, 1968). *Ascochyta fabae* Speg. lesions are also believed to be caused by a fungal toxin that impairs the selective permeability of the host cells. The fungus produces in culture ascochitine which causes lesions in broadbean similar to those caused by the pathogen. Ascochitine affects a wide range of plants from fungi to higher plants but in all of them a common feature is its effect on permeability, the leaking of metabolites from the cells (Oku and Nakanishi, 1966). In *Rhoeo discolor* L'Hér., for example, there is a marked loss of pigments from the cells. Citrinin, another fungal metabolite, also causes the leakage of cell components (Betina and Barathrova, 1968). It is likely that many other materials are synthesized by pathogens which are pathogenic primarily because such compounds act on cell membranes.

Cell Wall

Another vital site that could be susceptible to the action of toxic metabolites is the host cell wall. Compounds which act on cell walls could inhibit cell wall synthesis, interfere with functional elements of the wall, or act directly on the structural components of the host's cell wall to destroy its integrity. The most definitive investigations on the inhibition of synthesis have been done with metabolites from fungi that inhibit the growth of bacteria. *Penicillium chrysogenum* Thom produces penicillin, an antibiotic that inhibits the growth of Gram positive bacteria (Flynn and Godzeski, 1967). Protoplasts are often produced as well as elongated instead of the normal type cells. Penicillin inhibits one of the last steps in the formation of cell wall. It interferes with the transpeptidase that is responsible for cross linking of the mucopeptides that then form the complete wall. Other toxic metabolites interfere at different steps in cell wall synthesis. Cycloserine, for example, prevents the racemization of L-alanine to D-alanine and the formation of the D-alanine dipeptide which is essential in one of the early stages of wall formation. Interference with fungal wall synthesis has also been documented. The polyoxin antibiotics prevent growth and cause the formation of bulbous areas on hyphae, as well as the malformation of germ tubes. They act by preventing normal cell wall synthesis and interfere with the incorporation of N-acetyl glucosamine from UDP-N-acetylglucosamine into chitin (Endo and Misato, 1969).

Once the cell wall of the plant has been formed, pathogens can destroy it by the production of various hydrolytic and lyase enzymes which dissolve such structural elements as pectic substances and cellulose. These activities of bacteria and fungi are well documented and have been reviewed sufficiently in recent years. One must bear in mind that enzymes other than hydrolases and lyases can be classed as toxins and directly cause pathogenesis, and any discussion of toxin and plant disease should include not only these toxins but all other enzymes produced by pathogens that destroy host tissue.

References

Acs, G. and Reich, E. (1967). *In* "Antibiotics I. Mechanism of Action" (D. Gottlieb and P. D. Shaw, Eds), pp. 494–498. Springer-Verlag, New York.

Betina, V. and Barathrova, H. (1968). *J. Antibiot.* (*Tokyo*) **21**, 628–629.

Caajens, S. R. and Morris, A. J. (1965). *Biochim. biophys. Acta* **108**, 677.

Clark, J. M. Jr. and Chang, A. Y. (1965). *J. biol. Chem.* **240**, 4734.

DiMarco, A. (1967). *In* "Antibiotics I. Mechanism of Action" (D. Gottlieb and P. D. Shaw, Eds), pp. 191–210. Springer-Verlag, New York.

El-Nakeeb, M. A. and Lampen, J. O. (1964). *Bacteriol. Proc.* **60**.

El-Nakeeb, M. A., McLellan, W. L., and Lampen, J. O. (1965). *J. Bact.* **89**, 557–563.

Endo, A. and Misato, T. (1969). *Biochem. biophys. Res. Commun.* **37**, 718–722.

Feingold, D. S. (1965). *Biochem. biophys. Res. Commun.* **19**, 261–267.

Flynn, E. H. and Godzeski, C. W. (1967). *In* "Antibiotics I. Mechanism of Action" (D. Gottlieb and P. D. Shaw, Eds), pp. 1–19. Springer-Verlag, New York.

Gottlieb, D. and Inoue, Y. (1967). *J. Bact.* **94**, 844–849.

Gottlieb, D. and Nicolas, G. (1969). *Appl. Microbiol.* **18**, 35–40.

Gottlieb, D. and Shaw, P. D. (Eds) (1967). "Antibiotics I. Mechanism of Action". Springer-Verlag, New York, pp. 785.

Gottlieb, D., Carter, H. E., Wu, L., and Sloneker, J. H. (1960). *Phytopathology* **50**, 594–603.

Gottlieb, D., Carter, H. E., Sloneker, J. H., Wu, L., and Gaudy, Elizabeth. (1961). *Phytopathology* **51**, 321–330.

Hanka, L. J. (1960). *J. Bact.* **80**, 30–36.

Harding, R. W. and Mitchell, H. K. (1968). *Archs Biochem. Biophys.* **128**, 814–818.

Huang, K. T., Misato, T., and Asuyama, H. (1964a). *J. Antibiot.* (*Japan*) **17A**, 65–70.

Huang, K. T., Misato, T., and Asuyama, H. (1964b). *J. Antibiot.* (*Japan*) **17A**, 71–74.

Huber, F. M. and Gottlieb, D. (1968). *Can. J. Microbiol.* **14**, 111–118.

Hunter, F. E., Jr. and Schwartz, Lois S. (1967). *In* "Antibiotics I. Mechanism of Action" (D. Gottlieb and P. D. Shaw, Eds), pp. 631–635. Springer-Verlag, New York.

Inoue, Y. and Gottlieb, D. (1967). "Antimicrobial Agents and Chemotherapy 1966", Springer-Verlag, New York, pp. 470–479.

Kinsky, S. C., Luse, Sarah A., and Van Deenen, L. L. M. (1966). *Federation Proc.* **25**, 1503–1510.

Lamb, Adrian H., Clark-Walker, G. D., and Linnane, A. W. (1968). *Biochim. biophys. Acta* **161**, 415–427.

Lampen, J. O., Morgan, E. R., Slocum, Adelaid, and Arnow, P. (1959). *J. Bact.* **78**, 282–289.

Lardy, H. A., Johnson, D., and McMurray, W. C. (1958). *Archs Biochem. Biophys.* **78**, 587–597.

Lardy, H. A. and McMurray, W. C. (1959). *Federation Proc.* **18**, 269.

Lardy, H. A., Witonsky, P., and Johnson, Diane. (1965). *Biochemistry* **4**, 552–554.

Lynch, J. P. and Sisler, H. D. (1967). *Phytopathology* **57**, 367–373.

McMurray, W. and Begg, R. W. (1959). *Archs Biochem. Biophys.* **84**, 546–548.

Nathans, D. (1964). *Proc. Nat. Acad. Sci. U.S.* **51**, 585.

Nathans, D. (1967). *In* "Antibiotics I. Mechanism of Action" (D. Gottlieb and P. D. Shaw, Eds), pp. 259–277. Springer-Verlag, New York.

Oku, H. and Nakanishi, T. (1966). *Phytopathol. Z.* **55**, 1014.

Raw, W., Lindemann, I., and Rau-Hund, A. (1968). *Planta* **80**, 309–316.

Reich, E. (1966). *In* "Symposia of the Society for General Microbiology, Vol. XVII." Cambridge University Press, p. 266.

Reusser, F. (1969). *J. Bact.* **100**, 1335–1341.

Schnitzer, R. J. and Grunberg, E. (1957). "Drug Resistance of Microorganisms." Academic Press, New York, 395 pp.

Shaw, P. D. (1967). *In* "Antibiotics I. Mechanism of Action" (D. Gottlieb and P. D. Shaw, Eds), pp. 585–610. Springer-Verlag, New York.

Shemyakin, M. M., Ovchinnikov, Y. A., Ivanov, V. T., and Evstratov, A. V. (1967). *Nature* **213**, 412–413.

Shemyakin, M. M., Vinogradova, E. I., Geigina, M. Y., Aldanova, N. A., Loginova, N. F., Ryabova, I. D., and Pavlenki, I. A. (1965). *Experientia* **21**, 548–552.
Shiguera, H. T. (1967). *In* "Antibiotics I. Mechanism of Action" (D. Gottlieb and P. D. Shaw, Eds), pp. 360–365. Springer-Verlag, New York.
Siddhikol, C., Erbstoeszen, J. W., and Weisblum, B. (1969). *J. Bact.* **99**, 151–155.
Siegel, M. R. and Sisler, H. D. (1964a). *Biochim. biophys. Acta* **87**, 70–82.
Siegel, M. R. and Sisler, H. D. (1964b). *Biochim. biophys. Acta* **87**, 83–89.
Siegel, M. R. and Sisler, H. D. (1965). *Biochim. biophys. Acta* **103**, 558–567.
Szybalski, W. and Iyer, V. N. (1967). *In* "Antibiotics I. Mechanism of Action" (D. Gottlieb and P. D. Shaw, Eds), pp. 211–237. Springer-Verlag, New York.
Tomonaga, G. (1968). *J. gen. appl. Microbiol.* (*Tokyo*) **14**, 163–170.
Trakatellis, A. C. and Montjar, M. J. (1968). *Life Sci.* **7**, 413–419.
Tripathi, R. K. and Gottlieb, D. (1969). *J. Bact.* **100**, 310–318.
Vazquez, D. (1964). *Biochem. biophys. Res. Commun.* **15**, 464.
Vörös, J. (1965). *Phytopathol. Z.* **54**, 249–257.
Weber, M. M. and Kinsky, S. C. (1965). *J. Bact.* **89**, 306–312.
Wettstein, F. O., Noll, H. and Penman, S. (1964). *Biochim. biophys. Acta* **87**, 525–528.
Wheeler, H. and Black, H. S. (1963). *Am. J. Bot.* **50**, 686–693.
Wheeler, H. and Hanchey, Penelope. (1968). *In* "Annual Review of Plant Pathology" (J. G. Horsfall, K. F. Baker, and D. C. Hildebrand, Eds), Vol. 6, pp. 331–350.

Discussion

Paxton: You have spoken mainly about inhibition of growth and respiration by metabolites. But generally in plant diseases there is an initial increase in respiration followed by a decrease. Would you care to comment on this initial stimulation?

Gottlieb: The initial stimulation in oxygen uptake in plant disease probably is caused by a compensatory reaction of the plant to injury. If the injury remains at a low level, repair or limitation of the fungus growth follows. This requires more energy and oxygen consumption. As the invasion by a parasite progresses, the injury becomes too great and the respiratory activity of the host must fall because of extensive damage to the host.

Goodman: I concur with Gottlieb's comments on the dual or "oligo" effect of toxins and other substances. For example, IAA inhibits growth at high concentrations but stimulates at low concentrations.

Pringle: In response to the request for an explanation for how a toxin can stimulate respiration at low concentrations and inhibit respiration at high concentrations, I submit the following as a working hypothesis. At low concentrations the toxin may disturb the metabolism of the host in any one of a number of ways. The host attempts to neutralize this disturbance by its homeostatic mechanism. This leads to stimulation of metabolism in general and to increased respiration. As the dose of the toxin is increased homeostasis would be overwhelmed. Things would go from bad to worse until the plant was killed.

van Andel: The inhibitory action of auxin on higher plants is attributed by some authors to the increase in ethylene production which high concentrations of auxin induce. This, however, merely transfers the problem to another chemical and does little to explain the original effects of high concentration.

Durbin: Why do not toxins of types described by Gottlieb affect the micro-organisms which synthesize them?

Gottlieb: We do not always know. It might be that the toxins are synthesized in localized areas and never are in contact with the cell structure or mechanism on which they might otherwise act. In other instances these mechanisms might be resistant. Then, too, the organism producing the toxin might not need the susceptible system for growth nor have it. For example, *Streptomyces* species which produce antimycins have no mitochondria with which the antibiotic could react. Similarly, streptomycetes have no membrane sterols with which filipin could react. On the other hand, we know that *Sclerotium rolfsii* produces oxalic acid which eventually lowers the pH of the culture to a point where the hydrogen ion concentration inhibits further growth of the fungus.

Marré: I think that a possible partial explanation of the fact that many fungal or bacterial toxins, active on higher plants, are not toxic to micro-organisms depends on the presence, in the former and not in the latter, of complex protein systems involved in hormonal control, differentiation etc. Such systems might present specific targets for toxin action.

Ballio: I refer to Durbin's question. Some cases of specificity in antibiotics action are quite well understood. I shall quote two examples. Penicillins only act on biosynthesis of bacterial cell walls which contain a peculiar type of component, namely the mucopeptides; the macrocyclic polyenes mentioned in Gottlieb's lecture interfere with sterols which are present in membranes of fungi but absent in membranes of bacteria.

Goodman: Do the antibiotics pimaricin and nystatin antagonize cholesterol incorporation in membranes in the same way as filipin does?

Gottlieb: Yes, all such polyenes studied act in this manner though some also have other toxic mechanisms.

Kuć: Can you cite examples where microbial products exert their toxic effect by decreasing membrane permeability?

Gottlieb: Not at the moment.

Mirocha: I would like to comment on the fact that gibberellic acid was first discovered in plants and that its activity on plants was the first to be thoroughly investigated. Only recently it has been shown to have uterotropic effects on rats and will affect weight gain in swine. Conversely, F-2 or zearalenone was first discovered to affect animals, causing vulvovaginitis, hyperestrogenism and weight gain. Now it has been shown to have an effect on plants also by increasing their growth rate.

Morpurgo: You have stated that auxins have an effect on and are perhaps essential for micro-organisms. This surprises me because so far as I know, no one has yet established this.

Gottlieb: No exogenous indole acetic acid or other auxins are usually necessary because the fungi can make their own. When an analogue of indoleacetic acid, trichlorophenoxyacetic acid is added, which is an antagonist of the hormone, the growth of *Diplodia natalensis* ceases. This competitive action is the overcome by adding exogenous indoleacetic acid.

Matta: In general terms, and also from what we have heard here in the past few days, it seems that there is no positive correlation between pathogenicity and phytotoxin production. It is more likely that the production of phytotoxic substances is inversely correlated with the ability of a micro-organism to act as a perfect parasite, i.e. to establish with plants highly compatible combinations.

Page: I believe that we are constrained in our thinking about micro-organisms in that we pay homage each day to principles established by Koch and Pasteur, that is, we work with pure culture of single organisms. Yet, in the soil and elsewhere organisms grow and function in concert with other organisms. For example in examining micro-organisms associated with the Black Knot disease which is caused by *Dibotryon morbosum* we found another fungus, *Fusarium oxysporum* which in association with a *Brevibacterium* sp. produced an antibiotic toxic to an unidentified fungus also resident in the Black Knot stroma. The *Fusarium* or *Brevibacterium* above did not produce the antibiotic effect.

When a fungus and a higher plant interact we have a mixed culture complex. Interactions may result in disease which is conceived as an unusual response—perhaps because we are accustomed to thinking in terms of single organisms. It may be somewhat teleological then to regard the production of toxins by a fungus as an example of a special response. Instead, we should think of organisms co-existing and interacting as the normal state of affairs.

Meyer: I should like to raise the question as to what extent toxins are produced independently of the medium in which micro-organisms are grown, and how far production of toxins depends on constituents in the medium and conditions of culture?

Gottlieb: The function of toxins to microbes that produce them is unknown. Yet each might be necessary to the organism that produces it—but

8

in amounts so small that it is usually undetected. When placed in certain cultural conditions the microbe produces an excess over its need—let us say as a catalyst or hormone and it is only then that we detect them.

Durbin: For *Pseudomonas tabaci* and *P. angulatum*, there is no experimental evidence that toxin synthesis enhances the pathogen's ability to multiply in the host or exist as a soil saprophyte. For other *Pseudomonas* spp. there is a correlation between toxin production and multiplication. However, I think it remains to be shown that toxins enhance survival of an organism over long periods.

Backman: In relation to Durbin's comments, we find that *P. syringae* isolates that produce toxin have markedly wider host range than isolates that do not. We also find that the isolates most pathogenic on peach are toxin producers.

Rudolph: We have tested isolates of *P. phaseolicola* which do not produce toxin *in vitro*. Bean leaves infected with these isolates do not develop chlorotic haloes but only greasy looking translucent spots. These spots dry out after a few days. This suggests that bacterial multiplication is dependent on toxin production.

Marré: I wonder whether there are experiments that rule out the possibility that cholesterol reverses filipin action by simply forming an inactive complex with it. I would also like to ask about the possibility that filipin might act on some step of phenol biosynthesis, rather than directly on the membrane?

Gottlieb: It is possible that cholesterol forms an inactive complex with filipin since we have done no experiments that definitely rule out such a mechanism. I do not believe that the antibiotic acts on a step in sterol synthesis since much of our work has been under conditions in which synthesis is not taking place in whole cells and nevertheless filipin caused leakage.

Main: Cell free or pure enzyme systems are utilized to determine the site of action of a toxin against test organisms. The micro-organism that produced the toxin is usually insensitive to its own toxin. Once the site is pin-pointed, could you not test the toxin against its producer using the cell free approach? This might yield information on the nature of selectivity, e.g. is the producer insensitive because of permeability, detoxification, or other factors or does it lack the specific receptor site. This information could provide clues to the nature of selectivity of other micro-organisms to the toxin.

Gottlieb: I agree. If similar mechanisms are present in the producing organisms as are in the microbes which are inhibited this would be a good type of experiment.

12. Phytotoxins and Metabolism

CHESTER J. MIROCHA

Department of Plant Pathology, University of Minnesota, St. Paul, Minnesota 55101, U.S.A.

Introduction

Plants respond to infection in a variety of ways such as by the production of necrotic zones, chlorotic haloes, wilting, and disintegration of tissue. More fundamentally, the host undergoes an alteration in metabolism which can be demonstrated in synthesis of nucleic acids, change in nitrogen metabolism, change in respiratory ratios, more rapid CO_2 utilization, increase in glycolysis and turnover of the tricarboxylic acids. The living system is in a dynamic state but one which is balanced, and by altering one enzymatic sequence we disturb another and so on. For most of these metabolic responses, it is nearly impossible to demonstrate that a particular pathogen-produced toxin is directly involved in the stimulation of any one system although we can describe some of the more obvious effects associated with its application. Most often, the toxin that is used accounts for only one of the many signs associated with a disease, such as halo induction, and one is tempted to over-extend the involvement of a toxin in explaining pathogenesis. Perhaps the closest we have come to explaining pathogenesis by the activity of a single toxin is through experiments on oats with the toxin of *Helminthosporium victoriae* Meehan *et* Murphy. Even here, some plants other than oats are susceptible to certain isolates of this fungus, but are not affected by the toxin. This may mean that certain of the pathogen races carry genes that regulate pathogenicity but which are not operative or are masked when infecting oats.

In discussing the effect of plant toxins on disease, I shall use the following examples: (1) fumaric acid and the hull rot disease of almond, (2) the toxin of *H. victoriae* and (3) the toxin of *H. carbonum* Ullstr. These examples were chosen because the toxins associated with the diseases are either directly or indirectly involved in CO_2 fixation by the pathogen or the host and have an effect on the TCA cycle. Further, fumaric acid serves as an example of a toxin that is simple in structure, a common constituent of host metabolism and a secondary determinant of plant disease whereas the *Helminthosporium* toxins represent the other extreme and are examples of metabolites which are chemically complex, usually thought of as being foreign to host metabolism and are primary determinants of plant disease.

One of the objectives of this paper is to consider the aforementioned diseases from the standpoint of theoretical plant pathology, i.e. to take the known facts of the toxicology of these toxins and interweave them into well established metabolic pathways of plants and to propose a new metabolic pathway in diseased tissue.

Wood (1967) cautions that it is best to define toxins in terms of the effects they produce rather than on the basis of their origin or nature. He further states that the term toxins should not be limited solely to compounds which cause damage or kill protoplasts only at high dilutions. Toxins, then, could be normal host constituents that have been produced in pathogenesis to toxic amounts. Let us first consider toxicity by normal constituents of plants.

Hull Rot Disease of Almond

Symptoms and Development of Disease

The hull rot disease of almonds is most commonly found in the Sacramento and San Joaquin valleys of California, and is most often associated with infection by species of *Rhizopus*. Three symptoms are associated with the disease, rotting of hulls, leaf blighting and killing of twigs. Leaves of a cluster nearest the rotted hull curl up and later become brittle; some develop unilateral necrotic areas in the blades and veins. The blighted leaves do not abscise but remain attached to the spur. Brown to black streaks develop in the vascular tissue along the same side of the twig which bore the infected fruit. These spurs and twigs die later. In diseased twigs, gum pockets form near the vascular cambium and later enlarge into gum ducts between the rays of the xylem.

On a shoot with a rotted fruit at its base, necrosis occurs in leaves that arise from the same side of the shoot as the diseased fruit. Moreover, these leaves develop necrotic areas only along one side of the midvein. The unilateral distribution of the disease symptoms suggests the presence of a translocated toxin. Twigs that bear a number of infected fruits, however, may not show such unilateral symptoms.

Frequent isolations and examinations of tissue sections showed that the fungi do not invade the tissue of the adjoining twig suggesting the induction of twig and leaf symptoms by means other than direct invasion by fungi. A toxin is produced in the rotted hulls and transported to the twigs and leaves (Mirocha and Wilson, 1961).

Toxicology of the Disease

The toxin responsible for the symptoms described was found to be an anion and after further investigation identified as fumaric acid (Mirocha *et al.*, 1961). Tests were made of the toxicity to almond shoots of 0·05 M solutions of potassium fumarate (pH 4·1) and sodium succinate (pH 7·5). The latter was used as a control along with water buffered to the same pH as the fumarate. Current-year shoots (20–25 cm long) on Nonpareil almond seedlings were inserted into flasks containing the solutions. Within 21 h leaves on shoots treated with potassium fumarate showed a marked curling and twisting and a dark-brown discolouration of their mid and lateral veins. Necrosis also was visible in some leaf petioles. The plants treated with sodium succinate showed no symptoms and did not differ from controls in water. The symptoms caused by potassium fumarate resembled those of naturally blighted shoots.

Translocation of Toxin to Site of Action

Application of ^{14}C-labelled fumaric acid to the mesocarp of the dehiscing fruit resulted in rapid translocation of the ^{14}C fumaric acid into the shoot within 4 h. The distribution of the ^{14}C-fumarate in the leaves was unilateral, i.e., found in leaves on the same side as the treated fruit identical to that of symptoms occurring in natural infection. Symptom development in other treated shoots was evident after 2 days.

Copious amounts of fumaric acid (2300 ppm) were found in almond hulls rotted by *Rhizopus* sp. on orchard trees. After application of the metabolite to the mesocarp of fruit on trees, high concentrations (6090 ppm) were found in the leaves soon after application but before development of symptoms. After these had developed, only small to trace amounts of fumaric acid could be detected in the leaves.

Effect on TCA Cycle

Fumaric acid. As stated earlier, when ^{14}C-2,3-fumaric acid was applied to the mesocarp of fruit on orchard trees, an appreciable amount of radioactive fumarate was found in the leaves before symptom development. There was a considerable turnover of radioactivity from fumarate into malic, citric and tartaric acids. When ^{14}C-2,3-fumaric acid was applied to the mesocarp of the fruit of orchard trees and analyses not made until after symptom development, no fumaric acid was detected. Again, radioactivity was detected in citric, malic and tartaric acids. This supports the hypothesis that either fumaric acid is catabolized as a result of its toxic action or its metabolic derivatives are toxic.

In these experiments, the amount of incorporation of radiocarbon into malic and citric acids was nearly the same and about 2·5 times more than that incorporated into tartaric acid. Normally the equilibrium between malate and citrate as mediated by malate dehydrogenase is favoured in the direction of malic acid synthesis but this equilibrium is displaced through the continuous removal of oxalacetate in the synthesis of citric acid as mediated by the condensing enzyme. Conversion of fumarate to citrate is not surprising in that citrate has a central, regulating role in metabolism. Besides mediating synthesis in the TCA and glyoxylate cycle, citric acid is required for fatty acid synthesis, is a source of reducing power and is also an activator for acetyl CoA carboxylase, as well as a source of acetyl group generation. Synthesis of tartaric acid from fumarate in almond is more obscure as little is known about the biosynthesis of tartrate in higher plants. However, recent investigations by Saito and Kasai (1969) indicated that the C_1–C_2 moiety of L-ascorbic acid is an effective precursor of tartaric acid. Sucrose-U-^{14}C will yield uniformly labelled tartaric acid in the grape berry although the conversion into tartrate is not as high as with L-ascorbic acid. Similar studies in grapes show that when D-glucurono-lactone-6-^{14}C as well as L-ascorbic acid-1-^{14}C are used as precursors for the biosynthesis of tartaric acid, most of the radioactivity is localized in the carboxyl groups. Although tartaric acid may possibly be synthesized from ascorbic acid or sucrose in the almond hull rot syndrome, the data appear to favour another hypothesis, more directly involved with fumaric acid and the TCA cycle, which will be discussed later.

Martin and Foster (1955) reported the production of *trans*-L-epoxysuccinic acid by *Aspergillus fumigatus* Fres. and its biological conversion to *meso*-tartaric acid. They suggested malate as the precursor. In view of the data obtained from almond showing radioactivity in malic and tartaric acids, the formation of epoxysuccinic acid from fumaric acid seems feasible. Although the biological formation of epoxysuccinic acid from malic acid is chemically feasible, the production of epoxysuccinic acid from fumaric acid is more probable, because of the difficulty involved in removing a hydride ion (a strong bond) from malic acid in the formation of the epoxide. Wilkoff and Martin (1963) reached this same conclusion, and demonstrated that in *A. fumigatus* fumarate is the precursor. Other studies have shown that epoxysuccinic acid is converted to *meso*-tartaric acid, which in turn may be converted to oxalacetic acid (Martin and Foster, 1955; Shilo, 1957).

Cis- *and* Trans-*Epoxysuccinic Acid*

Although in biological systems, only the *trans* form of epoxysuccinate has been reported, in the laboratory, the *cis* form also exists and can be synthesized using normal techniques of organic chemistry. The *trans* isomer was prepared by treating an halohydrin of fumaric acid with NaOH while the *cis* isomer was prepared by the oxidation of hydroquinone with alkaline hydrogen peroxide. The two isomers differ from each other in reaction products in that the *trans* isomer will yield *meso*-tartaric acid after enzymatic or chemical reduction whereas under similar circumstances the *cis* form yields DL-tartaric acid.

The toxicity of both the *cis* and *trans* isomer of 2,3-epoxysuccinate was tested on young shoots of Nonpareil seedlings. Both isomers were toxic and incited the same symptoms as did fumaric acid but the symptoms were more severe and appeared sooner than with fumarate. Of the three acids, (fumarate, *cis* and *trans* isomers of epoxysuccinate) *cis*-epoxysuccinate induced the most severe toxicity, characterized by necrosis (brown discolouration) in the vascular tissue. The treated plants were also more odoriferous than the control plants.

The young almond shoots treated with the *cis* and *trans* isomer of epoxysuccinate were compared with shoots treated similarly with fumaric acid. The leaves were harvested and extracted after symptoms had developed. No tartaric acid was found in shoots used as control whereas the fumaric acid treated shoots had substantial amounts of DL-tartaric acid. No DL-tartaric acid nor the *meso* isomer was found in the *trans*-epoxysuccinate treated shoots. One would expect the *meso* isomer of tartarate to form from the *trans* isomer of epoxysuccinate but *meso*-tartaric acid is not stable in a biological system; it is oxidized rapidly and is therefore difficult to isolate. As expected, DL-tartarate was formed from *cis* epoxysuccinate, thus proving the feasibility of this metabolic route in almond.

The lability of *meso*-tartrate in a biological system is partly demonstrated by the fact that the source of *trans*-epoxysuccinate used in this test had a small amount of *meso*-tartrate present due to non-enzymatic degradation. The amount was sufficient to detect easily by paper and gas–liquid chromatography. None could be found in the extracts from treated plants. On the other hand, a small amount of DL-tartaric acid was also present as a non-enzymatic breakdown product of the *cis*-isomer of epoxysuccinate. A copious amount of the DL-stereoisomer was found in the extract from plants treated with the *cis*-isomer.

Biosynthesis of Fumaric Acid and Epoxysuccinic Acids

I have described above how the products of fungal metabolism i.e. fumaric and epoxysuccinic acids, are metabolized by or affect the host. Let me now consider the biosynthesis of these toxins by the fungus. This is important because certain nutritional requirements are necessary and perhaps unique for their synthesis which in this case involves CO_2 fixation, although the synthesis of fumaric acid is directly related to epoxysuccinic acid. I shall treat them separately and start with fumarate.

Species of *Rhizopus* have long been known to be capable of producing copious amounts of fumaric acid. The process of production has been patented (Waksman, 1943), and used industrially. For the purpose of this discussion it is more important to substantiate that hulls (mesocarp) of almond serve as an excellent source of carbon for the production of fumaric acid. This was demonstrated by Mirocha and DeVay (1961), by growing *R. arrhizus* Fisch., *R. stolonifer* (Ehrenb.) Lind and *R. circinans* v. Tiegh. on almond hulls with no added carbon source. Fumaric acid was detected in the medium within 24 h.

Foster *et al.* (1949) postulated that the high yields of fumaric acid in *Rhizopus* cultures could not be explained by operation of the TCA cycle. They favoured the hypothesis that fumarate arose via succinate formed by a Thunberg–Wieland 2-C_2 condensation. Foster *et al.* (1949) further concluded that CO_2 is a negligible factor in the synthesis of fumarate from alcohol. Foster and Carson (1950) postulated a net formation of succinate via the Thunberg reaction from two C_2-units derived from ethanol, coupled with exchange of the carboxyl groups formed with CO_2. Kornberg (1959) pointed out that the ^{14}C labelling pattern of fumaric acid obtained by Foster *et al.* (1949), (^{14}C incorporation predominantly in both methene carbons and some in the carboxyl groups), was consistent with the operation of the glyoxylate cycle, not considered previously by Foster. Kornberg (1959), resolved the labelling pattern by assuming that fumarate was the result of the metabolism of acetyl coenzyme A derived from ethanol via both the TCA and glyoxylate cycles (Fig. 1). For example, the cycling of 2-^{14}C-ethanol exclusively via the TCA cycle would result in isotope incorporation into the methene carbons of fumarate and some into the carboxyl groups after about five turns of the cycle. The methene carbons would have the same specific activity as the 2-carbon of ethanol but the carboxyl carbons would have half this activity. If the glyoxylate cycle were used exclusively, isotope from 2-^{14}C ethanol would be incorporated only into the methene carbons. If both pathways were involved simultaneously, then the result would be an isotope distribution in fumarate that was part way between that expected from the sole operation of the two cycles as observed by Foster.

Wegener and Romano (1964) found that the enzyme isocitritase (isocitrate lyase) of the glyoxylate cycle was repressed by glucose, yet fumarate synthesis was abundant in a medium containing high glucose (Fig. 1). They found that phosphoenolpyruvate also inhibits the key enzyme isocitritase. Romano *et al.* (1967), working with *Rhizopus* demonstrated that in the presence of glucose and aerobic conditions, fumarate production takes place through a CO_2 fixation mechanism through a C_3 compound such as pyruvic acid, phosphoenolpyruvic acid or possibly the malic enzyme.

In conclusion, the findings of Romano *et al.* (1967), most adequately describe fumarate synthesis by *Rhizopus*. In the presence of glucose, *Rhizopus* produce fumarate via a CO_2 fixation mechanism involving a 3 carbon metabolite. When this is nearly exhausted, the fungus then synthesizes isocitritase and can use for example, ethanol via the glyoxylate

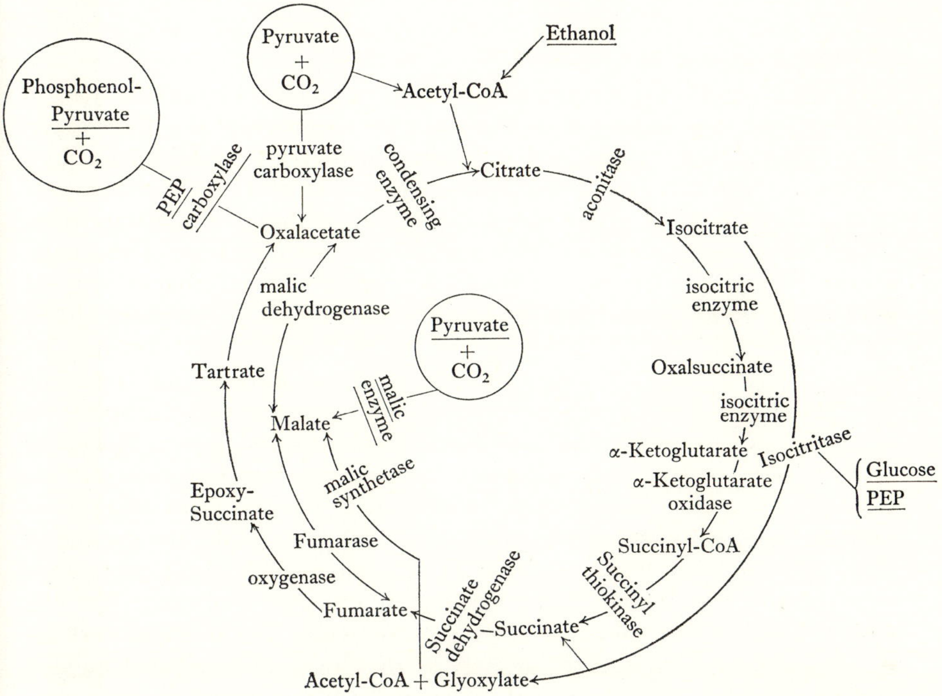

FIG. 1. Integrated metabolic routes comprising TCA, glyoxylate, epoxysuccinate cycles and CO_2 fixation mechanisms.

cycle for fumarate synthesis. Ethanol is beyond that point in metabolism where inhibition as obtained with glucose and phosphoenolpyruvate can occur.

Let me now proceed to the biosynthesis of epoxysuccinate by fungi. Evidence for the biosynthesis of epoxysuccinate by fungi is ample. For example, Martin and Foster (1955) found that *trans*-L-epoxysuccinic acid accumulated as a product of carbohydrate metabolism of *A. fumigatus*. Earlier, Sakaguchi *et al.* (1939) reported that *Paecilomyces varioti* Bain. (=*Monilia formosa* Sak., Inoue *et* Tada) and *Penicillium viniferum* Sak., Inoue *et* Tada produced the epoxy acid. Nomura *et al.* (1954) reported that

P. varioti (*M. formosa*) synthesized ethylene oxide-β-dicarboxylic acid from CO_2 and a 3 carbon acid. Recently Packter (1969) reported the biosynthesis of epoxysuccinate by *Lentinus degener* (Kalchr.).

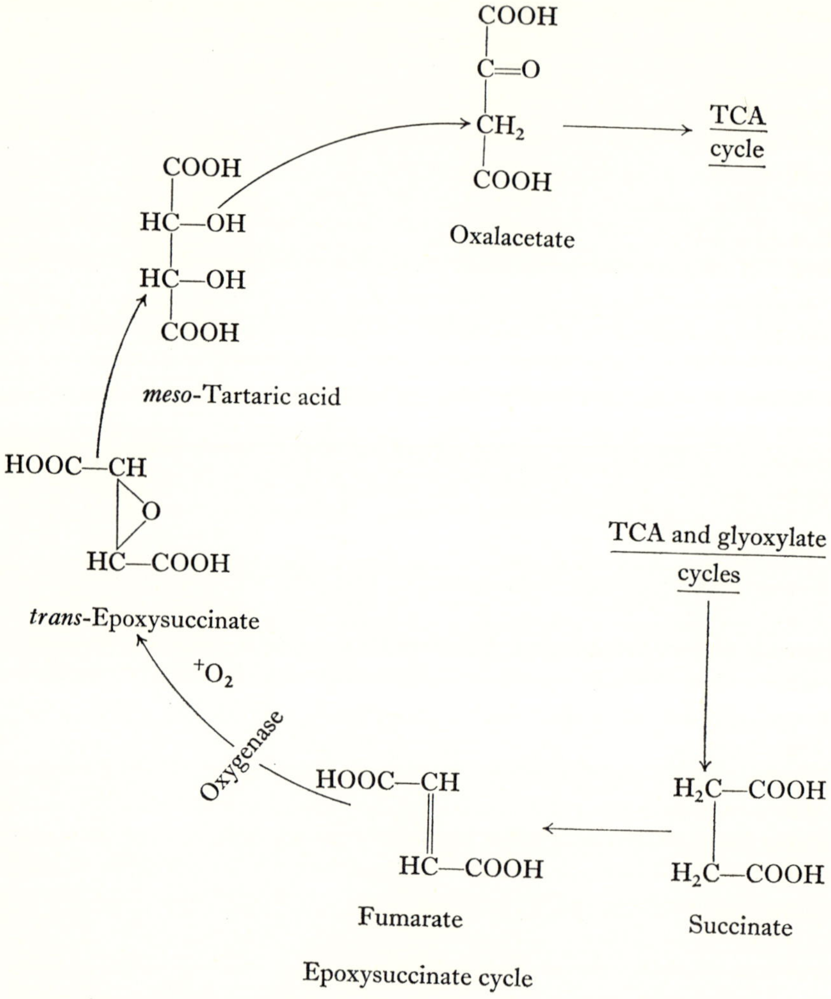

FIG. 2. Epoxysuccinate cycle as proposed by J. W. Foster (1958).

Foster (1958) postulated that the accumulation of epoxysuccinate indicates that the acid is formed as an intermediate in a new modification of a respiratory cycle. Strong evidence for this was obtained when it was found by Martin and Foster (1955) that *A. fumigatus* can metabolize epoxysuccinate to *meso*-tartaric acid. This is apparently the first report of the biological occurrence of *meso*-tartaric acid. The *meso* tautomer of tartaric acid does not accumulate in these cultures indicating its rapid utilization. Apparently, and as is known from bacterial and animal systems, *meso*-tartaric acid is rapidly oxidized. In mitochondria from animals, *meso*-tartrate is oxidized to oxaloglycolic acid (Shilo, 1957). The complete cycle as postulated is shown in Fig. 2.

Nomura *et al.* (1954) observed that the carboxyl groups of *trans*-epoxysuccinic acid produced in the presence of $^{14}CO_2$, have the same specific activity as carbons 1 and 6 of the citric acid produced in the same system. They postulated that oxalacetic acid is a common precursor for both acids. Aida and Foster (1962), demonstrated incorporation of atmospheric ^{18}O into epoxysuccinic acid. Wilkoff and Martin (1963) carried this investigation further and found that the intact carbon skeleton of fumarate is converted to epoxysuccinate and that the epoxide oxygen is derived directly from molecular oxygen. An uncharacterized oxygenase is proposed as the mechanisn of *trans*-L-epoxysuccinic acid biosynthesis. Recently, Albright and Schroepfer (1970) found that fumarase, in addition to hydrating fumarate to yield malate, can hydrate L-*trans*-2,3-epoxysuccinate to yield *meso*-tartrate.

Conclusions

In the light of the information derived from the literature on the physiology and metabolism of fumarate, epoxysuccinate and *meso*-tartrate, the following theory accounting for fumarate toxicity and its subsequent metabolism in almond is proposed. It centres around the epoxysuccinate cycle of metabolism originally introduced by Foster (1958) and shown in Fig. 2.

The fumarate is produced by *Rhizopus* spp. via a CO_2 fixation mechanism in the presence of glucose or its equivalent under aerobic condition. This fumaric acid, synthesized in the mesocarp, is translocated to the stems and leaves, the site of its toxic action, either in the form of fumarate or epoxysuccinate or a mixture of both. The fumarate is metabolized to the *trans* form of epoxysuccinate which is toxic and then further metabolized to the *meso* form of tartaric acid. The latter does not accumulate readily in host tissue but rather is quickly metabolized into oxalacetic acid and then taken up into various pathways. Although *cis*-epoxysuccinate is also toxic, it is doubtful that it is involved in the natural syndrome because fumaric acid has a *trans* configuration and is a precursor of only the *trans* isomer.

Figure 1 integrates the epoxysuccinate cycle to fit harmoniously into the TCA and glyoxylate cycles as well as into the various CO_2 fixation mechanisms which may be operative in the synthesis of fumarate. The proposed theory is supported by the facts that: (1) *Rhizopus* spp. readily synthesize fumaric acid in the mesocarp of dehiscing almond fruit; (2) the fumarate is readily translocated to its site of action, i.e. leaves and twigs; (3) the fumarate is rapidly metabolized to other dicarboxylic acids, most important of which is tartaric acid, it being presumed that fungal or host enzymes present in the invaded mesocarp or shoots convert fumarate to epoxysuccinate; (4) both the *cis* and *trans* form of epoxysuccinic acid are toxic to almond shoots; the *trans* form is metabolized to *meso*-tartrate which is rapidly oxidized whereas the *cis* form is metabolized to DL-tartrate; the DL forms are found in almond tissue when it is treated with the *cis* isomer of epoxysuccinate.

The isomers of epoxysuccinate are highly reactive by virtue of their epoxide group and hence are not expected to accumulate in the treated host. Fumaric acid, although stable, disappears at the time toxic symptoms appear, thus supporting the preceding thesis. Some of the carbon fragments of fumarate are found in tartaric acid, presumably via the proposed epoxysuccinate cycle.

Effect of the *Helminthosporium* Toxins on Metabolism

The host specific toxin produced by *H. victoriae* and other host-specific toxins are dealt with extensively elsewhere in this book. I will deal only with certain aspects of the effects this toxin has on metabolism in diseased plants and on CO_2 fixation by the host.

As discussed by Scheffer, the toxin from *H. victoriae* probably has the plasma membrane as its primary site of action and causes altered permeability. Besides the specific effect on the membrane and perhaps also on respiration, it is difficult to demonstrate a direct effect of this toxin on metabolism although this is altered during pathogenesis. For example, some of these metabolic changes noted may be a direct effect of the toxin on a key regulatory system but there is little evidence for this. Alternatively, the metabolic activity measured in infected plants may be a host response to infection or one induced by fungal metabolites other than the toxin. The binding of the toxin to the membrane may represent a very specific target site affecting both permeability and membrane integrity. Once leakage of host metabolites occurs, the chemical reactions resulting from the mixture of enzymes and their substrates may cause general metabolic alterations simply because of the increased availability of reaction sites. It must be remembered that metabolites and their enzymes are organized spatially within healthy cells in a way which maintains an orderly sequence of metabolic events. Once this organization is disrupted, the cell responds quickly and may do so in ways that lead to death of the cell. Altered respiration is one of the first measurable responses of living cells to metabolic disorganization.

Effect of Toxin on Tricarboxylic Acid Cycle of Oats

One of the first physiological signs of toxicity caused by *H. victoriae* is altered permeability as measured by conductivity of ambient solutions followed quickly by a sharp increase in respiration. Various metabolic changes have been reported in tissue treated with toxin from *H. victoriae* (HV) and *H. carbonum* (HC), including altered operation of the tricarboxylic acid (TCA) cycle in the cell. Luke and Freeman (1962) reported that treatment of susceptible oats with HV toxin decreased the concentration of malic, fumaric and succinic acids in host tissue. Later however, after more detailed study, Luke and Freeman (1965) stated that treatment of oats with HV toxin enhanced the synthesis and accumulation of malic and citric acids but had virtually no effect on aconitic and succinic acids. This discrepancy may have been due to the time difference in harvesting of tissue after treatment. Luke and Freeman (1965), proceeded to determine whether the toxin affects CO_2 fixation in the host. Their method was indirect in that they measured the concentration of TCA acids in the presence and absence of CO_2. In the presence of CO_2 malic acid accumulated, whereas in a CO_2 free system, citric acid accumulated but malic acid did not. This was interpreted as indicating that the CO_2 fixation capacity of the affected tissue is enhanced by treatment with toxin. Barring decarboxylation of other acids, CO_2 would be limiting and one would not expect the synthesis of malic or oxalacetic acid through any CO_2 fixation mechanism operative in the dark. Luke and Freeman (1965) postulated that the increase in malic acid could create a favourable buffering condition, i.e. to buffer the excessive amounts of cations that leak out of the vacuole. Citric acid accumulation in this experiment suggested that its synthesis is not dependent on any CO_2 fixation mechanism (source of oxalacetate) but rather occurs through the pyruvate—CoA enzyme system.

Kuo and Scheffer (1970) conducted more precise experiments measuring the direct effect of HV toxin on oat tissue and CO_2 fixation in the dark. They found that susceptible cuttings exposed to toxin (0·32 μg/ml) for 4 h, followed by 4-h incubation with $^{14}CO_2$, had 81% more CO_2 fixed than the control tissue.

In summary, Luke and Freeman (1965) concluded that the HV toxin initially causes a disruption of cell permeability, releasing cations from the vacuole and enhancing CO_2 fixation via a C_3 acid into a C_4 dicarboxylic acid. Whether CO_2 is fixed directly into malic acid by the malic enzyme or into oxalacetic acid by phosphoenolpyruvic (PEP) carboxylase or carboxykinase was not investigated. The excess acids (citric and malic) may be important as buffers in the presence of excess cations which have leaked out of the vacuole into the cytoplasm. Kuo and Scheffer (1970) showed that the effect of HV toxin on CO_2 fixation was direct.

Effect of Toxin on Tricarboxylic Acid Cycle of Corn

Malca and Zschiele (1963) investigating the effect of infection of corn seedlings by *H. carbonum*, found no consistent differences in the contents of fumaric, succinic, malic, aconitic and citric acids between non-inoculated healthy seedlings of susceptible and resistant plants. In inoculated plants however, they reported a large increase in isocitric acid in both susceptible and resistant varieties. The increased concentration of isocitrate could be due to a decrease in activity of enzymes such as isocitric dehydrogenase, isocitritase or malic synthetase. In analyses of TCA acids in culture, *H. carbonum* produced no isocitric acid but did produce malic acid in abundance.

Healthy and infected seedlings exhibiting a resistant reaction showed almost no difference in CO_2 fixation in the dark and no difference in malic acid synthesis. However, the amount of malic acid in seedlings exhibiting susceptible reactions was approximately half that of the non-inoculated controls. This reduction in malic acid content paralleled a reduction in $^{14}CO_2$ fixation in the dark and followed an initial brief period of increased CO_2 fixation.

The reduction in malic acid content in diseased corn leaves as reported by Malca and Zschiele (1963) indicates an inhibition of CO_2 fixation by the host either by action of the host-specific toxin or some other metabolic activity of the pathogen. A preliminary report by Kuo and Scheffer (1968) indicated that the action of the HC toxin (not infection) on corn tissue enhanced CO_2 fixation in the dark whereas the cumulative effect of the pathogen decreased CO_2 fixation in the dark when compared with the healthy controls. This was later confirmed by Kuo and Scheffer (1970) when they reported that in all cases studied, $^{14}CO_2$ fixation in the dark was decreased by infection (not toxin treatment) of corn with *H. carbonum*. When HC toxin was used on corn, however, without the pathogen, CO_2 fixation in the dark was increased. Similarly, HV toxin at low concentrations (0·32 μg/ml) stimulated CO_2 fixation in the dark in oat tissue. This agrees with the results of Luke and Freeman (1965) who obtained an increase in malic acid, presumably through enhanced CO_2 fixation when they treated oat tissue with HV toxin.

According to Malca *et al.* (1964) the inhibition of CO_2 fixation by infection of corn with *H. carbonum* can be explained by the fact that *H. carbonum* produces an inhibitor of CO_2 fixation in addition to the toxin obtained in culture; this inhibitor can be isolated from the toxin with charcoal. Malca *et al.* (1964) tried to explain the inhibition of CO_2 fixation in terms of the carboxylating enzymes phosphoenolpyruvate (PEP) carboxylase and those enzymes normally associated with the Calvin cycle of photosynthesis. He found a reduction in the carboxylation of PEP in diseased plants, but only after considerable injury had resulted from the invading fungus. Malca *et al.* (1964) further noticed a reduction in ribulose-1,5-diphosphate carboxylase activity about 24 h after inoculation which corresponded to about the time that symptoms of infection are expressed. Soon after penetration, but before symptoms were evident (24 h after inoculation), a stimulation of enzymes involved in the conversion of ribose-5-phosphate (R-5-P) to ribulose-1,5-diphosphate (Ru-1,5-DP) occurred. This coincides with the time at which they found a stimulation in CO_2 fixation by the infected host tissue. There appears then, a two fold effect of infection; a stimulation of CO_2 fixation during the early stages of pathogenesis (within 24 h) and an inhibition at later stages.

C_4-Dicarboxylic Acid Pathway of CO_2 Fixation

At this point, it is important to re-evaluate the early work on the effect of infection by *H. carbonum* on CO_2 fixation in corn in light of more recent information on the physiology of the corn plant.

The C_4-dicarboxylic acid pathway of photosynthesis unknown to Malca and Zschiele at the time of their report, operates in the dark as well as in the light, and is present in corn, a highly efficient photosynthetic plant (Hatch *et al.*, 1967). The first product of CO_2 fixation of this cycle as mediated by PEP carboxylase, is thought to be oxalacetic acid although most of the ^{14}C from $^{14}CO_2$ is only detected on the C-4 carbon of malic acid and C-1 carbon of 3-phosphoglyceric acid (Fig. 3). The above ^{14}C labelling pattern can be detected 3-4 sec after initial fixation (Hatch and Slack, 1966).

In the presence of light, there is a 250 fold increase over the rate of darkness in the fixation of CO_2 in the C_4-dicarboxylic acid cycle with a transfer of ^{14}C from the C-4 position of oxalacetic acid to the C-1 position of 3-phosphoglycerate (Hatch and Slack, 1966). Hatch and Slack (1966) postulated that ribulose-1,5-diphosphate by means of a transcarboxylation reaction in light, reacts with oxalacetic acid yielding 1 mol of 3-phosphoglyceric acid and two mol of pyruvic acid. Phosphopyruvate synthetase, demonstrated in corn leaves (Hatch and Slack, 1968), catalyses the conversion of pyruvate to phosphopyruvate thus ensuring a rapid turnover of pyruvate for the CO_2 fixation reaction.

Alternatively and according to Berry *et al.* (1970) the "malic" enzyme, which catalyses the oxidative decarboxylation of malic acid and is present in maize bundle sheath, could function to generate CO_2 from malic acid.

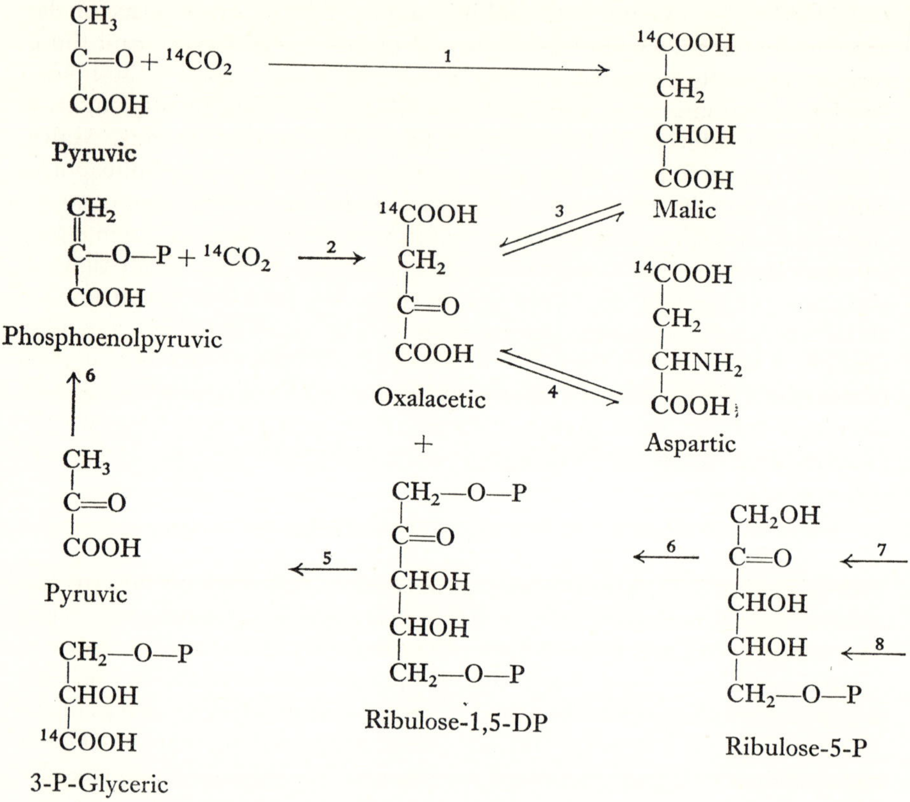

FIG. 3. Proposed scheme for the C_4 dicarboxylic acid cycle of CO_2 fixation: (1) "malic" enzyme; (2) phosphoenolpyruvic carboxylase: (3) malic dehydrogenase; (4) aspartic amino-transferase; (5) transcarboxylation between oxalacetate and R-1,5-DP; (6) phosphopentokinase; (7) phosphopentoisomerase; (8) phosphoketopentoepimerase.

If this reaction were coupled to that of R-1,5-DP carboxylase, then C-1 labelled 3-PGA would be formed from C4- labelled malic acid as in the C_4-dicarboxylic acid cycle. Similarly, as in the other scheme, inhibition of the formation of R-1,5-DP would inhibit CO_2 fixation. Both theories, i.e. transcarboxylation of R-1,5-DP with oxalacetate and coupled "malic" enzyme and R-1,5-DP carboxylase activity lend themselves to the interpretation that an inhibitor produced by *H. carbonum* in maize reduces CO_2 fixation.

The data of Malca and Zschiele (1963) on the decrease in CO_2 fixation and malic acid does allow the interpretation that the C_4-dicarboxylic acid pathway of photosynthesis, present in corn, is affected by infection with

H. carbonum. As mentioned earlier in the paper, Malca *et al.* (1964), reported a reduction in R-1,5-DP carboxylase in infected corn seedlings as early as 24 h after inoculation. The reduction in conversion to R-1,5-DP was apparently due to an undescribed inhibitor produced by the fungus. If the inhibitor affected the turnover of R-1,5-DP in the host, this would result in the inhibition of the proper cycling of the C_4-dicarboxylic acid pathway of photosynthesis and causes a net reduction in CO_2 fixation.

Kuo and Scheffer (1970) investigated the effects of HC toxin on carboxylation mechanisms and found that the phosphoenolpyruvate and 3-phosphoglycerate enzyme systems were not affected. When R-5-P was added to extracts from toxin-treated plants, a significant increase (*c.* 75%) in CO_2 fixation in the dark was noted. This is similar to the results obtained by Malca *et al.* (1964) when they analysed the R-5-P CO_2 reaction system 16 h after inoculation; 24 h after inoculation, they found a reduction of CO_2 fixation, presumably due to the synthesis of the inhibitor and not to the action of the toxins.

The data of Kuo and Scheffer (1970) indicate that the HC toxin does not affect the R-5-P CO_2 reaction system directly. Apparently the increase in dark fixation of CO_2 is due to the action of the toxin on some other factor present in diseased tissue. This is based on the observation that when enzymes are extracted from non-toxin-treated corn tissue, and ^{14}C-labelled bicarbonate is added and then HC toxin is added and incubated for 30 min, no effect on $^{14}CO_2$ uptake by R-5-P reaction system is noted. Further, the inhibition of CO_2 fixation in the dark is not due to the toxin but to another metabolite or condition imposed by infection.

Discussion and Conclusions

In summary, both *H. victoriae* and *H. carbonum* produce host-specific toxins which appear to affect their respective hosts in much the same fashion although physiologically the hosts differ appreciably. The host (oats) of *H. victoriae* has the Calvin cycle of photosynthesis whereas maize, the host of *H. carbonum*, is a highly efficient photosynthetic plant which has the C_4-dicarboxylic acid pathway of CO_2 fixation. Nevertheless, both toxins stimulate fixation of CO_2 in the dark in their respective hosts. This stimulation is apparent when fixation of $^{14}CO_2$ or the increase in the synthesis of malic acid is measured. Both HV toxin in oats and HC toxin in corn stimulate activity of the R-5-P and R-1,5-DP enzyme systems. This system is part of the dark reaction of the Calvin cycle and the C_4-dicarboxylic acid pathway of photosynthesis. This may explain why both toxins affect CO_2 fixation in the dark in plants having photosynthetic systems so different in nature. As an alternative to transcarboxylation reaction in maize, the CO_2

initially fixed by PEP carboxylase into oxalacetate, quickly finds its way to malic acid which is then decarboxylated by the "malic" enzyme and refixed into ribulose-1,5-diphosphate which then forms 2 mol of phosphoglyceric acid.

In addition to the effect of the toxins on CO_2 fixation mechanisms, there is inhibition of CO_2 fixation at the later stages of pathogenesis. This is more apparent in corn parasitized by *H. carbonum* and is due to an inhibitor synthesized during pathogenesis, at least 24 h after inoculation. Apparently, the initial stimulation of CO_2 fixation by the toxin is nullified by the synthesis of inhibitor later on in pathogenesis.

CO_2 fixation in the dark, by healthy plants exhibiting the Calvin cycle of photosynthesis as well as in those with the C_4-dicarboxylic acid photosynthetic pathway, is appreciable. This can be determined easily by exposing leaves of oats and maize to $^{14}CO_2$ in the dark and subjecting the leaves to autoradiography or by using direct combustion methods. The exact enzymatic mechanism operative in the dark and accounting for the fixation in both is not yet established. It is possible that PEP carboxylase or the "malic" enzyme may operate in both hosts under these conditions. Further, in the dark, carboxylation of R-1,5-DP to form 2 mol of 3-phosphoglyceric acid may function both in oats and maize or transcarboxylation of R-1,5-DP with oxalacetic acid to form 3-phosphoglyceric acid and pyruvic acid may occur in corn.

The effect of infection on the C_4-dicarboxylic acid pathway of CO_2 fixation has not been studied previously, mainly because its discovery is relatively recent. I am proposing it as an important pathway of metabolism to be studied in pathogenesis of plants which have this cycle, not only when infected by *Helminthosporium* spp. but by other pathogens as well. I wish to emphasize that the dark reaction concerned with the fixation of CO_2 is similar in certain higher plants and certain fungi. Thus, it occurs in highly efficient photosynthetic plants such as corn and sugarcane which fix CO_2 into malic acid and in the rust fungi (Mirocha and Rick, 1967). The latter fix CO_2 presumably by the "malic" enzyme whereas the former do so presumably by PEP carboxylase. No information is available on the ability of spores and mycelia of *H. victoriae* and *H. carbonum* to fix CO_2 in the dark by action of either of the carboxylating enzymes mentioned above but I would be surprised if they did not fix CO_2 independently of the host. If both the pathogen and the host contain similar enzymes capable of fixing CO_2 in the dark, it would be interesting to study whether their combined activity during pathogenesis results in a greater amount of CO_2 fixed in the dark than in the light.

The possible significance of accelerated CO_2 fixation during pathogenesis, whether it be by action of toxin or other secondary factors, lies in the increased metabolism during infection creating a demand or stress on the TCA cycle acids necessitating replenishment. The fixation of CO_2 directly into a C_4-dicarboxylic acid may serve as an anaplerotic (replenishment) mechanism as described by Kornberg (1966).

Acknowledgement

The author is indebted to Dr. W. R. Bushnell for reading the manuscript and the many contributions made to it through valuable criticism.

References

Aida, K. and Foster, J. W. (1962). *Nature* **196**, 672.
Albright, F., and Schroepfer Jr., G. J. (1970). *Biochem. biophys. Res. Commun.* **40**, 661–666.
Berry, J. A., Downton, W. J. S., and Tregunna, E. B. (1970). *Can. J. Bot.* **48**, 777–786.
Foster, J. W. (1958). *Texas Repts Biol. Med.* **16**, 79–100.
Foster, J. W., Carson, S. F., Anthony, D. S., Davis, J. B., Jefferson, W. E., and Long, M. V. (1949). *Proc. Nat. Acad. Sci. U.S.* **35**, 663–672.
Foster, J. W. and Carson, S. F. (1950). *Proc. Nat. Acad. Sci. U.S.* **36**, 219.
Hatch, M. D. and Slack, C. R. (1966). *Biochem. J.* **101**, 103–111.
Hatch, M. D., Slack, C. R., and Johnson, H. S. (1967). *Biochem. J.* **102**, 417–422.
Hatch, M. D. and Slack, C. R. (1968). *Biochem. J.* **106**, 141.
Kornberg, H. L. (1959). *A. Rev. Microbiol.* **13**, 49–78.
Kornberg, H. L. (1966). *In* "Essays in Biochemistry" (P. N. Campbell and G. D. Greville, Eds), Vol. 2. Academic Press, London and New York.
Kuo, M. S. and Scheffer, R. P. (1968). *Phytopathology* **58**, 1056 (Abstract).
Kuo, M. S. and Scheffer, R. P. (1970). *Phytopathology* **60**, 1391–1394.
Luke, H. H. and Freeman, T. E. (1962). *Phytopathology* **52**, 19 (Abstract).
Luke, H. H. and Freeman, T. E. (1965). *Phytopathology* **55**, 967–969.
Malca, I., Huffaker, R. C., and Zschiele, F. P., Jr. (1964). *Phytopathology* **54**, 663–669.
Malca, I. and Zschiele, F. P., Jr. (1963). *Phytopathology* **53**, 341–343.
Martin, W. R. and Foster, J. W. (1955). *J. Bact.* **70**, 405–414.
Mirocha, C. J. and DeVay, J. E. (1961). *Phytopathology* **51**, 274–276.
Mirocha, C. J. and Wilson, E. E. (1961). *Phytopathology* **51**, 843–847.
Mirocha, C. J., DeVay, J. E., and Wilson, E. E. (1961). *Phytopathology* **51**, 851–860.
Mirocha, C. J. and Rick, P. D. (1967). *In* "The Dynamic Role of Molecular Constituents in Plant Parasite Interaction", pp. 139–40 (C. J. Mirocha and I. Uritani, Eds). American Phytopathological Society, St. Paul, Minnesota.
Nomura, G., Takahashi, H., and Sakaguchi, K. (1954). *J. Agric. Chem. Soc. Japan* **28**, 376–382.
Packter, N. M. (1969). *Biochem. J.* **114**, 369–377.

Romano, A. H., Bright, M. H., and Scott, W. E. (1967). *J. Bact.* **93**, 600–604.
Saito, K. and Kasai, Z. (1969). *Phytochemistry* **8**, 2177–2182.
Sakaguchi, K., Inoue, T., and Tada, S. (1939). *Zentbl. Bakt. ParasitKde* Abt. II, **100**, 302–307.
Shilo, M. (1957). *J. gen. Microbiol.* **16**, 472–481.
Waksman, S. A. (1943). U.S. Patent 2, 326, 986.
Wegener, W. S. and Romano, A. H. (1964). *Bact. Rev.* **87**, 156–161.
Wilkoff, L. J. and Martin, W. R. (1963). *J. biol. Chem.* **238**, 843–846.
Wood, R. K. S. (1967). "Physiological Plant Pathology". Blackwell Scientific Publications, Oxford and Edinburgh, 570 pp. (see p. 191).

Discussion

Scheffer: Both *Helminthosporium victoriae* and *H. carbonum* toxins cause increased fixation of CO_2 in the dark. In the case of *H. carbonum* the effect of toxin mimics one of the effects of the fungus. I was confused by your reference to the difference between oats and corn in photosynthetic mechanisms. Do you suggest that this has something to do with action of toxin?

Mirocha: I indicated that in corn, the C_4-dicarboxylic acid pathway of CO_2 fixation is operative whereas in oats, the Calvin cycle is recognized as the major photosynthetic pathway. In both cases, the respective toxins from *H. carbonum* on corn and *H. victoriae* on oats, initially increase CO_2 fixation in the dark in host tissue. Both pathways use the enzymes involved in the formation of ribulose-5-phosphate (R-5-P) and ribulose-1,5-diphosphate (R-1,5-DP). I suggested that the toxins may have some physiological activity in common which may be expressed by the increased turnover of R-5-P and R-1,5-DP which in turn may affect CO_2 fixation rates.

Durbin: What is known about the translocation of toxic substances from the infected tissues to non-invaded areas of the host?

Mirocha: In the hull rot disease of almond, translocation of toxic substances has been demonstrated. I am not certain if the toxins from *Helminthosporium* are translocated apart from local diffusion; I suspect not. There is evidence that toxin from *Pseudomonas sojae* on soybean is translocatable.

Scheffer: The phytotoxic factor could be of host tissue or of parasite origin. Are there any facts or speculations bearing on this?

Mirocha: In the case of the disease in almond, fumaric acid is predominantly of fungal origin. The enzymes involved in the formation of epoxysuccinate, to my knowledge, have only been found in microorganisms and not higher plants.

Main: In your lecture, you distinctly distinguished between the terms "pathogenicity" and "toxin action" as applied to studies where the pathogen is or is not present. Perhaps this is only a matter of semantics, but it may also reflect on our concept of pathogenesis. In the case of host specific toxins, where the fungus is not necessary to reproduce all the physiological dysfunction and symptoms, are we not actually studying pathogenesis? "Toxic action" is just one facet of pathogenesis. Do we not refer to enzymatic imbalance in plant disease as pathogenesis?

Mirocha: In my lecture, I emphasized "pathogenesis" to distinguish between experiments where CO_2 fixation was studied in host tissue in the presence of added toxin apart from the fungus and those in which the toxin was synthesized *in situ* by the fungus. Pathogenesis in my example is more inclusive than "toxic action" because it includes among other things, synthesis of the inhibitor. I agree with you in that "toxic action" is one facet of pathogenesis.

Elgersma: In the case of the Dutch elm disease, the toxin can induce gummosis in the xylem vessels. What is the origin of the gums that you mentioned? If the gums are pectic in nature they may originate because the toxin induces the host to produce pectinases, or the toxin may cause autolysis in which pectinases are involved.

Mirocha: Gums as spoken in relation to plant diseases are of host origin.

Graniti: Generally, the reaction of a plant reacting to a parasite by the production of tyloses or gummosis in the xylem, largely depends on structural features of the vascular elements. Tyloses seem to develop only if the pore diameter of pits connecting vessels and ray cells is approximately more than 10 μm (Chataway (1949). *Aust. J. scient. Res.* ser. C. **2**, 227–240). If it is less, only gums are produced.

In the disease we are working with, e.g. the almond bud canker, a heavy gummosis occurs around infected tissues a few days after inoculation of twigs with *Fusicoccum amygdali*. This gummy reaction is thought to hinder the spreading of metabolites produced by the pathogen. However, information on whether and to what extent toxins produced by *F. amygdali* can stimulate gum formation in infected trees is still lacking.

13. Effect of Phytotoxins on the Permeability of Cell Membranes

ORVILLE T. PAGE

Department of Biology,
University of New Brunswick, Fredericton,
New Brunswick, Canada

Introduction

It is usual to consider a higher plant as a system of cells organized into tissues specialized to perform certain functions. Thus, from a plant pathologist's point of view, epidermal cells may be envisaged as functioning to inhibit invasion by bacteria, fungi, or viruses, or even by insects capable of transmitting pathogenic agents. In certain instances such diverse cells as root hairs, epidermal hairs of stems and leaves, mesophyll cells bounding a sub-stomatal chamber, and the cells of a lenticel may also serve to prevent invasion. This protective function of cells is enhanced frequently by structural modifications of the outer or exposed surface of epidermal cells (Goodman *et al.*, 1967; Mühlethaler, 1967; Franke, 1967). For example, the complexes of cuticular waxes, cutin, pectic substances, and primary and secondary cellulose lamellae, which constitute the cuticle and cell wall of foliar epidermal cells, may act as passive barriers to invasion by parasites. Clearly, the microfibrillar cellulose complex and matrix of non-cellulosic components of the walls of a living cell provide protection and support for the delicate protoplast within. The penetration of a plant cell wall, whether by forceful probe or by enzymic etching of wall substances by a parasite, is commonly thought of as the first step in plant infection (Albersheim *et al.*, 1969). Once the infection sequence has been initiated the pathogenic agent eventually may become established within the host. Inside the protective outer epidermis of a plant are other cells specialized for particular functions such as foliar mesophyll cells for photosynthesis, and the cells of the cortical and vascular tissues for storage and transport respectively. Cell to cell colonization of mesophyll cells may cause leaf spot disease symptoms, wilting may develop if vascular cells are invaded, or a soft rot may be caused by break-down of the living parenchymatous cells of cortex or pith.

This broadly sketched view of invasion of a plant by a parasite has a conceptual heritage based upon a three-hundred-year-old observation of structural compartments within a plant. The term "cell", first used in 1665 by the English physician Robert Hooke to describe the structure of fragments of cork, was used in a basic, everyday sense as, for example, a prison "cell". And it is still customary to conceive of plants as composed of aggregations of distinct microscopic cells, each bounded by rigid walls and, if the cells are living, each containing a viable protoplast.

Much information has been compiled about the composition and architecture of cell walls, particularly with respect to the presence of pectic substances, hemicelluloses, lignins, and structural proteins deposited within the lamellae of cellulose microfibrils (Albersheim *et al.*, 1969; Rogers and Perkins, 1968). It is still common practice, however, to think of the walls of a plant as sides of a cell separating adjacent protoplasts rather than as a unified skeletal wall system. Even such terms as "protoplast within", "invasion of living parenchymatous cells", and "cell to cell colonization", that I have already used, imply a concept of a discontinuous, compartmented plant structure. It is a fact, nevertheless, that a living plant is an integrated, multinucleate cytoplasmic system with a unified skeletal support for the internal medium.

The names of plant cell membranes were adopted when the sub-structure of plants was interpreted from observations made with optical microscopes. The plasma membrane, or plasmalemma, was considered to be the boundary layer enclosing the entire protoplast within a cell. The nuclear membrane and the vacuolar membrane, or tonoplast, were regarded as discrete membranes serving to separate functional zones within a cell. Concepts of membranes were developed wherein membranes were envisaged as layers bounding protoplasts, or functioning to delimit nuclei, vacuoles, and plastids, and to separate them from the surrounding colloidal hyaloplasm. These bounding membranes exhibited qualities of selective permeability and regulated the movement of ions and molecules between the compartments or zones within a cell, as well as between the protoplasts of contiguous cells.

It must be recalled, however, that the detailed structure of membranes could not be seen with optical microscopes which have a resolving power only down to about 0·2 μm. With the development of electron microscopes, which now have a resolving power down to 1 nm or less, much attention has been focused on the membrane system comprising the endoplasmic reticulum and its interrelationship with other membrane systems within a plant. Buvat (1969) has clearly illustrated the continuity of the nuclear membrane and the endoplasmic reticulum. The internal space of the endoplasmic reticulum is continuous with that enclosed by the nuclear membrane; the outer membrane of the nucleus has continuity with the diverticular processes of the endoplasmic reticulum. Thus the nuclear envelope (at least the outer membrane) is part of, and has apparent affinities with, the cytoplasm rather than the nucleus. Numerous pores of complex structure in the nuclear envelope would seem to ensure intimate contact for nucleo-cytoplasmic exchange whereby the regulation of nuclear and cytoplasmic activities might be correlated (Haemmerling, 1963; Yoo and Bayley, 1967). Changes in membrane profiles, which might be expected to conform with dynamic movements within living cytoplasm, have not been elucidated.

With the advent of the electron microscope it has been possible to determine that the plasma membrane of a cell forms a lining to the plasmodesmata. The plasma membrane extends as the bounding layer of tubules through the primary wall matrix, and is continuous with the plasma membrane of adjoining cells. The diverticula of the endoplasmic reticulum of adjacent cells are interconnected by filaments which extend through the tubular lumina of the plasmodesmata. It is also probable that hyaloplasm has continuity through the plasmodesmata (Buvat, 1969). A special kind of plasmodesmata called "ectodesmata" have been reported to occur in the outer walls of the epidermal cells of a number of plants. Evidence has been produced that in onion leaves liquids may be excreted via ectodesmata through the wall of the epidermis to the cuticle (Newcomb, 1963).

Although plasma membranes appear in electron micrographs as distinct, electron-dense tripartite layers, they are apparently bound to the surface of protoplasm. This can be demonstrated by plasmolysis of plant cells in hypertonic solution which shrinks the protoplast away from adjacent cell walls, yet the plasma membrane adheres to the surface of the protoplasm. Also, every mass of living protoplasm, including such organisms as amoebae, slime moulds, and bacterial spheroplasts, are bounded by a very thin elastic membrane. Presumably, molecular bonding exists between the colloidal protoplasmic body and its surface membrane; the plasma membrane is a differentiated outer region of the protoplasm. It is difficult, therefore, to avoid the view that damage to protoplasm in plants is communicated to the membrane system of the protoplasm, and, conversely, damage resulting from membrane penetration constitutes localized protoplasmic dysfunction.

Following the introduction of a virus into a plant through wounds (Black, 1969), the virus particles apparently have free passage from one viable region of the protoplasmic body of a plant to another region. Invasion of a plant by a fungal or bacterial pathogen usually involves penetration of the membrane bounding the protoplasmic body of the host. Because the protoplasmic body is an integrated system, it is reasonable to assume that metabolites released by a pathogen permeate through protoplasm, probably on a gradient basis. If such metabolites were toxic to protoplasm, it would be extremely difficult to isolate clearly these effects from those on the membranes referred to above.

The varied mechanisms by which microbial toxins damage a host are not clearly understood. However, toxins must eventually influence adversely the membrane system, whether these be the limiting cytoplasmic membranes, the internal membranous cytoskeleton or the diverse membranes of organelles, including vacuoles. Although it is premature to evaluate precisely the damage caused by toxins to plant membranes, it is still very worthwhile to assemble information on membrane structure and function and to explore the possible effects of microbial toxins on permeability.

Phytotoxins will be considered to be microbial metabolites toxic to higher plants (Wheeler and Luke, 1963). The concept of membranes will include all lipoprotein membranes of higher plants. These membranes will be viewed as dynamic systems which function to separate, yet allow communication, among micro-environments within plant cells.

Plant Membranes

Structure

The chemical and functional complexity of membranes and the difficulty of isolating membranes while retaining their structural integrity has led to the development of model systems in an attempt to understand relationship between membrane structure and function. For whereas the chemical composition of some membranes of biological origin is known with accuracy, the interactions between the main protein–lipid components is of importance in deducing their structure. Several models of membranes have been based on the Danielli–Davson concept of a bimolecular lipid layer bounded on either side by a mono-layer of protein. This protein–lipid–protein membrane has a hydrophobic interior and hydrophilic surfaces, possibly resulting from polar orientation of lipid–protein bonding (Branton, 1969). The clear-appearing band approximately 3·5 nm wide bounded by electron-dense dark bands approximately 2·0 nm wide, is consistent with the protein–lipid–protein hypothesis. This characteristic arrangement of image bands has been referred to as a "unit" membrane implying that the bands are part of a single membrane (Robertson, 1959).

In contrast, apolar bonding between the two major membrane components would support a lipid–protein–lipid model wherein enzymatically-active proteins would be in the interior of the membrane. However, it is still not clear whether the dark image bands in electron micrographs correspond to the hydrophilic ends of lipid molecules or to some substantial portion of the membrane protein (Branton, 1969; Thompson, 1965).

A third general model might consist of either protein–lipid–protein or lipid–protein–lipid layers organized into repeating, closely-packed subunits to form a membrane (Gross, 1967).

Or, a model recently proposed by Lesslauer *et al.* (1970) suggests a membrane geometry to include regions within a membrane which are of hydrocarbon-like nature. Proteins are bound to these regions, which are made up of phospholipids and other membrane lipids, by a combination of polar and hydrophobic interactions.

From the foregoing examples it is apparent that the spatial orientation of membrane components is still controversial (Hendler, 1968).

The relatively new freeze-etching techniques do not require chemical fixatives and agents for dehydrating tissue specimens. Instead, tissues are rapidly frozen and then fractured under vacuum, cleavage typically occurring along inherent planes of weakness. The surfaces exposed by freeze-etching techniques are three-dimensional; membranes are often split to expose their inner hydrophobic faces. Membrane cleavage consistently occurs parallel to the plane of the membrane and supports either a protein–lipid–protein or a lipid–protein–lipid unit structure, rather than a subunit concept in which cleavage perpendicular to the plane of a membrane might be expected. Electron microscope examination of freeze-etched replicas has shown small particles in the matrices of tonoplasts, plasmalemmae and chloroplast lamellae. The size and arrangement of particles varies with the source of a membrane. Small particles 6·5–13·0 nm in diameter, randomly distributed within the plane of the tonoplast matrix of root tip cells, are in contrast to the relatively large particles, 15·0–18·0 nm in diameter, which have been observed to be hexagonally arranged within yeast plasmalemma. Row patterns of small particles occur within the plasmalemmae of pea and onion roots whereas particles are randomly positioned within tonoplast membranes of these plant roots (Branton, 1969).

The plasma membrane is considered to control the difference in intra- and extracellular concentrations of inorganic salts and organic substances involved in the metabolism of a host plant. In addition, this membrane functions as the interface through which wall components must pass and upon which these components become organized as they are incorporated into wall matrix. It has been proposed that the microtubules observed in electron micrographs of glutaraldehyde-fixed plant tissue preparations, which lie more or less parallel to the plasmalemma, may convey materials for wall synthesis (Ledbetter and Porter, 1963).

In eucaryotic plants the photosynthetic and respiratory activities are carried out by separate organelles in contrast to procaryotes in which photosynthesis and oxidative phosphorylation occur in common membrane systems. It is of interest that both chloroplasts and mitochondria have an outer membrane system formed from two separate unit membranes. The thylakoids of chloroplasts apparently arise initially (in both a structural and phylogenetic sense) as membrane invaginations. Although mitochondria are usually structurally simpler than chloroplasts, their cristae also arise by invagination of the innermost of the bounding membranes of a mitochondrion (Allsopp, 1969). While the cristae of animal mitochondria are typically in parallel arrays, the cristae of plant mitochondria frequently connect with one another in closed loops (Bonner, 1965).

The structure of mitochondrial membranes is of considerable interest in regard to plant toxins because of the involvement of mitochondria in respiration and energy conservation within a plant. Although about 38–40% of the dry weight, e.g. of pea mitochondria, consists of protein, in a recent review it was concluded that only about half of the total mitochondrial protein was structural protein (Criddle, 1969). The remainder of mitochondrial protein is associated with the enzymic functions of membrane macromolecules. Of particular significance in disease is the fact that the greater part, if not all, of the enzymes and electron carriers that participate in oxidation-phosphorylation processes occur on the membranes of mitochondria.

It is probable that oxidation reactions which supply electrons to the interior of a mitochondrion are carried out in particles located on the outer membrane and that those concerned with transfer of electrons leading to ATP synthesis occur on the inner membrane (Green, 1964). The internal face of the inner membrane, including cristae, is covered with knob-like spheres, the "elementary particles" or "oxysomes", 9·0–11·0 nm in diameter and supported on a cylindrical stalk 3·5–4·0 nm wide and 4·5–5·0 nm long. Membranes free of the repeating knobs or headpieces have been shown to be fully active in electron transport; the components of the electron-transport chain are apparently within the membranes of the cristae and inner mitochondrial membranes and not in the projecting elementary particles (Hendler, 1968).

In concluding these brief comments on the structure of plant membranes, it is germane to subsequent discussion to consider the vacuolar membrane of a higher plant in the context of a lysosomal membrane. Although lysosomes are apparently absent from plant tissues, there is increasing evidence that plant vacuoles may function as lysosomes (Matile, 1969). A number of acid hydrolases including proteolytic enzymes, nucleases, glucosidases, esterases and phosphatases have been isolated from meristematic vacuoles from the root-tip cells of corn seedlings and from other plant vacuoles. The release of lysosomal enzymes into the cytoplasm occurs as the consequence of vacuolar membrane dysfunction.

"Particulate sites" which release acid phosphatase on freezing and thawing plant cells behave like lysosomes. These hydrolase-containing particles also yield lipases, esterases, acid deoxyribonuclease and β-glucuronidase. Since acid phosphatases have been demonstrated in small vacuoles, in spherosomes, and in the Golgi apparatus there may either be more than one site of enzyme localization or more than one type of lysosome (Gahan, 1969).

Mechanisms of Membrane Permeation

As recently pointed out by Stadelmann (1969), differences in permeability to glycerol, urea, methylurea, malonamide and erythritol, of 39 plant species, reflect variation in either composition and/or structure of membranes. These differences among species and the various hypotheses which have been proposed to account for mechanisms of membrane permeability render any brief review of little value. There are, however, several observations relevant to the present discussion which are worthy of note.

The term permeability will be used for the capacity of a lipoprotein membrane to allow the passive movement, or to regulate the passage of substances through it.

The maintenance of correct ionic composition appears to be an important function of plant plasma membranes. Correct ionic composition implies a selectivity in a membrane which permits certain species of ions to pass while the movement of other species is restricted. (This is apparent for Na^+ and K^+ which are maintained at levels within a plasma membrane against an external concentration gradient which is almost the reverse.) For example, a "pumping" action is required to offset leakage of Na^+ into a "cell" and the diffusion of K^+ out. Cells are also involved in import and export of complex organic compounds across a plasma membrane. It seems probable that a membrane interacts with molecules modifying their behaviour and permitting them in turn to modify membrane structure and function. Relatively large molecules do not permeate as would be expected from their low lipid solubility (Marsden, 1970; Sutton, 1969).

Functionally, cell walls and membrane systems may be very closely associated, particularly during the passage of wall substance through the plasma membrane during wall synthesis and also following nuclear division when endoplasmic reticulum is involved in the genesis of the phragmoplast (Albersheim, 1965). The phragmoplast is a zone along the equator of the spindle at the time of cell division consisting of highly branched elements of endoplasmic reticulum, spindle, fibres, dictyosomes, vesicles and mitochondria. The close association of cell wall and membranes is also apparent in the development of tyloses in certain wilt diseases and in the formation of the sheath and encapsulation associated with haustoria (Calonge, 1969).

Although several models have been proposed to account for ion transport, the nature of the connexion between metabolism and transport remains to be established. The concept of "carrier" molecules which form complexes with ions within membranes and transport them across the permeability barrier is a favoured model. The carrier operates in a cyclic fashion as a transport enzyme. Linkage between the operation of trans-membrane carriers and metabolism probably involves enzymatic phosphorylation of the carrier molecule by membrane ATP. Energy involved in pumping ions across membranes forms a considerable part of the total consumption of energy within the protoplast. Abolition of concentration gradients across a plasma membrane, or chemical inhibition of the pump leads to a marked drop in oxygen consumption (Epstein, 1965).

Permeability and Respiration

In a number of instances the most common sign of initial damage to a plant by a toxin is an alteration in water and ion permeability of the cytoplasmic membrane. Increased respiration has also been observed in a wide variety of plant species infected by pathogenic fungi. Non-enzymic substances which injure plant cells or disrupt their metabolism frequently act by stimulating synthetic processes or by uncoupling oxidative phosphorylation in the host. Because of the physicochemical interdependence of plasmalemma, endoplasmic reticulum, nuclear membrane and cytoplasm, a dysfunction in respiration spreads through host protoplasm. Thus it seems logical to suppose that permeability changes are the cause, rather than the result of increased respiratory activity (Wheeler and Black, 1963).

The term "permeability" has been defined recently as "the capacity of a membrane to allow substances to pass through it", and implies "nothing about the nature of the membrane" (Wheeler and Hanchey, 1968). A membrane was considered as being a "noncellular (cuticle), part of a cell (wall, plasmalemma, tonoplast), an entire cell, or one or more layers of cells (seed coat, coleoptile sheath)". This is to take such a gross view of the phenomenon of permeability and membranes as to make it difficult to comprehend what is involved in permeability. While the structure of membranes has not been explicitly resolved, electron microscopy has enabled a more definitive concept than that presented in this review.

As long ago as 1939, Thatcher (1939) employed a plasmolytic method to compare the influence of various fungi on the permeability and osmotic pressure of host cells. He observed that infection by certain rust fungi as well as by *Botrytis cinerea* Pers. *ex* Fr. and *Sclerotinia sclerotiorum* (Lib.) de Bary increased the permeability of plasma membranes of host cells. Wheeler and Black (1963) estimated permeability changes from differences in the electrical conductivity of water in which oat tissues were bathed. They observed that infected susceptible oat plants and susceptible oat tissues treated with victorin lost electrolytes more rapidly than did inoculated or victorin-treated resistant plants. Their results indicated that changes in permeability are responsible for changes in the respiratory activity of oats.

Rubin and Artsikhovskaya (1963) have reported a number of examples of increased permeability in diseased plants. Wheeler and Hanchey (1968) confirmed the now commonplace observation that changes in permeability are associated with disease. They state, "Every investigator who has looked, to our knowledge, has found changes in permeability associated with disease."

Changes in permeability may result from the action of toxin some distance from the location of a parasite within a host. In contrast, haustoria or specialized intracellular hyphae of certain pathogenic fungi such as *Exobasidium japonicum* Shir. establish an interface with the host which involves direct cytoplasmic connections (Calonge, 1969). The haustoria of other fungi such as *Phytophthora infestans* (Mont.) de Bary are enclosed in an encapsulation surrounded by a membrane which is continuous with the plasmalemma of the host. Considerable cytoplasmic continuity exists between the protoplasm of the intercellular hyphae and the intracellular anucleate haustoria owing to the absence of a narrow elongated haustorial neck. The encapsulation between the plasma membrane of the potato host and the parasite is considered to be a dynamic structure, actively engaged in the transfer of substances between host and parasite (Ehrlich and Ehrlich, 1966).

This leads to a concept of compartmentation by membranes which differs from the concept of continuity which characterizes endoplasmic reticulum and plasmalemma. The compartmentation of enzymes which when released by the action of a pathogen results in the formation of substances toxic to the host, has been a subject of much interest to us at the University of New Brunswick. Our interest developed from the differential conversion of adenosine by potato leaves. Leaves of apparently similar maturity were excised from greenhouse-grown Green Mountain plants and were selected for uniform fresh weight. During uptake of adenosine, leaves were maintained for 24 h under 40-ft-c artificial light, temperature 24°C, airflow 3 m.p.h. and relative humidity 25–30%. The volume of solution taken up was recorded and the concentration of the remaining solution was checked spectrophotometrically. No change in adenosine concentration was observed. Leaves were dried and the accumulated purines were leached with water. The leachate was diluted and the amount of adenosine taken up was determined.

During 24 h a mean of 12 μmol of adenosine per leaf was taken up. However, leachates from leaves treated with adenosine had a spectrum characteristic of inosine rather than adenosine, indicating that a substantial amount of adenosine had been deaminated. By differential spectrophotometry it was determined that an average of 78 mol% of adenosine had been deaminated, or up to 97 μmol of adenosine per mg dry weight of leaf tissue had been converted within 24 h. Relatively high levels of derived inosine and hypoxanthine, a trace amount of adenine, and residual adenosine were identified in leachates. Conversion pathways are as follows:

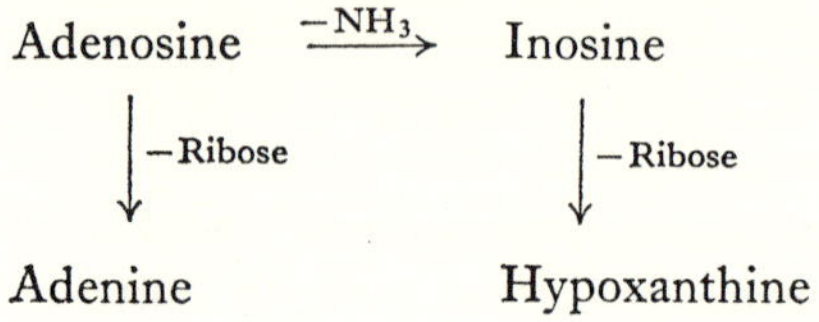

Intact Leaves

Leaves which had taken up adenosine often showed incipient drying of the margins of basal leaflets whereas control leaves which had taken up water remained turgid. The influence of adenosine derivatives, adenine, inosine and hypoxanthine on water loss from potato leaves was next examined. It was found that uptake of adenine resulted in severe drying of leaf tissues; the uptake of equivalent or larger amounts of inosine or hypoxanthine under environmental conditions described previously produced no measurable loss in fresh weight of leaves (Table I).

TABLE I. The effect of purine uptake on the loss of water from potato leaves (from Page, 1964)

Purine (4 mM)	Purine uptake (μmol)	Fresh wt (g)	Wt loss (%)
Adenine	9·6	1·65	62
Hypoxanthine	12·8	1·72	0
Inosine	15·7	1·36	0
Water	—	1·68	0

During 20 h leaves treated with 4 mM adenine transpired more than 4 ml of water in excess of that transpired by control leaves of equal initial weight. The adenine-treated leaves were irreversibly dehydrated having lost more than 60% of their original fresh weight. There was no loss of weight in control leaves or in leaves treated with inosine or hypoxanthine. Since adenine (6-aminopurine) and hypoxanthine (6-hydroxypurine) are otherwise similar in structure, presumably the altered leaf permeability is associated with the group attached to carbon-6 of the purine nucleus. The presence of ribose in the inosine molecule had no apparent effect on water loss.

In a further experiment, Green Mountain potato leaves were ground in a mortar with 0·5 ml of 4 mM adenosine or adenine for 0·1 g of fresh tissue, or with water alone. Samples were incubated at 30°C for up to 24 h. Solutions of adenosine were rapidly converted to adenine by ground leaf tissue as well as by the supernatant remaining after centrifugation of ground leaf tissue at 100,000 g. Solutions of adenine were not deaminated to hypoxanthine by ground leaf tissue; we have been unable to detect activity of adenosine deaminase in potato tissue.

From the foregoing it is apparent that intact potato cells degrade adenosine via inosine to hypoxanthine whereas cells ruptured by grinding converted adenosine to adenine. There is thus a differential degradation of adenosine. Direct conversion of adenosine to adenine was also achieved with potato leaf tissue which had been frozen and then thawed. Such conversion may result from the disruption of lysosomal-like compartmentation and the release of hydrolases observed by Gahan (1969) in freezing and thawing experiments.

The Fungus

In experiments reported in a series of recent papers we have established that the potato late blight fungus, *P. infestans* degrades ATP via adenosine and inosine to hypoxanthine (Page and Clark, 1964; Clark and Page, 1967; Page and Clark, 1967). The fungus is unable to convert adenosine to adenine. Experimental evidence indicated that the enzyme activity was intrahyphal. In contrast to the potato host, N-ribosyl-adenosine ribohydrolase, the enzyme responsible for the hydrolytic conversion of adenosine to adenine, appeared to be absent.

The capacity of the blight fungus to degrade an exogenous source of RNA has also been established. It is noteworthy that only two bases, uracil and hypoxanthine, were consistently recovered. The accumulation of a pyrimidine and a purine base as prominent cleavage end products depends upon prior hydrolysis of linkages between adjoining purine and pyrimidine nucleotides. Whereas pyrimidine mononucleotides derived from primary RNA degradation are accessible for further enzymatic hydrolysis, cleavage of inter-nucleotide ester linkages of purine oligonucleotide sequence is necessary before further breakdown of purine mononucleotides can proceed. The evidence is that the blight fungus possesses a complex of enzymes capable, *in vitro*, of cleaving nucleotide linkages of an exogenously supplied source of RNA. The fungus also degraded and converted derived nucleotides via nucleosides to the bases uracil and hypoxanthine. It is of interest to contemplate the spatial separation or binding of such degradative enzymes that prevents RNA breakdown within the fungus itself.

If intact mycelium of *P. infestans* is suspended in water or 0·01–0·5 M NaCl, there is a release of RNA or polynucleotides from the mycelium which is apparent after 3–4 min and continues for up to 1·5 h when such experiments were terminated. Perchlorate hydrolysis of this extrahyphal RNA or polynucleotides yielded adenine, guanine, cytosine, uracil and hypoxanthine. Thus, while yeast RNA was readily degraded by the fungus, polynucleotides leaked by the fungus apparently retained some degree of structural integrity. If the polynucleotides released by the fungus were of a transfer-RNA type it would be of considerable interest to know whether they would compete for specific host amino acids. Or, would the introduction of foreign nucleic acid into a host trigger an interferon-like or phytoalexin response? Or, if the nucleic acid functioned as a template would the host respond by proliferation of more, foreign nucleic acid as it does when infected with a virus.

Sixteen amino acids were also detected following suspension of mycelium of *P. infestans* in 0·01 M NaCl for 15 h at 20°C. There was no obvious correlation between molecular size or pH at isoelectric points relative to amino acid leakage through fungal plasma membranes.

From the foregoing and other information it has become apparent to us that membranes function to regulate a delicate interplay between the potato host and the late blight fungus. A number of interactions are apparently involved. The pathogen may trigger host auto-conversion of adenosine to adenine, the latter being toxic if accumulated in excess. Thus adenine functions as a toxin. A lysosomal or some similar membrane of the host confines the enzyme, N-ribosyl adenosine ribohydrolase; release of this enzyme results in the degradation of adenosine. The excretion of ribonucleic acid from the fungus, if of a transfer RNA type, could be competitive for host amino acids. Further, the release of RNAase by the fungus, demonstrated at present *in vitro*, is a potentially highly disruptive fungal product relative to the integrity of host RNA.

The effects of toxins on the permeability of membranes may involve many interlocking mechanisms. At this stage in our research it may be naïve to seek for simple cause and effect relationships. To generalize from one syndrome to another has little validity.

References

Albersheim, P. (1965). *In* "Plant Biochemistry" (J. Bonner and J. E. Varner, Eds), pp. 151–186. Academic Press, New York.

Albersheim, P., Jones, T. M., and English, Patricia D. (1969). *In* "Annual Review of Phytopathology" (J. G. Horsfall, Ed.), pp. 171–194. Annual Reviews, Inc., Palo Alto.

Allsopp, A. (1969). *New Phytol.* **68**, 591–612.

Black, L. M. (1969). *In* "Annual Review of Plant Pathology" (J. G. Horsfall, Ed.), pp. 73–100. Annual Reviews, Inc., Palo Alto.

Bonner, W. D. (1965). *In* "Plant Biochemistry" (J. Bonner and J. E. Varner, Eds), pp. 89–123. Academic Press, New York.
Branton, D. (1969). *In* "Annual Review of Plant Physiology" (L. Macklis, Ed.), pp. 209–238. Annual Reviews, Inc., Palo Alto.
Buvat, R. (1969). "Plant Cells." McGraw-Hill, Book Co., New York.
Calonge, F. D. (1969). *Arch. Mikrobiol.* **67**, 209–225.
Clark, M. C. and Page, O. T. (1967). *Can. J. Bot.* **46**, 267–270.
Criddle, R. S. (1969). *In* "Annual Review of Plant Physiology" (L. Macklis, Ed.), pp. 239–252. Annual Reviews, Inc., Palo Alto.
Ehrlich, M. A. and Ehrlich, H. G. (1966). *Can. J. Bot.* **44**, 1495–1503.
Epstein, E. (1965). *In* "Plant Biochemistry" (J. Bonner and J. E. Varner, Eds), pp. 438–466. Academic Press, New York.
Franke, W. (1967). *In* "Annual Review of Plant Physiology" (L. Macklis, Ed.), pp. 281–300. Annual Reviews, Inc. Palo Alto.
Gahan, P. B. (1969). *Proc. biochem. Soc.* **111**, 27–28.
Goodman, R. N., Király, Z., and Zaitlin, M. (1967). "The Biochemistry and Physiology of Infectious Plant Disease." D. Van Nostrand Co. Ltd. (Canada), Toronto.
Green, D. E. (1964). *Scientific American* **210**, 63–74.
Gross, L. (1967). *J. theoret. Biol.* **15**, 298–306.
Haemmerling, J. (1963). *In* "Annual Review of Plant Physiology" (T. Macklis, Ed.), pp. 65–92. Annual Reviews, Inc., Palo Alto.
Hendler, R. W. (1968). "Protein Biosynthesis and Membrane Biochemistry." John Wiley & Sons, Inc., New York.
Ledbetter, M. C. and Porter, K. R. (1963). *J. Cell. Biol.* **19**, 239–250.
Lesslauer, W., Wissler, F. C., and Parsons, D. F. (1970). *Biochim. biophys. Acta* **203**, 199–208.
Marsden, J. C. (1970). *New Scientist* 22 Jan., 152–155.
Matile, Ph. (1969). *Proc. biochem. Soc.* **111**, 26–27.
Mühlethaler, K. (1967). *In* "Annual Review of Plant Physiology" (L. Macklis, Ed.), pp. 1–24. Annual Reviews, Inc., Palo Alto.
Newcomb, E. H. (1963). *In* "Annual Review of Plant Physiology" (L. Macklis, Ed.), pp. 43–64. Annual Reviews, Inc., Palo Alto.
Page, O. T. (1964). *Can. J. Bot.* **42**, 951–952.
Page, O. T. and Clark, M. C. (1964). *Phytopathology* **54**, 1181–1186.
Page, O. T. and Clark, M. C. (1967). *Can. J. Bot.* **46**, 979–985.
Robertson, J. D. (1959). *In* "The Structure and Function of Subcellular Components" (E. M. Crook, Ed.), Biochem. Soc. Symp. no. 16, 3–43. University Press, Cambridge.
Rogers, H. J. and Perkins, H. R. (1968). "Cell Walls and Membranes." E. & F. N. Spon Ltd., London.
Rubin, B. A. and Artsikhovskaya, Y. V. (1963). *In* "Biochemistry and Physiology of Plant Immunity." Macmillan Co. New York.
Stadelmann, Ed. J. (1969). *In* "Annual Review of Plant Physiology" (L. Macklis, Ed.), pp. 585–606. Annual Reviews, Inc., Palo Alto.
Sutton, A. (1969). *New Scientist* 17 April, 125–127.

Thatcher, F. S. (1939). *Am. J. Bot.* **26**, 449–458.
Thompson, G. A. (1965). *In* "Plant Biochemistry" (J. Bonner and J. E. Varner, Eds), pp. 64–88. Academic Press, New York.
Wheeler, H., and Black, H. S. (1963). *Am. J. Bot.* **50**, 686–693.
Wheeler, H., and Hanchey, Penelope. (1968). *In* "Annual Review of Phytopathology" (J. G. Horsfall, Ed.), pp. 331–350. Annual Reviews, Inc., Palo Alto.
Wheeler, H. and Luke, H. H. (1963). *In* "Annual Review of Microbiology" (C. E. Clifton, Ed.), pp. 223–242. Annual Reviews, Inc., Palo Alto.
Yoo, B. Y. and Bayley, S. T. (1967). *J. Ultrastruct. Res.* **18**, 651–660.

Discussion

Rudolph: We found a decrease of permeability after treatment of Swiss chard leaves with the toxin of *Pseudomonas phaseolicola*. Since this observation is in contrast to all other reports on toxin action, we wonder if the measurement of electro-conductivity is sufficient for estimating permeability. The potential of ion inside the tissue should also be considered.

Goodman: What is the evidence you quoted that suggests vacuoles to be the sites of enzyme concentration? Does it extend to the major vacuole in mature leaf mesophyll cells? Our evidence with hypersensitive reactions induced by bacteria suggests that the vacuolar membrane (tonoplast) is the first vital membrane to lose its integrity.

Page: I refer to the conclusions of Matile and also of Gahan that plant vacuoles function as lysosomes. From our evidence which includes ultracentrifugation studies and also the relationships in spatial distribution of N-ribosyl adenosine ribohydrolase with certain phosphatases and other enzymes normally associated with lysosomal compartmentation, we believe that vacuoles can function as enzyme reservoirs comparable to lysosomes in animal systems.

Wood: A striking feature of infection particularly by obligate parasites is the proliferation of the plasmalemma that occurs to accommodate the invading hypha particularly when this is a haustorium. It would be interesting to know how the parasite stimulates the protoplast to behave in this way.

Hess: Membrane formation can be stimulated in onion root cells in the absence of the pathogen. Membranes are in a dynamic state of change in living systems. An electron micrograph has been defined as a frozen moment in the eternal nature of things, and does not adequately reveal the dynamic state of membranes in the living cells.

Strobel: We have demonstrated that changes in inorganic phosphate can be important in the yellow rust disease of wheat. The changes that occur during the infection process result in the activation of ADP-glucose pyrophosphorylase. This leads to an accumulation of starch in chloroplasts. It appears that upsetting the balance of metabolites in cells can be as important as the direct effect of toxins produced by the parasite.

Mussell: We have noticed that *in vitro* 6-benzyladenine enhances vegetative growth of *Phytophthora fragariae*, and that this effect is only evident when the compound is present at levels suggesting substrate not hormonal effects.

14. Interference with Plant Growth Regulation by Microbial Metabolites

O. M. VAN ANDEL

Institute of Biological and Chemical Research on Field Crops and Herbage, Wageningen, The Netherlands

AND

ADRIAAN FUCHS

Laboratory of Phytopathology, Agricultural University, Wageningen, The Netherlands

Introduction

Plant growth can be affected by a great number of chemical compounds including many microbial metabolites. Such substances may cause local injury or death of cells leading to a general inhibition of the growth of plants. In some cases, however, a more specific response of the plants occurs such as morphogenetic changes of leaves or stems, abnormal root or flower development, production of tumours, indicating specific interference with growth regulation. Such phenomena often form part of the symptoms of plant diseases; the causal agents, although involved in disease symptom expression, can hardly be called toxins though they may prevent optimal development or lead to damage of the plant.

It may sometimes be difficult to discriminate between gross non-specific inhibition of growth and specific interference with growth regulation. For instance, diacetoxyscirpenol, produced by *Fusarium equiseti* (Cda.) Sacc., causes development of axillary shoots in *Vicia sativa* L. (Brian *et al.*, 1961). This substance was, however, found to kill the apical buds of the plants, thus removing apical dominance rather than breaking the dormancy of the lateral buds directly. Because various stress effects can provoke comparable growth phenomena, conclusions as to specific growth regulatory activity should be drawn with all caution.

Interference with growth regulation of higher plants may be effected either directly by synthesis and secretion by micro-organisms of substances acting as plant hormones, or indirectly by induction of changes in synthesis or degradation or by inactivation of the growth hormones of the plant itself.

At present four compounds or groups of compounds are considered to play a major role in plant growth regulation: auxins, gibberellins, ethylene and cytokinins. Other natural growth regulators may be expected to be discovered. In fact abscisic acid is one but it will not be discussed separately here, because there is not sufficient known as yet about its synthesis by micro-organisms and its occurrence in diseased tissue. It might be mentioned, however, that a related substance, rishitin, also a sesquiterpene with growth regulatory properties, has been isolated from potatoes infected by *Phytophthora infestans* (Mont.) de Bary and is considered a phytoalexin (Tomiyama *et al.*, 1968).

Production of Plant Growth Hormones by Micro-organisms

Auxins (*indoleacetic acid* (*IAA*) *and other indole compounds*)

A list compiled by Gruen in his review of 1959 shows that most of the fungi investigated, about 80 species including plant pathogens and non-pathogens, produced compounds with auxin activity. Bacteria have also been shown to synthesize IAA (Risch, 1967). Usually, auxin activity was found in culture filtrates as well as in extracts of micro-organisms.

The response of the micro-organisms to exogenous auxin is often very slight compared to that of higher plants or plant organs, but some instances of slight growth promotion, growth inhibition, effect on spore germination and on perithecia formation are known (Gruen, 1959). The auxin concentration needed may be high. Considering the production of auxins by the micro-organisms themselves, more critically designed experiments are needed to establish whether auxins are of importance in the development of bacteria and fungi.

Gibberellins (*GA*, (I))

Contrary to what has been observed about auxin production the ability of micro-organisms to synthesize gibberellins seems to be limited. Gibberellin synthesis has been extensively studied in *Gibberella fujikuroi* (Saw.) Wollenw. or its imperfect form *Fusarium moniliforme* (Sheld.) Sny. *et* Hans.; in cultures of a few other *Fusarium* species and several bacteria, among them *Agrobacterium tumefaciens* (Sm. *et* Town.) Conn.; gibberellins, mainly GA_3, have been found, although in much smaller amounts than in cultures of *G. fujikuroi* (Katznelson and Cole, 1965; Lang, 1970). Curtis (1957) examined cultures of a large number of fungi and actinomycetes for gibberellin activity without positive results; unfortunately, only one bioassay and crude filtrates were used. Few data are available on the effect of gibberellins on growth of micro-organisms, but the fact that (2-chloroethyl)trimethyl-ammonium chloride inhibited almost completely synthesis of GA_3 and other gibberellins by *G. fujikuroi* without affecting dry weight production suggests that these compounds are not important in the growth of this fungus (Harada and Lang, 1965).

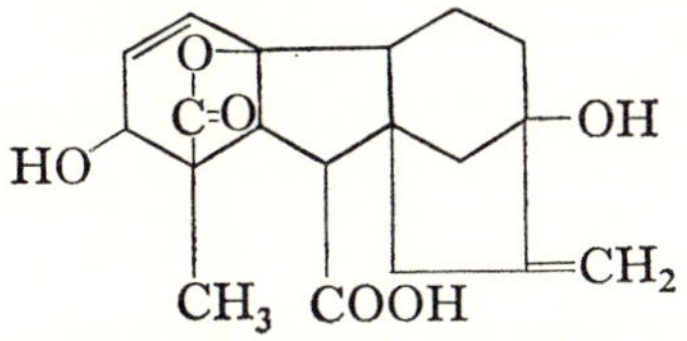

gibberellin A_3GA_3

I

Ethylene

Production of ethylene by micro-organisms has perhaps received less attention than that of auxins or gibberellins, but it has been demonstrated in several cases, for instance in cultures of *Penicillium digitatum* Sacc. (cf. Pratt and Goeschl, 1969), *Ceratocystis fimbriata* Ell. *et* Hallst. (Chalutz and DeVay, 1969) and *Pseudomonas solanacearum* (E. Sm.) E. Sm. (Freebairn and Buddenhagen, 1964). Ilag and Curtis (1968) found ethylene evolving cultures from no fewer than 58 out of 228 fungal species. Recent data on the influence of ethylene on growth of bacteria and fungi are lacking.

Cytokinins (II)

After the isolation of the first natural cytokinin, zeatin (6-(4-hydroxy-3-methyl-*trans*-2-butenylamino)purine) from maize kernels (Letham *et al.*, 1964) this and related compounds are being found in an increasing number of organisms. Isopentenyladenine (2iP, 6-(γ,γ-dimethyl-allylamino)purine) was found to occur as the free base in culture filtrates of *Corynebacterium fascians* (Fil.) Dows. (Helgeson and Leonard, 1966; Klämbt *et al.*, 1966). Table I shows the micro-organisms from which the cytokinins have been isolated and the structural formulae of the latter so far as established. Most species of fungi and bacteria mentioned produce 2iP, but in some cases a slightly different aminopurine analogue was found. According to Klämbt (1967) zeatin would be the common cytokinin of higher plants and isopentenyladenine that of micro-organisms but this view may now need correction, because cytokinins other than zeatin occur in higher plants (Skoog and Armstrong, 1970), and because zeatin has been found in at least one micro-organism, the mycorrhizal fungus *Rhizopogon roseolus* (Cda.) T. M. Fr. (Miller, 1967).

TABLE I. Cytokinins produced by micro-organisms

	R^1	R^2	R^3	Secreted in medium as free purine base	In tRNA	
Agrobacterium tumefaciens (Sm. *et* Town.) Conn.	$—CH_3$	$—CH_3$	—H	+	?	Klämbt, 1967; Romanov *et al.*, 1969
Azotobacter vinelandii Lipman	Structure not yet known				+	Skoog and Armstrong, 1970
Bacillus cereus Frankl. *et* Frankl.	Structure not yet known				+	Skoog and Armstrong, 1970
Corynebacterium fascians (Til.) Dows.	$—CH_3$	$—CH_3$	—H	+	+	Helgeson & Leonard, 1966; Klämbt *et al.*, 1966; Matsubara *et al.*, 1968
Escherichia coli (Mig.) Cast. *et* Chalm.	$—CH_3$	$—CH_3$	$—SCH_3$		+	Burrows *et al.*, 1968
Lactobacillus acidophilus (Moro.) Holl.	$—CH_3$	$—CH_3$	—H		+	Peterkofsky, 1968; Peterkofsky & Jesensky, 1969

Micrococcus roseus Flügge	Structure not yet known				+	Skoog and Armstrong, 1970
Rhizobium leguminosarum (Frank) Bald *et* Fred	—CH_3	—CH_3	—H	+	+	Coppola & Giannattasio, 1968; Giannattasio & Coppola, 1969
Staphylococcus aureus Rosenb.	Structure not yet known				+	Skoog and Armstrong, 1970
S. epidermidis (Winsl. *et* Winsl.) Evans	—CH_3	—CH_3	—H and —SCH_3		+	Skoog and Armstrong, 1970
Streptomyces flaveolus Waks. var. *rutilus*	Structure not yet known					Coppola & Giannattasio, 1969
Rhizopogon roseolus (Cda.) Th. M. Fr.	—CH_2OH	—CH_3	—H	+	(+)	Miller, 1967
Brewer's yeast	—CH_3	—CH_3	—H		+	Biemann *et al.*, 1966; Zachau *et al.*, 1966
Saccharomyces cerevisiae Meyen.	—CH_3	—CH_3	—H		+	Skoog *et al.*, 1966
Kluyveromyces lactis (Dombr.) Vandv.	—CH_3	—CH_3	—H		+	Skoog *et al.*, 1966

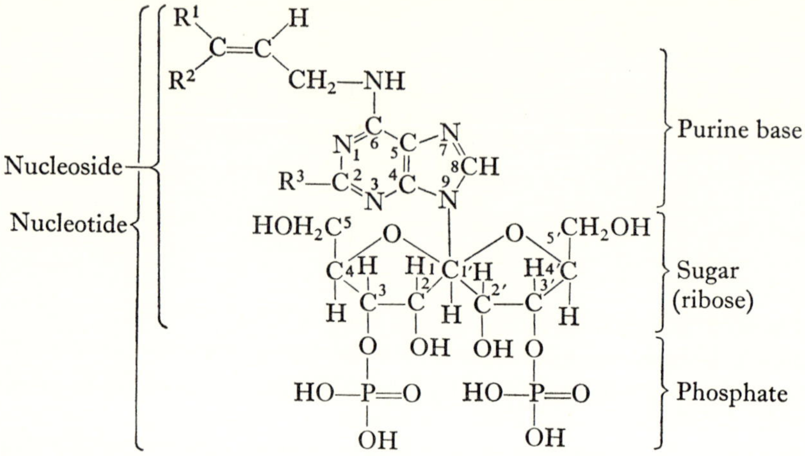

Cytokinins

R^1	R^2	R^3	
CH_3	CH_3	H	(2iP = N^6-($\triangle^2$-isopentenylamino) purine = 6-(γ,γ-) dimethylallylamino) purine = 6-(3-methyl-2-butenylamino) purine
CH_3	CH_3	SCH_3	6-(3-methyl-2-butenylamino)-2-methylthiopurine
CH_2OH	CH_3	H	zeatin = 6-(4-hydroxy-3-methyl-*trans*-2-butenylamino) purine

II

Cytokinins affect growth of bacteria and fungi, but here, too, their exact role in growth phenomena is not clear (Letham, 1967; Skoog and Armstrong, 1970). It is interesting that for the two phytopathogenic microorganisms and the two organisms associated with the root systems of higher plants, cytokinin activity was demonstrated in the culture fluids, indicating secretion of the free base. This must be a rather specific process considering the mode of synthesis of cytokinins (Zachau, 1969). In most other cases cytokinins have been isolated as the ribosides or ribotides from tRNA hydrolysates of the organisms (Table I).

Other compounds (III)

Besides the four groups of compounds mentioned above certain other substances are known which appear to exert a hormonal action in growth processes of higher plants, but which seem to be synthesized by only one species or even a limited number of strains of a particular fungus.

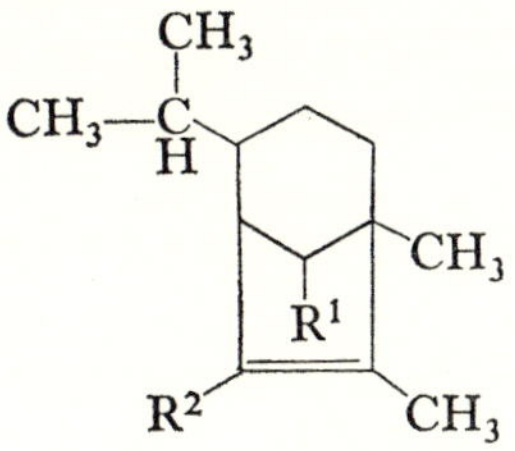

R_1	R^2	
CH_2OH	CHO	helminthosporol
CH_2OH	COOH	helminthosporic acid
CHO	CHO	helminthosporal

III

The ability to produce gibberellins is limited in fungi, but from cultures of strains of *Helminthosporium sativum* Pamm., King *et* Bakke other terpenes with growth regulating activity have been isolated. These are helminthosporol, helminthosporic acid (III) and dihydrohelminthosporic acid (Tamura *et al.*, 1963). The effect of these compounds resembles that of gibberellin in stimulating elongation of rice seedlings and inducing α-amylase synthesis in barley seeds, but they are inactive in other gibberellin bioassays (Kato and Shiotani, 1966; Hashimoto and Tamura, 1967; White and Taniguchi, 1969).

Abnormal growth on corn roots and bean shoots was induced by compounds present in filtrates of cultures of *Penicillium* and *Aspergillus* species (Curtis, 1958; 1961). After isolation and purification compounds named malformin A_1, A_2, B_1 and B_2, were identified as cyclopeptides: cyclo-L-isoleucyl(or alloisoleucyl)-D-cysteinyl-L-valyl-D-cysteinyl-D-leucyl (Table II), differing in the proportion of the various amino acids (Takeuchi *et al.*, 1967). Also, various phytotoxins have been identified as polypeptides, for instance victorin, which was found to inhibit IAA-induced cell elongation in oats susceptible to *H. victoriae* Meehan *et* Murphy (Wheeler, 1969). This suggests an indirect effect of victorin on cell growth, and one might wonder whether the same applies to the malformins and other compounds such as the amino acids involved in frenching of tobacco (Steinberg *et al.*, 1950), or yellow strap leaf of chrysanthemums (Woltz, 1963). It is worthwhile to note that Curtis (1968) recently demonstrated that ethylene induced the

TABLE II. Ratio of different amino acids in malformins

malformin = cyclo-(D,L)-(allo,iso)-leucyl-D-cysteinyl-L-valyl-D-cysteinyl-(D,L)-(allo,iso)-leucyl

A_1: $C_{23}H_{39}O_5N_5S_2$	cys: val: leu: ileu 2: 1: 1: 1
A_2: $C_{22}H_{37}O_5N_5S_2$	cys: val: leu or ileu 2: 2: 1
B_1: $C_{23}H_{39}O_5N_5S_2$	cys: val: leu: allo-ileu 2:1:1:1
B_2: $C_{22}H_{37}O_5N_5S_2$	cys: val: leu: 2: 2: 1

same effects in bean plants as the malformins, with CO_2 acting as an antagonist in both cases. This suggests that ethylene may be the growth regulator involved. On the other hand, the reaction of malformins with thiol compounds may be responsible for their effect on growth (Iriuchijima and Curtis, 1970).

Hadacidin, N-formylhydroxyaminoacetic acid (IV), has been isolated from cultures of *Penicillium purpurescens* (Sopp.) Rap. *et* Thom and

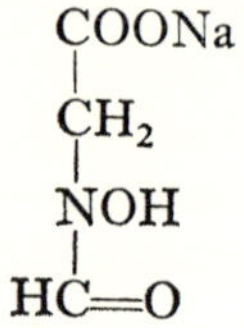

N-formylhydroxyaminoacetic acid
= hadacidin

IV

P. frequentans West (Gray *et al.*, 1964) and found to induce "dwarf" growth in pea, bean and other plants without any phytotoxic effects. Although preventing growth stimulation by gibberellins it seems to act as an antimetabolite rather than as a true antigibberellin.

Fusicoccin may be mentioned as a last substance of this kind. It is another diterpenoid, produced by *Fusicoccum amygdali* Del., the properties of which are discussed extensively elsewhere in this book.

In the next part of this paper little attention will be paid to helminthosporol and related substances, the malformins, hadacidin or fusicoccin. Their existence shows that micro-organisms may produce specific compounds that can act as plant growth regulators, but there is little evidence so far that this happens in plant–pathogen combinations except in almond wilt caused by *F. amygdali.*

Hormones in Plant Diseases and Interference of Microbial Metabolites with Plant Hormone Metabolism

It has already been suggested that microbial metabolites might interfere with plant growth regulation not because of their own regulatory properties, but because they affect hormone synthesis or degradation in plants. Direct and indirect interference with growth regulation may occur simultaneously and it may be impossible to distinguish which proportion each contributes to the overall effect.

Auxins

Fungal infection often produces phenomena in plants that indicate an increase in auxin content. Stem curvature of *Capsella bursa-pastoris* (L.)

Med. infected with *Albugo candida* (Hook.) Kuntze for instance, can be imitated by unilateral application of IAA (Kiermayer, 1958). Epinasty, tylosis, the formation of adventitious roots, symptoms of *Verticillium* wilt in tomatoes, are other examples (Pegg and Selman, 1959). Many more such cases are discussed in reviews by Sequeira (1963), Goodman *et al.* (1967) and Wood (1967). Crown gall is the best known example of a bacterial disease in which changes in auxin content are involved, although this may not be the primary cause of tumour formation (Wood, 1967; Schilperoort, 1969).

In extracts of both the infected *Capsella* and tomato plants mentioned above, auxin activity was higher than in extracts of healthy plants. Healthy as well as infected *Capsella* plants contained besides IAA another compound with high auxin activity, presumably indoleacetonitrile. It is not known whether *A. candida* can synthesize IAA; *Verticillium albo-atrum* Reinke *et* Berth. was found to produce IAA when grown on Czapek-Dox medium and might have contributed to the increase in auxin level but proof of this is lacking at present.

An increase in IAA content was also demonstrated in extracts of tobacco plants, in particular of stems, after infection with *Pseudomonas solanacearum* (E. Sm.) E. Sm. Application of IAA in amounts comparable to those found in the plants induced typical disease symptoms in tobacco cuttings: epinasty, yellowing and wilting (Phelps and Sequeira, 1968). Pathogenic as well as non-pathogenic strains of the bacterium synthesized IAA from tryptophan *in vitro*. This suggests that this property is not directly related to pathogenicity. In cell-free extracts of a less virulent strain tryptophan was first converted by transamination to indolepyruvic acid, which was subsequently decarboxylated to indoleacetaldehyde followed by oxidation to IAA. In contrast, IAA synthesis by the pathogenic strain proceeded along the aromatic pathway, with formylkynurenine, kynurenine and anthranilic acid as intermediates. In the plant tryptophan was first decarboxylated to tryptamine, followed by deamination to indoleacetaldehyde and oxidation to IAA. From data obtained with infected intact plants Phelps and Sequeira concluded that tryptamine was involved in the increased IAA synthesis following infection which indicates that the increased IAA content is due to an increase in synthesis by the host induced by infection, at least in the earlier stages of the disease. According to these authors pathogen-produced IAA might be present in plants in later stages of infection.

IAA synthesis from tryptophan by *P. savastanoi* (E. Sm.) Stev. provides an interesting example of the various ways in which IAA accumulation may be controlled (Kosuge and Kuo, 1968). In cultures of this bacterium IAA was first found to accumulate but later the concentrations decreased again with time. Tryptophan was converted to indoleacetamide, a reaction catalysed by tryptophan decarboxylase; indoleacetamide was subsequently hydrolysed to IAA. Both IAA and indoleacetamide appeared to inhibit tryptophan decarboxylase thus regulating the rate of IAA synthesis, whereas IAA, in the presence of the bacterial enzyme indoleacetyl-ϵ-L-lysine synthetase, was metabolized to indoleacetyl-ϵ-L-lysine. This compound showed less growth substance activity than IAA.

According to Gordon and Paleg (1961) a reaction of tryptophan with phenols under conditions leading to oxidation can also result in synthesis of indolepyruvic acid, which in its turn can be decarboxylated to IAA. Polyphenoloxidase from various plants was found to catalyse this process. Changes in polyphenol content or polyphenoloxidase activity are a common feature of diseased tissue (Kosuge, 1969). In this way increase in IAA synthesis might result indirectly from infection. The relation between IAA and polyphenols is, however, by no means clear. Another view is that IAA would stimulate polyphenol production, or that an increase in IAA content would parallel accumulation of polyphenols because the compounds originate from a common pathway.

So far, changes in the rate of synthesis of IAA have been dealt with. Next, changes in the rate of degradation will be discussed.

Many plants have been shown to have an enzyme system which, in the presence of certain monophenolic cofactors like *p*-coumaric acid and *p*-hydroxybenzoic acid, destroys exogenous IAA with the production of CO_2. Other polyphenols, for instance chlorogenic, caffeic or ferulic acids, may inhibit the degradation. The nature of the enzyme system is not known with certainty and there are conflicting data as to its specificity, but peroxidases and polyphenoloxidases are known to oxidize IAA *in vitro*. An intermediate of oxidative IAA degradation is probably 3-methyleneoxindole, a compound which may be involved in the induction of effects ascribed to IAA, in particular of growth inhibition, because of its high reactivity with SH-groups (Tuli and Moyed, 1967). 3-Methyleneoxindole and 3-hydroxymethyloxindole, the first intermediate in IAA oxidation, are bacteriostatic.

Summarizing, pathogens might affect IAA destruction by production of IAA-oxidizing enzymes, production of phenols acting as cofactors, thus increasing breakdown of IAA, production of phenols inhibiting IAA degradation, and by inducing host tissue to act in these ways.

In plants of *Euphorbia cyparissias* L. infection by *Uromyces pisi* (Pers.) Wint. induces morphological changes such as stem elongation, decreased branching, shortened leaves and abortive flowers, all of which indicate disturbance in growth substance metabolism. Pilet (1960) demonstrated an increase in free auxins in extracts of infected plants as well as decreased IAA oxidase activity. In the early stages of the disease there was good agreement between the ratios of IAA content in extracts of healthy and diseased plants and the rates of IAA oxidation; in later stages, however, the relative increase in IAA concentration was much higher than might be expected from the decreased rate of destruction. The surplus IAA was ascribed to synthesis by the pathogen. Evaluation of the change in IAA content needed for induction of disease symptoms was, however, complicated by the fact that beside IAA two unidentified substances with growth promoting activity and one inhibitor were found. The amounts of these also increased considerably after infection, the inhibitor even increasing to the same extent as IAA. The cause of this increase and the way in which IAA oxidase activity might be decreased were not explained. Decreased IAA oxidase activity was also found in tobacco infected with *P. solanacearum* (Sequeira and Kelman, 1962), presumably because of the high concentration of scopoletin in infected plants. Here, increased IAA content might be the result of increased synthesis as well as decreased destruction.

Sequeira and Steeves (1954) suggested that leaf drop of coffee caused by *Mycena citricolor* (Berk. *et* Curt.) Sacc. (*Omphalia flavida* Maubl. *et* Rang.) would result from a decrease in the IAA content through degradation by the fungus. Other pathogens are also known to produce IAA-oxidizing enzymes *in vitro* (Oaks and Shaw, 1960). Examining the degradation of IAA in extracts of healthy and rust or mildew infected susceptible and resistant barley leaves, Shaw and Hawkins (1958) demonstrated that infection produced a relative decrease in IAA in susceptible leaves during the first days after infection; but IAA increased in later stages. IAA oxidase activity in infected plants was first higher, then lower than in healthy plants. In infected resistant material the decrease in IAA oxidase activity started later and proceeded more slowly. Here, too, the enzyme might be provided by the pathogen in which case the different response of resistant leaves might be due to the more limited growth of the fungus. With disks of wheat leaves infected with *Puccinia graminis* Pers. f. *tritici* Erikss. *et* Henn., Antonelli and Daly (1966) obtained results similar to those of Shaw and Hawkins, but the final decrease in oxidation rate depended, at least partially, on decreased uptake of exogenous IAA.

From the foregoing it is obvious that measuring CO_2 production or the decrease in the concentration of IAA proper is an insufficient indication of loss of growth regulatory activity.

Oxidative decarboxylation is not the only way by which IAA is inactivated although it is the pathway most studied. Deverall and Daly (1964) established that most of the label of carboxyl-labelled IAA applied to wheat leaf tissue was found in four unidentified water soluble compounds one of which might have been indoleacetamide. Andreae and Van Ysselstein (1956) demonstrated in peas conjugation of IAA and aspartic acid to indoleacetyl-aspartic acid, a compound with no growth regulating activity in some plants, presumably because they cannot transform it into IAA. Conversion of IAA to indoleacetyl-ε-L-lysine by *P. savastanoi* has already been mentioned (Kosuge and Kuo, 1968). Leopold and Plummer (1961) demonstrated condensation of quinones and IAA *in vitro*, and suggested that similar reactions could occur in plants following polyphenoloxidase-mediated polyphenol oxidation. This might be another way in which polyphenol metabolism is related to IAA metabolism.

IAA is also known to combine with other compounds in plants, forming protein complexes, ascorbigen and glucose esters. At present, too little is known about the processes and conditions leading to the formation of such complexes and about their importance in IAA metabolism and activity to discuss them here at length. It should be noted, however, that it is not necessarily the concentration of free IAA which is most easily obtained in extracts, that determines the response of the plant cells. Certain products of IAA degradation such as methyleneoxindole may exert regulatory effects; some types of bound auxins may be inactive or act as storage reserves, whereas others would be physiologically active and perhaps even more important in plant growth than free IAA (Bentley, 1961).

It has been shown that auxin content of plants may be affected in various ways. Growth disturbances may result not only from changes in auxin content but also from interference with auxin action. The effect of several microbial metabolites may have to be explained in this way. The effect of victorin has already been mentioned. Chloramphenicol (Noodén and Thimann, 1965), and gougerotin (Burkett *et al.*, 1970) prevent the response of *Avena* coleoptile sections to IAA by inhibition of protein synthesis required for elongation; the similar effect of actinomycin is attributed to inhibition of RNA synthesis (Key, 1969). Finally, compounds interfering with auxin translocation, as for instance suggested for ethylene (Pratt and Goeschl, 1969), would also evoke disturbances in the normal growth pattern of plants.

Gibberellins (I, see p. 229)

Whereas a role of IAA has been implied in many plant diseases and changes in IAA content of infected tissue have been studied extensively, remarkably little attention has been paid in this respect to the gibberellins except for the famous *bakanae* disease of rice. Wood (1967) has already suggested that some disease symptoms ascribed to IAA, might be better explained by the action of gibberellins. Increases of elongation, for instance, in intact plants are rarely induced by application of IAA whereas gibberellins may have pronounced effects in this way. Moreover, the *bakanae* disease has perhaps always seemed so obvious an example of changes in growth of the host caused by a metabolite of a pathogen that the possibility of changes in the gibberellin metabolism of the host has not been considered. Rice has apparently not yet been investigated for its gibberellin content (Lang, 1970). Considering the fact that gibberellins have been found in many higher plants but so far in few micro-organisms (Lang, 1970), induction of changes in gibberellin synthesis, destruction and interconversion in infected plants deserves more attention.

Gibberellins are formed through the normal isoprene pathway of terpene biosynthesis (Cleland, 1969; Lang, 1970). Acetate is converted to mevalonate, isopentenylpyrophosphate, geranylgeranylpyrophosphate and kaurene, from which the gibberellane skeleton arises. It has not been established which gibberellin is first formed. Interconversion of certain gibberellins has been demonstrated, but little is known of the importance of interconversion or destruction for the regulation of the level of physiologically active gibberellins in the plant. It should be noted in this connexion, that plants differ in sensitivity to various gibberellins. There is some evidence of metabolism of exogenous gibberellin by tissues but the rate of degradation is low and details of the underlying mechanism are lacking (Lang, 1970). On the other hand gibberellin-like compounds can be isolated from plants and from them typical gibberellins may be obtained after treatment with acid, alkali or enzyme preparations. Some of these bound gibberellins have been identified as glucosides, and exogenous gibberellin may be transformed quickly into such compounds, for instance by ripening seeds (Rappaport *et al.*, 1967; Sembdner *et al.*, 1968). A change in β-glucosidase activity sometimes observed in diseased tissue might affect the state in which gibberellins are present and thus their activity. Formation of complexes or conjugates of gibberellins and proteins, peptides and amino acids has also been suggested (Cleland, 1969).

Inhibition of gibberellin action would also cause modifications in the plant's growth pattern. Helminthosporal, although itself inducing α-amylase synthesis in embryo-less barley seeds, was shown to inhibit GA_3 induced α-amylase production (White and Taniguchi, 1969).

Ethylene

According to Williamson (1950) diseased tissue may show increased ethylene production. Epinasty of *Fusarium*-wilted tomato plants was attributed to ethylene by Dimond and Waggoner (1953). Accelerated senescence as expressed in yellowing and abscission of leaves may also be a disease symptom caused by ethylene (Pratt and Goeschl, 1969), even where it seemed to be related to high auxin levels, because auxin has been found to stimulate ethylene production in plant tissue. Toxic or inhibitory effects of auxin might in fact be due to the increase in ethylene which it causes (Burg and Burg, 1966, 1968). Ethylene has been shown to be the cause of gummosis of *Fusarium*-infected tulip bulbs (Kamerbeek, 1970) but this effect does not seem to be directly related to its growth regulatory action.

The investigations of Stahmann *et al.* (1966) call attention to another aspect of the function of ethylene in pathogenesis. Treatment of sweet potato tissue with ethylene induced resistance to *Ceratocystis fimbriata*; further, more ethylene was evolved from slices inoculated with a non-pathogenic strain than from those infected with a pathogenic strain. Infection might thus induce the tissue to produce ethylene which would then stimulate metabolic changes leading to resistance. This conclusion was not confirmed by Chalutz and DeVay (1969) who found no effect of ethylene on resistance of sweet potato. A relation between ethylene and disease resistance is, however, indicated by certain other observations. Formation of the phytoalexin 3-methyl-6-methoxy-8-hydroxy-3,4-dihydroisocoumarin in carrot roots is induced by various treatments that could stimulate ethylene production. The relation is less clear for pisatin formation which was induced by ethylene only to a slight extent (Chalutz *et al.*, 1969; Chalutz and Stahmann, 1969). Here it should be mentioned that in sweet potato ethylene induced synthesis of phenylalanine ammonia lyase (Imaseki *et al.*, 1968), an enzyme that might be involved in pisatin synthesis (Hadwiger *et al.*, 1970). On the strength of observations of this kind and because auxins, gibberellins, and cytokinins, and stress effects such as wounding or mechanical stress affect ethylene production (Burg, 1962; Goeschl *et al.*, 1966; Pratt and Goeschl, 1969) it is tempting to attribute to ethylene a key position in the sequence: infection—changes in growth regulation—disease development. However, this may be an oversimplification.

Stahmann *et al.* (1966) found no ethylene synthesis in cultures of *C. fimbriata* so that in infected tissue it might be the host cells which produced the compound. However, both pathogenic and non-pathogenic strains of the fungus were shown to synthesize ethylene *in vitro* by Chalutz and DeVay (1969), so at present it is impossible to separate with certainty the roles of the pathogen and host in diseases caused by this fungus.

Various pathways of ethylene synthesis have been suggested, mainly based on data obtained with ripening fruits. The one most generally accepted at present is the methionine pathway, with γ-methylmercaptobutyric acid and methional as intermediates, and peroxidases as one of the enzyme systems (Yang, 1968; Pratt and Goeschl, 1969).

Cytokinins

Fasciation of dicotyledonous seedlings, caused by *Corynebacterium fascians*, has been found to be accompanied by an increased level of cytokinins, resulting, at least partially, from release of 2iP in the host cells by the pathogen (Helgeson and Leonard, 1966; Klämbt *et al.*, 1966; Thimann and Sachs, 1966). Formation of so-called "green islands" (Allen, 1942) in mildew and rust infected leaves, resulting from retention and resynthesis of chlorophyll around the infection court is also characteristic of cytokinin action. Mobilization and accumulation of metabolites at the site of infection, breaking of bud dormancy and stimulation of cell division, are other phenomena observed in diseased tissue (Yarwood and Jacobson, 1955; Shaw and Samborski, 1956; Sommer, 1961; Bushnell and Allen, 1962) which have been related to an increased cytokinin production resulting from infection (Király *et al.*, 1966, 1967; Poszár and Király, 1966; Dekhuijzen and Staples, 1968; Katsura *et al.*, 1969). The increased amount of cytokinin, presumably zeatin, found in *Plasmodiophora brassicae* infected turnips, is believed to be produced by the host cells (Dekhuijzen and Overeem, 1970). If tenable, Klämbt's hypothesis, already mentioned, might help in deciding whether secretion of microbial metabolites with cytokinin activity or increased hormone production by the host is involved.

Iospentenylpyrophosphate and mevalonate have been found to be converted to the isopentenyl group of isopentenyl adenosine (2iPA) of yeast and liver *in vitro* (Fittler and Hall, 1966) and in tobacco pith tissue (Chen and Hall, 1969). It may be noted that both compounds are also precursors of gibberellin. According to Wright (1968) abscisic acid too may be synthesized from mevalonic acid. A common precursor of three important plant growth regulators implies that interference with synthesis of one may also affect the other two.

Partial incorporation of synthetic exogenous cytokinin in tRNA has been demonstrated in tobacco and soybean tissues but the greater part was rapidly broken down, presumably by splitting off the side chain (Fox and Chen, 1968). Recently, Zachau (1969) has suggested that in tRNA synthesis there first forms a chain of unmodified nucleotides which are afterwards modified at several sites giving rise to the formation of the rarer nucleotides such as cytokinins.

Discussion

From the foregoing it may be concluded that there is little evidence to suggest that plant growth regulators of microbial origin are important in host–pathogen interactions. The changes in growth regulator level which are often characteristic of pathogenesis probably mainly result from changes in the hormone metabolism of the host induced by the pathogen. Not much can be said at the moment about the nature of the agents which induce these changes. Presumably, these agents may range from highly specific compounds such as bacterial DNA as suggested for crown gall caused by *A. tumefaciens*

(Schilperoot, 1969), to substances inducing stress conditions. For the latter the situation is perhaps comparable to that of phytoalexin production.

A discussion of the role of growth regulators produced by the host is outside the scope of this paper, but a few remarks seem appropriate.

To understand the possible role of growth substances in disease development their mode of action in normal plant metabolism clearly should be considered first. Notwithstanding much study over a long period many points are still far from clear. For all types of growth substances dealt with, interaction with nucleic acid and protein synthesis has been demonstrated or is suspected (Cleland, 1966; Key, 1969; Pratt and Goeschl, 1969; Thimann, 1969; Skoog and Armstrong, 1970).

Most is known perhaps about the cytokinins which are now known to be the rarer nucleotides in certain types of tRNA where they invariably occur next to the anticodon triplet which mediates in the translation of mRNA in protein synthesis (Peterkofsky and Jesensky, 1969; Zachau, 1969). As Skoog and co-workers (1966) have pointed out, however, it remains to be determined whether there is a functional relation between cytokinin nucleosides in tRNA and the morphogenetic changes that occur in response to the administration of the free bases.

According to Thimann (1969) the action of auxin on cell elongation of shoots results from the activation of a messenger type RNA. Armstrong (1966), to explain the manifold responses to auxin, concluded that it acted through tRNA. The induction of α-amylase synthesis in barley endosperm by gibberellin is caused by gene derepression, but other effects cannot be explained in this way (Cleland, 1969). In some cases interaction of gibberellin and IAA are suggested.

Obviously, changes in enzyme activity observed after treatment with the various growth substances, and their effect on cell metabolism, may be important for the development of the pathogen, even when the changes remain restricted to the subcellular level and do not result in visible disease symptoms. The fact that application of growth regulators has been found to affect the development of many diseases supports this view (Davis and Dimond, 1956; Corden and Edgington, 1960; Hale *et al.*, 1962; Farkas and Lovrekovich, 1963; Sinha and Wood, 1964; van Andel, 1968). The effect of IAA on pectinmethylesterase-mediated cell wall pectin methylation (Glasziou and Inglis, 1958; Datko and Maclachlan, 1968) was considered to be involved in induced resistance to *Fusarium* wilt in tomatoes (Corden and Edgington, 1960; Sequeira, 1963). Auxin might also influence cell wall structure by its effect on cellulase synthesis (Maclachlan *et al.*, 1968), or lignification (Sequeira and Kelman, 1962). Regulation of invertase synthesis by IAA and by gibberellin (Gayler and Glasziou, 1969) would affect reducing sugar content of tissue and thus, perhaps, susceptibility (Horsfall and

Dimond, 1957). Changes in cell membrane permeability (Powers and Cocking, 1970) may also be expected to be important in pathogenesis. Transport of nutrients to the infection site as a result of the creation of a metabolic sink by cytokinins may stimulate growth of the pathogens (Dekhuijzen and Staples, 1968).

There is little reason to expect that the hormone level in host tissue directly affects development of the pathogen, although some data do suggest that such effects may occur, for instance the decrease in polygalacturonase synthesis in *V. albo-atrum* cultures in the presence of cytokinins (Patil and Dimond, 1968), and the bacteriostatic activity of the IAA degradation product methyleneoxindole (Tuli and Moyed, 1967).

One more point should be considered. Most studies have investigated the relation between infection and only one of the growth substances. However, many processes involved in plant development seem to be controlled by a balance between the amounts of the various growth regulators rather than by the absolute amount of individual compounds. Thus the presence of IAA may be required for the action of gibberellin in cell elongation and cytokinin in cell division; gibberellin and ethylene both induce α-amylase synthesis but have opposite effects on elongation. Formation of adventitious roots is usually ascribed to high IAA concentrations, but it may depend on a high IAA/cytokinin ratio, and thus on a decrease in cytokinin concentration or an increase in abscisic acid concentration, the latter substance antagonizing the action of cytokinin. A complete picture of the role of growth regulators in plant diseases will be obtained only when in a particular case changes in the amounts of the various substances are considered. In a recent study of leaf abscission in *Verticillium*-infected cotton plants Wiese and DeVay (1970) compared ethylene production, IAA and abscisic acid concentration in leaves after infection with a defoliating and a non-defoliating strain. IAA concentration increased and rate of IAA oxidation decreased in leaves of infected plants; ethylene production increased to a greater extent after infection with the defoliating than with the non-defoliating strain; abscisic acid content only increased noticeably in the former instance. Induction of differential hormone imbalances in the host by the two strains is considered responsible for the difference in symptoms.

Summary

The ability to synthesize substances with plant growth regulatory activity is widespread among micro-organisms. These compounds have not been shown conclusively to be important for their growth. Changes in hormone content are very common in diseased plant tissue and they result probably mainly from alterations in host metabolism. The pathways of hormone biosynthesis or degradation may be altered at several points by microbial metabolites or by products resulting from disturbances in plant metabolism induced by infection such as changes in polyphenol metabolism. Some disease symptoms must be attributed to the hormone imbalance resulting from infection. The importance for disease development of changes in hormone content not accompanied by changes in the growth pattern of the plant remains speculative.

References

Allen, P. J. (1942). *Am. J. Bot.* **29**, 425–435.
Andreae, W. A. and Van Ysselstein, M. W. H. (1956). *Pl. Physiol.* **31**, 235–240.
Antonelli, E. and Daly, J. M. (1966). *Phytopathology* **56**, 610–618.
Armstrong, D. J. (1966). *Proc. Nat. Acad. Sci. N.Y.* **56**, 64–66.
Bentley, J. A. (1961). *In* "Encyclopaedia of Plant Physiology" (W. Ruhland, Ed.), Vol. XIV, pp. 609–619. Springer Verlag, Berlin.
Biemann, K., Tsunakawa, S., Sonnenbichler, J., Feldmann, H., Dütting, D., and Zachau, H. G. (1966). *Angew. Chem.* **78**, 600–601.
Brian, P. W., Dawkins, A. W., Grove, J. F., Hemming, H. G., Lowe, D., and Norris, G. L. P. (1961). *J. exp. Bot.* **12**, 1–12.
Burg, S. P. (1962), *A. Rev. Pl. Physiol.* **13**, 265–302.
Burg, S. P. and Burg, E. A. (1966). *Proc. Nat. Acad. Sci. U.S.* **55**, 262–269.
Burg, S. P. and Burg, E. A. (1968). *Pl. Physiol.* **43**, 1069–1074.
Burkett, A. R., Schlender, K. K., and Sell, H. M. (1970). *Phytochemistry* **9**, 545–547.
Burrows, W. J., Armstrong, D. J., Skoog, F., Hecht, S. M., Boyle, J. T. A., Leonard, N. J., and Occolowitz, J. (1968). *Science* **161**, 691–693.
Bushnell, W. R. and Allen, P. J. (1962). *Pl. Physiol.* **37**, 50–59.
Chalutz, E. and DeVay, J. E. (1969). *Phytopathology* **59**, 750–755.
Chalutz, E., DeVay, J. E., and Maxie, E. C. (1969). *Pl. Physiol.* **44**, 235–241.
Chalutz, E. and Stahmann, M. (1969). *Phytopathology* **59**, 1972–1973.
Chen, C. M. and Hall, R. H. (1969). *Phytochemistry* **8**, 1687–1695.
Cleland, R. E. (1969). *In* "Physiology of Plant Growth and Development" (M. B. Wilkins, Ed.), pp. 49–81. McGraw Hill Book Co. Inc., New York.
Coppola, S. and Giannattasio, M. (1968). *Ann. Fac. Sci. Agrar. Univ. Napoli* Portici, **IV**, 3 403–407.
Coppola, S. and Giannattasio, M. (1969). *Boll. Soc. ital. Biol. sper.* **44**, 1913–1914.
Corden, M. E. and Edgington, L. V. (1960). *Phytopathology* **50**, 625–626.
Curtis, R. W. (1957). *Science* **125**, 646.
Curtis, R. W. (1958). *Pl. Physiol.* **33**, 17–22.
Curtis, R. W. (1961). *Pl. Physiol.* **36**, 37–43.
Curtis, R. W. (1968). *Pl. Physiol.* **43**, 76–80.
Datko, A. H. and Maclachlan, G. A. (1968). *Pl. Physiol.* **43**, 735–742.
Davis, D. and Dimond, A. E. (1956). *Phytopathology* **46**, 551–552.
Dekhuijzen, H. M. and Overeem, J. C. (1970). *Acta Bot. Neerl.* **19**, 115.
Dekhuijzen, H. M. and Staples, R. M. (1968). *Contr. Boyce Thompson Inst.* **24**, 39–52.

Deverall, B. J. and Daly, J. M. (1964). *Pl. Physiol.* **39**, 1–9.
Dimond, A. E. and Waggoner, P. (1953). *Phytopathology* **43**, 663–669.
Farkas, G. L. and Lovrekovich, L. (1963). *Phytopath. Z.* **47**, 391–398.
Fittler, F. and Hall, R. H. (1966). *Biochem. biophys. Res. Commun.* **25**, 441–445.
Fox, J. E. and Chen, C. M. (1968). *In* "Biochemistry and Physiology of Plant Growth Substances" (F. Wightman and G. Setterfield, Eds), pp. 777–789. Runge Press Ltd., Ottawa.
Freebairn, H. T. and Buddenhagen, I. W. (1964). *Nature* **202**, 313–314.
Gayler, K. R. and Glasziou, K. T. (1969). *Planta* **84**, 185–194.
Giannattasio, M. and Coppola, S. (1969). *Giorn. Bot. ital.* **103**, 11–17.
Glasziou, K. T. and Inglis, S. D. (1958). *Austr. J. biol. Sci.* **2**, 127–141.
Goeschl, J. D., Rappaport, L., and Pratt, H. K. (1966). *Pl. Physiol.* **41**, 877–885.
Goodman, R. E., Király, Z., and Zaitlin, M. (1967). "The Biochemistry and Physiology of infectious Plant Disease." D. Van Nostrand Co. Inc., Princeton, N.J.
Gordon, S. A. and Paleg, L. G. (1961). *Pl. Physiol.* **36**, 838–845.
Gray, R. A., Ganger, G. W., Dulaney, E. L., Kaczka, E. A., and Woodruff, H. A. (1964). *Pl. Physiol.* **39**, 204–207.
Gruen, H. E. (1959). *A. Rev. Pl. Physiol.* **10**, 405–440.
Hadwiger, L. A., Hess, S. L., and von Broembsen, S. (1970). *Phytopathology* **60**, 332–336.
Hale, M. G., Roane, C. W., and Huang, M. R. C. (1962). *Phytopathology* **52**, 185–191.
Harada, H. and Lang, A. (1965). *Pl. Physiol.* **40**, 176–183.
Hashimoto, T. and Tamura, S. (1967). *Plant Cell Physiol.* **8**, 35–45.
Helgeson, J. P. and Leonard, N. J. (1966). *Proc. Nat. Acad. Sci. U.S.* **56**, 60–63.
Horsfall, J. G. and Dimond, A. E. (1957). *Z. Pflkrankh. PflPath. PflSchutz* **64**, 415–421.
Ilag, L. and Curtis, R. W. (1968). *Science* **159**, 1357–1358.
Imaseki, H., Asahi, T. and Uritani, I. (1968). *In* "Biochemical Regulation in Diseased Plants or Injury" (T. Hirai, Z. Hidaka and I. Uritani, Eds), pp. 189–209. Phytopath. Soc. Japan.
Iriuchijima, S. and Curtis, R. W. (1970). *Phytochemistry* **9**, 1199–1202.
Kamerbeek, G. A. (1970). *Acta bot. Neerl.* **19**, 115.
Kato, J. and Shiotani, Y. (1966). *Planta* **68**, 353–359.
Katsura, K., Egawa, H., Masuka, M., and Ueyama, A., (1969). *Proc. Kansai Plant Prot. Soc.* **11**, 23–27.
Katznelson, H. and Cole, S. E. (1965). *Can. J. Microbiol.* **11**, 733–741.
Key, J. (1969). *A. Rev. Pl. Physiol.* **20**, 449–474.
Kiermayer, O. (1958). *Oester. Bot. Z.* **105**, 515–528.
Király, Z., El Hammady, M., and Pozsár, B. I. (1967). *Phytopathology* **57**, 93–94.
Király, Z., Poszár, B. I., and El Hammady, M. (1966). *Acta Phytopath. Hung.* **1**, 29–38.
Klämbt, D. (1967). *Wiss. Z. Univ. Rostock* **16**, 623–625.
Klämbt, D., Thies, G., and Skoog, F. (1966). *Proc. Nat. Acad. Sci. N.Y.* **56**, 52–59.

Kosuge, T. (1969). *A. Rev. Phytopath.* **7**, 195–222.

Kosuge, T. and Kuo, T. T. (1968). *In* "Biochemical Regulation in diseased Plants or Injury" (T. Hirai, Z. Hidaka, and I. Uritani, Eds), pp. 313–351. Phytopath. Soc. Japan.

Lang, A. (1970). *A. Rev. Pl. Physiol.* **21**, 537–570.

Leopold, A. C. and Plummer, T. H. (1961). *Pl. Physiol.* **36**, 589–592.

Letham, D. S. (1967). *A. Rev. Pl. Physiol.* **18**, 349–364.

Letham, D. S., Shannon, J. S., and McDonald, I. R. (1964). *Proc. chem. Soc.* 1964, 230–231.

MacLachlan, G. A., Davies, E., and Fan, D. F. (1968). *In* "Biochemistry and Physiology of Plant Growth Substances" (F. Wightman and G. Setterfield, Eds), pp. 443–453. Runge Press Ltd., Ottawa.

Matsubara, S., Armstrong, D. J., and Skoog, F. (1968). *Pl. Physiol.* **43**, 452–453.

Miller, C. O. (1967). *Science* **157**, 1055–1057.

Noodén, L. D. and Thimann, K. V. (1965). *Pl. Physiol.* **40**, 193–201.

Oaks, A. and Shaw, M. (1960). *Can. J. Bot.* **38**, 761–767.

Patil, S. S. and Dimond, A. E. (1968). *Phytopathology* **58**, 868–869.

Pegg, G. F. and Selman, I. W. (1959). *Ann. appl. Biol.* **47**, 222–231.

Peterkofsky, A. (1968). *Biochemistry* **7**, 472–482.

Peterkofsky, A., and Jesensky, C. (1969). *Biochemistry* **8**, 3798–3809.

Phelps, R. H. and Sequeira, L. (1968). *In* "Biochemistry and Physiology of Plant Growth Substances" (F. Wightman and G. Setterfield, Eds), pp. 197–212. Runge Press, Ottawa.

Pilet, P.-E. (1960). *Phytopath. Z.* **40**, 75–90.

Poszár, B. I. and Király, Z. (1966). *Phytopath. Z.* **56**, 297–309.

Powers, J. B. and Cocking, E. C. (1970). *J. exp. Bot.* **21**, 64–70.

Pratt, H. K. and Goeschl, J. D. (1969). *A. Rev. Pl. Physiol.* **20**, 541–584.

Rappaport, L., Hsu, A., Thompson, R., and Yang, S. F. (1967). *Ann. N.Y. Acad. Sci.* **144**, 211–218.

Risch, H. (1967). *Wiss. Z. Univ. Rostock* **16**, 465–466.

Romanov, I., Chalvignac, M. A., and Pochon, J. (1969). *Ann. Inst. Pasteur* **117**, 58–63.

Schilperoort, R. A. (1969). Thesis, Leiden.

Sembdner, G., Weiland, J., Aurich, O., and Schreiber, K. (1968). *In* "Plant Growth Regulators", S.C.I. Monograph 31, pp. 70–86. Soc. Chem. Ind., London.

Sequeira, L. (1963). *A. Rev. Phytopath.* **1**, 5–30.

Sequeira, L. and Kelman, A. (1962). *Phytopathology* **52**, 439–448.

Sequeira, L. and Steeves, T. A. (1954). *Pl. Physiol.* **29**, 11–16.

Shaw, M. and Hawkins, A. R. (1958). *Can. J. Bot.* **36**, 1–16.

Shaw, M. and Samborski, D. J. (1956). *Can. J. Bot.* **34**, 389–405.

Sinha, A. K. and Wood, R. K. S. (1964). *Nature* **202**, 824.

Skoog, F. and Armstrong, D. J. (1970). *A. Rev. Pl. Physiol.* **21**, 359–384.

Skoog, F., Armstrong, D. J., Cherayil, J. D., Hampel, A. E., and Bock, R. M. (1966). *Science* **154**, 1354–1356.

Sommer, N. F. (1961). *Physiol. Plant.* **14**, 460–469.

Stahmann, M., Clare, B. G., and Woodbury, W. (1966). *Pl. Physiol.* **41**, 1505–1512.
Steinberg, R. A., Bowling, J. D., and McMurtray, J. E. (1950). *Pl. Physiol.* **25**, 279–288.
Takeuchi, S., Senn, M., Curtis, R. W., and McLafferty, F. (1967). *Phytochemistry* **6**, 287–292.
Tamura, S., Sakurai, A., Kainuma, K., and Takai, M. (1963). *Agr. biol. Chem.* **27**, 738–739.
Thimann, K. V. (1969). *In* "Physiology of Plant Growth and Development" (M. B. Wilkins, Ed.), pp. 3–45. McGraw Hill Book Co. Inc., New York.
Thimann, K. V. and Sachs, T. (1966). *Am. J. Bot.* **53**, 731–739.
Tomiyama, K., Ishizaka, N., Sato, N., Masamuna, T., and Katsui, N. (1968). *In* "Biochemical Regulation in diseased Plants or Injury" (T. Hirai, Z. Hidaka, and J. Uritani, Eds), pp. 287–292. Phytopath. Soc. Japan.
Tuli, V. and Moyed, H. S. (1967). *Pl. Physiol.* **42**, 425–430.
van Andel, O. M. (1968). *Neth. J. Pl. Path.* **74**, suppl. 1, 113–120.
Wheeler, H. (1969). *Phytopathology* **59**, 119.
White, G. A. and Taniguchi, E. (1969). *Can. J. Bot.* **47**, 873–884.
Wiese, M. V. and DeVay, J. E. (1970). *Pl. Physiol.* **45**, 304–309.
Williamson, C. E. (1950). *Phytopathology* **40**, 205–208.
Woltz, S. S. (1963). *Pl. Physiol.* **38**, 93–99.
Wood, R. K. S. (1967). "Physiological Plant Pathology." Blackwell Sci. Publ., Oxford.
Wright, S. T. C. (1968). *In* "Biochemistry and Physiology of Plant Growth Substances" (F. Wightman and G. Setterfield, Eds), pp. 521–542. Runge Press Ltd., Ottawa.
Yang, S. F. (1968). *In* "Biochemistry and Physiology of Plant Growth Substances" (F. Wightman and G. Setterfield, Eds), pp. 1217–1228. Runge Press Ltd., Ottawa.
Yarwood, C. E. and Jacobson, L. (1955). *Phytopathology* **45**, 43–48.
Zachau, H. G. (1969). *Angew. Chem.* **81**, 645–662.
Zachau, H. G., Dütting, D., and Feldmann, H. (1966). *Angew. Chem.* **78**, 392–393.

Discussion

Goodman: How important is the fact that the pathogen *per se* in the resistant plant does not grow in relation to the low accumulation of IAA in many resistant plants?

van Andel: This might be explained by a positive correlation between growth rate and accumulation of pathogen produced IAA. It has been demonstrated, however, that there is no such correlation *in vitro*. Another explanation would be that in this case IAA synthesis by the host is only weakly stimulated, but it may also be that the response of the resistant host is qualitatively different from the response of a susceptible host. Thirdly, it may suggest that IAA accumulation is unimportant in resistance, which does nevertheless not necessarily mean that IAA is not involved in the development of the disease.

Matta: I recall that Pegg and Sequeira have found more tryptophan in tobacco plants infected with an avirulent strain of *Pseudomonas solanacearum* than in plants infected with a virulent strain.

Kuć: The return to normalcy in an infected plant is of equal interest to me as is the activation of enzymes and accumulation of metabolites following infection. The organized chaos evident after infection often disappears with time both in resistant and susceptible hosts. There is a mechanism to initiate and stop the process.

It does not surprise me that organized chaos follows infection in a dynamic system, the alteration of concentration of a single metabolite can lead to profound changes in metabolism far distant from the original site of disturbance.

Durbin: Although *Agrobacterium rhizogenes* produces IAA it would appear that IAA cannot completely explain the production of roots. For example, on turnip slices none of the auxins we have tried will induce roots and in a bean hypocotyl bioassay the IAA produced by the bacterium is 100-fold too low to account for the rooting activity of the filtrate.

van Andel: Root initiation appears to depend on a high IAA/cytokinin ratio. If the increase in IAA content is insufficient to explain the induction of root formation after infection with *A. rhizogenes* perhaps a decrease in cytokinin activity might be involved, resulting from the production of a cytokinin antagonist such as abscisic acid which has been found to induce rooting in cuttings.

Maraite: Have you any idea of the nature of the metabolite of *Uromyces pisi* that induces in *Euphorbia* the changes in IAA oxidase and levels of IAA which are probably the factors that determine subsequent metabolic changes.

van Andel: Data on this subject are lacking. A shift in polyphenol content might be involved.

Patil: You distinguish between physiologically active and total IAA extracted from diseased tissue ? I am not clear on this point. Could you please comment on it ?

van Andel: Various results indicate that after application of auxins there is lack of correlation between the degree of their effect on growth phenomena and the "free" or ether extractable auxin content of the tissue. Such results have been explained by the concept of physiologically active "bound" (relatively ether insoluble) auxin: only this part of the auxin present in the tissue would affect growth.

Dimond: In reference to Patil's question, perhaps there is another way to interpret the fact that plant response is not always proportional to IAA levels.

IAA has multiple sites of action in the cell. The several reactions into which IAA enters are not independent of one another. If interrelated, one reaction may have a feedback relationship to the other. Under this condition the plant response will not be expected to be proportional to free auxin level in all cases. Perhaps it is more useful to look at the problem in this way rather than to consider free auxin as merely in surplus.

Ward: Dr. Stoessl of our laboratory has recently demonstrated that *Ceratocystis fimbriata* produces appreciable amounts of indoleacetic acid.* Do you think that this can be of importance during disease development by this fungus ?

van Andel: As has been mentioned many fungi, pathogens as well as non-pathogens, produce IAA. From this fact alone it is impossible to conclude whether IAA is involved in disease development.

Raggi: I am interested in tryptophan content of mycelium of obligate parasites as a source of precursor for IAA synthesis in diseased plants. Can mycelium autolysis contribute to the increase of IAA level of the infected host plant ? Is there any information about this topic ?

van Andel: So far as I am aware there are no data available.

* (A. Stoessl, 8-Hydroxy-6-methoxy-3-methylisocoumarin and other metabolites of *Ceratocystis fimbriata*. *Biochem. biophys. Res. Commun.* **35**, 186, 1969.)

15. Host-Specific Toxins and Selective Toxicity

ROBERT P. SCHEFFER AND OLEN C. YODER

Department of Botany and Plant Pathology,
Michigan State University, East Lansing, Michigan 48823, U.S.A.

Introduction

The host-specific toxins were introduced for this Institute by Dr. R. B. Pringle (pp. 139–154), who brought us up to date on the chemistry of these remarkable substances. The following will be an interpretative review of our knowledge of the action of host-specific toxins in biological systems. Genetic, physiological, histological, and biochemical data will be discussed in relation to selective toxicity. We especially recommend the genetic approach because we feel that this important tool is often neglected in phytotoxin studies. Three previous reviews have been published, in which detailed background of the subject is considered (Pringle and Scheffer, 1964; Scheffer and Pringle, 1967; Scheffer and Samaddar, 1970).

The roles of host-specific toxins in development of certain plant diseases are significant, because without exception such toxins show the same host-specificity as do the toxin-producing fungi. Some biologists find this incredible, or at best too unusual for further consideration. However, the list of known host-specific toxins is growing (Table I), and the subject may now have developed to the point that it can contribute to understanding of plant diseases in general. So far, only four cases have been examined in any detail: the host-specific toxins of *Helminthosporium victoriae* Meehan *et* Murphy, *H. carbonum* Ullstr., and *Periconia circinata* (Mang.) Sacc., as reported in papers by Pringle, Scheffer, Wheeler, and associates; and the selective toxins of *Alternaria kikuchiana* Tanaka as reported by Japanese workers (Hiroe *et al.*, 1954, 1958). Most of this paper will be concerned with *H. victoriae* (HV), *H. carbonum* (HC), and *P. circinata* (PC) toxins.

Nothing unusual is evident about these three plant diseases, except that host-specific compounds are known to be produced by the causal fungi. Histological studies (Comstock and Scheffer, 1970; Yoder and Scheffer, 1969a) show that infection, tissue colonization, and disease development are those typical of most fungal infections of plants. These studies, together with ecological, physiological, and genetic data show nothing to indicate

TABLE I. Host-specific toxins known to date (1970)

Fungus	Host*	Selective toxicity reported by	Toxin isolated by
1. *Alternaria kikuchiana* Tanaka	Japanese pear (*Pyrus serotina* Rehd.)	Tanaka (1933)	Hiroe *et al.* (1954, 1958)†
2. *Helminthosporium victoriae* Meehan *et* Murphy	Oats (*Avena sativa* L.)	Meehan and Murphy (1947)	Pringle and Braun (1957)
3. *Periconia circinata* (Mang.) Sacc.	Grain sorghum (*Sorghum vulgare* Pers.)	Leukel (1948); Scheffer and Pringle (1961)	Pringle and Scheffer (1963)
4. *H. carbonum* Ullstr.	Maize (*Zea mays* L.)	Scheffer and Ullstrup (1965)	Pringle and Scheffer (1967b)
5. *A. mali* Rob.	Apple (*Malus sylvestris* Mill.)	Sawamura (1966)	Sawamura (1966)†
6. *A. citri* (E. *et* P.) Bl. *et* Fawc.	Emperor mandarin (*Citrus reticulata* Blanco)	Pegg (1966)	
7. *H. maydis* Nis. *et* Miy.	Maize (*Z. mays*)	Smedegård-Petersen and Nelson (1969)	
8. *H. sacchari* Butler	Sugarcane (*Saccharum officinarum* L.)	Steiner and Byther (1969)	

* There are resistant and susceptible cultivars of each host species listed.
† Toxin apparently was not isolated in a highly purified state.

that the fungi involved are either more or less typical pathogens, or more or less saprophytic, than are the fungi that cause many other well-known plant diseases. The existence of host-specific toxins, at the very least, should bring attention to a fundamental concept, the chemical basis of pathogenicity.

The word "toxin" and associated terminology can have different meanings for different researchers. Therefore, each worker should define his terms, until some common usage is established. We have always reserved the term "toxin" to designate microbial products other than enzymes, that cause obvious damage to host plants, and are known with reasonable confidence to be involved in disease development. With this definition, such words as "vivotoxin" and "pathotoxin" appear to be superfluous. The host-specific toxins, as defined previously (Pringle and Scheffer, 1964), affect hosts of the toxin-producing micro-organism, but have little or no effect on resistant cultivars, or on other species, unless used in massive concentrations (Kuo *et al.*, 1970; Wheeler and Doupnik, 1969).

Unfortunately, all toxin definitions tend to channel our thinking. The danger of this restriction is clear if a primary aim is to understand the chemical basis of pathogenicity, or ability to cause disease. The toxin line of thought usually leads to a search for agents that cause necrosis, even though in most cases there is no assurance that substances involved in host–parasite interaction are directly toxic in the usual sense. Furthermore, some known toxins appear not to be required for colonization and pathogenicity, but account for development of certain special symptoms (Braun, 1955). Therefore, we were led to use of the terms primary (or essential) and secondary determinants of pathogenicity and disease (Scheffer and Pringle, 1967). Once these distinctions are clear, we can revert to use of the word "toxin" for its convenience.

The Host-Specific Toxin of *Helminthosporium victoriae*

HV-toxin is the best known and most active of the host-specific toxins, and has been the model for work on the others. Its most striking characteristic is extreme specificity (Fig. 1); this alone is enough to indicate a key role in host–pathogen interaction. In the standard bioassay (Pringle and Scheffer, 1964) all root growth of susceptible seedlings was prevented by as little as 0·0002 μg toxin/ml (Scheffer and Pringle, 1963). Seedlings of resistant oat cultivars tolerated >400,000 times higher concentration than did susceptible seedlings, without evident damage (Kuo *et al.*, 1970). All susceptible oat cultivars, without exception, were highly sensitive to the toxin. All resistant cultivars tested, plus all non-host plants, bacteria, and fungi were highly insensitive (Scheffer and Pringle, 1967).

Several other lines of evidence confirm the hypothesis that HV-toxin is necessary for pathogenicity of the fungus to certain cultivars of oats, and that the compound is the key factor in disease development. (1) All isolates of *H. victoriae* that produce toxin in culture are pathogenic. All isolates that fail to produce toxin are non-pathogenic to oats (Luke and Wheeler, 1955; Scheffer *et al.*, 1964 and 1967). The data cover wild-type isolates, mutants that have lost the ability to produce toxin, and progeny of

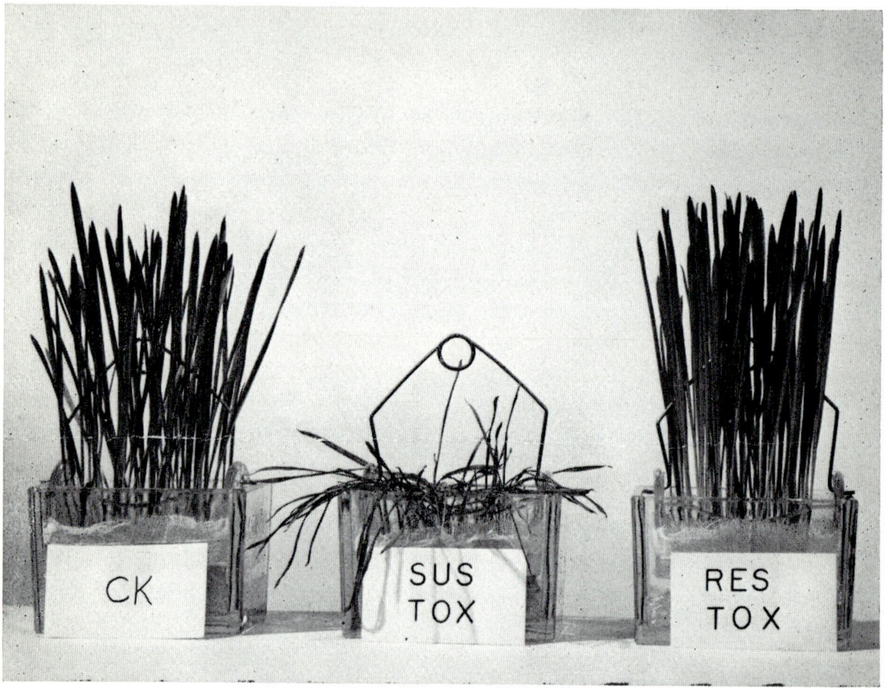

FIG. 1. Effect of *Helminthosporium victoriae* toxin on susceptible and resistant oat seedlings. A trace of toxin was added to the nutrient solution of the resistant (right) and susceptible (centre) seedlings three days before this picture was taken. The untreated control is on the left.

pathogenic isolates × non-pathogenic mutants. No exceptions have been found. (2) Oat cultivars that are intermediate in the susceptibility-to-resistance scale are intermediate in sensitivity to toxin (Luke and Wheeler, 1964; Wallace *et al.*, 1967). (3) The toxin is involved in initial colonization of host tissue by the fungus (Nishimura and Scheffer, 1965; Yoder and Scheffer, 1969a). (4) The known biochemical changes caused by the fungus in host tissue are also caused by the toxin (Scheffer and Pringle, 1967;

Scheffer and Samaddar, 1970; Yoder and Scheffer, 1969a). The evidence behind each category has been reviewed in some detail, but several points need bringing up to date and merit additional comment.

The most convincing evidence for a requirement of toxin for specific pathogenicity comes from genetic data. *Helminthosporium victoriae* and *H. carbonum* are sexually compatible, with the essential difference between them controlled by 2 gene pairs; one is for production of the oat-specific (HV) toxin, and the other is for production of a maize-specific toxin. Progeny of such matings segregate in a ratio of 1:1:1:1 for ability to produce oat toxin, maize toxin, both toxins, and neither toxin. Progeny isolates producing only oat toxin were pathogenic to oats but not maize; isolates producing only maize toxin were pathogenic to maize but not to oats; isolates producing both toxins were pathogenic to both oats and maize; and isolates producing neither toxin were not pathogenic to either oats or maize (Scheffer *et al.*, 1967). Thus pathogenicity was always correlated with toxin-producing ability.

Several lines of evidence show that toxin is involved in initial colonization of *H. victoriae* in oats. Spores contain toxin which is released on germination (Nishimura and Scheffer, 1965). If supplemental toxin is given at the site of penetration, non-pathogenic (non toxin-producing) mutants will colonize oat tissue just as they did before the mutation caused loss of pathogenicity (Yoder and Scheffer, 1969a). The first detectable physiological changes in host tissue after inoculation are the changes which toxin can cause. These considerations are discussed in more detail by Yoder (this volume, pp. 457–460).

We have attempted with limited success to study the mode of action as well as the role of toxin in control of pathogenicity. Toxin causes some well-known changes in cell physiology which are common to plant diseases in general; these changes will be discussed in a later section. However, most changes appear to be secondary effects of action of the toxin, as indicated by experiments with cell-free systems (Scheffer and Pringle, 1963; Scheffer and Samaddar, 1970). We think that the significant toxin lesion is in the plasma membrane, and that this leads to most or all of the secondary effects such as increased respiration and inhibition of protein synthesis.

There is considerable evidence that a plasma membrane lesion appears very quickly after exposure of susceptible cells to toxin. This evidence may be summarized as follows: HV-toxin destroys plasmolytic ability; stops active uptake of solutes; increases apparent free space; causes bursting and lysis of membrane-bound free protoplasts (Samaddar and Scheffer, 1968); causes rapid loss of electrolytes (Scheffer and Samaddar, 1970; Wheeler and Black, 1963) and organic materials (Luke *et al.*, 1969; Samaddar and Scheffer 1971); and its effects can be counteracted in part by certain compounds which bind with membranes and/or components of the protoplasm (Samaddar, 1968; Gardner and Scheffer, 1969 and 1970). None of these effects appears in cells of toxin-treated resistant plants. In each case, changes in the characteristics of susceptible cell membranes could account for such effects. Altogether the data show conclusively that the cell membrane is destroyed as a functional unit, at a very early time after exposure to toxin. However, the data do not prove conclusively that the membrane is the site of an initial lesion of toxin. The first effect could be elsewhere, with the membrane then destroyed because of physical and/or metabolic interconnections.

The earliest effect of toxin detected to date is an increase in loss of electrolytes from susceptible tissue (Fig. 2) (Scheffer and Samaddar, 1970). With carefully washed tissue, a response is evident within 2 min of exposure to toxin. Indeed, the curve of electrolyte loss extrapolates to zero time, indicating that the effect is almost instantaneous. An equally early response is loss of organic compounds (Samaddar and Scheffer, 1971). These experiments were based on earlier work by Wheeler and Black (1963), which was confirmed. Wheeler and Black showed that leaf tissues treated with toxin for 5 min took up electrolytes from the ambient solution, but the uptake was less than that of controls. Toxin treated and control tissues began to lose electrolytes after 5 and 15 min, respectively. The small but significant differences between our results (Fig. 2) and those reported earlier by Wheeler and Black (1963) can be attributed to differences in the conductivity of the ambient water and in the way tissues were handled.

If walls of oat coleoptile cells are digested carefully with hydrolytic enzymes, naked living protoplasts are released. The plasma membrane then becomes the outer surface of the cell. Such units will survive for some time, and show every indication of being functional (Ruesink and Thimann, 1965). When a trace of HV-toxin was added, the protoplasts from susceptible oats began to burst almost immediately. The ruptured plasma membrane soon lysed, while the mitochondria floated away without obvious damage. Protoplasts from resistant oats and other plants were not affected by HV-toxin (Samaddar and Scheffer, 1968). Again, these results indicate a prompt effect of toxin on the outer membrane of the cell.

There are interesting results with several compounds known to bind with components of the plasma membrane or the cytoplasm. These experiments were suggested by the observation that toxin breakdown products will give partial protection to tissues against toxin (Scheffer and Pringle, 1964). The rationale is that if the sites of toxin attack are covered, toxin cannot act and do its damage. We hoped that this might give clues to the nature of toxin receptors.

Very few compounds tested affected toxicity, when applied briefly before toxin treatment, or at the same time with toxin. Uranyl acetate and uranyl nitrate (which bind to the membrane), or semicarbazide, bisulphite and hydroxylamine (carbonyl binding reagents) gave partial to almost complete protection. The uranyl ion complexed with toxin in solution, so we do not know that it protects by way of its membrane-binding properties (Samaddar, 1968; Samaddar and Scheffer, 1971; Hanchey, 1969). Certain sulphydryl-binding compounds (arsenite, iodoacetate, dinitrofluorobenzene,

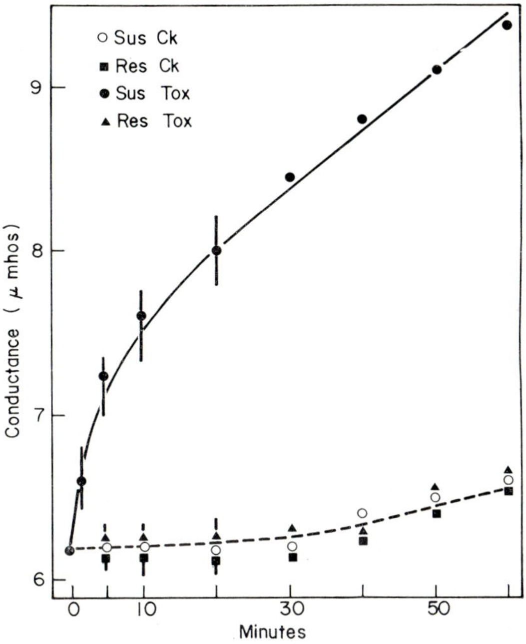

FIG. 2. Effect of *Helminthosporium victoriae* toxin on loss of electrolytes from oat tissue (Sus = susceptible tissue; Res = resistant tissue; CK = control; Tox = toxin-treated; "minutes" refers to time after exposure of tissue to toxin) (From Scheffer and Samaddar, *Recent Advances in Phytochemistry*, **3**, 1970).

and N-ethylmaleimide) also gave partial to striking protection. The protective effects of arsenite and dinitrofluorobenzene were reversed by mercaptoethanol, a sulphydryl donor (Gardner and Scheffer, 1969 and 1970). These results must be interpreted with caution, but they appear to reflect a change in toxin receptor sites or substances.

Pretreatment of oat tissues for 12 h with cycloheximide caused a striking decrease in sensitivity to HV-toxin. The effect was freely and repeatedly reversible; tissues regained sensitivity about 48 h after removal from cycloheximide solutions (Gardner and Scheffer, 1970). Sterol-affecting compounds did not change toxin sensitivity. The data with various protective compounds support the hypothesis that an initial lesion occurs after toxin exposure, that the lesion may be in the plasma membrane, and that a protein may be involved as a receptor substance. The possibility that HV-toxin has its initial effect on protein synthesis has not been eliminated entirely. Toxin did not affect amino acid incorporation by ribosomes from oat tissues and red blood cells (Scheffer and Samaddar, 1970), although incorporation by the oat preparation was too low for a firm conclusion. Damage to the plasma membrane could lead to inhibition of apparent protein synthesis or turnover, which has been reported (Scheffer and Pringle, 1967). Further tests of these hypotheses are under way.

Host-Specific Toxins of *Periconia circinata*

P. circinata infects the roots and crowns and blights the foliage of certain cultivars of grain sorghum. Other cultivars are essentially immune. The fungus produces in culture two or more closely related compounds (PC-toxins), each with a high degree of host specificity (Pringle and Scheffer, 1967a). All growth of susceptible seedling roots was prevented by 0·1 μg toxin/ml, whereas resistant seedling tolerated >26,000 times higher concentration without deleterious effects (Mansour, 1968). There was little or no effect of the toxins on other non-host plants, except with massive concentrations (Mansour, 1968; Scheffer and Pringle, 1961). Comparative work reported here was with one of the host-specific compounds. There is no reason to suspect that the others behave differently.

It has been said that *P. circinata* does not invade sorghum roots, but exerts all its pathogenic effects by action of its extracellular toxin (Flentje, 1965). If this were true, *P. circinata* would be a "saprophytic pathogen" but not a parasite. However, such was not found to be the case. Histological studies showed that the fungus penetrated young seedling roots to the second cell below the surface within 24 h after inoculation. By 48 h, the fungus had penetrated almost to the pericycle; by 4 days after inoculation, it had invaded tissues extensively, but was not found within the stele. No disintegration or disruption of host tissue was evident before the fourth day (Mansour, 1968). The role of PC-toxin in colonization of sorghum tissue appears to parallel the role of HV-toxin in oat tissue. We have not found a basis for the notion that *P. circinata* and *H. victoriae* are "saprophytic pathogens".

PC-toxin causes the same kinds of physiological changes in susceptible sorghum that HV-toxin causes in oats. These effects, which will be discussed later, include increased respiration, loss of cellular materials, and decreased ability to incorporate amino acids and uridine into TCA-insoluble substances (Mansour, 1968). PC-toxin did not affect activities of cell-free preparations of mitochondria or nuclei, but it did have profound effects on membrane functions (Mansour, 1968; Gardner, 1971). Thus there are many parallels between HV-toxin action and PC-toxin action, although PC-toxin affects a different host species, is somewhat less drastic in its action, and is less toxic on a dry weight basis. However, there must be still other differences, because most of the compounds which protected oat tissues against HV-toxin did not protect susceptible sorghum against PC-toxin (Gardner and Scheffer, 1969; Gardner, 1971).

The Host-Selective Toxin of *Helminthosporium carbonum*

HC-toxin specifically affects maize lines which are susceptible to *H. carbonum* race 1. Evidence for the critical role of HC-toxin in pathogenesis is similar to that described previously for HV-toxin. Plants susceptible to *H. carbonum* are sensitive to low levels (0·2 μg/ml) of toxin; resistant plants can tolerate approximately 100 times this concentration (Pringle and Scheffer, 1967b). Maize selections intermediate in susceptibility to *H. carbonum* have a corresponding intermediate sensitivity to toxin (Kuo *et al.*, 1970). All pathogenic isolates of *H. carbonum* race 1 produce the host-specific toxin; all isolates not pathogenic to susceptible maize do not produce HC-toxin (Scheffer *et al.*, 1967). Physiological changes which occur in infected maize can be reproduced by toxin treatment (Kuo and Scheffer, 1970).

Our estimate of importance of a toxin is enhanced if the toxin is known to determine the success or failure of an infection, as is the case with HV-toxin (Yoder and Scheffer, 1969a). The involvement of HC-toxin in initial interactions between *H. carbonum* and maize tissue has been examined in detail by Comstock and Scheffer (1970). Spores germinated on the leaf surface and penetrated the epidermis of susceptible, intermediate, and resistant leaves within 12 h after inoculation. Fungal growth was confined to the epidermal cells of resistant leaves but continued throughout susceptible leaf tissues. In leaves of intermediate resistance the fungus grew beyond the epidermis into the mesophyll but growth was slower and less extensive than in susceptible leaves. The sequence of fungal development and host response by susceptible and resistant tissue is characteristic of fungal infections in general.

Spores of *H. victoriae* (pathogenic to oats but not to maize) germinated and penetrated *H. carbonum* susceptible, intermediate and resistant maize leaves in 12 h but fungal growth generally did not advance beyond the epidermal cells. However, when HC-toxin was added to *H. victoriae* spores the fungus invaded and grew in maize leaves just as did *H. carbonum*. A toxin concentration of 2μg/ml in the infection droplet allowed *H. victoriae* to make abundant growth in susceptible leaves, limited growth in intermediate leaves, and little or no growth beyond the epidermis in resistant leaves. In the presence of 40 μg HC-toxin/ml, *H. victoriae* advanced into the mesophyll of resistant leaves in 24 h but after 48 h growth was restricted to relatively small areas. Furthermore, growth of *H. carbonum* in resistant maize tissues is a function of spore concentration. Five spores/mm^2 on the leaf surface resulted in a typical resistance sequence described above. When 20 spores/mm^2 were applied, *H. carbonum* invaded resistant maize leaves to the same extent as did *H. victoriae* in the presence of 40 μg HC-toxin/ml (described above) (Comstock and Scheffer, 1970). When more spores are present on the leaf surface, more toxin is released. Resistant cells are known to be affected if toxin concentration is high enough (Pringle and Scheffer, 1967b), and this appears to make resistant tissue a substrate for the fungus.

Some of the physiological changes caused by HC-toxin in susceptible maize are similar to those caused by HV-toxin in susceptible oats. When sensitive tissues are exposed, both toxins cause increases in respiration, electrolyte leakage, and dark fixation of CO_2 whereas synthesis (Scheffer and Samaddar, 1970; Yoder, 1970), and amino acid and uridine incorporation (Scheffer and Samaddar, 1970) decrease.

HC- and HV-toxins differ in certain respects in their effects on susceptible tissues. HC-toxin is less active on a dry weight basis (Kuo *et al.*, 1970) and acts more slowly than HV-toxin (Kuo and Scheffer, 1970). Relatively low levels of HC-toxin (*c.*–45 μg/ml) inhibit growth of resistant maize and non-host plants. In contrast, HV-toxin is extremely specific for susceptible oats and massive doses (3600 μg/ml) are required to affect resistant oats or non-hosts (Kuo *et al.*, 1970). Also, HV-toxin appears to act by a simple physical process which is independent of host cell metabolism (Scheffer and Pringle, 1964) whereas HC-toxin requires metabolic activity to exert its effects (Kuo and Scheffer, 1969). At the proper concentration and/or exposure time, HC-toxin stimulates certain processes which HV-toxin invariably inhibits. HC-toxin can cause increased root elongation of susceptible and resistant maize seedlings at 0·125 and 3·125 μg/ml, respectively (Kuo *et al.*, 1970). Relatively brief exposure to toxin results in increased influx of certain materials across the plasma membrane and an apparent increase in synthetic capacity (Yoder and Scheffer, 1969b). HC-toxin treatment of maize seedling roots for 2–4 h increased the rate of uptake of leucine-^{14}C into the free leucine pool, the size of the free leucine pool, and the rate of incorporation into the ethanol insoluble fraction. Activity of nitrate reductase is increased 30–70% if toxin is present during the induction period (Yoder, 1970). This increase appears to be caused by increased NO_3^- uptake.

The capacity of HC-toxin to be both inhibitory and stimulatory suggested that it may have a hormone-like effect similar to that of gibberellic acid (GA). Helminthosporol, a biologically active compound produced by certain *Helminthosporium* spp., is known to mimic many GA effects (Katsumi *et al.*, 1967). Tests confirmed that GA promoted dark germination of tobacco seeds, elongation of cucumber hypocotyls, and synthesis of α-amylase by barley endosperm. However, HC-toxin has a different effect because it had no activity in any of these tests (Yoder, 1970).

Comparative Physiological Changes Induced by Infections and Host-Specific Toxins

Infection of plants by bacteria and fungi causes some well-known changes in host cell and tissue physiology (Table II). These changes are the subjects of a large descriptive literature, most of which is inconclusive as to cause

TABLE II. Comparative physiological changes induced by pathogenic bacteria and fungi (in general) and by host-specific toxins

Response	Infections*	HV-toxin	PC-toxin	HC-toxin
Loss of electrolytes	+	+	+	+
Respiration	+	+	+	+
Dark fixation of CO_2	+	+	?	+
Free amino acid content	+	+[1]	?	+
Indoleacetic acid content	+	?	?	?
Phenol and/or ascorbic acid oxidases	+	+[2]	?	?
Growth	+ or −†	−	−	+
Protein synthesis	+ or −†	−	−	+
Uptake of solutes	+ or −†	−	−	+
Phytoalexin production	+	0	?	?

* + = increase; − = decrease; 0 = no effect; ? = not examined.
† Increase or decrease, depending on the particular host-pathogen combination.
[1] Luke and Freeman, 1964.
[2] Krupka, 1959; and Novacky and Wheeler, 1970.

and effect relationships. The same kinds of changes are induced by the host-specific toxins. A notable exception is phytoalexin production, which is not induced by HV-toxin.

The earliest detectable changes in host physiology after infection with *H. victoriae* and *H. carbonum* can be attributed to toxin released from the invading fungus. *H. victoriae* infection results in increased loss of electrolytes from host cells within 6 h. Toxin is capable of inducing similar losses almost immediately. The parallel between effects of toxin and infection is evident, considering the time required for spore germination and toxin production. Comparable but less drastic effects were found for *H. carbonum*-infection and HC-toxin treatment (Comstock and Scheffer, 1970). Inoculation of susceptible oats with *H. victoriae* also caused, within 8 h, a decrease in apparent synthesis of protein, another effect comparable to that of HV-toxin treatment (Yoder and Scheffer, 1969a).

Increased respiration is induced in sensitive tissues by HV, HC, and PC-toxins. The stimulations are quantitatively and qualitatively comparable to those caused by plant infections in general. Exposure to HV-toxin caused a rise in oat tissue respiration within minutes (Scheffer and Pringle, 1963). PC-toxin-induced respiration in sorghum tissues appears somewhat more slowly (Mansour, 1968), while stimulation of corn tissue by HC-toxin is evident only after >5 h exposure (Kuo and Scheffer, 1970). Oxygen and phosphorus uptake by mitochondria from sensitive tissues was not affected by HV or PC-toxins (Scheffer and Pringle, 1963; Scheffer and Samaddar, 1970; Black and Wheeler, 1966), which led to the conclusion that respiratory disturbance is an indirect or secondary effect of toxic action. Some loss of respiratory control and slightly decreased P:O ratios in mitochondria isolated from tissues previously treated with HV-toxin (Wheeler and Hanchey, 1966) may result from host cell breakdown products or other secondary effects. The respiratory effects of toxin on susceptible plants were given much attention in earlier work, and the results have been reviewed in some detail (Pringle and Scheffer, 1964; Scheffer and Pringle, 1967).

Increased growth and protein synthesis, especially near the zone of infection, have been observed for some plant diseases (Daly and Sayre, 1957); with others, only decreases are evident. HV and PC-toxins cause inhibition of growth and protein synthesis; stimulatory effects have not been observed at any concentration. The inhibitory effects on protein synthesis may be the secondary result of some other toxin lesion, possibly in the plasma membrane. In contrast, low concentrations of HC-toxin stimulated maize seedling growth (Kuo *et al.*, 1970). Low concentrations and relatively short exposure times also increased the incorporation of amino acids and uridine into TCA-insoluble components of cells (Kuo, 1968; Yoder and Scheffer, 1969b). Higher concentrations and longer exposure to HC-toxin led to inhibition. Resistant maize gave similar responses, but higher levels of toxin were required. The growth and protein synthesis effects of these toxins appear to parallel the effects of certain pathogens on susceptible plant tissues (Daly and Sayre, 1957; Pegg and Sequeira, 1968).

The toxins affect cell permeability, including both influx and efflux of substances, and these effects also parallel known responses to these and certain other plant infections (Wheeler and Hanchey, 1968). Toxin-induced changes in permeability were discussed above, in the sections for each toxin.

Another interesting parallel between infections and toxin is the effect on dark fixation of CO_2 by plant tissues. Increased dark fixation has now been observed to follow infection by several fungi, including rusts (Mirocha and Rick, 1967). *H. carbonum* infection is known to cause a brief phase of increased CO_2 fixation in the dark, followed by a decrease in this capacity (Malca *et al.*, 1964). An inhibitor which collects in the area of infection was shown to cause the decrease; when the inhibitor was removed, the tissue was shown to retain an increased capacity for CO_2 fixation. HC-toxin-treated tissue was found to fix as much as 2·9 fold more CO_2 than healthy control tissue (Kuo and Scheffer, 1970). Cell-free preparations from infected (Malca *et al.*, 1964) and HC-toxin treated (Kuo and Scheffer, 1970) leaves gave more active carboxylation of ribose-5-P (forming 3-phosphoglyceric acid) than did similar preparations from healthy leaves. However, the effects on carboxylation do not appear to be direct effects on the enzymes and reactions involved, because addition of toxin after extraction of the enzymes did not result in increased carboxylation. HV-toxin also causes increased carboxylation by intact oat tissues (Kuo and Scheffer, 1970).

It is clear that the host-specific toxins can cause the changes in host physiology that are typical of plant infections in general. The effects of the several toxins differ somewhat but they all seem to cause a change in characteristics of the plasma membrane. Presumably this is to the advantage of the pathogen at the site of infection.

Resistance to the Host-Specific Toxins

Several lines of evidence indicate that resistance of plants to infection by each of the fungi considered herein is the same as resistance to the pathogen-produced, host-specific toxin. The genetic evidence is especially significant. Resistance or susceptibility to infections by *H. victoriae*, *H. carbonum*, and *P. circinata* is controlled in each case by one gene pair (Scheffer and Pringle, 1967). Resistance or susceptibility to toxin is controlled by the same gene pair, in each case. Oat (Luke and Wheeler, 1964; Luke and Wallace, 1969), and maize (Nelson and Ullstrup, 1964) selections have been found which are intermediate between resistant and susceptible. The intermediate reaction is controlled by modifying genes which affect the gene for susceptibility (Wallace *et al.*, 1967; Nelson and Ullstrup, 1964). The host lines that are intermediate in susceptibility to the fungus are also intermediate in sensitivity to toxin (Kuo *et al.*, 1970; Luke and Wheeler, 1964).

If we accept a toxin receptor or sensitive site in the susceptible cell, as discussed previously, it is logical to suggest that the resistant cell simply lacks such a site or substance. Other possibilities have been suggested; among them, that resistant tissue inactivates HV-toxin more rapidly than does susceptible tissue (Romanko, 1959). This hypothesis was rejected because our data show that toxin taken up by both resistant and susceptible oat cuttings was completely inactivated (Scheffer and Pringle, 1964). Wheeler (1969) has confirmed host inactivation of toxin by using a high toxin to tissue ratio, with the toxin solution on the plant surface within the hollow of coleoptiles. He found that resistant and susceptible coleoptiles inactivate toxin at equal rates for the first 8–12 h after exposure. By then, susceptible cells are disorganized and leaking badly, while resistant cells continue to function normally. Since resistance or susceptibility to HV-toxin is evident within less than a minute of exposure, it seems probable that resistance and ability to inactivate toxin are independent characteristics.

HC- and PC-toxins were used in further tests of inactivation by plant tissues. HC-toxin could not be recovered from resistant or susceptible cuttings (Kuo and Scheffer, 1969). However, PC-toxin appears to be much more stable in host tissues than are HC- and HV-toxins; 25–72% of PC-toxin taken in by cuttings was recovered after exposure times ranging from 5 to 21 h. There were no significant differences between the amounts recovered from resistant and susceptible tissues (Mansour, 1968; Gardner, 1971). Thus the fundamental picture with HC- and PC-toxins is the same as with HV-toxin; there is no good evidence that resistance depends on superior ability of resistant tissue to inactivate toxin.

Resistance was expressed equally by active vegetative cells, by resting seeds, and by resting aleurone cells. Seeds in the resting condition, with very low metabolism, were exposed briefly to toxin or to water, re-dried, and stored until no toxin could be detected. These seeds were later placed under conditions conducive to germination. The water-exposed controls germinated and grew normally, but the seeds exposed to toxin while in the resting condition did not germinate. There were similar experiments with half-seeds devoid of embryos. In such preparations only the aleurone cells are capable of synthetic activity. Aleurone cells are triploid, fully differentiated, non-dividing, and do not synthesize α-amylase until they are activated by an exogenous source of gibberellic acid (GA). Toxin blocked synthesis and secretion of amylase by susceptible but not by resistant aleurone cells. Inactive aleurones can be allowed to imbibe water or toxin solution, dried, and activated days or weeks later with GA. Control aleurones exposed to water produced α-amylase after GA treatment. Susceptible aleurones exposed to toxin did not synthesize α-amylase after GA treatment, although no residual toxin was detected (Samaddar and Scheffer, 1970).

The conclusion from the seed germination and aleurone experiments is that toxin will act on resting tissue as well as on tissue with an active metabolism. Germination of resistant seeds and activation of resistant aleurones was not affected by toxin, whether the exposure occurred during resting or active metabolism. Active metabolism thus is not required for expression of resistance; the data are compatible with the hypothesis that susceptible cells possess preformed receptor or sensitive sites not present in resistant cells.

We have considered still other possible explanations of resistance to *H. victoriae*, *H. carbonum*, and *P. circinata*. The available data indicate no significant role for certain post-invasion-produced fungal inhibitors or phytoalexins (Nishimura and Scheffer, 1965; Samaddar and Scheffer, 1970), in resistance to *H. victoriae*. The single gene pair control of resistance and susceptibility is not easily rationalized in multi-factor hypotheses of resistance. Wheeler and Pirone (1969) have suggested a further interesting possibility, that HV-toxin can induce resistance in non-host plants to other infections. Bean cuttings were allowed to take up solutions containing HV-toxin or inactivated toxin. The concentrations used caused no obvious symptoms in the bean shoot. Cuttings thus treated with active toxin were protected against subsequent virus infection to a significant degree, whereas the controls with inactive toxin were not. This was considered possible evidence for toxin activation of a defensive self-repair mechanism which might account for toxin resistance. It is not clear at this time whether or not bean is unusual in this respect. In similar experiments, HV-toxin gave no protection of barley against infection by powdery mildew (Gardner and Scheffer, unpublished). Furthermore, a defensive self-repair mechanism does not seem to fit the data obtained with resting cells (Samaddar and Scheffer, 1970).

In summary, experiments with three host-specific toxins tell us that resistance to some fungi is the same as lack of sensitivity to certain products of these fungi. The best available hypothesis at this time is that resistance of this type is a constitutive or passive characteristic; that is, absence of receptor sites or substances that are present in susceptible cells. Whether or not this type of resistance is widespread is a matter for speculation. Other types of plant disease resistance seem likely, but little is known about them at this time.

General Discussion

There seems to be general agreement that pathogenicity, host-selectivity, and disease resistance have chemical bases. Work with the host-specific toxins has application to each of these phenomena. We do not know how widespread such toxins may be in nature, but it seems likely that we will be able to isolate significant determinants from many other plant pathogens. This may be only a matter of learning the exact techniques and conditions required for each case. To date, all substances known with full confidence to be essential for colonization and pathogenicity are host-specific toxins. The potential significance of these factors became evident when the early studies were shown to have predictive value. Several essential determinants were discovered on the basis of knowledge gained with HV-toxin (Scheffer and Pringle, 1961; Scheffer and Ullstrup, 1965; Smedegård-Petersen and Nelson, 1969; Steiner and Byther, 1969). It seems logical that there are entirely different mechanisms of pathogenicity, but other essential determinants or mechanisms have not been established to date.

High degrees of toxicity and selectivity are striking features of the essential determinants of pathogenicity found so far. However, neither of these characteristics should be expected of all determinants. Clearly, many pathogenic micro-organisms have only limited specificity of host range. More than one substance may be essential for a given pathogen, and determinants could include both host-specific and non-specific factors, either toxic or non-toxic. The determinants are not necessarily diffusable factors from the pathogen, and they do not necessarily account directly for all or any of the major disease symptoms. The important thing to remember is that pathogenicity and host-specificity must have chemical bases. If one or more host-selective compounds are not involved, then how can we explain host-specific pathogenicity?

There is theoretical significance for the host-specific toxins in addition to their significance for understanding disease development and disease resistance. We have in these substances low molecular weight compounds that can "recognize" one type of one species among all others tested. Is it theoretically possible to find or to make molecules that could "recognize" any other individual species or type? If a host-specific toxin is shown unequivocally to have its primary lesion in the plasma membrane, then the toxins may be valuable tools for further study of this structure. Once chemical structures are determined for a toxin and its hypothetical receptor we may have further clues to a knowledge of membrane composition. We could hope to ask whether or not there are heritable differences in structure of the plasma membrane in closely related species and cultivars. There have even been suggestions that HV-toxin affects the plasma membrane without a direct effect on the tonoplast (Luke *et al.*, 1966).

Finally, the toxin studies have had some practical applications to agriculture. Wheeler and Luke (1955) first showed that HV-toxin can be used to eliminate susceptible seedlings in plant breeding programmes, making mass screening rapid and efficient. PC-toxin now has similar uses in breeding for resistance to *P. circinata* in sorghum (Schertz and Tai, 1969). As a result of the work of Steiner and Byther (1969), geneticists of the Hawaii Sugar Planters' Association are using the host-specific toxin of *H. sacchari* in their sugarcane breeding programme. Other known and potential host-specific toxins may have similar uses.

General Summary

Three different fungi (*Helminthosporium victoriae* affecting oats, *H. carbonum* affecting corn, and *Periconia circinata* affecting sorghum) and their host-specific toxins were used in comparative studies. In each case, resistance or susceptibility to the fungus and to the toxin is controlled at a single gene locus. The toxins are low molecular weight peptides that cause the same biochemical changes found in infected plants. Non-host plants, including resistant cultivars, tolerate from 100 to 1,000,000 times higher levels of toxins than do suscepts. Extensive data show that in each case toxin is required for pathogenicity by the fungus producing it. Data indicate that one of these, the toxin of *H. victoriae*, may cause an initial biochemical lesion in the plasma membrane of susceptible oat cells. Resistant oat cells appear to lack a receptor or sensitive site, or to have a receptor with less affinity for toxin. Other host-specific toxins are known from species of *Alternaria* and *Helminthosporium*.

Acknowledgements

The work of R. P. Scheffer and associates at Michigan State University, discussed as part of this chapter, was supported by grants from the National Science Foundation (G 13913, GB 1448, and GB 6560) and by the Michigan Experiment Station (Journal Article No. 5239).

References

Braun, A. C. (1955). *Phytopathology* **45**, 659–664.

Black, H. S. and Wheeler, H. (1966). *Am. J. Bot.* **53**, 1108–1112.

Comstock, J. C. and Scheffer, R. P. (1970). *Phytopathology* **60**, 1288 (Abst.).

Daly, J. M. and Sayre, R. M. (1957). *Phytopathology* **47**, 163–168.

Flentje, N. T. (1965). *In* "Ecology of Soil-Borne Plant Pathogens" (K. F. Baker and W. C. Snyder, Eds), pp. 255–268. University of California Press, Berkeley.

Gardner, J. M. (1971). Ph.D. Thesis, Michigan State University, East Lansing 48823, U.S.A.

Gardner, J. M. and Scheffer, R. P. (1969). *Phytopathology* **59**, 1027 (Abst.).

Gardner, J. M. and Scheffer, R. P. (1970). *Phytopathology* **60**, 1292–1293 (Abst.).
Hanchey, Penelope (1969). *Phytopathology* **59**, 1960–1962.
Hiroe, I. and Aoe, S. (1954). *J. Fac. Agr. Tottori Univ.* **11**, 1–26.
Hiroe, I., Nishimura, S., and Sato, M. (1958). *Trans. Tottori Soc. agr. Sci.* **11**, 291–299 (in Japanese, English summary).
Katsumi, M., Tamura, S., and Sakurai, A. (1967). *Pl. Cell Physiol.* **8**, 399–407.
Krupka, L. R. (1959). *Phytopathology* **49**, 587–594.
Kuo, M. S. (1968). Ph.D. Thesis, Michigan State University, East Lansing 48823, U.S.A.
Kuo, M. S. and Scheffer, R. P. (1969). *Phytopathology* **59**, 1779–1782.
Kuo, M. S. and Scheffer, R. P. (1970). *Phytopathology* **60**, 1391–1394.
Kuo, M. S., Yoder, O. C., and Scheffer, R. P. (1970). *Phytopathology* **60**, 365–368.
Leukel, R. W. (1948). *J. agr. Res.* **77**, 201–222.
Luke, H. H. and Freeman, T. E. (1964). *Nature* **202**, 719–720.
Luke, H. H., Freeman, T. E., Garrard, L. A., and Humphreys, T. E. (1969). *Phytopathology* **59**, 1002–1004.
Luke, H. H., and Wallace, A. T. (1969). *Phytopathology* **59**, 1769–1770.
Luke, H. H., Warmke, H. E., and Hanchey, P. (1966). *Phytopathology* **56**, 1178–1183.
Luke, H. H. and Wheeler, H. (1955). *Phytopathology* **45**, 453–458.
Luke, H. H. and Wheeler, H. (1964). *Phytopathology* **54**, 1492–1493.
Malca, I., Huffaker, R. C., and Zscheile, F. P. (1964). *Phytopathology* **54**, 663–669.
Mansour, I. S. (1968). Ph.D. Thesis, Michigan State University, East Lansing, 48823.
Meehan, F. and Murphy, H. C. (1947). *Science* **106**, 270–271.
Mirocha, C. J. and Rick, P. D. (1967). *In* "The Dynamic Role of Molecular Constituents in Plant-Parasite Interaction" (C. J. Mirocha and I. Uritani, Eds), pp. 121–143. Bruce Publishing Co., Minneapolis, Minn.
Nelson, O. E. and Ullstrup, A. J. (1964). *J. Heredity* **55**, 195–199.
Nishimura, S. and Scheffer, R. P. (1965). *Phytopathology* **55**, 629–634.
Novacky, A. and Wheeler, H. (1970). *Phytopathology* **60**, 467–471.
Pegg, K. G. (1966). *Qd. J. Agr. Animal Sci.* **23**, 15–28.
Pegg, G. F. and Sequeira, L. (1968). *Phytopathology* **58**, 476–483.
Pringle, R. B. and Braun, A. C. (1957). *Phytopathology* **47**, 369–371.
Pringle, R. B. and Scheffer, R. P. (1963). *Phytopathology* **53**, 785–787.
Pringle, R. B. and Scheffer, R. P. (1964). *A. Rev. Phytopath.* **2**, 133–156.
Pringle, R. B. and Scheffer, R. P. (1967a). *Phytopathology* **57**, 530–532.
Pringle, R. B. and Scheffer, R. P. (1967b). *Phytopathology* **57**, 1169–1172.
Romanko, R. R. (1959). *Phytopathology* **49**, 32–36.
Ruesink, A. W. and Thimann, K. V. (1965). *Proc. Nat. Acad. Sci.* (*U.S.*) **54**, 56–64.
Samaddar, K. R. (1968). *Phytopathology* **58**, 1065–1066 (Abst.).
Samaddar, K. R. and Scheffer, R. P. (1968). *Pl. Physiol.* **43**, 21–28.
Samaddar, K. R. and Scheffer, R. P. (1970). *Pl. Physiol.* **45**, 586–590.
Samaddar, K. R. and Scheffer, R. P. (1971). *Physiol. Pl. Path.* **1**, 319–328.

Sawamura, K. (1966). *Bull. Hort. Res. Sta., Japan* (Morioka), series C, No. 4 (in Japanese, English summary).
Scheffer, R. P., Nelson, R. R., and Pringle, R. B. (1964). *Phytopathology* **54**, 602–603.
Scheffer, R. P., Nelson, R. R., and Ullstrup, A. J. (1967). *Phytopathology* **57**, 1288–1291.
Scheffer, R. P. and Pringle, R. B. (1961). *Nature* **191**, 912–913.
Scheffer, R. P. and Pringle, R. B. (1963). *Phytopathology* **53**, 465–468.
Scheffer, R. P. and Pringle, R. B. (1964). *Phytopathology* **54**, 832–835.
Scheffer, R. P. and Pringle, R. B. (1967). *In* "The Dynamic Role of Molecular Constituents in Plant-Parasite Interaction" (C. J. Mirocha and I. Uritani, Eds), pp. 217–236. Bruce Publishing Co., Minneapolis, Minn.
Scheffer, R. P. and Samaddar, K. R. (1970). *In* "Recent Advances in Phytochemistry", Vol. 3, pp. 123–142. Appleton-Century-Crofts, New York.
Scheffer, R. P. and Ullstrup, A. J. (1965). *Phytopathology* **55**, 1037–1038.
Schertz, K. F. and Tai, Y. P. (1969). *Crop Sci.* **9**, 621–624.
Smedegård-Petersen, V. and Nelson, R. R. (1969). *Can. J. Bot.* **47**, 951–957.
Steiner, G. W. and Byther, R. S. (1969). *Phytopathology* **59**, 1051 (Abst.).
Tanaka, S. (1933). *Mem. College Agr. Kyoto Imp. Univ.* **28**, 1–31.
Wallace, A. T., Singh, R. M., and Browning, R. M. (1967). *In* "Induzierte Mutationen und Ihre Nutzung." Edwin-Baur Gedachtnisvorlesungen IV. Akademie-Verlag, Berlin.
Wheeler, H. (1969). *Phytopathology* **59**, 1093–1097.
Wheeler, H. and Black, H. S. (1963). *Am. J. Bot.* **50**, 686–693.
Wheeler, H. and Doupnik, Jr., B. (1969). *Phytopathology* **59**, 1460–1463.
Wheeler, H. and Hanchey, P. (1966). *Science* **154**, 1569–1571.
Wheeler, H. and Hanchey, P. (1968). *A. Rev. Phytopath.* **6**, 331–351.
Wheeler, H. and Luke, H. H. (1955). *Science* **122**, 1229.
Wheeler, H. and Pirone, T. P. (1969). *Science* **166**, 1415–1417.
Yoder, O. C. and Scheffer, R. P. (1969a). *Phytopathology* **59**, 1954–1959.
Yoder, O. C. and Scheffer, R. P. (1969b). *Phytopathology* **59**, 1058–1059 (Abst.).
Yoder, O. C. (1970). *Phytopathology* **60**, 1320–1321 (Abst.).

Discussion

Marrè: If I understood correctly, the concentration of the toxin giving a significant response is indeed very low. Can you put it in number of molecules per cell? If this number is exceedingly low, also the number of target molecules in the cell should be very low. Then, one might wonder about the possibility that an amplification mechanism mediates the action of the toxin from its primary site (possibly in the cytoplasm) to its final physiological effects at cell membrane level.

Scheffer: My guess, based on calculations, is that <100 toxin molecules per cell can cause damage. We have considered possible amplification effects, but have no experiments that bear on this question. The amplification could be from target sites in the plasma membrane or in the cytoplasm.

Page: May I have comments on the mechanisms of de-toxification of host-specific toxins in resistant plants?

Scheffer: All available data indicate that resistant and susceptible tissues inactivate host-specific toxins at approximately equal rates. Our results were confirmed by other workers. We do not know the mechanisms of de-toxification; there does not appear to be a relationship between ability to inactivate toxin and resistance to toxin or to the fungus.

Pringle: It is very difficult to design experiments to study detoxification of HV toxin because of the great chemical instability of the molecule. The property of host-specific toxicity is quickly lost under very mild conditions at pH 8·4 with sodium bicarbonate for example.

We first became aware of the possibility of amplification of toxicity of HV toxin, something like an autocatalytic effect, many years ago when we were developing a quantitative bioassay for the toxin. We found that we did not get a typical sigmoidal dose—response curve, that it was difficult to find a point of 50% inhibition of root growth. This is why we measure inhibition as the minimum quantity to produce complete inhibition (root growth less than 1 cm).

Kuć: We have observed that *H. carbonum*, *H. maydis*, *H. turcicum* and other *Helminthosporium* spp. cause symptoms on many non-hosts including apple, cucumber, tomato, bean and potato leaves, carrot roots and potato tubers. Symptoms range from the collapse of one or two cells to a readily visible necrotic fleck. Resistant reactions to *H. victoriae* and *H. carbonum* in their respective hosts are as distinct as symptoms of susceptibility and include collapse of cells, necrosis or chlorosis and containment of the fungus. I therefore suggest that there may be two toxins, one responsible for resistance and one for susceptibility or that there is a mechanism in addition to resistance to toxin to explain containment of the fungus in the resistant host.

Scheffer: Your hypothesis could be correct; certainly, I cannot cite conclusive evidence to the contrary. However, resistance to *H. victoriae* and to its toxin is genetically the same thing. This is true also for *P. circinata* and *H. carbonum.* The genetic analyses do not indicate a second system. Also, maize plants susceptible to *H. carbonum* show the resistant-type necrotic response when inoculated with a non-pathogenic (non-toxic-producing) mutant of the fungus. This response does not occur when a trace of toxin is applied along with the inoculum.

Wood: Many pathogens occur as races highly specific in their parasitism on closely related cultivars of a host species. Is this type of specificity likely to be explained in terms of host-specific toxins such as HV toxin?

Also, many other pathogens are much less specific in their parasitism than is *H. victoriae* though, of course, comparatively few species are still attacked by them. Here, presumably, toxins *per se* have a different role in determining whether or not a host is parasitized.

Scheffer: The essential difference between *H. victoriae* and *H. carbonum* is controlled by two gene pairs; thus the two may be considered as races of one species. In this case host specificity is determined by toxins.

Some fungi have wide host ranges. If their pathogenicities depend on toxins, then the toxins must have wider ranges of sensitive plants than do the host-specific toxins described to date.

Pringle: In attempting to answer the question of non-specific pathogens —fungi that attack more than one host—I think of the example of crosses between *H. victoriae* and *H. carbonum* where some of the progeny are non-specific in that they attack oats and corn. Perhaps a naturally-occurring, non-specific pathogen such as *H. sativum*, which attacks barley and wheat, is a similar example.

Strobel: Is there any evidence that the *Periconia* toxin is produced in the soil before entry of the parasite?

Scheffer: There is no evidence. Your question reminds me of the widely held notion that *P. circinata* is a "pathogenic saprophyte" which does not invade living tissue. We could find no basis for this belief. On the contrary, *P. circinata* invades susceptible tissue in an orthodox manner.

Turner: What is the effect of concentration of HV-toxin on the speed of electrolyte loss in susceptible oats?

Scheffer: Within a certain range, which covers approximately a thousand-fold dilution, the higher the toxin concentrations, the more rapid is the rate of electrolyte loss from exposed tissues. One of our toxin assays is based on this. Less than 0·1 μg toxin/ml will cause an almost instantaneous increase in electrolyte loss.

Morpurgo: I should like to have additional information on the genetic basis of resistance to HV toxin, in particular whether there is only one gene for resistance?

Scheffer: Resistance to HV-toxin is controlled by a single recessive gene. Reaction to PC-toxin is also controlled by a single gene pair, with susceptibility dominant over resistance. HC-toxin differs in that resistance is dominant over susceptibility. (This question and answer are covered adequately in the text.—ED.)

Matta: Is there any evidence for localization of HV toxin in cells penetrated by the pathogen? Could not this be obtained with labelled material?

To what extent does *H. victoriae* grow in host tissue after cells have been killed by the toxin?

Scheffer: We have used ^{14}C-labelled HV- and HC-toxins without conclusive results. Both resistant and susceptible tissues inactivate toxin.

It is unlikely that we can differentiate between the fate of labelled toxin and the fate of labelled breakdown products.

In the initial phase of colonization, *H. victoriae* penetrates and grows in tissues that are alive, as far as we can tell (see Yoder's paper). In later stages of disease, host cells are killed beyond the area of fungal growth. The fungus may continue to grow in killed tissues for some time.

Goodman: Can simple diffusion of the HV toxin be demonstrated as movement across a membrane of epidermal cells and cuticle?

Scheffer: We have not tried this exact experiment. However, we know from indirect evidence that toxin can pass through susceptible root tissues, and move upward with the transpiration stream. Apparently the movement occurs after roots cells are disorganized. Active toxin does not move through resistant tissues, as far as we can tell.

Main: Granting you do not know the exact structure of your host specific toxin, would it be possible to use some method of "protecting" the toxin itself from degradation or detoxification? Perhaps some chemical treatment could cover or protect the sensitive site on the toxin and not materially affect its biological activity. The path of entry, localization via labelling, and fate of the intact, *active* molecule could then be followed.

Of course this might complicate purification and characterization, but again it might give information on the structure by indicating the degradation sensitive site on the toxin molecule itself.

Scheffer: We have not tried systematically to do this. Toxin is relatively stable in crude preparations at pH 3·5. Perhaps a stabilizing substance could be found in such preparations.

Lebeau: Can you compare the concentrations of the toxins you describe with those of lycomarasmin that are needed to produce disease symptoms?

Do you consider lycomarasmin a phytotoxin?

Scheffer: The best preparation of HV-toxin used to date completely inhibited growth of susceptible seedlings roots at 0·0002 μg/ml. The other host-specific toxins are not so active, but they give complete inhibition at $<1{\cdot}0$ μg/ml. Lycomarasmin is relatively innocuous by comparison. The question of lycomarasmin as a "phytotoxin" depends on your definition of the term. There is no conclusive evidence that lycomarasmin is involved in disease development in plants infected with *Fusarium* wilt fungi.

16. The Killing of Plant Cells by Soft Rot Parasites

RONALD K. S. WOOD

Department of Botany and Plant Technology,
Imperial College, London, S.W.7, England

Introduction

Soft rots are diseases of higher plants caused by bacteria and fungi. They are typical of parenchyma especially of storage tissue such as potato tubers and of chlorenchyma in which they take the form of rapidly spreading lesions which may or may not be discoloured. Rotted tissue usually shows no shrinkage in volume and is water soaked with little or no coherence so that on transfer to water it readily breaks up into a suspension of individual cells or small clumps of cells. Individual cells retain their identity for some time after the tissue has lost its coherence. This loss of coherence is usually described as maceration. However, this term is also used for the much more drastic effects on tissues in which cell and cell parts lose their identity. It has been suggested with good reason that the term "cell separation" be used instead of maceration for the early, critical stages in soft rots in which the tissue loses coherence but in this paper I shall continue to use the term maceration.

The other striking feature of soft rots is that protoplasts of cells in macerated tissue rapidly die. Why they do so is the subject of this paper.

Death of cells in soft rots was recognized by de Bary in 1886 working with a thermolabile principle extracted from tissue rotted by the fungus *Sclerotinia libertiana* Fckl. (=*S. sclerotiorum* (Lib.) de Bary) which readily macerated plant tissues. But de Bary did little to establish the relationship between maceration and killing. This had to await the classical work of Brown (1915) on how the fungus *Botrytis cinerea* Pers. *ex* Fr. affected plant tissues. Until recently, most of what was known about maceration and cell death derived from this and later work by Brown and his colleagues at Imperial College. But during the past ten years or so there has been a renewed interest in the subject from another viewpoint when cell physiologists began to use cell wall degrading enzymes from commercial sources to isolate protoplasts, largely in ignorance, so it would seem, of the work done by pathologists on the killing of protoplasts by the types of enzymes that they were using.

A small number of plant pathologists have also once again become interested in this problem, stimulated no doubt by the recent review by Brown in 1965, and encouraged by the increasing availability during the past decade or so of better and more easily used techniques for separating the many enzymes that act on cells, and for studying how they act.

In this paper I shall begin by surveying the subject generally, continue by giving a short account of work done recently at Imperial College and end with some speculations about what happens to protoplasts when cells separate.

Types of Tissues Attacked

Soft rots by their nature are confined to parenchyma and chlorenchyma, the cell walls of which are little altered by deposition of substances such as suberin or lignin, and in which the protoplast is active and surrounds a more or less well developed vacuole. Meristematic tissue with cell walls of different composition and in which the vacuole is less well developed may also be attacked by soft rot parasites, but this has been very little studied and will not be dealt with here though there are some interesting problems connected with it.

Nature and Composition of Material between Protoplasts

Middle Lamella

By this is generally meant the zone between primary walls of adjacent cells but the nomenclature is a little confused. I am not certain whether the middle lamella is to be regarded as part of the cell wall but for convenience I shall refer to it as such so that half of it belongs to the wall of one cell and half to the next cell.

The middle lamella does not contain cellulose microfibrils. It starts as the phragmoplast at the equator of the spindle, between two daughter nuclei of a dividing cell, the zone in which is formed the cell plate which later extends from the spindle to the existing side walls of the cell. The cell plate is initiated by a large number of 100 nm vesicles probably derived from the Golgi bodies and which probably contain precursors of pectic substances. The evidence for this is that the cell plate stains strongly with ruthenium red which is, however, specific only for carboxyl groups. Also, the middle lamella stains strongly after treatment with alkaline hydroxylamine which reacts with methyl esters to give hydroxamic acids. These bind ferric ions tightly to reveal sites of former methyl esters. Carboxyl groups and methyl esters are typical of pectic substances but are not confined to them.

Outside the middle lamella is the primary wall dominated structurally by its network of cellulose microfibrils which may be several nm long, 5–10 nm thick and 10–30 nm wide. Each is a bundle of parallel unbranched cellulose (β-1,4-glucan) molecules. About one-third of the microfibril is crystalline with regular H-bonding between OH groups of adjacent chains. The remainder is less well ordered with no regular bonding and occurs within and at the surface of the microfibril; the OH groups of these paracrystalline zones are important for H-bonding between microfibrils. The primary wall is regarded as that part of the cell wall produced while the cell is growing in surface. In it the microfibril meshwork is relatively loose.

Outside the primary wall is the secondary wall formed after the wall has stopped growing. Here the microfibril meshwork is more compact.

For reasons which will become apparent we are mainly concerned with the primary wall as the zone next to the middle lamella.

Microfibrils of both walls are embedded in a matrix which presumably is continuous with the middle lamella. The matrix of the primary wall contains pectic substances and hemicelluloses of types and in proportions that differ considerably in different tissues of the same plant, in tissues of different ages, and in different species. The matrix of the secondary wall is dominated by hemicelluloses. Both groups of substances are more hydrophilic, some very much more so than cellulose.

In vivo both middle lamella and matrix are highly hydrated; a figure of *c.* 60% water is given for collenchyma and corn coleoptile cell walls. A cellulose content of *c.* 30% dry weight would mean that the microfibrils occupied about 10% of the volume of the cell wall with a mean interfibrillar distance *c.* 4 × diameter of the microfibrils (Roelofsen, 1963). The density of meshwork of microfibrils is much less in the outer than in the inner zones (Setterfield and Bayley, 1958). This means that the microfibrils near the middle lamella would be closer together with increased chances of H-bonding through substances of the matrix (Roelofsen, 1963).

The general picture is, therefore, of a highly hydrated, very firm gel continuous between adjacent protoplasts, changing in composition from the centre outwards, with a thin central zone free of the cellulose microfibrils that form a complex meshwork through the much thicker zones on either side.

Plasmodesmata, the protoplasmic connexions between cells that break the continuity of the cell wall will be referred to later.

Substances of the Middle Lamella and Matrix

These are amorphous, isotropic and show little or no structure under the electron microscope. There are two main groups, the pectic substances and hemicelluloses. Much less quantitatively though possibly of considerable importance functionally are proteins and lipids. The proportions of these groups vary tremendously. Some general figures are: hemicelluloses 35–50%, pectic substances 10–20%, proteins 3–10% and lipids 2–7% (with cellulose at 20–30%).

This paper is concerned mainly with the pectic substances and to a much less degree with the hemicelluloses. The classification and characterization of both groups of substances is difficult because almost always in the past they have been isolated from plants by arbitrary extractions with acid and alkali. Now it is becoming increasing recognized, by biochemists and, I hope, also by pathologists, that the degraded products of such extractions give a poor picture of the various polymers as they occur *in vivo* (Albersheim, 1965). With these limitations in mind we can for the present regard the pectic substances as complex polymers dominated by α-D-(1,4) linked galacturonic acid residues, mainly linear but containing as side chains or within the main chains significant amounts of arabinose and galactose, less frequently rhamnose, and still smaller amounts of other sugars or sugar acids.

Many of the galacturonic acid carboxyl groups may be esterified with methanol, transforming pectates (polygalacturonates) to pectinates (polymethylgalacturonates). The predominance of carboxyl and methyl ester groups is the basis of the ruthenium red and hydroxylamine-ferric ion tests for this group of substances. The proportions of these groups have profound effects on the properties of the polymers.

The pectic substances are extracted from plants with water, dilute acid and solutions of chelating agents. Arabans and galactans, branched and linear polymers of arabinose and galactose, may also be extracted by these procedures.

The hemicelluloses are usually extracted by hot alkaline solutions. They are a complex and heterogeneous group of polysaccharides occurring usually as linear, less frequently branched polymers of xylose, arabinose, mannose, galactose and glucose. In some, the polymer is dominated by one type of residue, in others there is a mixture of two or more. Glucuronic acid or its methyl ester may occur as minor components usually in side chains.

Protein is now regarded as an integral component of cell walls though its role does not yet seem to have been established. It has been suggested that protein is covalently linked to cell wall carbohydrate to give a complex called "extensin" because it has been found that after removal of some 70% of the cell wall by enzyme treatment, the wall shaped structure remaining still contained 80–90% of the original protein. It also contained a large amount of arabinose and galactose. On this sort of evidence "extensin" is envisaged as playing a part in cell wall extension.

Linkages between Microfibrils and Components of Matrix

These are the linkages that will be in part responsible for the structural integrity of the middle lamella and cell wall. They augment those between the units of the different polymers and they are well described by Roelofsen (1965). Between polysaccharides there will be H bonds between two hydroxyl groups or between a hydroxyl group and a O-atom of pyranose or furanose rings or of glycosidic bridges.

Carboxyl groups of pectic substances or hemicelluloses may bind OH groups, ionized or not.

Relatively strong double bonds may occur between two non-ionized COOH groups in an acid medium. Proteins with peptide groups and H-donating or accepting groups in side chains may form complexes with polysaccharides.

Acid groups may be bound through metal ions, especially calcium and magnesium, which may also form bridges between uronic acid or other anionic groups and also between these and OH groups of polysaccharides.

This multiplicity of possible linkages between the components of the cell wall makes possible very large complexes and explains the insolubility of the cell wall and why slices of parenchyma remain coherent in water for so long in the absence of contamination.

The Process of Maceration

The first major study of maceration was by Brown (1915) for the fungus *Botrytis cinerea* in which he demonstrated the close connexion between cell separation caused by extracts of young hyphae and death of protoplasts as indicated by loss of ability to plasmolyse. It is only comparatively recently that any appreciable advances have been made beyond the stage reached by Brown and all have emphasized as did Brown's work the importance of cell separation in causing cell death. A short account of maceration will, therefore be given first.

It has long been established that enzymes of one sort or another are responsible for cell separation. Because cells separate along the line of the middle lamella it is not surprising that pathologists and others have emphasized the importance in maceration of the enzymes that degrade the main structure of pectic substances; these have usually been considered as the most important components of this zone between primary cell walls. Earlier, these enzymes were called "pectinases" or "protopectinases", the latter referring to protopectin, an insoluble parent substances yielding soluble pectin on limited acid hydrolysis.

In fact, until relatively recent direct evidence that either of these ill-defined groups of enzymes was responsible for maceration was indeed quite slight, and even now, some would not regard it as conclusive.

However, during the past fifteen years or so the pectic substances and the enzymes that degrade them have been studied in considerable detail to reveal a picture of considerable complexity. The more important conclusions from this work are as follows.

(1) As mentioned earlier, the main structure of pectic substances extracted from plants and, therefore, presumably *in vivo* too, is dominated by galacturonic acid residues the carboxyl groups of which may or may not be esterified with methanol.

(2) The forms in which pectic substances occur *in situ* and the reasons for their insolubility are still not known with certainty but considerable importance is usually attached to the possible formation of large complexes through calcium between carboxyl groups.

(3) The enzymes that act upon pectic substances are classified as follows:

- (a) Pectinmethylesterase (PME) acting upon ester bonds between carboxyl groups and methanol and controlling the number of free carboxyl groups available for cross linkage and other reactions.
- (b) Chain-splitting enzymes that act upon α-(1,4) glycosidic linkages between adjacent galacturonic acid residues. These enzymes can be divided into the following groups.
  - (i) Glycosodic bonds ruptured by hydrolysis (polygalacturonases), or by *trans*-elimination of the proton from C5 of one residue to give an unsaturated bond between C4 and 5 of the reaction product which absorbs strongly at 235 mμ (*trans*-eliminases, lyases).
  - (ii) Enzymes attacking preferentially esterified or non-esterified parts of the chain.
  - (iii) Enzymes attacking chains terminally (exo-), or at random (endo-).

On this basis, eight major groups of chain-splitting enzymes are possible.

Most *trans*-eliminases so far studied are activated by calcium with a pH optimum at *c.* 5·0, or more usually *c.* 9·0. Polygalacturonases usually have pH optima *c.* 5·0, are not activated, indeed are usually inhibited by calcium.

The above means that the same end result, the breaking of long chain polymers of pectic substances into smaller units may be brought about in many different ways by the multitude of pectic enzymes produced by plant parasites. These enzymes all rupture linkages between galacturonic acid residues. The role of enzymes rupturing linkages between these and other residues known to occur in pectic substances has been little investigated though it is known that the chain-splitting enzymes referred to above yield substantial amounts of substances other than galacturonic acid or its unsaturated derivative.

For many years attempts have been made to equate maceration with one or more of the enzymes that degrade the main components of cell walls. It is now clear that maceration can occur quite rapidly in extracts of rotted tissue or in cultures that have a low or not readily detected proteolytic or cellulolytic activity. And there is, as yet, no good evidence that hemicellulases are critical for the process. This is not to say that one or more of these enzymes do not play a part in maceration *in vivo* especially in the later stages. But all the good evidence available so far suggests that it is one or other of the chain-splitting pectic enzymes that are essential for the early and important stages of maceration. The following are among those implicated so far; polygalacturonase, *Rhizoctonia solani* Kühn (Bateman, 1963); *Corticium praticola* Kotila (Hall and Wood, 1971); *Rhizopus tritici* Saitô (Zaitlin and Coltrin, 1964); *Rhizopus stolonifer* (Ehrenb. *ex* Fr.) Lind. or *Rhizopus* sp. (Susuki *et al.*, 1967); pectate *trans*-eliminase, *Erwinia aroideae* (Towns.) Hall. (Dean and Wood, 1967); *E. atroseptica* (van Hall) Jenn. (Hall and Wood, 1971); pectin methyl *trans*-eliminase, *Sclerotinia fructigena* Aderh. *et* Ruhl. (Byrde and Fielding, 1968). This list is not a complete one.

All the above strongly suggest that it is the rupture, by hydrolysis or *trans*-elimination, of the main chain of pectic substances occurring in the middle lamella or in the matrix of the cell wall immediately adjacent that causes cell separation, though the precise mechanism is still largely unknown. In fact the work of the past few years has confirmed what has been assumed for very many years.

At this point it might be as well to warn that in spite of the evidence provided by the papers referred to above we may not yet be quite out the wood because in a recent paper, Karr and Albersheim (1970) claim that Pectinol R-10 a commercial preparation from *Aspergillus niger* van Tiegh. cultures contains a "wall modifying enzyme" which must act before cell walls prepared from slices of bean hypocotyl can be degraded by other enzymes in the mixture, or by an α-galactosidase from *Colletotrichum lindemuthianum* (Sacc. *et* Magn.) Br. *et* Cav. This enzyme releases water soluble, 70% ethanol insoluble polymers from cell walls but is unable to degrade a wide variety of polysaccharide, glycoside and peptide substrates though it does degrade polygalacturonic acid to a limited extent. Whether this enzyme plays an important part in maceration remains to be established.

For this paper it will be assumed that at least in some diseases chain-splitting pectic enzymes with polygalacturonase, *trans*-eliminase or other, undetermined, activities, are critical in cell separation. At the same time it must be remembered that the activity of these enzymes *in vivo* may be profoundly altered by the activity of enzymes acting on other components of the cell wall and by factors that affect the substrate. A good example is provided by calcium which may increase the activity of *trans*-eliminase, inhibit polygalacturonase, and affect the susceptibility of the substrate to both.

The connexion between cell separation and killing of protoplasts will now be considered.

Killing of Protoplasts

The early work is substantially that of Brown (1915) who established the close relationship between maceration and cell death. It was followed much later by Tribe (1955) and Fushtey (1957) who confirmed Brown's work and extended it to include maceration and killing by the soft rot bacterium *Erwinia aroideae*. Tribe also made the significant discovery that death of protoplasts as shown by inability to plasmolyse and retain neutral red stain was greatly retarded if tissues were macerated in enzyme solutions containing one of a variety of substances in plasmolysing concentrations. Fushtey found that the protoplasts became sensitive again on de-plasmolysis, that cells treated with macerating enzymes are also more susceptible to the action of other toxins, and that substances produced through the action of macerating enzymes are not toxic in concentrations likely to be found *in vivo*.

I shall now deal with more recent work by Spalding (1969) and by Mrs. J. A. Hall at Imperial College. Spalding used extracts from sweet potatoes rotted by *Rhizopus stolonifer* which had no caseinase activity, only low gelatinase activity, no lipase and no phosphatidase activity. But they macerated sweet potato tissue and were also active on pectate and cellulose substrates. An earlier paper (Spalding, 1963) had shown that cellulase was unlikely to be important in maceration so we are left with the pectic enzymes for this process. Toxicity as indicated by a decrease in respiration of tissue disks of sweet potatoes, potato tuber, turnip root, cucumber fruit and beet root was related to maceration. A variety of other evidence suggested a close relation between decrease in respiration, maceration and polymethylgalacturonase activity. Spalding also found that KNO_3 which at high concentrations had little effect on hydrolysis of pectate greatly retarded maceration of sweet potato disks at 0·5 M but respiration was little affected. Potassium nitrate had little effect on maceration caused by enzymes of *Erwinia carotovora* (Jones) Holland a result obtained earlier for *E. aroideae* by Tribe (1955).

I now turn to the work of Mrs. Hall started in 1967 on the rotting of potato tubers by the bacterium *E. atroseptica* and the fungus *Corticium praticola*. Parallel studies with extracts of rotted tubers and culture filtrates gave essentially similar results. Evidence along a number of lines showed that maceration by *C. praticola* depends on an endopolygalacturonase most active at pH 4·0–5·0 and inhibited by Ca^{++}. In contrast, *E. atroseptica* macerates by producing a pectate *trans*-eliminase with an optimum above pH 9·0 and activated by Ca^{++}. There is a close connexion between the effects of different factors on the activity of these enzymes on pectate substrates in solution and on maceration of tuber disks.

For *E. atroseptica* there is also a close relation between maceration and ability of cells to plasmolyse and retain neutral red stain. In an extract diluted to macerate disks in *c.* 60 min cells begin to lose their ability to plasmolyse in 15 min and on this criterion over half are dead in 60 min and all are dead in 120 min. The neutral red test has given anomalous results with *C. praticola* extracts which need to be re-examined.

The effects of macerating preparations on permeability have also been studied. Disks of tubers were placed in buffered enzyme extracts (untreated or boiled) for known times, removed, washed, placed in water and the increase in conductivity measured over 15 min to give a measure of increase of permeability of protoplasts to electrolytes. This technique was adopted at first to avoid the high background conductivity of extracts but interpretation of results obtained with it are not easy so in later work changes in permeability will be measured in terms of labelled potassium ions.

From the conductivity measurements it is clear that electrolytes begin to leak from cells long before they are separated from each other. Thus with preparations of *C. praticola* and *E. atroseptica* macerating tuber disks in 50 min at optimum pH, increase in conductivity of the ambient solution is readily detectable after 10 min and reaches a peak at about 50 min or earlier. For *E. atroseptica* the increase in permeability is low at pH 4·0–5·0 high at pH 9·0; for *C. praticola* it is high at pH 4·0–5·0 low at pH 9·0.

For *E. atroseptica* with a calcium dependent *trans*-eliminase loss of electrolytes from cells is much lower in dialysed than in non-dialysed extracts; the rate of loss is restored on addition of calcium. Here it is interesting to note that it is easier to demonstrate a calcium effect on changes in permeability than on maceration.

Rot extracts can be readily fractioned with $(NH_4)_2SO_4$. For *E. atroseptica* almost all *trans*-eliminase activity is in the 60–70% saturation fraction which contains only a very small proportion of the total protein. This is the fraction which causes the most rapid increase in permeability. It has no phosphatidase activity.

With *C. praticola* the highest polygalacturonase activity is in the 70–90% saturation fraction and this too is the one causing the largest increase in permeability; again this fraction has no phosphatidase activity. But other fractions with phosphatidase activity and much lower polygalacturonase activity also cause pronounced increases in permeability indicating that phosphatidases of *C. praticola* may have a role in cell death.

The macerating enzymes of *E. atroseptica* also have other interesting effects on potato tuber cells. Thus if disks are placed in an extract which macerates in about 40 min there are striking losses in fresh weight which start soon after immersion and reach some 25% of the initial fresh weight when disks begin to lose coherence. In boiled extracts the loss of weight is negligible.

If the enzyme solution is replaced shortly before maceration fresh weight increases slightly for a short time, then continues to fall. This loss of weight cannot be explained in terms of loss of electrolytes or of non-electrolytes, neither can it be explained by loss of cells, because there is virtually none if disks are not agitated.

Another interesting point is that disks lose about 30% fresh weight if placed in 10^{-2} M IAA for 60 min. If the IAA is then removed and replaced by a macerating solution there is a further loss of 25% fresh weight in the 40-min period required for maceration. Again this does not occur when boiled enzyme solutions are used (Webb, 1970).

These experiments and other evidence suggest strongly that cells rapidly lose water under the influence of macerating enzymes.

From this work it seems clear that extracts containing polygalacturonase or *trans*-eliminase as the main enzymes cause changes in permeability of cells that allow electrolytes to escape and that this occurs well before separation. And it would seem that it is these changes that quite soon cause protoplasts to lose their ability to plasmolyse and then to die.

Isolation of Viable Protoplasts

The killing of protoplasts in soft rots and by cell wall degrading enzymes has attracted the interest of plant pathologists for many years. In contrast it is only comparatively recently that cell physiologists have tried to obtain free protoplasts in quantity and they have resorted to cell wall-degrading enzymes for this purpose. But because one assumes that they are interested primarily in living protoplasts one sees here an apparent contradiction. Some of the more prominent workers in this field have been Cocking and his colleagues in the U.K., Thimann, Ruesink and Hildebrandt in the U.S.A., Eriksson and Jonasson in Europe, and Takebe and Toyama in Japan. Much of the recent work has been done with "Macerozyme" a commercial product based on *Rhizopus* sp. which has a low cellulase activity. Extracts derived from cultures of *Trichoderma viride* Fr. have also been used, though here in some cases the active principles are, in fact, contaminating pectic enzymes. The usual technique is to place plant tissue in high concentrations of the commercial products before or after partial purification, and in plasmolysing solutions e.g. 0·6 M sucrose. Protoplasts released on cell wall degradation are collected by filtration, washed and kept in buffer in sucrose or some other plasmolyticum. Under suitable conditions protoplasts are stated to remain alive for some days and are even reported as regenerating thin cell walls (Pojner *et al.*, 1967).

The apparent contradiction between the aims of the cell physiologists and the findings of plant pathologists is to some extent resolved by Tribe's discovery that cells retain their ability to plasmolyse long after cell separation, if tissues are macerated in solutions that plasmolyse the protoplasts. And we have found recently that if potato tuber disks are macerated in slightly hypertonic solutions of sucrose the expected increase in permeabiliy to electrolytes is considerably delayed. The successful use of cell wall-degrading enzymes in obtaining isolated protoplasts would seem to depend therefore on the fact that a plasmolyticum is used to prevent protoplasts from bursting.

Conclusions

These will be based on the assumption that chain-splitting enzymes acting upon α-(1,4) glycosidic linkages between galacturonic acid residues are mainly responsible for cell separation caused by certain bacteria and fungi. Enzymes with other activities either play no part or are far less important, though some such as cellulases will obviously have a role in later degradation of the cell wall such as may be required, for example, to obtain wall free protoplasts.

Before cells obviously separate from each other the pectic enzymes cause changes that lead to loss of electrolytes and this increase in permeability of the protoplasts culminates in an inability to plasmolyse and, presumably, death. Some of our most recent work with *E. atroseptica* suggests that increase in permeability is accompanied by a decrease in respiration. Increase in permeability to electrolytes and loss of ability to plasmolyse are both retarded in hypertonic solutions. The problem now remains as to how polygalacturonases, or *trans*-eliminases highly specific for substrates considered to be confined to the middle lamella and matrix of the primary cell wall, can so drastically alter permeability which is presumably based on the plasmalemma and the tonoplast. So far as I am aware there is in current thinking no place for galacturonic acid-dominated polymers in the structure of these or any other membranes, and indeed it would be difficult to imagine one. Explanations other than those based on direct effects on the limiting membranes must therefore be sought. Some time ago I suggested that the primary effect of maceration on the protoplasts may be largely physical.

It seems possible that dissolution of substances in the middle lamella or in the outermost layers of the matrix of the primary cell wall may remove a mechanical barrier to the expansion of the plasmalemma under turgor pressure, and that the protoplast ruptures at these points of weakness. Albersheim (1965) points out that the matrix of the primary wall with a polysaccharide content of 10–20% is probably a very dense gel; the middle lamella continuous with this matrix may also be a dense gel, and it is presumably these gels that play a large part in containing the plasmalemma under normal conditions of turgor. Where then are the points of weakness likely to be exposed by the action of the pectic enzymes? Could they be the plasmodesmata? The structure of these connexions through middle lamella and cell walls is still not clearly established because of discrepancies between the results of light and electron microscopy. The latter reveals them for the most part as straight tubes *c.* 25 nm diameter with expanded central regions within the primary wall in some species. The tubes are considered to be lined with plasmalemma continuous with that of the parent protoplasts. They may contain thin hollow threads which are sometimes seen to be connected to ER membranes within the main bodies of cytoplasm. Estimates of the numbers of plasmodesmata between cells made by various workers differ tremendously but it has been claimed by Krull (1960) for cortical parenchyma of *Viscum album* that a fully mature cell may have 6,000–24,000, and figures of the same order are given for cells of onion root tip (Strugger, 1957).

In the normal cell one can assume that the plasmalemma lining plasmodesmata is contained laterally by the middle lamella and the matrix of the cell wall. What happens when the middle lamella is degraded even if cells are not separated mechanically? Does the plasmalemma of the plasmodesmata become exposed to the hypotonic ambient solution and is it then ruptured by osmotic forces? This, if it happened, could account for the rapid increase in permeability before cell separation, the subsequent death of the cells and why both are retarded in hypertonic solutions. What is needed now is a critical examination of what happens to plasmodesmata when cells are treated with enzymes that destroy the integrity of part or the whole of the substance between primary cell walls.

References

Albersheim, P. (1965). *In* "Plant Biochemistry." (J. Bonner and J. Varner, Eds), pp. 151–186, 298–321. Academic Press, New York.

de Bary, A. (1886). *Bot. Z.* **44**, 377–474.

Bateman, D. F. (1963). *Phytopathology* **53**, 1178–1186.

Brown, W. (1915). *Ann. Bot.* (*Lond*). **29**, 313–348.

Byrde, R. J. W. and Fielding, A. H. (1968). *J. gen. Microbiol.* **52**, 287–297.
Dean, M. and Wood, R. K. S. (1967). *Nature* **214**, 408–410.
Eriksson, T. and Jonasson, K. (1969). *Planta* **89**, 85–89.
Fushtey, S. G. (1957). *Ann. Bot.* (*Lond*). **21**, 273–286.
Gregory, D. W. and Cocking, E. C. (1965). *J. Cell Biol.* **24**, 143–146.
Hall, J. A. and Wood, R. K. S. (1970). *Nature* **227**, 1266–1267.
Karr, A. L. and Albersheim, P. (1970). *Pl. Physiol.* **46**, 69–80.
Krull, R. (1960). *Planta* **55**, 598–629.
Pojner, E., Willison, J. H. M., and Cocking, E. C. (1967). *Protoplasma* **64**, 460–480.
Powers, J. B., Cummins, S. E., and Cocking, E. C. (1970). *Nature*, **225**, 1016–1022.
Roelofsen, P. A. (1963). *Adv. bot. Res.* **2**, 67–149.
Ruesink, A. W. and Thimann, K. V. (1966). *Science* **154**, 280–281.
Schenk, R. U. and Hildebrandt, A. C. (1969). *Crop. Sci.* **9**, 629–631.
Setterfield, G. and Bayley, S. T. (1958). *Can. J. Bot.* **35**, 435–444.
Spalding, D. H. (1963). *Phytopathology* **53**, 929–931.
Spalding, D. H. (1969). *Phytopathology* **59**, 685–692.
Strugger, R. (1957). *Protoplasma* **48**, 231–236.
Susuki, H., Abe, T., Urade, M., Nishizawa, K. and Kuroda, A. (1967). *J. Ferment. Technol.* **45**, 73–85.
Takebe, I., Otsuki, Y., and Aoki, S. (1968). *Pl. Cell. Physiol.* **9**, 115–124.
Toyama, N. (1965). *J. Ferment. Technol.* **43**, 683–689.
Tribe, H. L. (1955). *Ann. Bot.* (*Lond*). **19**, 351–368.
Webb, L. R. (1970). Ph.D. Thesis, University of London.
Zaitlin, M. and Coltrin, D. (1964). *Pl. Physiol.* **39**, 91–95.

Discussion

Mirocha: How does a fungus get to the middle lamella to hydrolyse it? You have disregarded the complex of enzymes and stressed instead the pectinases (polygalacturonase and *trans*-eliminase). A complex of enzymes is necessary to hydrolyse both the cellulose complex and the middle lamella. Furthermore, how pure a preparation of *trans*-eliminase can you obtain for experimental use and although we say that this enzyme is highly specific how do we know that it will not attack polymers other than those of the middle lamella.

Wood: I do not know that our *trans*-eliminase preparations are pure. All I can say is that preparations which cause cell separation actively degrade suitable pectic substrates but have little or no activity on other substrates of types that occur in cell walls, hemicelluloses, cellulose and proteins in particular. I do not assert that cell separation involves only degradation of the middle lamella. What I do say is that it occurs along the line of the middle lamella. This implies degradation of the middle lamella or zones in the matrix of the primary cell wall immediately next to the middle lamella. I would remind you of the probable continuity of both. Degradation of cellulose is not important in the early, critical stages of cell separation. Preparations with no cellulase activity can cause cell separation. Rapid degradation of the middle lamella or adjacent matrix *in situ* is somewhat difficult to conceive because when cells are stuck together, little would seem to be exposed to the ambient solution. The fact is that it happens.

Page: I would like to make the observation that many soft rot organisms are wound parasites so that initial penetration through cellulose is unnecessary.

Patil: How pure were the pectic enzyme preparations used in your studies of electrolyte leakage from protoplasts? We know that peptides have effects on membrane permeability. Is it possible that membrane leakage resulted from such peptides or possibly other enzyme preparations?

Wood: The short answer is that we do not know how pure our pectic enzyme preparations are. But they are active only on pectic of the many substrates tested.

If peptides or substances tested other than the pectic enzymes cause electrolyte leakage then all I can say is that they must have a host of properties in common with these enzymes as I have stated in my lecture. This would seem most unlikely.

Matta: My colleague Dr. A. Garibaldi has worked with Dr. Bateman at Cornell on the problem of maceration and cell death caused by bacterial enzymes. They have separated and purified by electrofocusing 6 distinct polygalacturonic acid *trans*-eliminase fractions from two strains of *Erwinia chrysanthemi* which were free of phosphatidase, proteinase and RNAase.

Five out of the six isolated fractions, induced loss of electrolytes, macerated tissue and killed plant cells. The fractions enriched in exo-type enzymes tended to kill at a faster rate than they macerated tissues. No cell death occurred with the fraction unable to produce maceration.

Backman: Electron microscopists have not been able to demonstrate plasmodesmata in callus tissue culture cells. Have you studied callus tissue regarding killing by pectolytic enzymes?

Wood: Not yet but I suspect that callus tissue might be rather resistant to maceration compared with parenchyma of storage type tissue. On this basis I would predict that the protoplasts would not be killed by solutions of pectic enzymes. It would be interesting to find out.

Dimond: Mirocha has asked how a fungus macerates when middle lamella and pectic substances are seldom exposed. Albersheim proposes and has experimental evidence for successive and rapid induction in the fungus of enzymes that can attack the exposed portion of the wall. As he stresses the heterogeneity in cell wall substances is far greater than has been realized in the past. Pectins are rarely pure but contain sugar residues in their chains and the production of various kinds of enzymes would seem to be essential for degradation of cell wall middle lamella.

Wood: I agree with your comments though I think that the significance in pathogenesis of induction of enzymes with specific cell wall components as substrates remain to be appraised. Also, in the past, pathologists have too readily assumed commercially available substrates from cell walls such as pectin to be well-defined, homogeneous products which they certainly are not. There is too the problem that it is still a far cry from degradation of a soluble extractive from a cell wall *in vitro* to the making soluble of certain parts of middle lamella and cell wall that are insoluble *in vivo*.

Dimond: Your ideas on the mechanism of injury and death following cell separation would seem to rule out the possibility of separation and recovery of uninjured protoplasts. Zaitlin has separated tobacco leaf cells infected with tobacco mosaic virus with chelating agents and has maintained these cells for studies on multiplication of the virus. That the virus can multiply suggests the cells are not injured. Moreover, leaf cells have been separated by a purified polygalacturonase preparation. The enzyme showed highly specific properties in not separating all cells in a given leaf and in being inactive in leaves of different varieties. These kind of results suggest that cells may be separable without injury if sufficiently pure enzymes of the correct type for the specific tissue being studied are used.

Wood: I should be surprised if Zaitlin did not separate his protoplasts in a plasmolyticum otherwise they would have burst rapidly. And this is the burden of my argument. I agree about the specificity of cell separating enzymes for some types of tissues. I found many years ago that red beet tissue is very difficult to macerate with preparations that readily disintegrate potato tuber tissue. But I should like to see someone produce a preparation that *will* macerate these tissues that are resistant to maceration by some preparations. Having mentioned specificity, I should also like to emphasize too the *lack* of specificity of cell separating enzymes. Many who have worked with these enzymes from all sorts of bacteria and fungi end up by using potato tubers as the test tissue!

Paxton: Would you care to comment on the fact that the soft rot bacteria almost exclusively attack storage organs rather than other parts of the plant? If the pectin degrading enzymes are the primary determinants of virulence, why is there this tissue specificity?

Wood: The soft rot pathogens are not confined to storage tissue. But their effects on this type of tissue are usually spectacular and it is, of course, convenient to have a large mass of homogeneous tissue to work with. I do not think that I have ever asserted that cell separating enzymes are "primary determinants" of pathogenicity though I am not sure that I accept this concept anyway. But that they are involved in pathogenesis and in a critical way is surely self-evident. In no other group of diseases is the role of a particular group of substances so easily demonstrated.

Durbin: Does the pH of potato tuber attacked by *Corticium praticola* drift downward and that of tubers attacked by *E. atroseptica* drift upwards?

Wood: Yes.

Durbin: You have indicated that ruthenium red is not a specific stain for pectin. What other chemical structures found in plant cells react with ruthenium red?

Wood: I believe that ruthenium red is a stain for free COOH groups.

Mussell: Have you considered the possibility that the products of pectin degradation may in fact be toxic or deleterious to host membranes?

Wood: Fushtey showed some years ago that the products referred to are not toxic to tuber disks in the concentrations likely to occur in rotted tissue.

17. The Origin of Symptoms of Vascular Wilt Diseases

ALBERT E. DIMOND

The Connecticut Agricultural Experiment Station, New Haven, Connecticut 06504, U.S.A.

Introduction

A discussion of how symptoms arise in the vascular wilt diseases is highly appropriate to the general theme of the role of phytotoxins in plant disease. It was for a wilt disease that the toxin theory was first proposed and Hutchinson (1913) did so to account for the wilting of apical tobacco leaves when only the base of the stem was infected by *Pseudomonas solanacearum* (E. Sm.) E. Sm.

For many years thereafter, toxic substances were described from culture filtrates of various pathogens. The idea that toxins are responsible for wilting in the vascular diseases of plants was intensively pursued by the Gäumann school. Phytotoxins from cultures of *Fusarium oxysporum* Schl. f. sp. *lycopersici* (Sacc.) Sny. *et* Hans. were shown to have characteristic effects on water economy of cuttings and other tissues exposed to them. Such compounds as lycomarasmin and fusaric acid were called wilting toxins and their effects on tomato cuttings were called pathological wilting (Gäumann, 1951, 1958). Characteristics of some of these compounds have already been discussed in the presentations of Professor Kern and Mme. Naef-Roth (pp. 35–45 and 49–67).

The term "vivotoxin" was proposed in an effort to clarify a situation in which phytotoxins were being implicated in pathogenesis before they were detected in wilt-diseased plants (Dimond and Waggoner, 1953a). Although fusaric acid has since been detected in *Fusarium*-infected cotton (Lakshminaranayan and Subramanian, 1955) and tomato (Kluepfel, 1957), its role in development of symptoms of *Fusarium* wilt is not yet apparent.

A variety of other compounds is formed in the host–pathogen interaction, each of which may be involved in the development of different symptoms. Some of these compounds, such as ethylene and indoleacetic acid (IAA), occur in the healthy plant, but are present in abnormal amounts in infected plants. Others, such as polysaccharides of high molecular weight, are products of the pathogen and act physically. Some compounds have specific modes of biochemical action. Enzymes that attack cell walls are also involved. Whether any or all of these are toxins is a matter of definition, which cannot be attempted in this chapter.

Pathogenesis in the wilt diseases has been summarized from differing viewpoints (Dimond, 1955, 1967, 1970; Goodman *et al.*, 1967; Graniti and Matta, 1969; Sadasivan, 1961; Subramanian and Saraswathi-Devi, 1959; Wood, 1967). Reviews on special aspects of this subject or on specific wilt diseases are also available (Beckman, 1964; Buddenhagen and Kelman, 1964; Kelman, 1953; Sequeira, 1963).

The Nature of Vascular Wilt Diseases

The typical wilt diseases are restricted to angiosperms in which the functional conductive element for transporting water is the vessel. Vessels are of fairly large diameter, as compared with the tracheids that transport water in gymnosperms, and are of considerable length. Many angiosperms suffer from wilt diseases. These diseases are systemic. A single infection often kills the plant or renders it unproductive. Some wilt pathogens are highly specific and attack only a single species of host; others, such as *Verticillium albo-atrum* Reinke *et* Berth., have a broad host range.

General Characteristics

The pathogens causing wilt diseases are typical facultative parasites, having few unique characteristics other than the ability to grow and reproduce in xylem vessels. Some are bacteria, others are fungi. Both can grow in the dilute sap of the transpiration stream (Kessler, 1966). This sap is low in sugars (Wood, 1961), but contains amino acids (Elgersma, 1969; Singh and Smalley, 1969) that serve as a carbon source for the pathogen. It also contains inorganic nutrients absorbed by the plant from the soil. Oxygen levels are low (Dimond, 1962a). Additional food is obtained from cell walls that are attacked during pathogenesis (Husain and Kelman, 1958b).

What makes the vascular wilt diseases unique is the establishment of the pathogen in the water transport system of the host. Whatever the pathogen produces extracellularly or frees from the host in attacking it may be transported apically through vessels with water. Even propagules of the pathogen are transported. The symptoms of the wilt diseases are unique because of the locus of infection and because of the nature of the tissues that respond to stimuli during pathogenesis. In what other group of diseases can xylem and the tissue to which it delivers water respond selectively?

Symptoms

The vascular wilt diseases share many symptoms in common. The infected host often is epinastic. Other growth abnormalities may appear, such as the shepherd's crook symptom of the Dutch elm diseases. Soon afterwards, leaves wilt by day and recover their turgor by night. Eventually, leaves wilt permanently, then die, and, in hosts that form an abscission layer, leaves fall from the plant. During the wilting phase, leaves may show marginal desiccation and withering. Less commonly, areas of leaf tissue between veins wither and dry. Yellowing of leaves is common. In the infected zone, xylem usually turns brown, but in certain wilt diseases, it may be a bright green, e.g., *Verticillium* wilt of maple. Adventitious roots may be formed at the base of the plant. In the elm disease, adventitious buds become active and their growth gives the trunk of the infected tree a shaggy appearance. Infected plants are usually stunted, and leaves may fail to expand to their normal size.

The Water Economy of Infected Plants

Water shortage in wilt diseases does not result from inability of roots to absorb water from soils, even when wilt pathogens invade the root system (Ludwig, 1952; Subramanian and Saraswathi-Devi, 1959). Usually, when the root is damaged in wilt diseases, this occurs after above ground parts have succumbed. Even then if water is plentiful, it moves into damaged roots by capillary forces.

Xylem vessels of stem, branches, shoots and petioles are plugged as a consequence of vascular infection. The resistance of xylem to flow of water is increased. The magnitude of this increase is large. *Fusarium*-infected stems and petioles of tomato transport water at about 2% of the rate of healthy ones under the same pressure head (Ludwig, 1952; Dimond and Waggoner, 1953b). A comparable increase in resistance to flow has been found in other wilt diseases, such as oak wilt (Beckman *et al.*, 1953), *Cephalosporium* wilt of soybean (McAlister and Chamberlain, 1951), and *Verticillium* wilt of tomato (Threlfall, 1959).

In at least some wilt diseases, changes in permeability of parenchyma cells and osmotic absorption of water by leaves have been measured. In other cases, such changes have not yet been demonstrated. The glycopeptides produced by species of *Corynebacterium* induce wilting, then withering of leaves and leaf margins, and these symptoms are regularly seen in systemically infected plants. They are attributable to damaged membranes and permanently changed differential permeability of exposed cells (Strobel and Hess, 1968). In other wilt diseases, infection causes a reversible increase in permeability of osmotically functional cells. Gottlieb (1944), for example, found that tracheal fluid from *Fusarium*-infected tomato plants caused small and reversible changes in permeability of parenchyma cells. Collins and Scheffer (1958) observed more loss of electrolytes from leaf cells of diseased than of healthy plants; conductivity of leachates increased with severity of symptoms.

Yet other evidence based on behaviour of tissue from diseased plants suggests that permeability is not altered permanently. For example, completely flaccid green leaves from *Fusarium*-infected tomato plants readily recover their turgor when floated on water or when allowed to absorb water through cut ends of stems or petioles from which dysfunctional conductive tissue has been removed. Cells in such leaves are still osmotically functional (Dimond and Waggoner, 1953b).

Hammel and Waggoner (1964) failed to detect changes in permeability of elm leaves from trees infected with *Ceratocystis ulmi* (Buism.) Moreau. The leafy portions of twigs were placed in a metal cylinder with the cut ends of the twigs exposed to air through a gas-tight seal. Gas applied under pressure to leaves caused displacement of sap at the cut surface of the twig. Sap from both healthy and diseased twigs was almost pure water and froze at the same temperature. Appropriate plots of the gas pressure applied versus the volume of sap displaced were linear and curves based on healthy and diseased tissues were parallel. More pressure was required to displace the same volume of sap from a diseased twig than from a healthy one. The evidence suggests that resistance to flow through conductive elements of xylem is high in diseased twigs, but that membranes of parenchyma cells in leaves from healthy and diseased trees behave very similarly. These experiments offer no evidence of membrane damage.

Water Transport

Through Vessels

Vessels, in which wilt pathogens live, transport water and inorganic nutrients from roots to leaves. Water moves in response to a gradient in water potential, extending from transpiring surfaces to absorptive cells at the root surface. Vessels are long capillaries, devoid of protoplasm, through which water flows in accordance with Poiseuille's law to a close approximation (Ludwig, 1952).

It is worthwhile to look at Poiseuille's law in relation to the changes that wilt diseases cause in plants. This law states that water of viscosity, η, will flow through a vessel of radius, r, and length, l, at a rate of v ml/sec when a difference in water potential amounting to p is present over its length, according to the relation $v = (\pi p r^4)/(8 l \eta)$. Wilt disease may reduce the rate of flow of sap through a vessel or through a stem as a consequence of a reduction in the gradient of water potential. It may also be reduced by a reduction in the net radius of a vessel. Finally, if viscosity of tracheal fluid is increased, the rate of flow is reduced. One notes that a relative change in net radius has a far greater effect on flow rate than a corresponding change in either viscosity or water potential.

In the wilt diseases, a change in the factor p is likely to be associated with vivotoxins that alter differential permeability of leaf cells. A change in η will be related to formation of extracellular polysaccharides or other products of high molecular weight that are liberated by the pathogen into the transpiration stream. Decrease in net radius of vessels is caused by any material that occupies space in lumens of vessels or by collapse of vessels.

Through Bundles of Vessels

In stems and branches, vessels are organized into functional units. In young, herbaceous stems, xylem consists of vascular bundles in which vessels of different radius are associated with each other and with other non-conductive xylem cells. In the tomato, vascular bundles of primary xylem are of two types. The large bundles that traverse the stem form a network. Small bundles of the stem and petiolar bundles have no connexion with each other from their point of origin, except at their termini in leaf tissue. The impact of a plug on flow to tissues is very different in these two cases. In a network bundle, a plug has little effect, because of existing by-passes. In an independent bundle, such as one that traverses a green shoot or petiole, a change in net radius to one-half of the initial value causes flow to drop to one-sixteenth of the initial rate if p remains constant. In a sheath of secondary xylem, the association between vessels is comparable to that in a network, but greater numbers of cells are involved. The effect of an occlusion in this circumstance can be no greater than in a network bundle (Dimond, 1966).

Water is redistributed in plants according to localized demand. Redistribution can be demonstrated by notching experiments (Ludwig, 1952; Dimond and Waggoner, 1953b). The magnitude of redistribution in a transpiring tomato plant has been evaluated in a typical circumstance (Dimond, 1966). Redistribution of water occurs as a result of its movement through pits in vessels at points where they are opposite to one another. These pits occupy sites of former plasmodesmata when the cells were living. Here secondary cell wall material fails to be deposited and the two cells are separated only by a pit membrane that is a derivative of the middle lamella. Pit membranes contain very small apertures or pores through which water and solutes can move in solution. The pit membrane is thin as compared with the secondary cell wall that divides vessels from each other elsewhere.

Pits, in addition to serving in the redistribution of water, serve as effective valves against the passage of gas bubbles (Dimond, 1962b). They serve as valves because of the tiny size of the openings in pit membranes. When gas fills one vessel, it cannot pass through the membrane into the adjacent vessel. To do so, a bubble with radius no larger than that of the aperture in the pit membrane would have to pass through it. The internal pressure of a bubble is inversely proportional to its radius. A very small bubble is unstable because of its high internal pressure. Thus, as water shortage develops in diseased plants, when water columns fracture in vessels and they become vapour locked, the vapour lock remains confined to the affected vessels and cannot pass to adjacent ones.

Pit membranes serve yet another purpose, that of ultrafilters that prevent the passage of large molecules from one vessel to another. Large molecules, such as polysaccharides, will be filtered out of sap that passes through a pit membrane. As a consequence, the pit membrane eventually becomes plugged and, if this occurs on an extensive scale, redistribution of water in the plant ceases to occur. Then localized differences in water potential cannot be equalized by means of redistribution of water, but instead continue to increase.

Products of Wilt Pathogens in their Hosts

Mycelium

Once established, wilt-inducing pathogens grow in xylem vessels. Not all vessels are invaded, and the amount of mycelium in a vessel is often sparse, sometimes dense (Chambers and Corden, 1963). Opinions have varied as to whether or not mycelium interferes seriously with flow of water. A hydraulic analysis has shown that mycelium contributes to but is not solely responsible for increased resistance to flow (Waggoner and Dimond, 1954).

Propagules and their Migration

In wilt diseases, the host is invaded more rapidly than can be accounted for by mycelial growth alone. Wilt pathogens multiply in xylem and bacterial cells or propagules of fungi are transported with water through vessels. Originally described by Yoshii (1933) for *F. oxysporum* f. sp. *niveum* (E. F. Sm.) Sny. *et* Hans. in watermelon, this process is known to occur regularly in all other wilt diseases. Migration of conidia in the host may occur in short steps or in long ones, depending upon the obstructions present in vessels. In the plate region of banana, propagules of *F. oxysporum* f. sp. *cubense* (E. F. Sm.) Sny. *et* Hans. move only short distances before they become lodged. These then germinate and produce more spores, which repeat the process (Beckman *et al.*, 1961; Trujillo, 1963). In species of oak that form tyloses readily, spores of *Ceratocystis fagacearum* (Bretz) Hunt become lodged against tyloses, germinate, and may invade the tylose or produce more spores (Nair *et al.*, 1967). In Dutch elm disease, propagules are carried through vessels for long distances in a single step (Banfield, 1968), and may even be found in petioles and leaf veins (Pomerleau, 1966). In fact, seasonal susceptibility of elms to *C. ulmi* depends upon the time when vessels mature in the spring. Only then can conidia migrate to newly developed terminal tissues (Pomerleau, 1968).

Glycopeptides and Polysaccharides

In addition to their own structures, bacteria and fungi liberate materials into tracheal fluid. Among these are the polysaccharides and glycopeptides that Professor Salemink has so ably discussed (p. 127). Likewise the glycopeptides produced by *Corynebacterium sepedonicum* (Spieck. *et* Kotth.) Skapt. *et* Burkh. have been discussed by Professor Strobel with respect to their molecular architecture and how they produce wilting (p. 379). These groups of compounds may be produced extracellularly or be liberated into tracheal fluid as cells of the pathogen autolyse.

Such compounds are non-selective in toxicity. Their toxic action depends upon their molecular weight. If less than 50,000, the compounds can be transported through xylem, and accumulate in margins of leaves, where they produce marginal desiccation and withering. Compounds of higher molecular weight remain in xylem and produce a physical plugging of vessels which is expressed by flaccidity of stem and petioles (Hodgson *et al.*, 1949; Strobel and Hess, 1968).

The glycopeptide produced by *C. sepedonicum* has a molecular weight in the range of 21,400 and produces marginal withering in leaves and flaccidity of stems as a result of its accumulation in cell membranes of mesophyll and pith parenchyma (Strobel, 1970). Studies by Strobel and Hess (1968) have clearly shown that the glycopeptide enters cells and damages membranes permanently. Properties of the glycopeptide described by Salemink *et al.* (1965) from *Ceratocystis ulmi* appear to be similar.

Other vascular wilt pathogens produce polysaccharides of much higher molecular weight, among them *P. solanacearum* (Husain and Kelman, 1958a), *C. fagacearum* (Gregory, 1969), and *C. ulmi* (Salemink, p. 127). These polysaccharides appear to be retained in xylem vessels where, by their entrapment, they physically increase the resistance of the stem to flow of water. These are among the materials that are caught on pit membranes, plugging the pores and reducing lateral movement of water to other vessels and to parenchyma of xylem and pith. For example, the transport of radiophosphate into cabbage leaf tissue was prevented by the polysaccharides present in vessels of leaves, infected by *Xanthomonas campestris* (Pamm.) Dows. (Sutton and Williams, 1968).

Wall-Attacking Hydrolytic Enzymes

Manner of action. The wilt pathogens produce extracellular cellulolytic and pectolytic enzymes together with a variety of other enzymes that free neutral sugars from host cell walls. Of these, the endopolygalacturonases and endopolymethylgalacturonases may have the most direct effect on water transport. These enzymes cleave large fragments from pectic substances in cell walls and free them into the transpiration stream. These may be deposited on pit membranes or with other debris at narrow orifices in vessels. Pectolytic action also weakens pit membranes and cellulolytic activity weakens cell walls. Such hydrolytic action permits leakage of melanoid pigments from xylem parenchyma, and contributes to collapse of vessels that are subjected to compression when hyperplasia occurs.

Evidence for their role in pathogenesis. (a) Presence in diseased plants. Endopolygalacturonase activity has been detected in vascular sap from diseased plants but not in sap from healthy ones. This has been established for tomato infected by *F. oxysporum* f. sp. *lycopersici* (Deese and Stahmann, 1962; Mussell and Green, 1970; Patil and Dimond, 1968) or by *V. albo-atrum* (Deese and Stahmann, 1960; Mussell and Green, 1970), for wheat infected with *Cephalosporium gramineum* Nisik. *et* Ikata (Spalding *et al.*, 1961) and both endopolygalacturonase and cellulase activity were detected in tomato infected with *P. solanacearum* (Husain and Kelman, 1957, 1958b).

The amount of polygalacturonase activity produced by virulent strains of *F. oxysporum* f. sp. *lycopersici* is greater than for avirulent ones (Paquin and Coulombe, 1962). In *V. albo-atrum*, virulence of strains was also correlated with polygalacturonase production (Mussell and Green, 1970).

More polygalacturonase was produced by *Fusarium* or *Verticillium* on susceptible than on resistant tomato tissues, according to Deese and Stahmann (1960). In fact, on resistant tomato, no polygalacturonase was produced, although both pathogens grew without producing symptoms on the resistant tissue (Mussell and Green, 1970).

(b) Damage to host cell walls. Cell walls of oak xylem in late stages of oak wilt have shown damage from invasion by mycelium of *C. fagacearum*. Sachs *et al.* (1967) reported degradation of the compound middle lamella and secondary cell walls in which cellulose, lignin, and, presumably, pectic substances were affected during penetration of walls by mycelium.

Microchemical and microscopic examination by Husain and Kelman (1958b) of tomato and tobacco tissues from plants infected by *P. solanacearum* showed that both cellulosic and pectic substances of the host were altered by the pathogen *in vivo*. Kelman and Cowling (1965) have also demonstrated degrading of cellulose in infected tobacco and tomato tissue.

(c) Presence of cleavage products in diseased plants. The presence of cleavage products of pectolytic activity or of entrapped products related to cell walls in diseased tissue argues strongly for involvement of enzymes in pathogenesis. Both pectic fragments and galacturonic acid were found by McIntyre (1964) in preparations of *Verticillium*-infected tomato.

Pectic materials and lignins are associated with plugging material in vessels of elms infected by *C. ulmi* (Gagnon, 1967). The vascular gels associated with plugged vessels in *Fusarium*-infected banana are composed of pectins, calcium pectate, and hemicelluloses, according to Beckman (1969) and Beckman and Zaroogian (1967).

(d) Effect of inhibitors or repressors of pectolytic activity on wilt diseases. *Fusarium* wilt of tomato is reduced in severity when pectolytic activity of the pathogen is inhibited by rufianic acid, as Grossmann (1962, 1963) has shown. Professor Grossmann's excellent presentation on control (p. 347) makes further elaboration of this idea unnecessary here.

The normal development of symptoms in *Fusarium*-infected tomato can be arrested by the presence of glucose or 2-deoxy-D-glucose in vessels. These measurably repress the synthesis of polygalacturonase in infected plants (Patil and Dimond, 1968). Thus both inhibitors and repressors of enzyme activity reduce severity of disease.

(e) Relation of ease of hydrolysis of substrates to susceptibility of the host to infection. The polygalacturonase of *F. oxysporum* f. sp. *lycopersici* attacks pectic acid readily but hydrolysis of calcium pectate is very slow. Tomato plants grown on a normal supply of calcium are less susceptible to wilt than calcium-deficient plants (Edgington and Walker, 1958). When plants are treated with naphthaleneacetic acid, they temporarily become resistant to wilt and treatment increases the amount of pectic acid in tissues. The increased resistance resulting from treatment with growth-regulators is calcium-dependent (Corden and Edgington, 1960). Under a normal calcium supply calcium pectate predominates and it is enzyme resistant. Hence resistance of pectic substances to hydrolysis by enzymes is correlated with resistance of plants to *Fusarium* wilt (Edgington *et al.*, 1961).

Growth Regulating Compounds

IAA is produced in cultures of *F. oxysporum* f. sp. *cubense* (Mace, 1965) and of *P. solanacearum* (Sequiera and Williams, 1964). Ethylene is produced by the tomato wilt *Fusarium* (Dimond and Waggoner, 1953c) and by *P. solanacearum* (Freebairn and Buddenhagen, 1964). Growth regulatory activity of undetermined nature has been reported in cultures of a number of other vascular wilt pathogens. The role of growth regulators in pathogenesis will be considered when their production by the infected host is discussed.

Phytotoxins

Wilting phytotoxins. The role of wilting toxins in pathogenesis of wilt diseases remains unsettled. Gottlieb (1944) showed vascular fluid from *Fusarium*-infected tomato plants to affect the permeability of parenchyma cells reversibly. Gnanam (1956) noted that in *Fusarium*-infected cotton leaves, conductivity of sap was higher than in healthy tissues, starting 15 days after inoculation. Talboys (1958) and Selman and Pegg (1957) have considered that toxins must have a role in wilt pathogenesis. Their evidence is indirect.

Most studied of the wilting toxins are lycomarasmin and fusaric acid. The former has never been identified in or isolated from diseased plants. Fusaric acid, on the other hand, occurs in plants infected by various *formae* of *Fusarium oxysporum* and its effect on the water economy of cuttings is well described (Sivadjian and Kern, 1958). Its mode of action has been proposed in detail (Gäumann, 1958). However, the effects of fusaric acid on plant tissues are not observed in infected plants. Fusaric acid causes a burst in transpiration of treated cuttings whereas diseased plants do not show this (Scheffer and Walker, 1953). Fusaric acid competitively inhibits polyphenol oxidase (Bossi, 1959), but in naturally infected plants, activity of polyphenol oxidase is markedly increased (Matta and Dimond, 1963). Fusaric acid causes a marginal withering of leaves and desiccation in interveinal areas of leaf tissue. Both of these symptoms are seen only occasionally in infected plants. Kuo and Scheffer (1964) noted that fusaric acid injury does not resemble the usual symptoms of *Fusarium* wilt of tomato and that, because its effects on respiration are the opposite of what occurs in infected plants, fusaric acid can have no more than a secondary role in pathogenesis.

Other phytotoxins. Respiration increases sharply in infected tissues but this increase is not ascribable to fusaric acid or to other known products of infection. The respiratory rate of infected tissues is greater than that of the host and pathogen together. On this basis, respiratory toxins have been postulated (Weir, 1962; Collins and Scheffer, 1958).

Evidence for toxins of another sort has also been adduced from grafting experiments. Davis (1954) used tomato root stocks in intergeneric grafts with a number of other solanaceous species. Stocks inoculated with *Fusarium* became invaded but immune scions did not. The scions, however, showed severe vascular discolouration and wilting, which suggests the presence of a toxin that is transported from the infected stock to the scion.

Scions of resistant varieties of tomato respond similarly to the presence of *F. oxysporum* f. sp. *lycopersici* in a susceptible stock. The response of resistant scions to toxins is greater than of susceptible ones, according to Keyworth (1964). He has concluded that genetically resistant tissue is hypersensitive to a toxin that is produced in inoculated, susceptible stocks. The nature of this toxin is not known.

Host Responses to Invasion

Increased Respiration

Soon after a susceptible host is inoculated with a wilt pathogen, respiration of host tissue increases. This increase is of fundamental significance. In resistant plants, it occurs as a burst, then subsides as the pathogen subsides. In susceptible hosts, the activity increases more slowly and persists for a longer time. Increased respiration even occurs in leaf tissue that is not yet invaded by the pathogen (Collins and Scheffer, 1958). This also suggests the action of a toxin.

In *Fusarium*-infected tomatoes, respiratory rates may double, reaching a peak from 9 to 14 days after inoculation (Collins and Scheffer, 1958). In *Ceratocystis*-infected elms, the peak occurs between 11 and 22 days after inoculation (Landis and Hart, 1968).

Activity of polyphenol oxidase increases as much as ten-fold (Matta and Dimond, 1963). Ascorbic acid oxidase, other oxidases, and peroxidase also increase in activity (Heitefuss *et al.*, 1960). Activity of indoleacetic acid oxidase decreases as scopoletin accumulates and causes inhibition (Sequeira, 1963). Oxidation rates of substrates is high in mitochondria of diseased stems and mitochondrial protein N increases in them (Wu and Scheffer, 1962).

Increased Aromatic Syntheses

A number of aromatic syntheses are stimulated in plants soon after infection. Pegg and Sequeira (1968) report an increase in phenylalanine, tryptophan, tyrosine, dihydroxyphenylalanine, and aromatic amino acids in tobacco 24 h after inoculation with *P. solanacearum*. These are precursors for IAA and for phenolic compounds. Scopolin and scopoletin accumulate within 48 h (Sequeira, 1968). Phenolic compounds increase in other wilt diseases, e.g., *Fusarium* wilt of tomato (Matta *et al.*, 1967), and *Verticillium* wilt of cotton (Bell, 1969). The increase is greater in resistant varieties than in susceptible ones.

These changes, together with activation of such systems as polyphenol oxidase, initiate two processes in the infected plant. The one leads to vascular browning and the other to hyperauxiny and resulting hyperplasia.

Development of vascular browning. Vascular discolouration first appears in xylem parenchyma and results from oxidation and polymerization of dihydroxyphenols (Davis *et al.*, 1953). According to Mace (1963) 3-hydroxy-tyramine exists in scattered cells of xylem parenchyma in roots of banana and these are the cells that show discolouration when roots become infected by *F. oxysporum* f. sp. *cubense*. Laccase, phenol oxidase, and peroxidase of *Fusarium* and cytochrome-c oxidase and polyphenol oxidase of the host are all participants in the oxidation according to Mace (1964) and Mace and Wilson (1964). Vascular browning originates similarly in xylem parenchyma in other wilt diseases. With the increase in phenolic substrates, and of polyphenol and related oxidase activity, dark melanoid pigments are formed.

During invasion of vessels and adjacent cells, mycelium penetrates through pits of vessels and hydrolytic enzymes cause some maceration. The melanoid pigments leak from parenchyma cells into vessels, where they are transported as particles and become trapped on pit membranes, as shown for *Fusarium* wilt of tomato by Corden and Chambers (1966). Thus, the pigments causing vascular browning contribute to the occlusion of vessels and decreased water flow.

Accumulation of IAA. How pathogens interfere with growth regulation in plants has been ably presented by Drs. van Andel and Fuchs (p. 227). In addition to what the pathogen does, the host also produces IAA in response to effects of the pathogen. Gordon and Paleg (1961) have shown how *o*-dihydroxyphenols, under the influence of polyphenol oxidase, react with tryptophan to form IAA *via* indolepyruvic acid. These are the conditions that prevail in infected plants. Matta and Gentile (1965, 1968, 1970) have called attention to how levels of IAA in tomato plants increase with severity of symptoms of *Fusarium* wilt. They have demonstrated the dependence of IAA biosynthesis in diseased tissue upon phenolase activity in the presence of chlorogenic acid or other *o*-diphenols.

That wilt diseased plants contain greater than normal amounts of auxin has been recognized for some time (Sequeira, 1963). Tobacco and banana, infected with *P. solanacearum*, tomato suffering from *Verticillium* wilt are but two examples in which the build-up of auxin in diseased plants is well documented (Sequeira and Kelman, 1962; Sequeira, 1965; Pegg and Selman, 1959). Indirect evidence exists for hyperauxiny in Dutch elm disease and oak wilt (Beckman *et al.*, 1953; Beckman, 1956).

The diseased host responds to higher than normal levels of auxin in a variety of ways. IAA increases the plasticity of cell walls, making them more extensible through altering the nature of the substances deposited in them. Tyloses involve an extension of the middle lamella of xylem parenchyma cells into lumens of vessels. Beckman *et al.* (1953) have suggested that the heavy development of tyloses in oak infected with *C. fagacearum* is a response to hyperauxiny. Tyloses have been experimentally induced by Mace and Solit (1966) in banana roots following treatment with IAA.

Hyperplasia of xylem parenchyma and pith occurs in a number of wilt diseases, e.g., *Verticillium* wilts of hops and tomato (Talboys, 1958; Pegg and Selman, 1959). The lateral growth pressures resulting from such cell division leads to crushing and collapse of xylem vessels, a symptom seen in *Fusarium* wilt of tomato. By administration of IAA or *Fusarium* culture filtrates to tomato cuttings, Chambers and Corden (1963) have experimentally duplicated the hyperplasia of xylem parenchyma and collapse of vessels in tomato plants.

Production of Ethylene

Ethylene is produced in wilt-diseased tissues. It has been identified both in *Fusarium*-infected tomatoes and in banana infected by *P. solanacearum* (Dimond and Waggoner, 1953c; Freebairn and Buddenhagen, 1964). According to Abeles and Rubinstein (1964), plants treated with IAA are stimulated to produce ethylene. Thus, production of ethylene may result from the increase in IAA that occurs in infected plants.

Three symptoms of wilt diseases are attributable to ethylene. Hosts that form abscission layers become partially or completely defoliated as infection progresses. Elm and cotton characteristically form abscission layers; tomato and oak do not. Ethylene breaks the dormancy of buds. Dormancy of adventitious buds along the trunks and branches of elms is broken when trees are infected with *C. ulmi* in such a manner that the tree is not rapidly killed. Adventitious roots are symptoms of *Fusarium* wilt of tomato and of southern bacterial wilt of tomato and tobacco. Epinasty is an early symptom of *Fusarium* wilt of tomato. These are probably caused by ethylene, IAA, or both.

Production of Vascular Gels

Vascular gels appear promptly in infected plants and they appear, together with tyloses, more promptly and in greater quantity in resistant tissues of root and stem than in susceptible ones. Beckman (1964) has proposed that this response confers resistance by preventing migration of propagules of the pathogen, so that infection does not become systemic. This view is compatible with the idea that symptoms of water shortage arise in wilt diseases as a result of increased resistance of conductive tissues to flow of water. If an infection remains localized, the host responses and the water shortage in the host remain localized.

Vascular gels stain with ruthenium red and have other properties of pectic substances (Husain and Kelman, 1958b; Ludwig, 1952). However, the occluding material in infected plants is not removed by pectolytic enzymes (Saaltink and Dimond, 1964; Beckman and Zaroogian, 1967). Current evidence of Beckman (1969) and Beckman and Zaroogian (1967) indicates that the gels in *Fusarium*-infected banana roots are composed of a mixture of pectins, calcium pectate, hemicelluloses, and traces of protein. Beckman (1969) has suggested that this gel is stabilized through cross-bonding between the components, and, by mixing these components in suitable ratios, has produced highly stable gels.

Whether the components of gels are formed from partially cleaved products as a result of the action of cell wall-attacking enzymes of the pathogen or are formed primarily as a host response (Beckman, 1966) is not yet known, but their formation above constrictions in vessels, rather than below, is a characteristic that must not be ignored.

Consequences of Infection

Water becomes deficient in wilt-diseased plants as a result of a number of events. In oak wilt vessels become completely plugged by tyloses and gums (Struckmeyer *et al.*, 1954). Resistance to flow of water through vessels increases as a result of development of tyloses, vascular gels, the presence of mycelium, the lodging of propagules in constrictions. Resistance to transfer of water from vessel to vessel and to redistribution of water in the plant also increases in diseased plants as a result of the plugging of pit membranes by compounds of high molecular weight. These include polysaccharides excreted by the pathogen, the resinous melanoid pigments associated with vascular browning, and the partially hydrolysed cleavage products of cell walls freed by hydrolytic enzymes of the pathogen.

Glycopeptides produced by some pathogens accumulate at termini of the vascular system in leaf margins, where they accumulate in the membranes of parenchyma cells and damage differential permeability.

The plant shows that water is limiting its processes in a variety of ways. Non-symptomatic leaves on infected plants have the characteristics of leaves on drought-hardened plants (Dimond and Waggoner, 1953b). Their transpirational losses may be but one-third of those on healthy plants receiving a normal supply of water. Stomata remain closed. Tissues are subjected to an ever-increasing shortage of water.

Photosynthesis requires water. Infected plants have a lower photosynthetic rate than healthy plants; measurements by Mathre (1968) on *Verticillium*-infected cotton show leaves to have less starch than healthy ones, the Hill reaction is less efficient, and the chlorophyllase content is lower. Beckman *et al.* (1962) measured both photosynthesis and transpiration in banana plants infected by *P. solanacearum* and found that both processes fluctuated together before wilting appeared. When water was supplied through saturating the atmosphere and water loss was reduced by shielding plants from light, the fluctuation disappeared. Abnormally high rates of transpiration were never observed. Depressed rates of photosynthesis and transpiration are attributed to a limited supply of water.

Leaf expansion is reduced in *Verticillium* wilt of tomato (Selman and Pegg, 1957) as well as in other wilt diseases. Leaf expansion is primarily a matter of hydration; cell number does not change.

Both terminal and cambial growth are reduced in infected plants. Kozlowski *et al.* (1962) have measured changes in diameter of oaks suffering from oak wilt with a dendrometer. Reduced cambial expansion is attributed to change in hydration of the stem as water becomes limiting, rather than from growth inhibition of cambium.

Internal tissues receive what water is available and can respond as soon as auxin levels rise to critical heights. This will account for how the growth responses in diseased plants are internal in nature and do not tend to involve terminal tissues.

From the foregoing, it would appear that the symptoms of the wilt diseases have a multiplicity of causes rather than a single cause. The role of toxins in these processes may be great but has yet to be delimited.

References

Abeles, F. B. and Rubinstein, B. (1964). *Pl. Physiol.* **39**, 963–969.
Banfield, W. M. (1968). *Phytopath. Z.* **62**, 21–60.
Beckman, C. H. (1956). *Phytopathology* **46**, 605–609.
Beckman, C. H. (1964). *A. Rev. Phytopath.* **2**, 231–252.
Beckman, C. H. (1966). *Phytopathology* **56**, 821–824.
Beckman, C. H. (1969). *Phytopathology* **59**, 837–843.
Beckman, C. H., Brun, W. A., and Buddenhagen, I. W. (1962). *Phytopathology* **52**, 1144–1148.
Beckman, C. H., Kuntz, J. E., Riker, A. J., and Berbee, J. G. (1953). *Phytopathology* **43**, 448–454.
Beckman, C. H., Mace, M. E., Halmos, S., and McGahan, M. W. (1961). *Phytopathology* **51**, 507–515.
Beckman, C. H. and Zaroogian, G. E. (1967). *Phytopathology* **57**, 11–13.
Bell, A. A. (1969). *Phytopathology* **59**, 1119–1127.
Bossi, R. (1959). *Phytopath. Z.* **37**, 273–316.
Buddenhagen, I. W. and Kelman, A. (1964). *A. Rev. Phytopath.* **2**, 204–230.
Chambers, H. L. and Corden, M. E. (1963). *Phytopathology* **53**, 1006–1010.
Collins, R. P. and Scheffer, R. P. (1958). *Phytopathology* **48**, 349–355.
Corden, M. E. and Chambers, H. L. (1966). *Am. J. Bot.* **53**, 284–287.
Corden, M. E. and Edgington, L. V. (1960). *Phytopathology* **50**, 625–626.
Davis, D. (1954). *Am. J. Bot.* **41**, 395–398.
Davis, D., Waggoner, P. E., and Dimond, A. E. (1953). *Nature* **172**, 959.
Deese, D. C. and Stahmann, M. A. (1960). *Phytopathology* **50**, 633.
Deese, D. C. and Stahmann, M. A. (1962). *Phytopathology* **52**, 255–260.

Dimond, A. E. (1955). *A. Rev. Pl. Physiol.* **6**, 329–350.
Dimond, A. E. (1962a). *Phytopathology* **52**, 7.
Dimond, A. E. (1962b). *Phytopathology* **52**, 7.
Dimond, A. E. (1966). *Pl. Physiol.* **41**, 119–131.
Dimond, A. E. (1967). *In* "The Dynamic Role of Molecular Constituents in Plant-parasite Interaction" (C. J. Mirocha and I. Uritani, Eds), pp. 100–120. Am. Phytopath. Soc., St. Paul, Minn.
Dimond, A. E. (1970). *A. Rev. Phytopath.* **8**, 301–322.
Dimond, A. E. and Waggoner, P. E. (1953a). *Phytopathology* **43**, 229–235.
Dimond, A. E. and Waggoner, P. E. (1953b). *Phytopathology* **43**, 619–623.
Dimond, A. E. and Waggoner, P. E. (1953c). *Phytopathology* **43**, 663–669.
Edgington, L. V., Corden, M. E., and Dimond, A. E. (1961). *Phytopathology* **51**, 179-182.
Edgington, L. V. and Walker, J. C. (1958). *Phytopathology* **48**, 324–326.
Elgersma, D. M. (1969). *Meded. Phytopath. Lab. Willie Commelin Scholten* **77**, 84 pp., Baarn, Netherlands.
Freebairn, H. T. and Buddenhagen, I. W. (1964). *Nature* **202**, 313–314.
Gagnon, C. (1967). *Can. J. Bot.* **45**, 1619–1623.
Gäumann, E. (1951). *Adv. Enzymol.* **11**, 401–435.
Gäumann, E. (1958). *Phytopathology* **48**, 670–686.
Gnanam, P. (1956). *Proc. Indian Acad. Sci.* **44**, 125–129.
Goodman, R. N., Király, Z., and Zaitlin, M. (1967). "The Biochemistry and Physiology of Infectious Plant Disease." Van Nostrand, Princeton, N.J.
Gordon, S. A. and Paleg, L. G. (1961). *Pl. Physiol.* **36**, 838–845.
Gottlieb, D. (1944). *Phytopathology* **34**, 41–59.
Graniti, A. and Matta, A. (1969). *Ann. Phytopath.* **1**, 77–120.
Gregory, G. F. (1969). *Phytopathology* **59**, 399.
Grossmann, F. (1962). *Naturwiss.* **49**, 138–139.
Grossmann, F. (1963). *Meded. LandHoogesch. OpzoekStns Gent* **28**, 604–610.
Hammel, H. T. and Waggoner, P. E. (1964). Unpublished.
Heitefuss, R., Stahmann, M. A., and Walker, J. C. (1960). *Phytopathology* **50**, 371–375.
Hodgson, R., Peterson, W. H., and Riker, A. J. (1949). *Phytopathology* **39**, 47–62.
Husain, A., and Kelman, A. (1957). *Phytopathology* **47**, 111–112.
Husain, A. and Kelman, A. (1958a). *Phytopathology* **48**, 155–165.
Husain, A. and Kelman, A. (1958b). *Phytopathology* **48**, 377–386.
Hutchinson, C. M. (1913). Pusa Agr. Res. Inst. Dept. Agr., Bact. Ser., Mem. 1, 67–83.
Kelman, A. (1953). N. C. Agr. Exp. Station Tech. Bull. 99, 194 pp.
Kelman, A. and Cowling, E. B. (1965). *Phytopathology* **55**, 148–155.
Kessler, K. J., Jr. (1966). *Phytopathology* **56**, 1165–1169.
Keyworth, W. G. (1964). *Ann. appl. Biol.* **54**, 99–105.
Kluepfel, D. (1957). *Phytopath. Z.* **29**, 349–379.
Kozlowski, T. T., Kuntz, J. E., and Winget, C. H. (1962). *J. Forestry* **60**, 558–561.
Kuo, M. S. and Scheffer, R. P. (1964). *Phytopathology* **54**, 1041–1044.
Lakshminarayanan, K. and Subramanian, D. (1955). *Nature* **176**, 697–698.

Landis, W. R. and Hart, J. H. (1968). *Phytopathology* **58**, 1057.
Ludwig, R. A. (1952). Tech. Bull. 20, 40 pp., Macdonald Coll., Montreal.
Mace, M. E. (1963). *Physiol. Plantarum* **16**, 915–925.
Mace, M. E. (1964). *Phytopathology* **54**, 1033–1034.
Mace, M. E. (1965). *Phytopathology* **55**, 240–241.
Mace, M. E. and Solit, E. (1966). *Phytopathology* **56**, 245–247.
Mace, M. E. and Wilson, E. M. (1964). *Phytopathology* **54**, 840–842.
Mathre, D. E. (1968). *Phytopathology* **58**, 137–141.
Matta, A. and Dimond, A. E. (1963). *Phytopathology* **53**, 574–578.
Matta, A. and Gentile, I. (1965). *Phytopath. medit.* **4**, 129–137.
Matta, A. and Gentile, I. A. (1968). *Neth. J. Pl. Pathol.* **74**, (1968 Suppl.) 47–51.
Matta, A. and Gentile, I. A. (1970). In Ms.
Matta, A., Gentile, I. A., and Giai, I. (1967). *Ann. Fac. Sci. Agr. Univ. Torino* **4**, 17–32.
McAlister, D. F. and Chamberlain, D. W. (1951). *Plant Dis. Rep.* **35**, 318–319.
McIntyre, G. A. (1964). *Diss. Abstr.* **25**, 779–780.
Mussell, H. W. and Green, R. J., Jr. (1970). *Phytopathology* **60**, 192–195.
Nair, V. M. G., Kuntz, J. E., and Sachs, I. B. (1967). *Phytopathology* **57**, 823–824.
Paquin, R. and Coulombe, L. J. (1962). *Can. J. Bot.* **40**, 533–541.
Patil, S. S. and Dimond, A. E. (1968). *Phytopathology* **58**, 676–682.
Pegg, G. F. and Selman, I. E. (1959). *Ann. appl. Biol.* **47**, 222–231.
Pegg, G. F. and Sequeira, L. (1968). *Phytopathology* **58**, 476–483.
Pomerleau, R. (1966). *Can. J. Bot.* **44**, 109–111.
Pomerleau, R. (1968). *Phytopath. Z.* **63**, 301–327.
Saaltink, G. J. and Dimond, A. E. (1964). *Phytopathology* **54**, 1137–1140.
Sachs, I. B., Nair, V. M. G., and Kuntz, J. E. (1967). *Phytopathology* **57**, 827–828.
Sadasivan, T. S. (1961). *A. Rev. Pl. Physiol.* **12**, 449–468.
Salemink, C. A., Rebel, H., Kerling, L. C. P., and Tchernoff, V. (1965). *Science* **149**, 202–203.
Scheffer, R. P. and Walker, J. C. (1953). *Phytopathology* **43**, 116–125.
Selman, I. W. and Pegg, G. F. (1957). *Ann. appl. Biol.* **45**, 674–681.
Sequeira, L. (1963). *A. Rev. Phytopathol.* **1**, 5–30.
Sequeira, L. (1965). *Phytopathology* **55**, 1232–1236.
Sequeira, L. (1968). *Phytopathology* **59**, 473–478.
Sequeira, L. and Kelman, A. (1962). *Phytopathology* **52**, 439–448.
Sequeira, L. and Williams, P. H. (1964). *Phytopathology* **54**, 1240–1246.
Singh, D. and Smalley, E. B. (1969). *Phytopathology* **59**, 891–896.
Sivadjian, J. and Kern, H. (1958). *Phytopath. Z.* **33**, 241–247.
Spalding, D. H., Bruehl, G. W., and Foster, R. J. (1961). *Phytopathology* **51**, 227–235.
Strobel, G. A. (1970). *J. biol. Chem.* **245**, 32–38.
Strobel, G. A. and Hess, W. M. (1968). *Pl. Physiol.* **43**, 1673–1688.
Struckmeyer, B. E., Beckman, C. H., Kuntz, J. E., and Riker, A. J. (1954). *Phytopathology* **44**, 148–153.
Subramanian, D. and Saraswathi-Devi, L. (1959). *In* "Plant Pathology" (J. G. Horsfall and A. E. Dimond, Eds), Vol. 1, pp. 313–348. Academic Press, New York.

Sutton, J. C. and Williams, P. H. (1968). *Phytopathology* **58**, 1069.
Talboys, P. W. (1958). *Trans. Br. mycol. Soc.* **41**, 249–260.
Threlfall, R. J. (1959). *Ann. appl. Biol.* **47**, 55–77.
Trujillo, E. E. (1963). *Phytopathology* **53**, 162–166.
Waggoner, P. E. and Dimond, A. E. (1954). *Am. J. Bot.* **41**, 637–640.
Waggoner, P. E. and Dimond, E. (1955). *Phytopathology* **25**, 79–87.
Weir, G. M. (1962). *Mycopathol.* **18**, 194–198.
Wood, R. K. S. (1961). *Ann. appl. Biol.* **49**, 120–139.
Wood, R. K. S. (1967). "Physiological Plant Pathology." Blackwell Sci. Publ., 570 pp., Oxford and Edinburgh.
Wu, L.-C. and Scheffer, R. P. (1962). *Phytopathology* **52**, 354–358.
Yoshii, H. (1933). Bul. Sci. Fac. Terkultura, Kyushu Imp. Univ. 5, 578–589.

Discussion

Page: What is your basis for considering flow to be laminar when the secondary wall thickenings, whether spiral, annular or reticulate, are internal to the primary wall?

Dimond: Whether flow is laminar or turbulent is not primarily a result of wall roughness. The lift of an aircraft wing is caused by differential turbulent flow around the airfoil, which is smooth. The type of flow is determined by viscosity of the fluid η, its density d, and its velocity of flow v, through a tube of radius r. These factors are combined in Reynold's number, according to the relation $R = (2dvr)/\eta$. Many physical studies have shown that when Reynold's number has a value less than 2000, flow is laminar. In a healthy tomato stem, R was shown by Waggoner and Dimond (1954) to have a value of about 0·1 and is even smaller in a diseased stem.

Page: There is a gradual reduction of flow with time through the vascular system of banana fruit stem. Do you consider the primary walls of xylem vessels to be inflexible or do they respond to liquid flow?

Dimond: The walls of xylem vessels do respond to the volume of water they contain. Kozlowski *et al.* (1962) have measured with a dendrograph the changes in diameter of oak trees suffering from oak wilt and have related these to changes in rate of flow and to changes in hydration of diseased tissues. Vascular tissue of diseased banana should behave similarly.

Page: There is a striking reduction in volume of liquid in diseased banana pseudostems in contrast to the amount of fluid in healthy pseudostems. Would you care to comment on this difference considering that the vascular system has rigidity so that the volume should remain constant in diseased and healthy plants?

Dimond: As the amount of water in diseased banana pseudostem decreases, the volume of the tissue must also decrease or become filled with air. One would suppose that longitudinal walls of vessels would yield first to a decrease in volume. In other wilt diseased woody tissues, a decrease in radius has been measured. This reflects a yielding of walls to stress. We cannot consider walls to be rigid in any absolute sense.

Graniti: In investigations on wilt diseases, I found it very useful considering the plant body as composed of the two Münch's (1930) systems, namely symplast and apoplast.

In the concept of Crafts (1968), the symplast is represented by the interconnected continuum of all living cells, including phloem sieve tubes; the apoplast is formed by the cell wall continuum that surrounds and contains the symplast, and includes the non-living xylem elements. Water and solutes absorbed by plants move in *via* the apoplast and substances within the symplast move throughout this system without passing permeability barriers.

Energy and other changes occurring in any point of one of these systems are consequently expected to influence the whole system.

Generally speaking, one could say that toxins excreted by a pathogen within a plant affect the symplast, and aggressins affect the apoplast.

In some vascular diseases, such as fusarial wilt of tomato, the free spread of toxins through the apoplastic structures is often hindered by plugging materials. The parasite itself contributes to occlusion by the production of aggressins. As you have outlined, wilt in this case sets in more as a result of vascular dysfunction than as a consequence of a direct disruption of the mechanisms controlling the cell turgor in leaves. However, if plugging of vessels and occlusion of cell wall micropores are prevented e.g. by grafting resistant varieties on infected susceptible root-stocks (Keyworth, 1963, 1964), the toxic syndrome can manifest itself.

Vascular diseases, therefore, are not the best examples of toxigenic diseases.

Dimond: Yes. I agree with the thoughts you have expressed.

Wood: Your schemes for water transport in plants attach great importance to the functioning of pit membranes. Is there any evidence that these are significantly degraded in diseased plants through the action of cell wall degrading enzymes.

Dimond: In oak wilt Nair has shown penetration of pits by mycelium and in Dutch elm disease Pomerleau has seen mycelium passing from vessel to vessel through pits only. The pathogen would damage pit membranes in passing through them. However, one would expect colloidal materials or slightly larger ones to accumulate on intact pit membranes.

Goodman: A recent report by Dr. McDonald of Iowa State University, based on electron micrographs, bases resistance to *Ceratocystis ulmi* between species of *Ulmus* on diameter of xylem vessels, number of contiguous vessels, and area occupied by groups of vessels. *U. americana*, the most susceptible, has the largest vessels, the most contiguous vessels, and vessel groups occupying the largest areas.

Dimond: My comments on origin of wilting in vascular diseases are entirely consistent with what Dr. McDonald has found. The more a tree is constructed like a conifer the more resistant it should be to wilt diseases. In such a structure, neither propagules nor products of colloidal size can move freely or cause extensive dysfunction of the transport system.

Meyer: I find it difficult to accept that tyloses interfere materially with induction of wilt, because in hop plants apparently healthy vessels are completely plugged with tyloses.

Dimond: Tyloses alone, like mycelium of the pathogen in vessels, probably are not solely responsible for increased resistance to flow of water. But they surely contribute to the increase. Moreover, when vessels contain many tyloses, the tyloses form a structural filter bed on which particulate matter or gels can lodge to form a more effective plug than could be in the absence of tyloses.

Page: Could you offer an explanation for the presence of *Fusarium oxysporum* in such plants as tobacco, banana and red clover in the absence of visible wilt symptoms?

Dimond: One can think of a number of reasons why *F. oxysporum* might be present in a plant and not produce visible wilting: (a) A clone of a pathogen that did not produce propagules might remain restricted to a zone in the stem around which water could flow readily to aerial tissues. (b) In trees that have recovered naturally from a vascular wilt disease, the pathogen is walled off by a ring of healthy conductive wood, yet the pathogen continues to live for years without causing wilting. (c) A resistant host, according to Beckman (1964), responds by rapid basal vascular plugging with gels and tyloses that prevent migration of propagules and spread of infection. (d) A *forma specialis* of *F. oxysporum* can enter the vascular system of a plant that is not a host. Then the vascular response of the plant is similar to that of a resistant plant.

Main: You referred to research showing reduced water supply to leaves affected photosynthesis and resulted in chlorosis. I wonder if pathological inhibition of water transport across leaf parenchyma cells could account in part for the chlorotic halos commonly observed surrounding foliar lesions?

Dimond: During the course of this Institute, we have heard of a number of chlorosis-inducing toxins. Although such compounds may be active in wilt diseases, I know of no persuasive evidence that photosynthesis is reduced as a result of the action of such a toxin in these diseases. Rather, because the photosynthetic reaction requires water, it is likely that the reaction is directly affected as water becomes limited in the wilt diseased plant. If this be the mechanism, then the situation that leads to reduced photosynthesis in leaf-spotting diseases must be quite different.

Durbin: In what bacterial diseases do you think wilting is a consequence of increased viscosity in the vessels, rather than of occlusion of pit membranes?

Dimond: In advanced stages of the bacterial wilt of cucumber, caused by *Erwinia tracheiphila*, the viscosity of tracheal fluid is used as a diagnostic symptom. When infected stems are broken and drawn apart, the fluid may be drawn out in strands for some distance. In this advanced stage of disease, bacterial polysaccharide has accumulated and viscosity has increased. Possibly at this very late stage, water shortage is related to viscosity. But in early stages, viscosity is not high and occlusion of pit membranes must occur. After this happens, flow decreases, and viscosity of tracheal fluid increases as polysaccharide accumulates. In some other diseases the increase in viscosity is not so striking.

18. Phytotoxin-Induced Ultrastructural Modifications of Plant Cells

ROBERT N. GOODMAN,

Department of Plant Pathology,
University of Missouri,
Columbia, Missouri 65201, U.S.A.

Introduction

Published accounts of ultrastructural modifications induced by toxic substances produced by plant pathogens are few (Goodman and Burkowicz, 1970; Goodman and Plurad, 1971; Hanchey *et al.*, 1968; Luke *et al.*, 1966; and Strobel and Hess, 1968). Nevertheless, from the visual symptoms attributed to phytotoxins and from the electron micrographs that have been published it would appear that the damage they cause at the ultrastructural level is profound.

I have chosen to describe in this paper the effects of three toxic substances on plant cells at the ultrastructural level. Although they cause markedly differing visual symptoms, their effects at the cellular level have some basic similarities and also some which are unique to the specific host-toxin interaction.

Effects of Victorin

Victorin (Woolley *et al.*, 1955) or the HV-toxin (Scheffer and Pringle, 1963) produced by *Helminthspoorium victoriae* Meehan *et* Murphy has been reported by Luke *et al.* (1966) to cause changes in permeability and respiration of diseased tissues. Attempting to correlate permeability changes with ultrastructural modifications, they observed a general disruption of internal membrane systems (exclusive of the mitochondria) in root cap cells and green leaf tissues of susceptible oat varieties exposed to the toxin for 24 h. The mitochondria seemed to remain intact until cellular disorganization was complete. The first effect on root cap cells was the appearance of densely stained material between the cell wall and plasma membrane. The material was believed to be a product of a reaction between the toxin and either cell wall *per se* or the outer surface of the plasma membrane. A localized separation of the plasma membrane from the cell wall with the formation of blister-like structures containing fragments of densely stained material was also

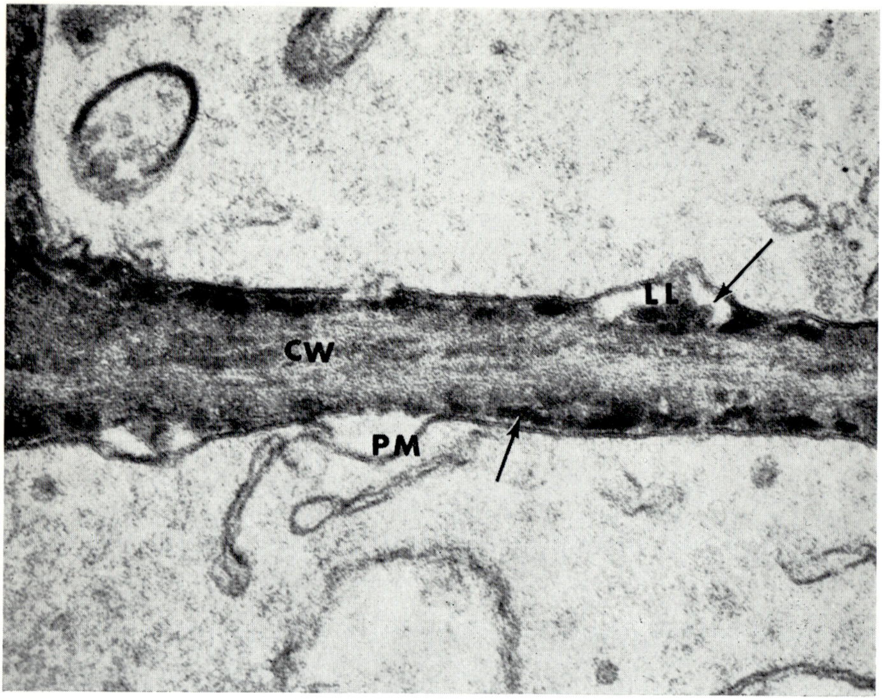

FIG. 1. Oat root tissue treated with 0·1 units of victorin for 24 h. Upper arrow points to Golgi vesicles in a lomasome-like structure whereas the lower arrow locates dark staining material in the cell wall adjacent to the plasma membrane (120,000 ×). (After Hanchey *et al.*, 1968.)

KEY TO ABBREVIATIONS: AC—aggregated cytoplasm, B—bacterium, CM—chloroplast membrane, CW—cell wall, EN—enzyme, ER—endoplasmic reticulum, G—grana, GB—Golgi body, IL—intergrana lamellae, IS—intercellular space, LL—lomasome-like structure, M—mitochondrion, Mb—microbody, N—nucleus, NM—nuclear membrane, Nu—nucleolus, PG—plastoglobuli, PM—plasma membrane, R—ribosomes, S—chloroplast stroma, SG—starch grain, T—tonoplast, V—vacuole, VG—vesiculated grana.

observed in toxin-treated root cap cells. Oat leaf tissue exposed to the toxin for 24 h displayed swelling of the plastids and disorganization of the granal and intergranal fretwork. This disruption of the chloroplasts was interpreted as being the cause of yellowing and striping symptoms characteristic of

Victoria blight of oats. None of these symptoms were apparent in electromicrographs of toxin-treated root cap or leaf tissue from a resistant variety.

In a subsequent study, Hanchey *et al.* (1968) suggested that the dense staining material was a carbohydrate component of the wall and represented the earliest indication of a change in its chemical and physical properties.

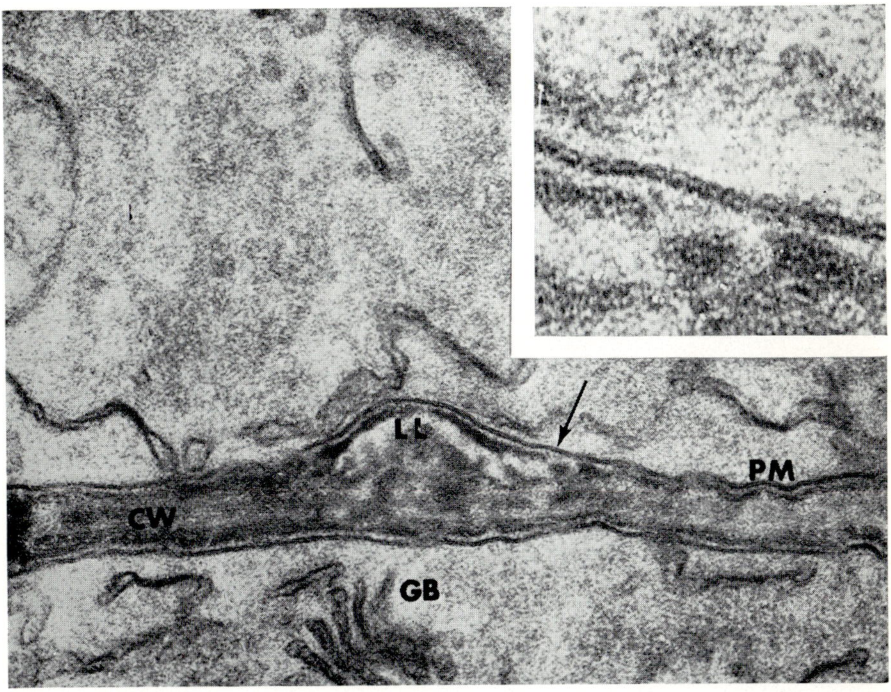

FIG. 2. Oat root tissue treated with 1·0 units of victorin for 24 h. Note Golgi body and larger lomasome-like structure caused by the separation of the plasma membrane from the cell wall. Arrow locates the enlarged segment of the plasma membrane. The apparent "unit structure" actually reflects a damaged membrane. (66,000 ×) (insert 200,000 ×). (After Hanchey *et al.*, 1968.)

In this study it was observed that the Golgi bodies, which are responsible for cell wall synthesis (Whaley *et al.*, 1964), produced abnormally large and densely stained vesicles in external root cap cells treated with victorin. Neither dense wall staining nor Golgi vesicle staining (Fig. 1) was detected in toxin-treated tissue that was fixed in glutaraldehyde and post-fixed in OsO_4. Only potassium permanganate fixation revealed these effects. The blister-like structures (invaginations of the plasma membrane) induced by victorin, when observed at higher magnification (Fig. 2), revealed a unit structure configuration which was interpreted by Hanchey *et al.* (1968) as reflecting both a structural and functional change in the membrane. They related this change to the rapid leakage of electrolytes from victorin-treated tissue. The blister-like structures (Fig. 1 and 2) induced by victorin, and termed lomasome-like, were believed to arise from dissociation of the plasma membrane from its normal position which is closely appressed to the cell

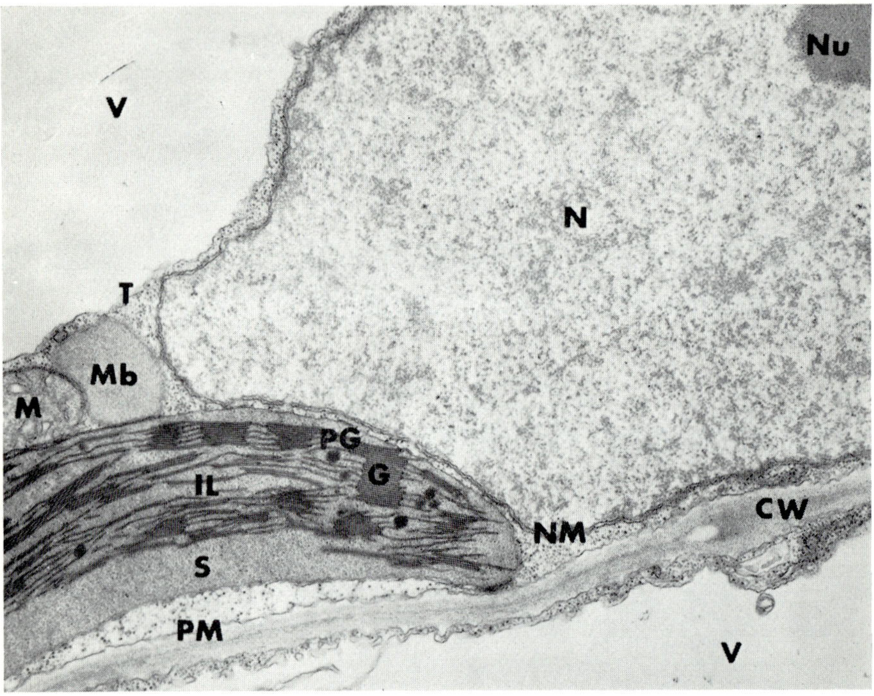

FIG. 3. Control (7 h after buffer infiltration) tobacco leaf tissue with normal appearing nucleus, nucleolus, nuclear membrane, cell wall, plasma membrane, chloroplast stroma, grana, intergrana lamellae, plastoglobuli, mitochondrion, microbody, tonoplast and vacuole (12,500 ×). (After Goodman and Plurad, 1971.)

wall. Victorin seems to cause the plasmalemma to migrate from the cell wall giving rise to an electron lucid region within which densely stained Golgi secretion products aggregate. This suggests an apparent alteration in cell wall synthesis where Golgi vesicle products are not integrated into existing wall material. It would appear, therefore, that victorin alters the structure and function of both the cell wall and plasma membrane.

Halo-inducing *Pseudomonas tabaci* Toxin

The halo-inducing toxin of *Pseudomonas tabaci* (Wolf *et* Foster) Stevens has been the most intensively studied toxin produced by a bacterial plant pathogen. Its biochemical composition has been described (Woolley *et al.*, 1955) and its possible mode of action has recently been redefined (Sinden and Durbin, 1968). Nevertheless, its effects in tobacco leaf tissue at the subcellular level have never been reported.

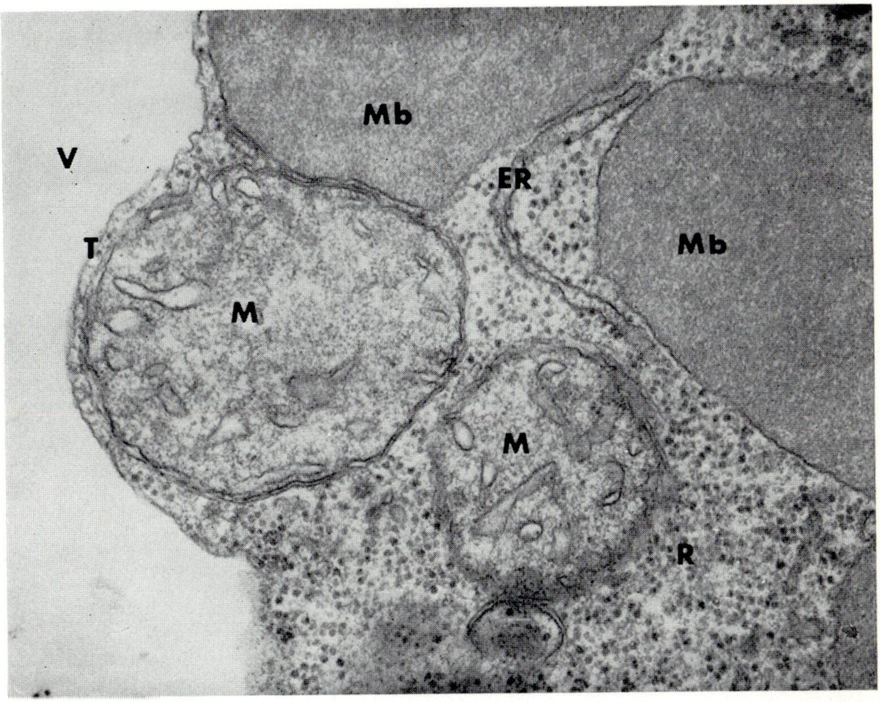

FIG. 4. Control (7 h after buffer infiltration) tobacco leaf cell revealing normal mitochondria, microbodies, tonoplast, ribosomes, and endoplasmic reticulum (38,500 ×). (After Goodman and Plurad, 1971.)

For the purpose of this Advanced Study Institute I have made a rapid and admittedly cursory electron microscopic examination of tobacco tissue showing the halo-symptom caused by *P. tabaci*.

The yellowing or chlorosis caused by the *P. tabaci* toxin is similar in appearance to the yellowing which occurs in naturally senescing tobacco leaves or in yellowing caused by leaf detachment. Chlorosis in all three instances can be reversed for a short time by an exogenous application of cytokinin (Farkas and Lovrekovich, 1963; Lovrekovich and Farkas, 1964; Lovrekovich *et al.*, 1964). It has been suggested, therefore, that the observed yellowing occurs as the result of a decrease in RNA and subsequently protein synthesis (Lovrekovich *et al.*, 1964; Osborne, 1962).

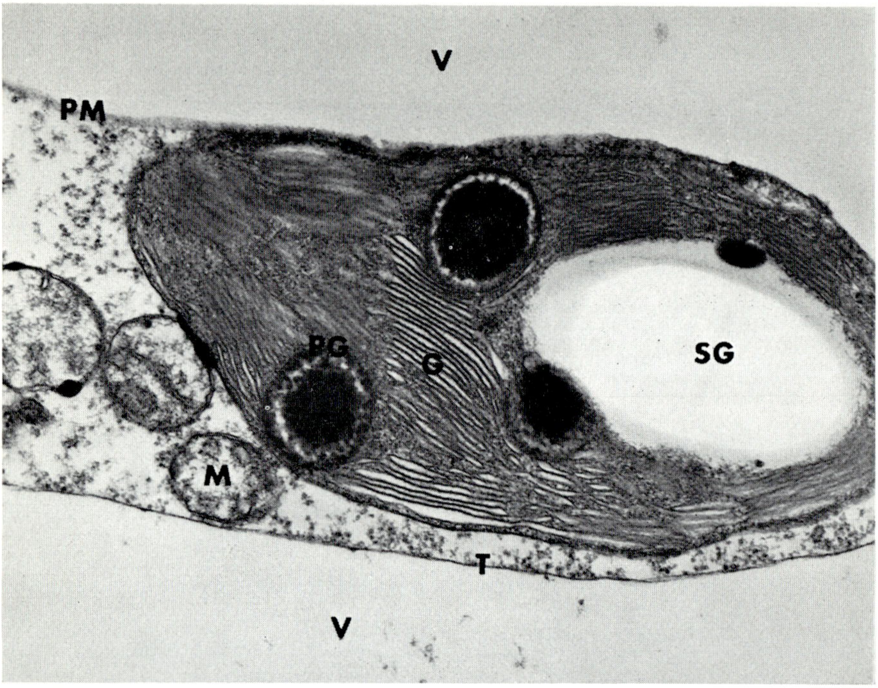

FIG. 5. Tobacco leaf tissue senescing (yellowing) normally (leaf attached). Note particularly the few cristae in mitochondria, the paucity of ribosomes in the cytoplasm, thickening of the plasma membrane and its separation from the cell wall (30,000 ×).

Tissue samples were obtained from leaves showing early signs of chlorosis (faint yellowing) as the result of inoculation with *P. tabaci* (10^8 cells/ml), natural ageing or detachment. In addition, tobacco leaves were treated with 10 μg/ml of actinomycin-D, which suppresses RNA and in turn protein synthesis (Osborne, 1962), and sampled for electron microscopic examination at the same stage of chlorosis induced in leaves either by infection or natural or induced senescence. The tissues were fixed in glutaraldehyde and post fixed in OsO_4 and embedded in Epon 812. Thin sections were stained in uranyl acetate for 20 min and in lead citrate for 5 min. The electron micrographs of yellowing tissues (Figs 5–8) in all instances revealed strikingly similar ultrastructural modifications and differed considerably from control tissue (Figs 3, 4). The major effects of "senescence", whether induced (pathologically, chemically or physically) or natural, include swelling of the chloroplasts and reduction in cytoplasmic and chloroplast ribosomal content. There is a pronounced decrease in chloroplast stroma

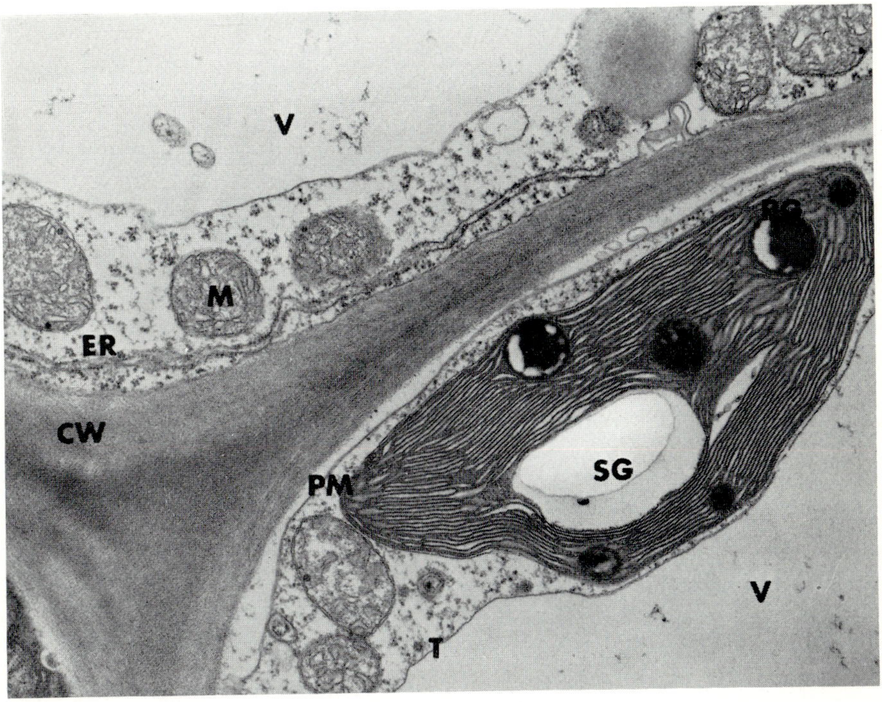

FIG. 6. Tobacco leaf tissue senescing (yellowing) as a consequence of leaf detachment. Note almost complete absence of staining in the cytoplasm (soluble protein) and the comparatively few ribosomes there as well as limited amount of stroma in the chloroplast (24,000 ×).

and in groundplasm of mitochondria (indications of reduction in soluble protein). The plastoglobuli (plastoquinone pool which is chlorophyll free) increase in size and number. This is in good agreement with a recent report on ageing chloroplasts in spinach (Lichtenthaller, 1969). The globuli also lose their osmiophilic quality and become less electron dense at their periphery. Attention is called to starch grains in Figs 5–7. These do not appear in either control (Figs 3, 4) or inoculated tissue (Fig. 8). Starch grains generally appear in tissue fixed soon after maximum photosynthetic activity (end of the daylight cycle). Their presence does not indicate that the treatments result in a reduction of metabolism of starch. Note also the distortion in chloroplasts due to the accumulated starch. It would appear that even these senescent chloroplasts are capable of photosynthesizing and storing starch.

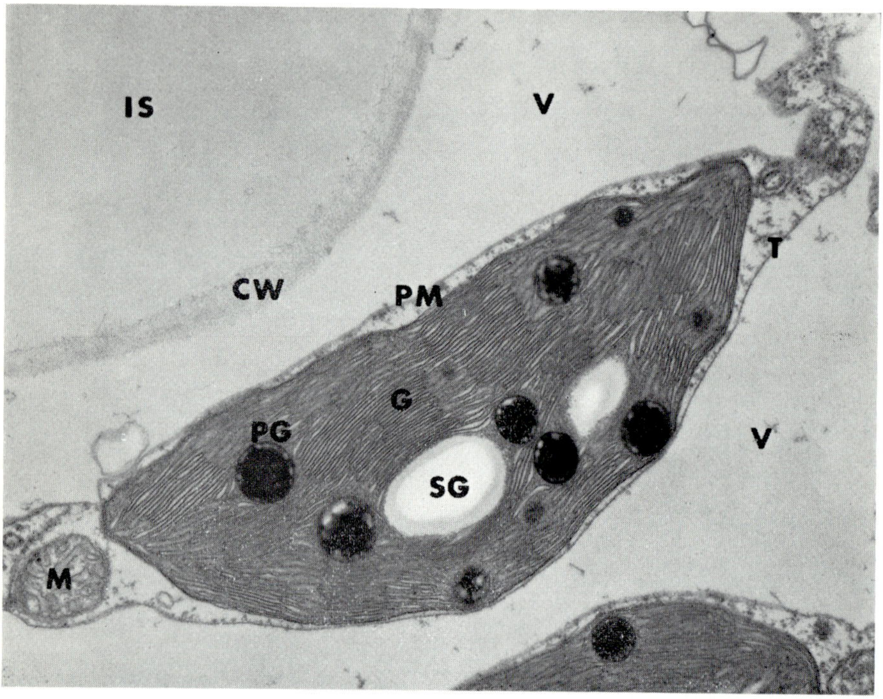

FIG. 7. Tobacco leaf tissue senescing (yellowing) 5 d after infiltration of the intercellular spaces with 10 μg/ml actinomycin-D. The cytoplasm is virtually devoid of stainable material and the plasma membrane has separated from the cell wall along its entire length (24,000 $\times$).

The appearance of crystalline structures in microbodies is also more frequent in ageing cells (Fig. 8). The crystalline lattice structures found in microbodies (also referred to as peroxisomes) (Frederick and Newcomb, 1969a) are more clearly seen in Fig. 12. These crystals are reported to be catalase by Frederick and Newcomb (1969b) which is responsible for the elimination of H_2O_2 produced in the oxidation of glycolate excreted from the chloroplasts.

Another consequence of senescence observed in all treatments is the separation of the plasma membrane from the cell wall. This is as clearly an effect of the toxin as it is of ageing.

It would seem that the treatments resulted in two primary effects: (a) a reduction in the nucleic acid and protein synthesizing capability of the cell;

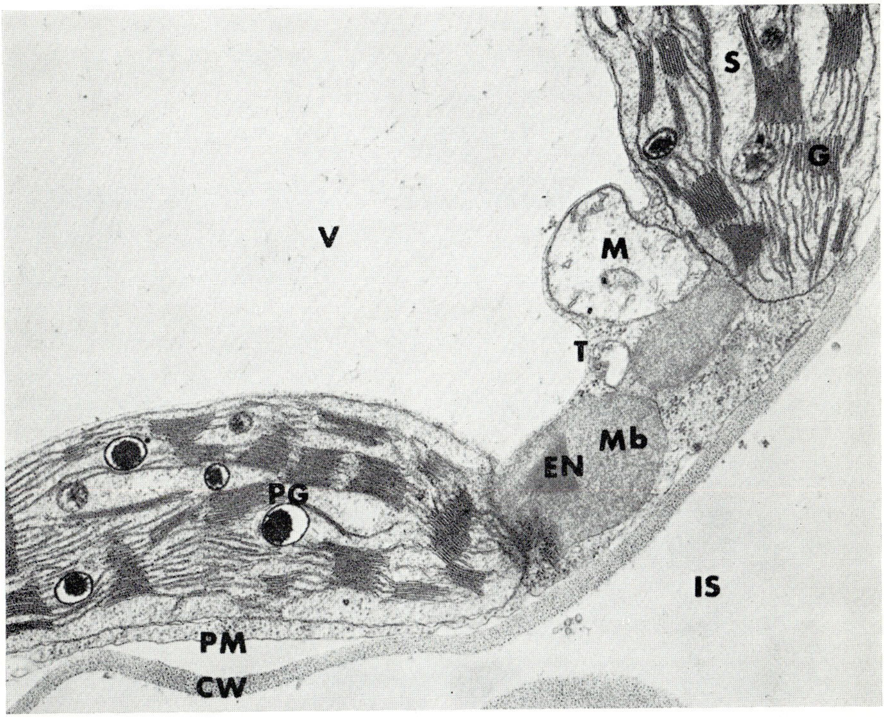

FIG. 8. Tobacco leaf tissue at the periphery of the yellow halo caused by a needle inoculation with *P. tabaci* (taken 5 d after inoculation). Plastoglobuli, as in Figs 4–6, are much enlarged and numerous, mitochondrion appears empty, one of the microbodies reveals an enzyme crystal and the plasma membrane has largely separated from the cell wall (12,500 ×).

and (b) a loss of at least some of the dynamic or functional activities of the plasma membrane which normally proceed at its interface with the cell wall.

Ammonia as a Necrotoxin

Recent reports by Lovrekovich *et al.* (1969, 1970a–c) have implicated ammonia as a toxic substance in plants inoculated with bacterial plant pathogens. High levels of ammonia accumulated within 36 h in tobacco leaf tissue inoculated with either *Pseudomonas tabaci* or within 18 h with a number of "incompatible" species capable of causing the hypersensitive reaction (HR). These data were interpreted as suggesting that necrosis was caused by the ammonia that accumulated and that pathogenesis and HR differed only in the rate at which these two phenomena proceeded following

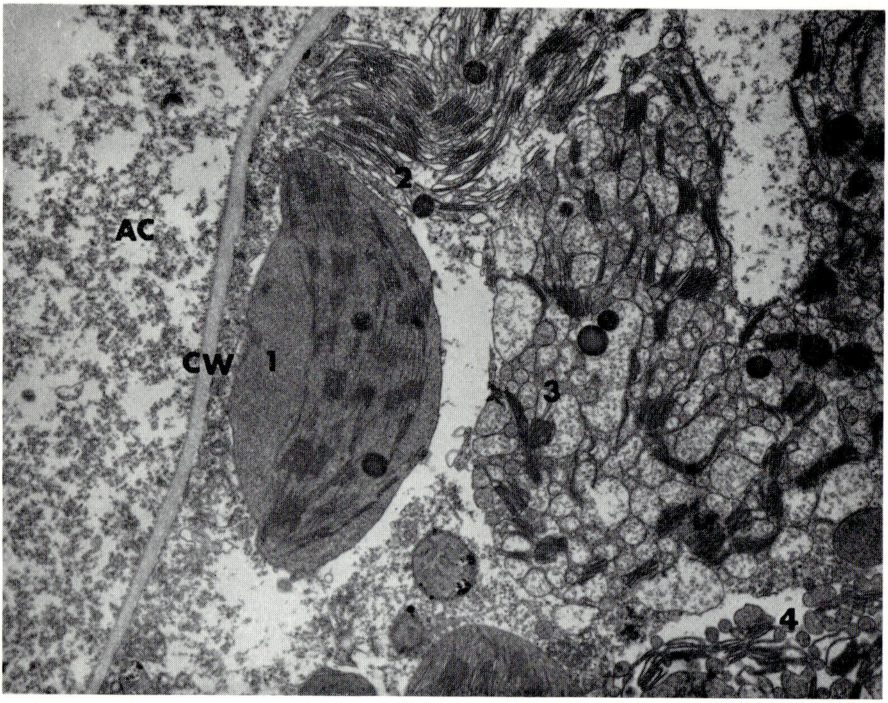

FIG. 9. Tobacco leaf tissue 7 h after infiltration with 10^8 cells/ml *Erwinia amylovora* showing aggregated cytoplasm, and 4 conditions of chloroplast degeneration: (1) nearly normal, (2) spreading, (3) branched, and (4) vesiculated (11,500 ×). (After Goodman and Plurad, 1971.)

inoculation. Briefly, it was concluded that HR was a rapid pathogenesis (Lovrekovich *et al.*, 1970b). Essentially the same results were obtained in cotton plants inoculated with *Xanthomonas malvacearum* (E. Sm.) Dows. (the compatible combination) and *P. tabaci* as the HR-inducing pathogen (Lovrekovich *et al.*, 1970a). That ammonia played a role in pathogenesis of *Erwinia amylovora* (Burr.) Winsl. *et al.* in apple stem tissue was also reported (Lovrekovich *et al.*, 1970c).

Although these bacterial plant pathogens produce copious amounts of ammonia *in vitro*, Lovrekovich and co-workers were unable to present evidence that ammonia accumulating in bacteria-inoculated plant tissues was produced by the bacteria *per se*. Furthermore, their data did not establish ammonia accumulation as the initial process in the sequence of events culminating in either the expression of disease symptoms or HR.

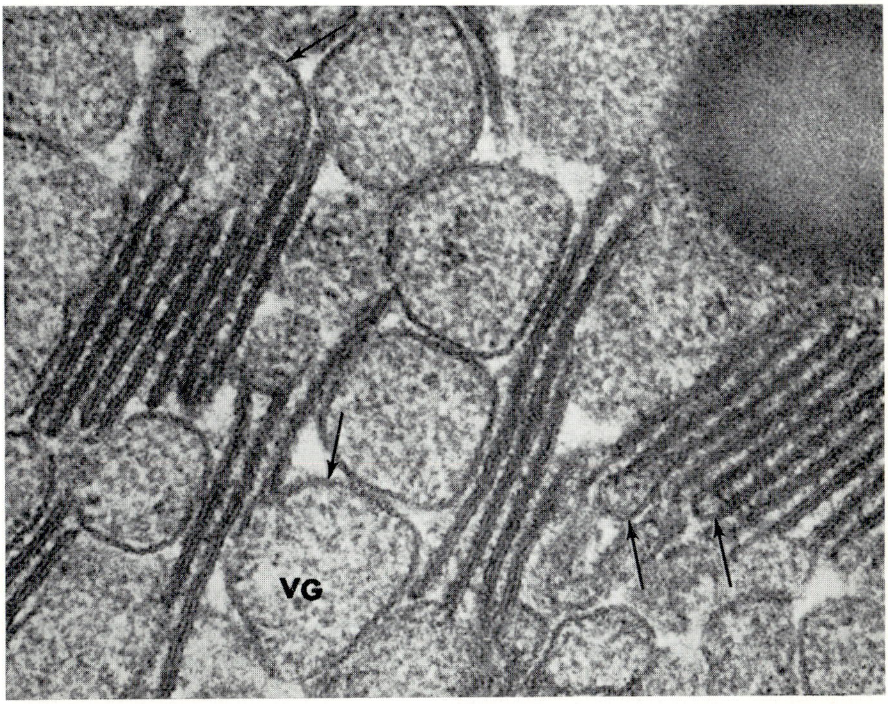

FIG. 10. Vesiculated grana and intergrana lamellae 7 h after infiltration with 10^8 cells/ml *Erwinia amylovora* with membrane stretching apparent at granal margins (arrows) (108,400 ×). (After Goodman and Plurad, 1971.)

Several years earlier Klement and Goodman (1967) clearly demonstrated that HR was irrevocably set into motion within 30 min after inoculation of tobacco tissue with 5×10^7 *E. amylovora* cells/ml. This suggested that the induction period for HR occurs shortly after inoculation and that a considerable period elapses before either symptoms are expressed or electrolyte leakage occurs (Goodman, 1968), and long before ammonia accumulates in the inoculated tissues (Lovrekovich *et al.*, 1970b).

Goodman (1968) has presented indirect evidence that HR resulted in permeability changes that reflected profound modifications in plant cell membranes. He noted that these changes were detected on tobacco leaf tissue either visually or conductimetrically 6–8 h after inoculation with 10^8 cells/ml of an incompatible bacterial species (*E. amylovora*). Evidence has also been obtained in our laboratory that the target in the membrane was the protein (Goodman, 1968) and not the phospholipid component (Huang and Goodman, 1970). Hence, it would seem that ammonia is not the *initial* cause of either disease symptoms or HR.

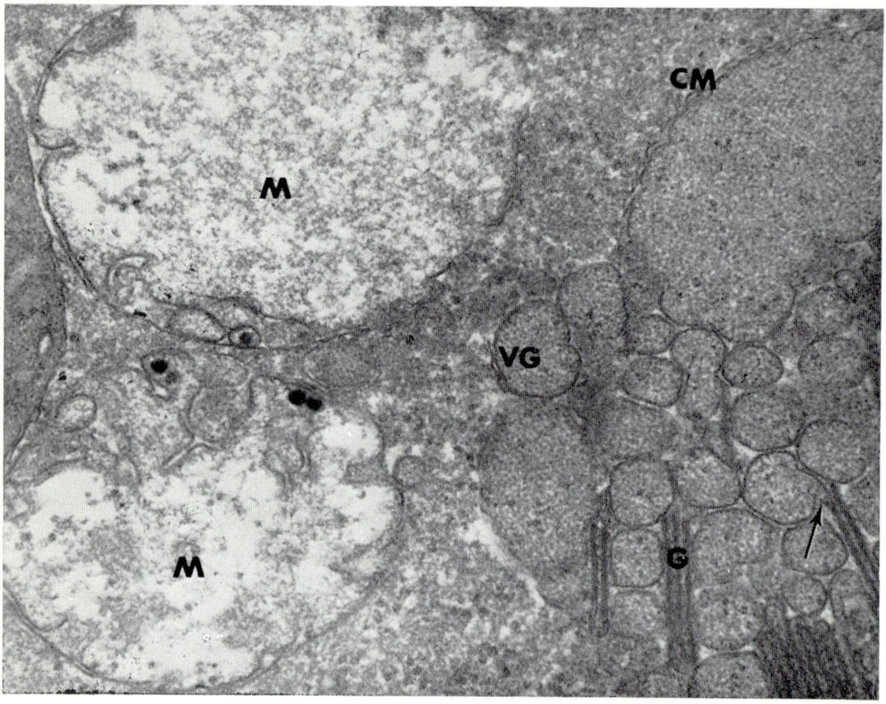

FIG. 11. Disintegrating mitochondria and vesiculated grana in tobacco leaf chloroplast 7 h after infiltration with 10^8 cells/ml *Erwinia amylovora* (43,750 ×). (After Goodman and Plurad, 1971.)

Recently Goodman and Plurad (1971) have presented direct evidence that as a consequence of HR, induced by inoculation of tobacco with 10^8 cells/ml of either *E. amylovora* or *Pseudomonas pisi* Sackett, the membrane systems of most subcellular organelles are drastically altered. The pattern of membrane damage is presented in Figs 9–13. The plasma membrane, tonoplast, as well as the bounding and internal membranes of chloroplasts and mitochondria, and the external membrane of microbodies are profoundly deranged within 7 h after inoculation. Cytoplasm, groundplasm of chloroplasts, mitochondria, microbodies and ribosomes, both membrane-bound and free, are similarly affected. The damage coincides with and is probably the cause of the observed tissue collapse and rapid loss of electrolyte-laden water. The presence of few bacterial cells (not more than 4×10^6 cells/leaf

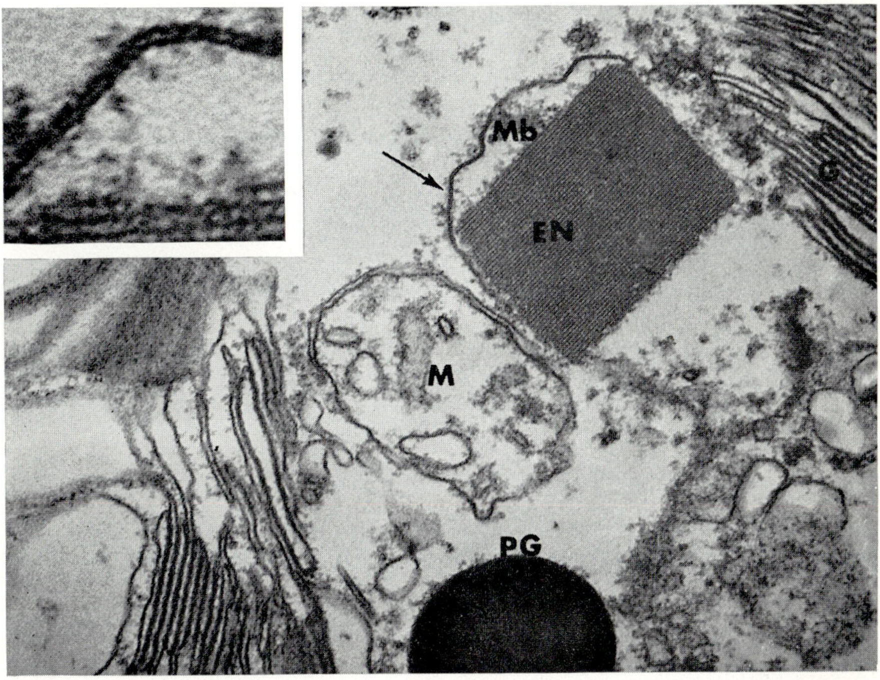

FIG. 12. Severely disintegrated tobacco leaf cell 7 h after infiltration with 10^8 cells/ml *Pseudomonas pisi*. Chloroplast grana have spread and the bounding membranes and stroma have been destroyed. The outer mitochondrial membrane has disintegrated, the inner one has large perforations and the vestigial cristae are distorted. The microbody membrane which only partially surrounds a crystalline inclusion (catalase enzyme) has an apparent unit structure. Note similarity between structure of the membrane in this insert and the one in Fig. 2 (66,000 ×) (insert 250,000 ×).

disk 1 cm in diameter) cause the observed symptoms which appear, at the ultrastructural level, to be unique. A similar array of membrane modifications was reported by Goodman and Burkowicz (1970) for *E. amylovora* in apple leaf tissue (a compatible combination).

That ammonia may have caused the observed damage at the subcellular level was surmised following the application of exogenous ammonia which

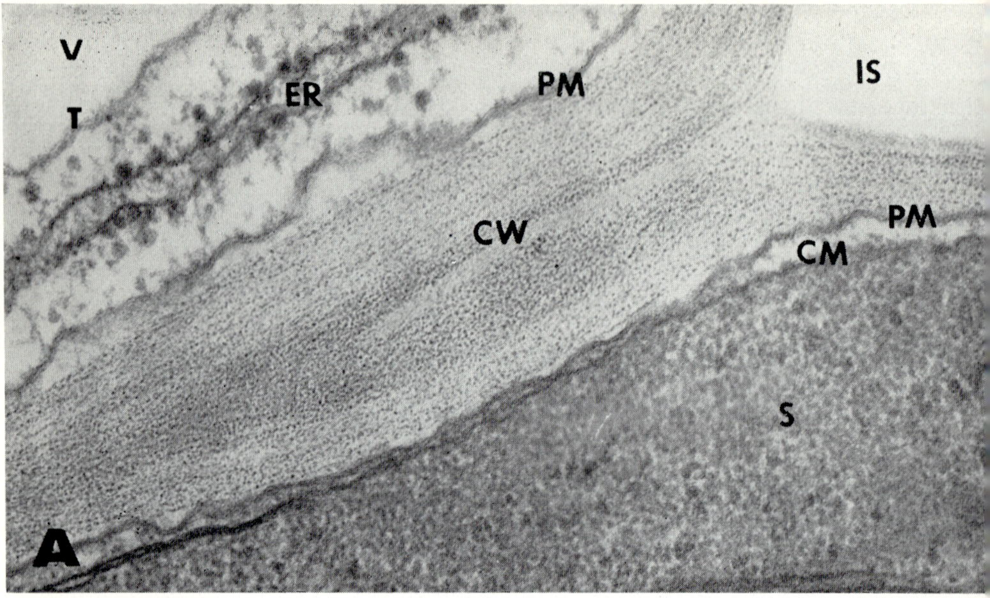

FIG. 13 (A). Control (7 h after buffer infiltration) tobacco leaf tissue showing apparently sound plasma membrane and tonoplast. (115,000 ×).

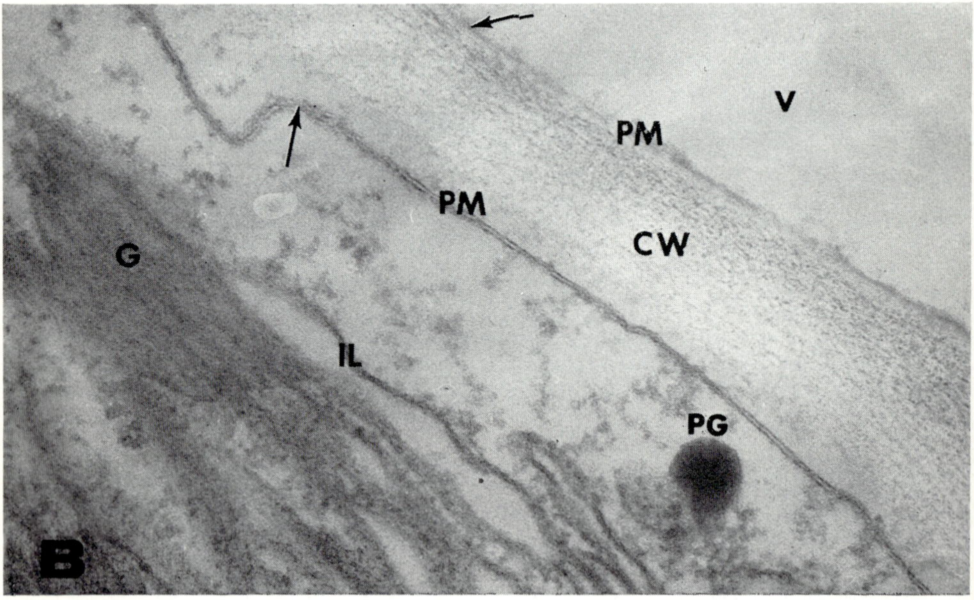

FIG. 13 (B). Tobacco leaf tissue 7 h after infiltration with 10^8 cells/ml *Erwinia amylovora* showing only a vestigial plasma membrane in the upper cell (arrow) and a swollen, interrupted plasma membrane (arrow) in lower cell. Upper cell is completely devoid of cellular organelles and particulate matter (93,150 ×).

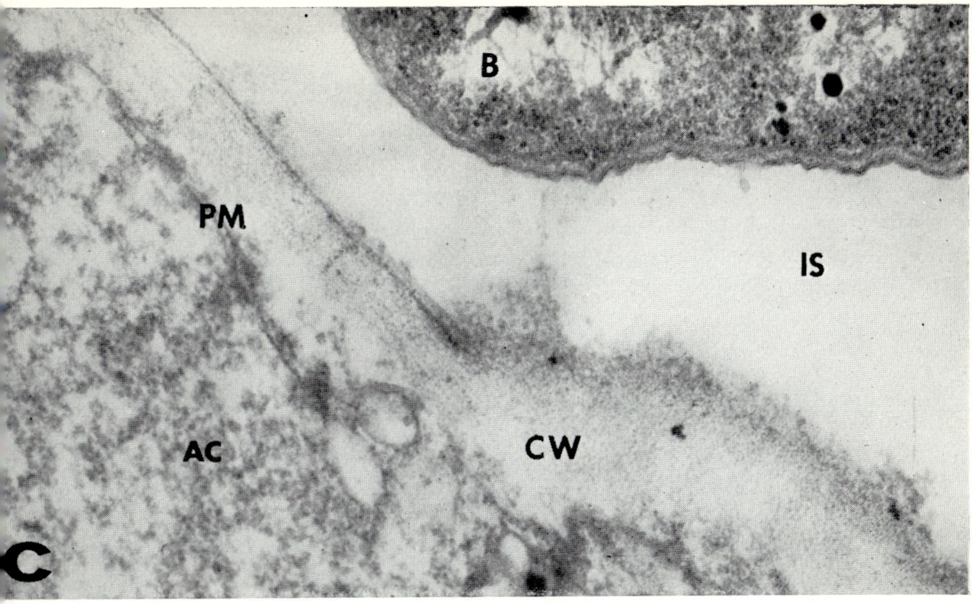

FIG. 13 (C). Tobacco leaf tissue 7 h after infiltration with 10^8 cells/ml *Pseudomonas pisi* showing bacterium in the intercellular space. Plasma membrane is discontinuous and vesiculated and the cytoplasm has aggregated (50,000 ×). (After Goodman and Plurad, 1971.)

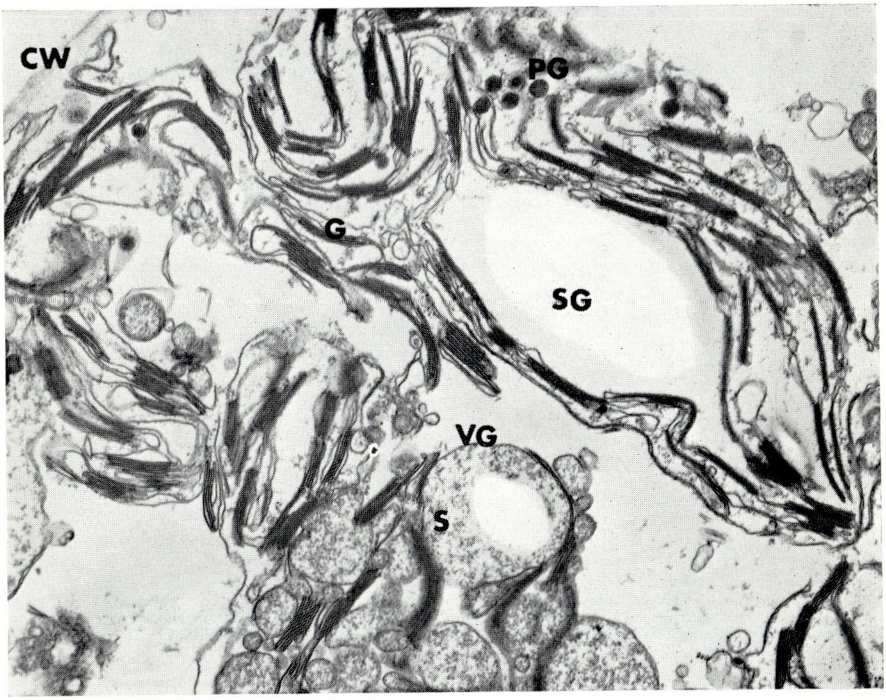

FIG. 14. Tobacco leaf tissue fixed at the first visual signs of an HR-like reaction (30 min after a 5 min exposure to NH_3 vapours from ammonium carbonate). Several chloroplasts are spreading because of destruction of their bounding membranes and chloroplast shows vesiculated grana as in Fig. 9 (16,500 ×).

mimicked the ultrastructural modifications caused by HR-inducing bacteria. The level of ammonia detected in tobacco tissue exposed to ammonia vapours from ammonium carbonate that showed HR-like symptoms is of the same order of magnitude as that detected in tobacco showing HR symptoms induced by *P. pisi* (Lovrekovich *et al.*, 1970c). The ultrastructural changes in tobacco leaf tissue induced by ammonia liberated from ammonium carbonate are shown in Figs 14 and 15.

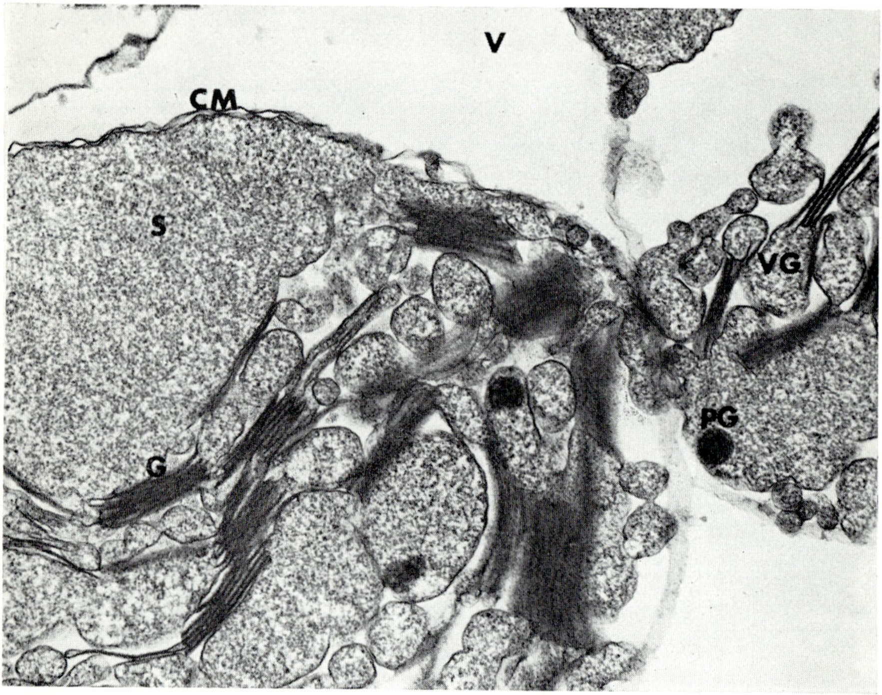

FIG. 15. An enlargement of two chloroplasts revealing vesiculated grana caused by NH_3 liberated from ammonium carbonate. Conditions of the experiment are the same as in Fig. 14. Compare vesiculated grana with those in Fig. 10 (25,600 ×).

These are similar to the modifications shown in Figs 9 and 10 where the HR was induced by *E. amylovora*. Furthermore these modifications of subcellular organelles, particularly those of the chloroplast caused either by ammonia vapours or by incompatible plant pathogenic bacteria, seem unlike any induced by either light (Crofts *et al.*, 1967) or osmotic (Deamer *et al.*, 1967) effects.

Nevertheless, the precise role of ammonia in the hypersensitive reaction is at this time uncertain. Experiments performed in my laboratory since this paper was presented suggest that membrane damage occurs more than 18–24 h prior to the time that Lovrekovich *et al.* (1970b) reported ammonia to accumulate in inoculated tissues.

Summary

In summarizing the observed ultrastructural modifications induced by the three toxins it would appear that victorin acts rapidly upon both cell wall and plasma membrane structure and function; ammonia also acts quickly upon membranes in general and the *P. tabaci* toxin exerts its effects more slowly by gradually suppressing nucleic acid and protein synthesis.

Acknowledgement

Research of the author reported herein was supported by a research grant from the National Science Foundation, GB-17729.

References

Crofts, A. R., Deamer, D. W., and Packer, L. (1967). *Biochim. biophys. Acta* **131**, 97–118.
Deamer, D. W., Crofts, A. R., and Packer, L. (1967). *Biochim. biophys. Acta* **131**, 81–96.
Farkas, G. L. and Lovrekovich, L. (1963). *Phytopath. Z.* **47**, 391–398.
Frederick, S. E. and Newcomb, E. H. (1969a). *Science* **163**, 1353–1355.
Frederick, S. E. and Newcomb, E. H. (1969b). *J. Cell Biol.* **43**, 243–253.
Goodman, R. N. (1968). *Phytopathology* **53**, 872–873.
Goodman, R. N. and Burkowicz, A. (1970). *Phytopath. Z.* **68**, 258–268.
Goodman, R. N. and Plurad, S. B. (1971). *Physiol. Pl. Path.* **1**, 11–15.
Hanchey, Penelope, Wheeler, H., and Luke, H. H. (1968). *Am. J. Bot.* **55**, 53–61.
Huang, J. S. and Goodman, R. N. (1970). *Phytopathology* **60**, 1020–1021.
Klement, Z. and Goodman, R. N. (1967). *Phytopathology* **57**, 322–323.
Lichtenthaller, H. K. (1969). *Protoplasma* **68**, 315–320.
Lovrekovich, L. and Farkas, G. L. (1964). *Nature* **198**, 710.
Lovrekovich, L., Klement, Z., and Farkas, G. L. (1964). *Science* **145**, 165.
Lovrekovich, L. and Lovrekovich, H. (1970a). *Plant Dis. Reptr* **54**, 713–716.
Lovrekovich, L., Lovrekovich, H., and Goodman, R. N. (1969). *Phytopathology* **59**, 1713–1716.
Lovrekovich, L., Lovrekovich, H., and Goodman, R. N. (1970b). *Can. J. Bot.* **48**, 167–171.
Lovrekovich, L., Lovrekovich, H., and Goodman, R. N. (1970c). *Can. J. Bot.* **48**, 999–1000.
Luke, H. H., Warmke, H. E., and Hanchey, P. (1966). *Phytopathology* **56**, 1178 1183.
Osborne, D. J. (1962). *Pl. Physiol.* **37**, 595–602.
Scheffer, R. P. and Pringle, R. B. (1963). *Phytopathology* **53**, 465–468.

Sinden, S. L. and Durbin, R. D. (1968). *Nature* **210**, 379–380.
Strobel, G. A. and Hess, W. M. (1968). *Pl. Physiol.* **43**, 1673–1688.
Whaley, W. G., Kephart, J. E., and Mollenhauer, H. H. (1964). *In* M. Locke (Ed.) "Cellular membranes during development." Academic Press, Inc., New York, pp. 135–173.
Woolley, D. W., Schaffner, G., and Braun, A. C. (1955). *J. biol. Chem.* **215**, 485–493.

Discussion

Wood: Our recent work on the fine structure of *Phaseolus vulgaris* infected by *Colletotrichum lindemuthianum* has emphasized the specificity, rapidity and the localization of the hypersensitive response. Cells next to those that have been killed following penetration by an avirulent race of the pathogen remain alive though their constituents become different in a number of ways. In contrast, in the earliest stages of infection a hypha of a virulent race passes through the cytoplasm of a cell without killing it.

Goodman: I remind the audience that fungi act on the host in the disease cited by Wood only intracellularly whereas bacteria operate extracellularily.

Durbin: Do you believe that the wildfire toxin is involved in producing chlorosis *in vivo*, or is chlorosis due entirely to ammonia produced by the bacterium?

Goodman: From our cursory examination of subcellular modifications caused by the wildfire toxin it would seem that the toxin *per se* is responsible for the chlorosis observed. Although ammonia may play a role in producing these symptoms, no evidence has been presented that would suggest that ammonia accumulation causes the symptoms observed. It would be of interest to determine ammonia accumulation in tissues becoming senescent normally and where actinomycin-D has induced senescence.

Kuć: A message is apparently transmitted to cells adjacent in those that have collapsed in the hypersensitive reaction. These and more distant cells behave as resistant cells when inoculated with a race of *C. lindemuthianum* to which they would be expected to be susceptible.

Durbin: Does Wood know what happens when one cell dies and the adjacent cell is alive? How does this affect the continuity of membrane and E.R. *via* plasmodesmata?

Wood: This has not yet been determined.

Scheffer: Tomiyama in Japan has shown that spores of *P. infestans* penetrate cell walls of both resistant and susceptible cells. Within minutes after germ tubes contact the resistant protoplast, protoplasmic streaming stops suddenly. This does not happen in the susceptible cell which accepts and tolerates the fungus for several days. Thus a difference between resistant and susceptible cells is evident within minutes, possibly seconds. Any explanations of disease resistance or hypersensitivity must take this into account. Otherwise, the proposed mechanism is not satisfactory.

Rudolph: Klement has developed a method for differentiating between saprophytic and plant pathogenic pseudomonads. Certain concentrations of phytopathogenic pseudomonads infiltrated in tobacco leaves, induce hypersensitive reactions, whereas the same concentrations of saprophytic pseudomonads do not. If the ammonia hypothesis is valid, it has to be assumed that the phytopathogenic pseudomonads liberate high amounts of ammonia in the host plant, whereas the saprophytic pseudomonads do not. The question becomes why the phytopathogenic pseudomonads are able to liberate ammonia.

Increases in respiration, for instance glucose-6-phosphate dehydrogenase and isocitric-dehydrogenase, also appear following infection of bean leaves by *Pseudomonas phaseolicola*. There are no important differences in the respiration increases found in bacterial and fungal plant diseases. When bean leaves are infected with very dilute suspensions of *Pseudomonas phaseolicola*, so that each leaf contains only one or two infection centres, often a greasy spot is surrounded by a large halo and between the halo and the normal green of the leaf there appears a green circle indicating high metabolic activity.

Wood: I should like to refer to the rapid killing of cells following the action of enzymes that degrade cell walls. There is now good evidence that chain-splitting pectic enzymes are the most important in this killing. But these are highly specific for pectic substrates and so far as I know these substances are not constituents of the plasmalemma. Also, they are concentrated mainly in the middle lamella and primary cell wall and not in that part of the wall in contact with the plasmalemma. It is difficult to see, therefore, how protoplasts are killed other than by physical, osmotic effects.

Goodman: The membrane theory is consistent with mine with one modification. I believe that the plasmalemma actually extends at intervals along its length into the wall *per se*. Hence pectin-degrading enzymes actually free, at least partially, the membrane from the wall. Protoplasts so prepared would have stress points which would make them osmotically sensitive.

Page: I would be interested in comments on the formation of microtubules in wall synthesis. Is it necessary for the microtubules to be in intimate contact with walls for synthesis to proceed?

Goodman: In my experience with spores of the alga *Cladophora rupestris* microtubules are usually apparent at times when wall synthesis has not yet been initiated. When the wall is being laid down and fibrils are being oriented the microtubules are not in evidence. So here microtubules do not seem to play a role in wall synthesis or microfibril orientation.

19. The Ecological Significance of Toxin Production by Micro-organisms

J. A. MEYER

Université Catholique de Louvain,
Laboratoire de Phytopathologie,
Heverlee, Belgium

Introduction

During their growth *in vitro*, micro-organisms secrete into the culture medium many metabolites that can be classified broadly as toxins, antibiotics, enzymes, growth factors and others that are biologically inactive, at least in low concentrations. Conditions for the production of these metabolites are often well known, but the *raison d'être* of many of them and their significance remain unknown.

In nature, however, for plant pathogens and other micro-organisms, very little is known about the production of metabolites and because of this the study of their ecological significance is all the more difficult although we often assume that they play a part in the survival of the micro-organisms and in the infection and colonization of plants by pathogens.

The behaviour of micro-organisms in nature and in the laboratory may be very different. The media used *in vitro* never reproduce and are often far removed from the natural environment of a parasite in the soil or in a root, a stem, a leaf, or a xylem vessel of a host plant. A typical example of the discrepancy between *in vitro* and *in vivo* conditions for growth of a pathogen can be seen with *Fusarium oxysporum* (Schl.) Sny. *et* Hans. a typical agent of vascular wilts. *In vitro* studies of the physiology of the fungus aimed at explaining specificity and pathogenicity have often used highly oxygenated shaken cultures on a rich nutrient medium although the fungus as a parasite lives in a xylem vessel, at a very low oxygen tension, in a substrate which is a continuously renewed, dilute solution of many substances, presumably very poor in organic compounds especially at the start of infection. These characteristics of the natural environment profoundly influence the metabolism of the fungus and, thus, the metabolites excreted in that environment. In these circumstances, deductions about the ecological significance of metabolites produced *in vitro* would be dangerous.

It is often difficult to determine whether a bacterial or a fungal product has some effect under natural conditions. Also, failure to demonstrate toxin production *in vitro* does not mean necessarily that the organism is unable to produce that toxin because it may not be easy to find the correct conditions for the production of a particular substance. These may not be the same as those for good growth. Another relevant question about the ecological significance of toxins is whether a micro-organism in the soil or in a plant produces enough toxin to be detected or to be harmful to plants.

Although in plant pathology a toxin is generally defined as a non-enzymic substance (Owens, 1969), in this paper the word will be used in a broader sense, not only because of the many different types of metabolites secreted by micro-organisms, but also because in only a few diseases is the substance causing symptoms known with certainty so that it can or cannot be called a toxin in the more restricted use of the word. The different uses of the word toxin are discussed by Wood (1967).

During the life of most pathogens, there are two phases during which the ecological significance of toxins may differ fundamentally, outside and inside the plant. These will be dealt with separately.

Toxins Produced Outside the Plant

Two groups of toxins are distinguished according to their origin: toxins resulting from the decomposition of organic debris by saphrophytes and metabolites secreted by saprophytes or pathogens.

Toxins Released during the Decay of Plant Residues or by Living Roots

Toxic action on plants. Toxins in this group explain the poor growth of corn in stubble-mulch farming or in tropical forest soils the first season after felling the forest. Water extracts of such soils strongly inhibit root development and several organic compounds have been isolated from them, the most toxic being coumarin, vanillin, salicylic aldehyde, cinnamic acid; many are derived from the degradation of lignin material.

The toxins liberated from decomposing plant residues have been extensively studied by Winter (1955), and more recently by McCalla and Haskins (1964), Patrick *et al.* (1964) and Linderman (1970). McCalla and Haskins investigated these toxins in relation to stubble-mulch farming, a practice which reduces yields, particularly of corn. When this occurs, plants are smaller and often show chlorotic symptoms resembling those caused by nitrogen deficiency. Carley and Watson (1967) found that aqueous extracts of crop residues caused death of root meristems. Corky root rot of lettuce is also claimed to be caused by substances coming from decomposing tissues (Amin and Sequeira, 1966).

Patrick *et al.* (1964) believe that the toxic substances released from decaying plant material which inhibit germination of seeds or discolour or kill roots, are most probably formed during the early stages of decomposition.

Other disorders which may be induced by toxins from decomposing residues are the replant problems of peaches and apples. Characteristics of these diseases are their high specificity, the persistence of the causal factor (at least two years), and their control by treatments with substances with a broad spectrum of activity against soil organisms. However, with further study of these diseases, the toxin hypothesis is becoming less tenable (Hoestra, 1968; Savory, 1969), and it now seems probable that soil organisms are an essential factor in the etiology of replant problems.

Toxic action on micro-organisms. Toxins from decomposing root material are also invoked to account for the treeless prairies in some regions of the U.S. Prairie soils depress the activity of certain soil fungi that are essential for the growth of trees. This adverse effect is believed to be caused by harmful substances released from living and decaying roots of certain prairie grasses; after many years they appear to accumulate in the humus layer in concentrations sufficient to eliminate certain free living fungi needed by trees. But it has never been shown that these fungi were originally present in these prairie soils.

A similar hypothesis has been proposed by Theron (1951) in South Africa to explain the absence of nitrate under grass fallow. He claimed that root excretions inhibit nitrifying bacteria. But when samples from these soils are removed and incubated, they immediately begin to accumulate nitrate in the absence of living, absorbing roots. These roots are the main reason why nitrate does not accumulate in the field (Meyer, 1959).

Increase of the incidence and the severity of root diseases. Toxins may act on the parasite or on the host plant. They may stimulate chlamydospore germination. This may be beneficial for the fungus, as has been shown for *Fusarium solani* (Mart.) Sny. *et* Hans. f. sp. *phaseoli* (Burk.) Sny. *et* Hans. (Toussoun *et al.*, 1963), or it may be detrimental as it is for *F. oxysporum* (Toussoun *et al.*, 1969). But more often, extracts of plant residues seem to act directly on the host tissues, probably by altering cell permeability so that exudation of materials is increased. This may enhance the activity of the parasite in various ways. Linderman and Toussoun (1968) showed that water extracts of decomposing plant residues in non-sterile soils do not stimulate germination of chlamydospores of *Thielaviopsis basicola* (Berk. *et* Br.) Ferr. But exposure of roots to phytotoxic extracts increases susceptibility of bean plants to *F. solani*, *Rhizoctonia solani* Kühn and *T. basicola* (Toussoun and Patrick, 1963), and the susceptibility of tobacco to *T.*

basicola (Patrick and Koch, 1963). Roots treated with the extracts are much more extensively colonized because the toxins induce higher exudation of nutrients from roots. In these experiments the toxins were used at levels not visibly damaging to the plant.

Beneficial effects of toxic compounds. Decomposition is not always detrimental. Clearly, *inter alia*, it provides plants with essential nutrients in a readily assimilable form. But I would like to mention here ammonia as a decomposition product of green manures which may affect disease control by its toxicity to fungi and higher plants (Gilpatrick, 1969). Ammonia may injure *Persea indica* Spreng. roots, but roots so injured were not so easily invaded as were uninjured roots when both were inoculated with zoospores of *Phytophthora cinnamomi* probably because they do not attract zoospores or cause the growth of germ tubes to be orientated. Furthermore, high concentrations of ammonia kill mycelium and zoospores of *Phytophthora* species.

Also interesting are the numerous toxic compounds excreted into soil by growing plants. These may damage other plants and may play a role in certain soil sickness problems. This subject was reviewed recently by Tukey (1969).

Toxins Secreted by Micro-organisms

1. *Beneficial substances*

Antibiosis in the soil. Various authors have demonstrated that specific antibiotics may be present in soil that has been inoculated with organisms known to produce these antibiotics. But unfortunately most of the work has been done with sterilized soil which usually has been heavily enriched with readily decomposable organic matter at concentrations far higher than would occur in normal agricultural practice. Brian (1957a) concluded that only very rarely will the organic matter content of soil be of a quality and quantity suitable to stimulate appreciable production of antibiotics if one considers the bulk of soil as a whole. I would also add that only rarely will the proportion of antibiotic producing organisms in a natural soil be high enough to produce detectable quantities of antibiotics even after enrichment. There is in fact a complete lack of convincing evidence to support the thesis that antibiotics produced under natural conditions can exert a significant and general effect in soils (Pramer, 1965).

However, some workers believe that antibiosis may be significant in disease control generally on the basis of a correlation between reduction or suppression of disease and an increase in the total microflora which will include antibiotic producing organisms. There are a few special ecological situations or conditions where such correlations might appear as when organic material or fertilizers are added to the soil (Ghaffar, 1969; Grossmann, 1967; Huber and Watson, 1970). Also an antagonistic organism may increase naturally in the rhizosphere and in some diseases, resistance of varieties has been attributed to the presence around their roots of more antagonistic organisms which in turn reduce the number of infection units of a pathogen at the surfaces of the roots. This was claimed for the resistance of some banana varieties to *F. oxysporum* f. sp. *cubense* Sny. *et* Hans. (Harper, 1950) but disproved a few years later by Rombouts (1953). Agnihothrudu (1955) stated that the rhizosphere of a wilt resistant variety of pigeon pea compared with susceptible varieties, contained larger numbers of actinomycetes antagonistic to *F. episphaeria* (Tode) Sny. *et* Hans. Spraying leaves with substances such as 2,4-D and urea may increase the proportion of antagonistic organisms in the rhizosphere (Horst and Herr, 1962).

Microbial antagonism in the rhizosphere of subterranean clover has also been invoked to explain the absence of nodulation following inoculation with a potentially effective strain of *Rhizobium.* Here again the factor was believed to be microbial in origin, transferable to other soils and disappearing with sterilization or after adding charcoal to the seed when it was sown (Hely *et al.*, 1957). Another situation in which microbial antagonism may be important is given by Pugh and Van Emden (1969) who explained the natural reduction of *Ophiobolus graminis* and *R. solani* in polder soil by the gradual increase in the proportions of antagonistic organisms with continuous cultivation.

I should also like to recall the impetus given to the use of induced antagonism caused by partial sterilization and which is associated with the growth of *Trichoderma* spp., bacteria and other antagonists (Olsen and Baker, 1968). Another approach aims to promote the development of an antagonistic flora by adding to the soil an organism known to produce an antibiotic *in vitro.* This, however, is unlikely to succeed unless conditions in the soil are altered to permit the establishment and growth of the new organism. This can be done by inoculating seed with an organism preferably derived from the rhizosphere and whose antagonism has been demonstrated in the laboratory (Baker, 1968). However, except for partial sterilization these promising methods have rarely been developed for practical use, and antibiotic substances have never been implicated in the control that has been obtained. The great difficulty with these methods is that their results are usually unpredictable.

Mycostasis in the soil. Mycostasis or fungistasis in soil means that the spores of many fungi do not germinate so long as they remain in the soil. The cause of this inhibition is still not known with certainty; toxins have never been detected, but there is some evidence that volatile substances are involved. Irrespective of the nature of the inhibition, its activity is nullified by adding to soil various readily metabolized organic substances. Experiments in my laboratory (Mpungi, 1967) suggest that the most plausible explanation of fungistasis is that spores do not germinate in soil because they lack readily available sources of energy. Other workers have come to similar conclusions (Ko and Lockwood, 1967; Lockwood, 1964; Steiner and Lockwood, 1970).

Antagonism in the phyllosphere. Antagonistic phenomena have also been reported in the phyllosphere at the surface of the leaf. A few examples can be given. *Erwinia herbicola* (Geilinger) Dye, a saprophytic yellow pigmented bacterium, which is often associated with *E. amylovora* (Burr.) Winsl. *et al.* in fireblight diseased tissues, possesses a β-glucosidase which cleaves arbutin to yield hydroquinone and glucose; this hydroquinone may inhibit the metabolic activities of *E. amylovora* and reduce its effectiveness as a pathogen (Chatterjee *et al.*, 1969). Inoculation of dwarf bean leaves with a suspension of spores of *Aureobasidium pullulans* (de Bary) Arn. one day before or simultaneously with *Alternaria zinniae* Pape significantly decreased the number of lesions caused by the latter, the result being explained by the fact that *Aureobasidium pullulans* restricted the germination of *Alternaria zinniae* on bean leaves (Van den Heuvel, 1969).

In the most important ecological situations so far examined, antibiotics would be expected to have a role; unfortunately this has only rarely been demonstrated and present evidence only suggests rather than proves that there are significant interactions between parasites and other organisms. The mechanism regulating these interactions may well be predominantly competition.

All the above considerations, however, do not exclude the use of antibiotics as such in plant disease control. Applied to plants, they may be absorbed and protect them against parasites (Dekker, 1963).

2. *Phytotoxic substances produced in the soil by micro-organisms*

Micro-organisms often secrete in their environment low molecular weight substances such as organic acids and amino acids which may at certain concentrations be toxic to plants. However, even for the same compound various values for the lowest level for toxicity are given. For example, for tryptophan figures in the range from 10 to 1000 ppm are reported by McCalla and Haskins (1964). But in natural environment, products such as these will never accumulate because they will be rapidly metabolized by the microflora. Thus, Putnam and Schmidt (1959) reported only small but detectable amounts of free amino acids in the soil, never sufficiently high to be toxic to plants.

However, toxic material may be produced by micro-organisms in the immediate vicinity of the plant, causing necrosis without physical contact between host and pathogen. This happens in necrosis induced by *R. solani* on roots of Chippewa soybean seedlings (Wyllie, 1962), and probably also in crown rot of groundnut seedlings where oxalic acid produced by *Aspergillus niger* van Tiegh. causes necrosis (Gibson, 1953).

Another example of action at a distance is that of some *Marasmius* spp. producing HCN that damages tissues of the crown of several plants; only after this damage has occurred are tissues invaded by the fungus (Ylimäki, 1967).

Antibiotics may also damage plants. Brian (1957b) stated that *in vitro* most antibiotics exhibited some degree of phytotoxicity. However, in nature, again examples seem to be rare. But recently, patulin has been found to be produced by *Penicillium expansum* Link *em.* Thom. on apple tree root and leaf debris and has been correlated with apple soil sickness (Börner, 1963). Patulin, produced by *P. patulum* Bain. has also been suggested as a possible source of toxicity in stubble-mulching (Norstadt and McCalla, 1969).

Another example of toxic substances produced outside but damaging the plant, is provided by certain fungi that induce albinism in seedlings. Albinism develops in citrus seedlings when seeds are soaked in a suspension of spores of *Aspergillus flavus* Link *ex* Fr. (Durbin, 1959) and when seeds are germinated in extracts from cultures of certain clones of *Alternaria tenuis* Nees (Rayan *et al.*, 1961). The metabolites of *A. tenuis* inhibit chlorophyll formation in seedlings of citrus and many other plants (Fulton *et al.*, 1965). But not all plants tested were affected; chlorosis occurs only if the fungal metabolites come in contact with plant tissues before they are exposed to light.

Toxins Produced Within the Plant

During the parasitic phase of the life of a parasite, we have to distinguish clearly two stages in considering the ecological significance of toxic compounds. The first is penetration or "aggressiveness", related to specificity and the second is the period during which pathogenesis occurs. If toxins produced in the environment outside plants are generally interpreted as beneficial to the micro-organisms secreting them, what is the function of these metabolites when produced inside plants? They may help the pathogen to become established by overcoming the resistance that living tissue presents naturally to invasion. But at the same time, these substances may often induce increase in the resistance of the host to a point where the environment becomes unsuitable for the pathogen so that its growth is limited or it is eliminated. With the aid of biochemical, genetical and physiological techniques, pathologists have persistently tried to analyse the potential mechanisms of pathogenicity *in vitro* in the hope of determining those that would explain specificity and pathogenicity. But we are still a long way from understanding the real nature of both. These will now be considered.

Toxins and Specificity

First we have to recall that specificity of a pathogen should be regarded as the result of a complex of interacting factors and can only be expressed in relation to a particular host; it is impossible to determine it *in vitro* and to separate specificity from the mechanisms of resistance or immunity. The following are some examples.

Several phytopathogenic fungi produce highly specific compounds that only affect the host of the pathogen (Scheffer, p. 251). It seems clear that we can accept that these toxins are responsible for specificity; pathogen and toxin have the same host spectrum, and virulent strains that have lost their toxin producing ability also lose their aggressiveness; the toxin is necessary for the fungus which will infect only when it "recognizes" the host receptor sites. Specific toxins are so far described from species of the genera *Helminthosporium*, *Periconia* and *Alternaria*.

In a second group of examples, determinants of specificity are only partially known. Such examples have been discussed by Mahadevan (1969). According to him, inhibitory substances in the plant govern the specificity of pathogenic micro-organisms. When the inhibitors are removed or their activity is overcome by its metabolites, a fungus can then infect the plant. This happens when the β-glucosidase of *Septoria lycopersici* Speg. detoxifies α-tomatine (Arneson and Durbin, 1967); it remains to be proved, however, that α-tomatine is the resistance factor as has been pointed out by Kern (1952). The enzyme of *S. lycopersici* was found to have an unusually high degree of substrate specificity (Durbin and Uchytil, 1969). Another example is given by avenacinase which destroys avenacin which makes oats resistant to *Ophiobolus graminis* Sacc. (Turner, 1961). However, this does not explain why *O. graminis* is pathogenic for wheat and some grasses but not others. Host toxicants (Davis, 1964) would seem to be responsible, at least in part, for selective pathogenicity.

Finally are the many diseases for which we have no idea about the factors responsible for specificity. Whether or not metabolites of the parasites are at all involved remains to be investigated.

Other hypotheses for specificity that have been proposed are not related to toxins, as for example the inhibition–nutrition hypothesis of Garber (1956) which he deduced from his work with nutrient deficient mutants.

To conclude this brief reference to the specificity problem, it is clear that only for a very few of the many diseases studied has it been shown that a metabolite of the parasite, a phytoirritant, is responsible for specificity (host-specific toxins). In a few other diseases there are some reasons to believe that metabolites of the parasite contribute partially to specificity.

Toxins and Pathogenesis

The role of toxins in pathogenesis has been discussed in recent reviews (Deverall, 1964; Wheeler and Luke, 1963), and in books (Mirocha and Uritani, 1967; Wood, 1967). At this Institute, it has been mentioned several times that many pathogens produce one or several substances toxic to plants. These toxins alter more or less profoundly the tissues within which the pathogen lives, alterations which can result in the death of these tissues with all the consequences that this implies for the parasite.

The host specific toxins (Scheffer, p. 251) are the primary determinants of diseases in which they are involved; in their absence there is no tissue colonization and thus no appearance of symptoms.

More generally, however, symptoms are not attributable to one metabolite but are the result of a complex action of several compounds. Among the many metabolites secreted by the pathogen into host-plant tissues or resulting from the host–parasite interaction it is usually impossible to establish which one is the primary phytoirritant. A few examples will illustrate the confusion that still exists on the role of toxins in pathogenesis. Some strains of *Alternaria solani* (Ell. *et* Mart.) Sor. producing alternaric acid are, however, unable to cause the disease in tomato or potato shoots; also it has not always been possible to isolate the toxin from virulent strains (Brian *et al.*, 1952). Stelar reddening in strawberries resistant or susceptible to *Phytophthora fragariae* Hickman can be induced by infection or by sterile filtrates of cultures of *P. fragariae*, and also by sterile basal medium and a streptomycin-captan mixture (Otterbacher *et al.*, 1969). In *Verticillium* wilt disease, the polygalacturonase activity within stems correlated well with the onset and development of disease symptoms (Mussell and Green, 1970), but the importance of polygalacturonase in pathogenesis is questioned by Wiese *et al.* (1970). Paquin and Coulombe (1962) claimed that *in vitro* synthesis of pectic enzymes by *F. oxysporum* f. sp. *lycopersici* (Sacc.) Sny. *et* Hans. appears to be correlated with the virulence of the organism; Meyer (1964, 1967) found no relation between enzymes and fusaric acid production and virulence.

Toxins and Host Reaction

A very important ecological role of toxins during the incubation period of the disease is the induction in the host plant of a reaction that is often paralleled with the appearance of a resistance mechanism. Any infection alters the metabolism of the host (Wood, 1967). These changes, whatever their mechanisms, may often be a defence reaction against invading organism so that infection is prevented or limited in extent. This type of resistance has the following characteristics.

The induction and resistance are not specific and the substance(s) responsible for the resistance may be induced and their synthesis increased by many agents, living, dead parts or metabolites of the pathogen, non-pathogenic fungi, different chemical substances or even mechanical injury (Bell, 1967a and 1967b; Bell and Presley, 1969; Cruickshank, 1963; Cruickshank and Perrin, 1965; Deverall and Vessey, 1969; Hadwiger and Schwochau, 1969). The acquired resistance is thus not related to a pathogen but seems to be only host specific. A typical example is pisatin, induced by biological, chemical or mechanical injuries but only in peas (Kuć, 1966).

The defence reaction is not only active against the inducer organism but also often against many other parasites, leading to a delay in symptom appearance, and pronounced, even total protection (cross protection). Protection has been observed between fungi (Cheng *et al.*, 1968; Langton, 1969; Littlefield 1969; Meyer and Maraite, 1970; Rahe *et al.*, 1969), between viruses and fungi (Dias-Polance *et al.*, 1969; King *et al.*, 1964; Mandryk, 1963) and even between a fungal toxin and viruses by Wheeler and Pirone (1969) who speculate on a defensive, self-repair mechanism which once activated by one pathogenic agent might render plants resistant to other agents.

Roots bearing ectotrophic mycorrhiza may also be less susceptible to infection by root pathogens than are non-mycorrhizal roots; possible mechanisms of resistance are competition, mechanical protection or antibiosis (Marx and Davey, 1969; Zak, 1964).

Generally, not one but many substances are produced by the host in response to infection (Frank and Paxton, 1970; Rahe *et al.*, 1969; Weber *et al.*, 1967), and the resistance response of the plant develops early. Rahe *et al.* (1969) working with anthracnose of *Phaseolus vulgaris* L. and several parasites, found that high levels of protection to pathogenic races of *Colletotrichum lindemuthianum* (Sacc. *et* Magn.) Briosi *et* Cav. were induced by all three fungi (*Alternaria* sp., *Helminthosporium carbonum* Ullstr. and non-pathogenic races of *C. lindemuthianum*), when the inducer preceded the pathogen by 24 h. At the time of cell collapse, 60–120 h after infection, inhibition of an invading organism paralleled phaseollin concentration in the extracts but this is not the primary cause of resistance. Frank and Paxton (1970) also claimed that in Harosoy soybeans the reactions governing resistance or susceptibility take place 4–8 h after inoculation. Hecht and Bateman (1964) working on acquired resistance in tobacco, claimed that non-specific resistance induced by *T. basicola* appears to be similar to that brought about by infection with lesion-causing viruses, emphasizing the necessity of lesions because leaves systemically infected with TMV showed no resistance against *T. basicola.*

In these examples, the reaction of the plant against infection in some way creates an inhospitable environment for the pathogen. However, substances synthesized by the plant may sometimes aggravate the disease. Thus, ipomeamarone which is considered by some to be responsible for the resistance of sweet potatoes to *Ceratocystis fimbriata* Ellis *et* Halst. (although this has been questioned by Hyodo *et al.*, 1969) may also be a vivotoxin that damages and may kill tissues of sweet potato roots (Suzuki and Toyoda, 1957).

Pre-, or simultaneous infection with different parasites may also sometimes accentuate symptoms; this is well known for viruses X and Y of potatoes but may happen also with fungi (Alconero and Santiago, 1969; Goujon, 1967), and between viruses and fungi (Farley and Lockwood, 1964; Futrell and Scott, 1969; Nitzany, 1966).

References

Agnihothrudu, V. (1955). *Naturwiss.* **42** (12), 1–2.
Alconero, R. and Santiago, A. G. (1969). *Phytopathology* **59**, 1521–1524.
Amin, K. S. and Sequeira, L. (1966). *Phytopathology* **56**, 1047–1061.
Arneson, P. A. and Durbin, R. D. (1967). *Phytopathology* **57**, 1358–1360.
Baker, R. (1968). *A. Rev. Phytopath.* **6**, 263–294.
Bell, A. A. (1967a). *Phytopathology* **57**, 759–764.
Bell, A. A. (1967b). *Phytopathology* **57**, 1111–1115.
Bell, A. A. and Presley, J. T. (1969). *Phytopathology* **59**, 1141–1151.
Börner, H. (1963). *Phytopath. Z.* **48**, 370–376.
Brian, P. W. (1957a). *In* "Microbial Ecology" (R. E. O. Williams and C. C. Spicer, Eds), pp. 168–188. Cambridge University Press.
Brian, P. W. (1957b). *A. Rev. Pl. Physiol.* **8**, 413–426.
Brian, P. W., Elson, G. W., Hemming, H. G. and Wright, J. M. (1952). *Ann. appl. Biol.* **39**, 308–321.
Carley, H. E. and Watson, R. D. (1967). *Phytopathology* **57**, 401–404.
Chatterjee, A. K., Gibbins, L. N., and Carpenter, J. A. (1969). *Can. J. Microbiol.* **15**, 640–642.
Cheng, Y. H., Yu, C. and Lin, Y. N. (1968). *Pl. Prot. Bull. Taiwan* **10**, 63–69.
Cruickshank, I. A. M. (1963). *A. Rev. Phytopath.* **1**, 351–374.
Cruickshank, I. A. M. and Perrin, D. R. (1965). *Aust. J. biol. Sci.* **18**, 803–828.
Davis, D. (1964). *Phytopathology* **54**, 290–293.
Dekker, J. (1963). *A. Rev. Microbiol.* **17**, 243–262.
Deverall, B. J. (1964). *In* "Microbial behaviour 'in vivo' and 'in vitro'" (H. Smith and J. Taylor, Eds), pp. 165–186. Cambridge University Press.
Deverall, B. J. and Vessey, J. C. (1969). *Ann. appl. Biol.* **63**, 449–458.

Dias-Polance, C., Smith, S. H., and Hancock, J. C. (1969). *Phytopathology* **59**, 18–22.

Durbin, R. D. (1959). *Pl. Dis. Reptr* **43**, 922–923.

Durbin, R. D. and Uchytil, T. F. (1969). *Biochim. biophys. Acta* **191**, 176–178.

Farley, J. D. and Lockwood, J. L. (1964). *Phytopathology* **54**, 1279–1280.

Frank, J. A. and Paxton, J. D. (1970). *Phytopathology* **60**, 315–318.

Fulton, N. D., Bollenbacher, K., and Templeton, G. E. (1965). *Phytopathology* **55**, 49–51.

Futrell, M. C. and Scott, G. E. (1969). *Pl. Dis. Reptr* **53**, 600–602.

Garber, E. D. (1956). *Am. Nat.* **90**, 183–194.

Ghaffar, A. (1969). *Mycopath. Mycol. appl.* **38**, 101–127.

Gibson, I. A. S. (1953). *Trans. Br. mycol. Soc.* **36**, 198–209.

Gilpatrick, J. D. (1969) *Phytopathology* **59**, 973–978.

Goujon, M. (1967). *Cah. ORSTROM, sér. Biol.* **4**, 43–51.

Grossmann, F. (1967). *Z. PflKrankh. PflPath. PflSchutz.* **74**, 143–149.

Hadwiger, L. A. and Schwochau, M. E. (1969). *Phytopathology* **59**, 223–227.

Harper, J. L. (1950). *Pl. Soil* **2**, 374–394.

Hecht, E. I. and Bateman, D. F. (1964). *Phytopathology* **54**, 523–530.

Hely, F. W., Bergersen, F. J., and Brockwell, J. (1957). *Aust. J. agric. Res.* **8**, 24–44.

Hoestra, H. (1968). *Meded. LandbHoogesch. Wageningen* **63** (13), 1–105.

Horst, R. K. and Herr, L. I. (1962). *Phytopathology* **52**, 423–427.

Huber, D. M. and Watson, R. D. (1970). *Phytopathology* **60**, 22–26.

Hyodo, H., Uritani, I., and Akai, S. (1969). *Phytopath. Z.* **65**, 332–340.

Kern, H. (1952). *Phytopath. Z.* **19**, 351–382.

King, L., Hampton, R. E., and Diachun, S. (1964). *Science* **146**, 1054–1055.

Ko, W. and Lockwood, J. L. (1967). *Phytopathology* **57**, 894–901.

Kuć, J. (1966). *A. Rev. Microbiol.* **20**, 337–370.

Langton, F. A. (1969). *Ann. appl. Biol.* **62**, 413–427.

Linderman, R. G. (1970). *Phytopathology* **60**, 19–22.

Linderman, R. G. and Toussoun, T. A. (1968). *Phytopathology* **58**, 1571–1574.

Littlefield, L. J. (1969). *Phytopathology* **59**, 1323–1328.

Lockwood, J. L. (1964). *A. Rev. Phytopath.* **2**, 341–362.

Mahadevan, A. (1969). *Phytopath. Z.* **65**, 358–363.

Mandryk, M. (1963). *Aust. J. agric. Res.* **14**, 315–318.

Marx, D. H. and Davey, C. B. (1969). *Phytopathology* **59**, 549–557.

McCalla, T. M. and Haskins, F. A. (1964). *Bact. Rev.* **28**, 181–207.

Meyer, J. A. (1959). Proc. 3rd. Inter. African soil conf. **1**, 517–524.

Meyer, J. A. (1964). *Meded. LandbHoogesch. Gent* **28**, 832–838.

Meyer, J. A. (1967). *Ann. Epiphyties* **18**, 241–247.

Meyer, J. A. and Maraite, H. (1970). *Trans. Br. mycol. Soc.* (In press.)

Mirocha, C. J. and Uritani, I. (1967). "The dynamic role of molecular constituents in plant-parasite interaction." A.P.S. St Paul.

Mpungi, P. (1967). MS thesis, Univ. Louvain, 54 pp.

Mussell, H. W. and Green, R. J. (1970). *Phytopathology* **60**, 192–195.

Nitzany, F. (1966). *Phytopathology* **56**, 1386–1389.

Norstadt, F. A. and McCalla, T. M. (1969). *Pl. Soil.* **30**, 129–133.

Olsen, C. M. and Baker, K. F. (1968). *Phytopathology* **58**, 79–87.
Otterbacher, A. G., Powell, D., and Zych, C. C. (1969). *Pl. Dis. Reptr* **53**, 794–797.
Owens, L. D. (1969). *Science* **165**, 18–25.
Paquin, R. and Coulombe, L. J. (1962). *Can. J. Bot.* **40**, 533–541.
Patrick, Z. A. and Koch, L. W. (1963). *Can. J. Bot.* **41**, 747–758.
Patrick, Z. A., Toussoun, T. A., and Koch, L. W. (1964). *A. Rev. Phytopath.* **2**, 267–292.
Pramer, D. (1965). *In* "Ecology of soil-borne plant pathogens" (K. F. Baker and W. C. Snyder, Eds), pp. 17–18. University California Press.
Pugh, G. J. F. and Van Emden, J. H. (1969). *Neth. J. Pl. Path.* **75**, 287–295.
Putnam, H. D. and Schmidt, E. L. (1959). *Soil Sci.* **87**, 22–27.
Rahe, J. E., Kuć, J., Chien-Hei Chuang, and Williams, E. B. (1969). *Phytopathology* **59**, 1641–1645.
Rayan, G. F., Greenblatt, G., and Al-Delaimny, K. A. (1961). *Science* **134**, 833–834.
Rombouts, J. E. (1953). *Pl. Soil* **4**, 276–288.
Savory, B. M. (1969). *Ann. appl. Biol.* **63**, 225–231.
Steiner, G. W. and Lockwood, J. L. (1970). *Phytopathology* **60**, 89–91.
Suzuki, N. and Toyoda, S. (1957). *Bull. natl. Inst. agr. Sci. Japan, sér. C.* **8**, 131–173.
Theron, J. J. (1951). *J. agric. Sci.* **41**, 289–296.
Toussoun, T. A., Menzinger, W., and Smith, R. S. (1969). *Phytopathology* **59**, 1396–1399.
Toussoun, T. A. and Patrick, Z. A. (1963). *Phytopathology* **53**, 265–270.
Toussoun, T. A., Patrick, Z. A., and Snyder, W. C. (1963). *Nature* **197**, 1314–1316.
Tukey, H. B. (1969). *Bot. Rev.* **35**, 1–16.
Turner, E. M. C. (1961). *J. exp. Bot.* **12**, 169–175.
Van den Heuvel, J. (1969). *Neth. J. Pl. Path.* **75**, 300–307.
Weber, D. J., Clare, B., and Stahmann, M. A. (1967). *Phytopathology* **57**, 421–424.
Wheeler, H. and Luke, H. H. (1963). *A. Rev. Microbiol.* **17**, 223–243.
Wheeler, H. and Pirone, T. P. (1969). *Science* **166**, 1415–1417.
Wiese, M. V., DeVay, J. E., and Ravencroft, A. V. (1970). *Phytopathology* **60**, 641–646.
Winter, A. G. (1955). *Z. PflErnähr. Dung.* **69**, 224–233.
Wood, R. K. S. (1967). "Physiological Plant Pathology." Blackwell Scientific Publications, Oxford.
Wyllie, T. D. (1962). *Phytopathology* **52**, 202–206.
Ylimäki, A. (1967). *Ann. Agric. fenn.* **6**, 1–59.
Zak, B. (1964). *A. Rev. Phytopath.* **2**, 377–392.

Discussion

Wood: May I point out that Dr. B. J. Deverall was to have attended the Institute to talk about toxins in disease resistance with particular reference to the induction of phytoalexins. In his absence, this aspect of the action of toxins has been rather neglected but we should not underrate its importance particularly in light of Meyer's approach to the role of toxins in plant diseases.

Schlösser: Here is an example where a toxin might be involved in soil microflora ecology: an apple orchard in Lebanon where the trees were dying rapidly; the only micro-organism found on the decayed roots and in the soil (!) was a *Trichoderma* species. Owing to favourable conditions for its development this fungus which is known to produce powerful antibacterial and antifungal toxins, becomes dominant to such an extent that the soil was virtually free of other micro-organisms. This extreme shift in the otherwise multi-species soil microflora appears possible only through the action of toxins.

Meyer: The presumption that the death of the trees is caused by *Trichoderma* is plausible; I would however be very cautious because the great saprophytic competitive ability of this fungus allows rapid and early colonization of roots killed in other ways. Also, only a very low percentage of strains of *Trichoderma* produce the antibiotics; we find it less than 1% even among strains belonging to *Gliocladium virens* (*Revue mycol.* **29**, 307, 1964) and Webster (*Trans. Br. mycol. Soc.* **47**, 535–540) finds also that the proportion of strains producing antibiotics is very low among the true tufted species of *Trichoderma*.

Durbin: Glucoside linkages are widely distributed in plants. Recent work has shown that certain of these in tomato plants may be important in resistance. *Septoria lycopersici* produces an enzyme that attacks only a specific linkage in certain glucosides.

Meyer: As suggested in the lecture, it is still necessary to prove that tomatine is the factor responsible for resistance. Its role in this respect has been questioned by Kern (1952), at least in vascular wilts.

Wood: Apple trees die if planted in old apple orchards even many years after the old trees have been removed. The problem is eliminated by soil sterilization. This suggests that one or more pathogens are responsible. Have any been isolated?

Meyer: The replant problem of apple trees has been referred to in the lecture (see also the work of Hoestra (1968) and Savory (1969)).

Page: The technique of flood fallowing to control *Fusarium oxysporum* f. sp. *cubense* was not a successful treatment. It would also appear that extremely low levels of carbon substrate e.g. 2 mg/litre of glucose or magnesium carbonate will trigger chlamydospore production when *F. oxysporum* is grown on 2 salts ($MgSO_4.7H_2O$ and KH_2PO_4) in glass distilled water. Chlamydospore production was absent above approximately 32 mg of glucose and a slightly higher level of magnesium carbonate.

Feldman: In referring to Page's comment on the practice of flooding for the control of *Fusarium* in banana, I should like to inquire whether flooding also eliminates or reduces the population of *Radopholus similis*, the nematode associated with the Panama disease of banana? I mention this because our problem on spreading decline of citrus in central Florida is also caused by *R. similis* (burrowing nematode). In this area of Florida, citrus is grown on deep sands where roots of citrus will often reach to depths of 3 to 10 m. Control by flooding would not be possible and even eradication by high rates of nematocides is virtually impossible.

Page: I am unaware of the association of *Radopholus* and *Fusarium* in banana wilt. However, with respect to the association of animals in the soil, an insect *Scaptocorus* sp., produced several complex volatile alcohols. These were so toxic to *Fusarium* that a single gland fixed to the inside of a Petri dish lid prevents spore germination and mycelial growth.

Feldman: I was especially interested in Meyer's comment on "disease control" through the release and perhaps the accumulation in the soil of specific phenolics from crop residues. If I may again refer to citrus, we have found that essentially the same kinds and amounts of phenolics are found in both the healthy tolerant (to *R. similis*) and the healthy susceptible citrus cultivars. However, after infection, the phenolic composition in both plant groups is quantitatively changed. There is a 25–50% reduction of phenolics in the susceptible roots and a 27–300% increase (depending on the cultivar) in the tolerant roots with subsequent elimination of the pathogen in the latter in *c.* 6 months. If the susceptible cultivars are treated with phenolics, particularly vanillic acid (often found in residues of some crops) certain phenolics, especially *p*-hydroxybenzoic acid, sinapic acid, salicylic acid, vanillic acid, gentisic acid, ferulic acid, scopoletin, and umbelliferone are increased. Most of these phenolics are also increased in the *R. similis*-infected tolerant citrus cultivars but unlike the roots from these tolerant hosts, total phenolics in the susceptible citrus cultivars are not increased above the amount of those phenolics found in the healthy roots. Infected susceptible-seedlings treated with vanillic acid had plant weights comparable to the healthy controls but the nematode was not eliminated from the host. In effect, then, certain phenolics appear to modify the plant sufficiently so that near normal development can be sustained in spite of the presence of the pathogen.

Page: Such results are very interesting and are also known in certain diseases caused by fungi.

20. Chemical Control of Toxigenic Plant Diseases

FRIEDRICH GROSSMANN

Institut für Phytopathologie, Justus Liebig-Universität, Giessen, German Federal Republic

The subject, about which I am to speak presents some difficulties. First of all, what is a "toxigenic plant disease"? Gäumann (1954) who pioneered work in the field of phytotoxins, has stated "Microorganisms are only pathogenic if they are toxigenic". If this be true, then all plant diseases so far as they are caused by micro-organisms are, in fact, "toxigenic plant diseases". But this, undoubtedly, is a somewhat extreme view of plant diseases, because there are certainly factors other than toxins which participate in plant diseases caused by micro-organisms. Brian (1958) has mentioned mechanical effects, deprivation of nutrients and excretion of enzymes which degrade host tissue. And to these I would add pathogenic information submitted to the host cells by genetic material from viruses and, perhaps, other pathogens. Nevertheless, there can be little doubt that toxins play a dominant role in many plant diseases (Owens, 1969). There are, however, only a few diseases in which they have been isolated and where their mode of action is known. I shall restrict my paper to such diseases, and more particularly to chemical control in which it is likely that the effect of a chemical is to prevent production of a toxin or to nullify its effects. A better title for my paper would, perhaps, have been, "Chemical control of plant diseases by interfering with toxins." But this title also raises difficulties. What is a toxin? It will be one of the main purposes of this Institute to answer this question and Professor Graniti in his introductory paper has dealt with it thoroughly. But when I compiled my paper I did not know what the views of this Institute would be and I therefore decided on the following, rather simple definition. Toxins, especially phytotoxins, are products of living organisms which damage plants. This definition is independent of whether toxins are produced by the pathogen itself or as a result of a host–pathogen interaction. It also includes the extracellular enzymes which destroy plant tissue. I know that substances with enzymic properties are not usually considered as toxins. Thus, Owens (1969) specially emphasizes that "In plant pathology a toxin is generally defined as a non-enzymic substance which injures plant cells or disrupts their metabolism." I ask you for your indulgence when I nevertheless include extracellular enzymes in my considerations, as phyto-aggressins in the sense of Graniti (1972). I shall show that there are parallels between toxins and extracellular enzymes so far as their production, action and inactivation are concerned so that the inclusion of enzymes in this discussion will be profitable.

One of the first examples in which the effect of a chemotherapeutant was attributed to its antidoting of a toxin, is to be found in the work of Howard (1941), who confirmed that *Phytophthora cactorum* (Leb. *et* Cohn) Schr., the pathogen of the bleeding canker disease of maples, produces a toxin in culture on different media. Maples and tomato cuttings wilted when placed in culture filtrates. The toxic effect was neutralized by adding to filtrates a solution of the dihydrochloride salt of diaminoazobenzene. Injection of the toxic filtrate into healthy maple trees killed them whereas injection of the filtrate containing 0·5% of the diaminoazobenzene salt did not harm the trees. Furthermore, many naturally infected trees reacted to an injection of the diaminoazobenzene salt with cessation of bleeding and a significant improvement of vegetative growth. Howard suggested that this chemotherapeutant acted by counteracting the toxin but he did not present any views about the mode of action of the antidote. This is understandable because he knew neither the chemical nature of the toxin nor how it worked.

The ways in which chemicals could counteract the effects of toxins produced by micro-organisms are outlined in simplified form in Fig. 1. The following are some of the possibilities.

1. Chemicals may suppress biosynthesis of toxins or enzymes (*Suppression of biosynthesis*)
2. The host plants frequently are able to degrade toxins biologically or to neutralize them in other ways; chemicals may promote this property of the plant tissue (*Stimulation of biodegradation*)
3. Chemicals may directly modify the toxin molecule, e.g. by forming an inactive complex (*Modification of the toxin molecule*)
4. Certain toxins are effective only in a specific environment, e.g. within a certain pH-range; chemicals may change the milieu so that it becomes unfavourable for the toxin (*Change of milieu*)
5. Chemicals may alter the substrate on which the toxin or the enzyme acts, e.g. by influencing receptor sites or the chemical structure of the substrate (*Alteration of the substrate*)
6. Toxins may act by forming blocks in plant metabolism; chemicals may allow the plant to circumvent such blocks (*Removal of metabolic blocks*)

I shall now discuss in more detail these possibilities with the help of a few examples. The discussion will often be speculative because for the most

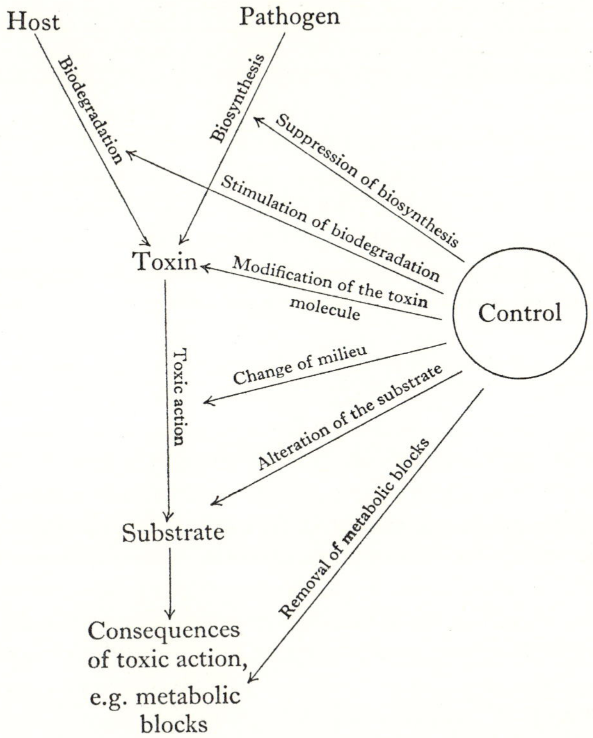

FIG. 1. Possibilities of antidoting pathogen-produced toxins.

part our knowledge of the subject is still fragmentary. We are still unable to describe, in molecular terms, the mechanisms by which most toxins damage plant cells (Owens, 1969). I shall, therefore, frequently have to rely on hypotheses or model conceptions. In so doing I hope I shall stimulate your thoughts in this field.

Suppression of Biosynthesis

This method should be mentioned first, because if it were successful the production of toxins would be prevented. The method is directed not against the secondary consequences of infection but against the production of the toxin. The ability of a micro-organism to produce a toxin or an enzyme is, of course, fixed genetically, but certain chemicals may be able to repress activation of genetic information or to prevent it from functioning in cell metabolism. According to Pouteau-Thouvenot *et al.* (1968), a commercial extract from the blossoms of *Helichrysum arenarium* Moench (Compositae), inhibits biosynthesis of lycomarasmin by *Fusarium oxysporum* Schl. f. sp. *melonis* Sny. *et* Hans. *in vitro*. Three terpenic acids, abietic, agathic and eburicoic acid have the same effect. The authors suggest, therefore, that the active principles in the blossom extract are the terpenic acids.

The production of fusaric acid by *F. oxysporum* f. sp. *vasinfectum* (Atk.) Sny. *et* Hans. depends on the type of nitrogen available to the fungus (Kalyanasundaram and Braun, 1958). It is also controlled by the zinc content of the medium (Kalyanasundaram and Saraswathi Devi, 1955). Apart from these examples little seems to be known about the suppression of the biosynthesis of phytotoxins. Here then is an interesting field for future investigations.

Substantially more has been done on the biosynthesis of extracellular enzymes and its regulation. Patil and Dimond (1968c) found that the synthesis of a polygalacturonase by *F. oxysporum* f. sp. *lycopersici* (Sacc.) Sny. *et* Hans. *in vitro* is repressed by high concentrations of phosphate and that production of polygalacturonase by *Verticillium albo-atrum* Reinke *et* Berth. in culture was reduced by certain phenols (chlorogenic acid, catechol, rufianic acid) and cytokinins (kinetin, benzyladenine) (Patil and Dimond, 1968b). With rufianic acid and cytokinins, although synthesis of polygalacturonase was reduced, secretion of total protein into the culture medium increased. These results suggest how naturally occurring phenols could act in defence reactions and are also of interest in connexion with the therapeutic effects of rufianic acid about which more will be said later. It has been known for some time that several amino acid analogues, e.g. ethionine and phenylserine, may suppress the formation of enzymes (van Andel, 1966). L-Canavanine, an antimetabolite for L-arginine, inhibits the induced biosynthesis of polygalacturonase by *Sclerotinia* (*Monilinia*) *fructigena* Aderh. *et* Ruhl. (Byrde, 1963). This is possibly because canavanine displaces arginine from the active site of the enzyme. In this connexion it is interesting that canavanine decreases the development of rust on detached wheat leaves (Samborski and Forsyth, 1960) and that Van Sumere *et al.* (1957) claim that rust uredospores secrete polygalacturonase as well as other enzymes during germination. It may be possible that canavanine inhibits polygalacturonase synthesis by the rust, thereby preventing its establishment in the host plant.

Another frequently observed effect is the suppression of enzyme synthesis by various sugars and their derivatives. Patil and Dimond (1967) found that monogalacturonic acid, the basic constituent of pectic polymers, inhibits the production of polygalacturonase by *Verticillium albo-atrum* in a glucose nitrate medium. In contrast production was stimulated by addition of glucose, sucrose, ribitol and 2-amino-D-glucose. Synthesis of polygalacturonase by *F. oxysporum* f. sp. *lycopersici* was reduced in unbuffered pectin shake cultures by glucose as well as by monogalacturonic acid, but the biggest reduction was caused by 2-deoxy-D-glucose (Patil and Dimond, 1968a). Certain sugar alcohols, like sorbitol, have a similar but smaller effect. When tomato plants infected with *F. oxysporum* f. sp. *lycopersici* were continuously supplied with glucose, 2-deoxy-D-glucose, or mono-galacturonic acid through exposed bundles at the bases of stems, the wilt symptoms on the leaves and the polygalacturonase activity in stem extracts were decreased, although, at least with glucose, the growth of mycelium in the stems was increased. From these results, Patil and Dimond suggest that the different sugars and sugar alcohols repress synthesis of polygalacturonase by the pathogen in culture as well as in the tomato plants.

Some years earlier Feldman and Caroselli (1951) had reported that absorption of sugars by foliage and trunks of elms retarded development of wilt symptoms for 3–4 weeks. Horsfall and Dimond (1957) classified wilt diseases including Dutch elm disease as "low sugar diseases". Perhaps they belong to this group because the biosynthesis of extracellular enzymes important in pathogenesis is suppressed at higher sugar levels.

Stimulation of Biodegradation

Toxins produced by pathogens may be inactivated by the plant tissue. There are probably many biological mechanisms for inactivating toxins but

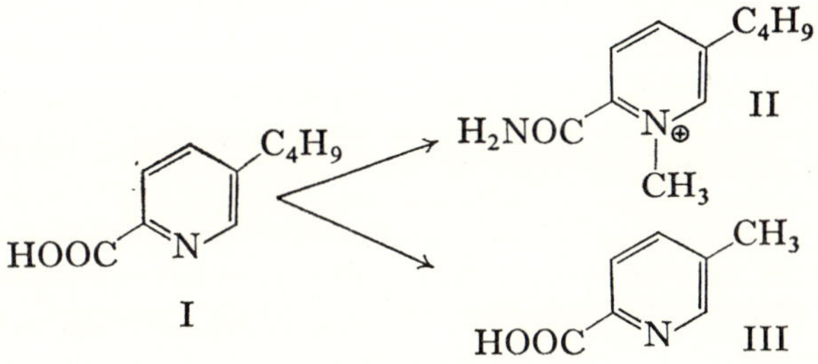

so far they have attracted little attention. I shall confine myself to one example. Fusaric acid (5-n-butyl-picolinic acid (I)), produced by a number of *Fusarium* species (Gäumann, 1957), is probably one of the best known and most studied phytotoxins. Jost (1965) investigated its conversion and detoxification in pea, tomato, and cotton plants. In tomato plants it is, *inter alia*, changed to the amide of N-methylfusaric acid (II), and to 5-methyl-picolinic acid (III). The tomato plant may also split the pyridine nucleus of fusaric acid. Addition of indoleacetic acid (IAA) enhances the degradation of the toxin in tomato cuttings and also increases their resistance to fusaric acid. These results recall earlier work of Davis and Dimond (1953), in which IAA and other growth-regulating substances increased the resistance of tomato plants to *F. oxysporum* f. sp. *lycopersici*.

Extracellular enzymes of pathogens may also be inactivated by plant tissue. Phenolic compounds may play an important role as natural inhibitors of pectolytic and cellulolytic enzymes (Byrde, 1963), so that chemicals which influence the phenolic composition of plants may interfere with biodegradation of these enzymes (Dimond, 1963). The oxidation products of phenols are usually even more effective and compounds which activate oxidases, especially phenol oxidase and peroxidase, often increase inhibition. Thus, Matta (1963) stimulated the polyphenol oxidase in roots and stems of tomato plants by application of naphthaleneacetic acid and at the same time increased resistance to *F. oxysporum* f. sp. *lycopersici*. The resistance of sweet potatoes to *Ceratocystis fimbriata* Ellis *et* Halst. induced by exposure to ethylene is also connected with a higher activity of peroxidase and polyphenol oxidase (Stahmann *et al.*, 1966). In these cases the chemically induced resistance may depend, partially at least, on inactivation of the extracellular enzymes of the pathogens.

Modification of the Toxin Molecule

The preceding section has discussed how chemicals can stimulate inactivating systems in plants. Applied chemotherapeutants may also directly modify the toxin molecule and thereby make it ineffective.

Fusaric acid, already mentioned, forms chelates with metals between the N-atom of the pyridine nucleus and the carboxyl group in the α-position (IV). Of the heavy metals the ferric ion is most strongly chelated, followed by the cupric ion (Tamari *et al.*, 1963). Fusaric acid, therefore, may affect the iron and copper enzymes which play important roles in the respiratory systems of plant tissues. The capacity to form complexes can be destroyed, e.g. by methylation at the N-atom (see preceding section) or esterification

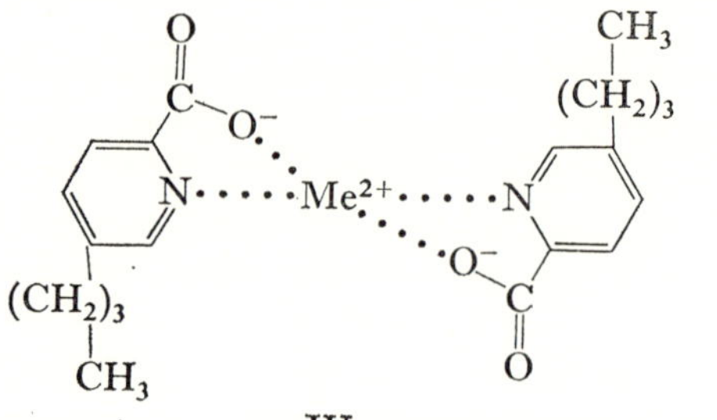

IV

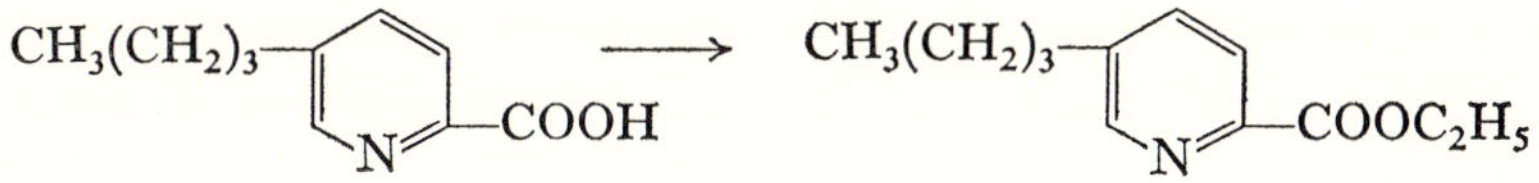

V

of the carboxyl radical (V). In this way the activity of the toxin is greatly reduced. This also happens if the compound is saturated with heavy metal ions. Thus fusaric acid is inactivated, especially at higher pH, by addition of iron and, to a lesser degree, of cobalt, copper, nickel, zinc and manganese (Braun, 1960; Bär, 1963).

Heavy metal ions may also detoxify other phytotoxins. Lycomarasmin, a second wilt toxin produced by *F. oxysporum* f. sp. *lycopersici*, is also a chelating agent. Although addition of iron increases its toxicity to tomato cuttings about 10-fold, cobalt, nickel and copper, added in equimolar concentrations, form stable complexes with lycomarasmin thereby inactivating it (Gäumann *et al.*, 1955; Gäumann and Naef-Roth, 1956).

Fusarium solani and related species secrete in culture and in infected pea plants certain toxic naphthazarin derivatives including marticin and isomarticin (VI) (Kern and Naef-Roth, 1965). In the presence of copper, aluminium or iron the marticins lose their phytotoxicity. The same metals, added to water culture, reduce the sensitivity of pea tissue to the toxins and inhibit pathogenesis in infected plants (Kern and Naef-Roth, 1966).

Ascochitine (VII), a metabolite from *Ascochyta fabae* which is phytotoxic and antibiotic, is also antagonized by equimolar concentrations of ferric ions (Oku and Nakanishi, 1963).

Pyricularia oryzae, causal organism of the rice blast disease, produces at least two toxins, α-picolinic acid which is closely related to fusaric acid and apparently has a similar mode of action, and piricularin, a nitrogen-containing compound of as yet unknown structure (Tamari *et al.*, 1963). Piricularin, which appears to be the main component in pathogenesis, is detoxified by chlorogenic acid and ferulic acid, presumably because inactive complexes

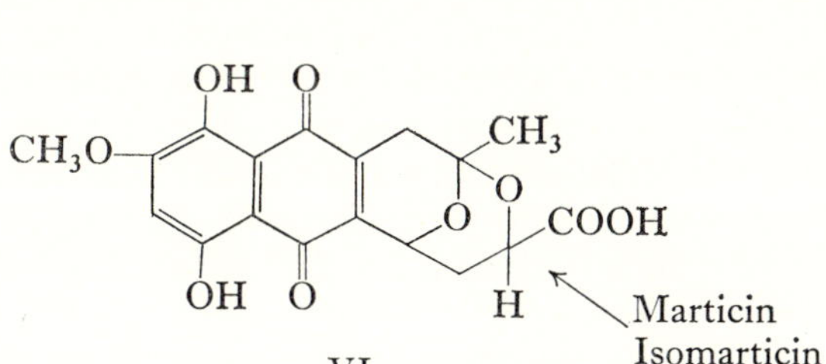

VI

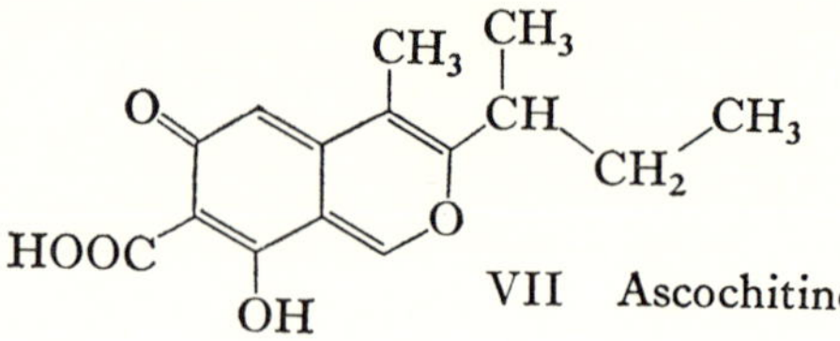

VII Ascochitine

are formed. Application of ferulic acid to rice plants increases resistance to infection by the blast fungus. The oxidation product of ferulic acid also detoxifies piricularin.

As mentioned earlier, phenolics and their oxidation products may also inactivate extracellular enzymes of microbial pathogens; this has been considered to be an important factor in the inactivation of these enzymes *in vivo*. It may be possible to obtain the same effect by adding phenols or quinones which probably denature the enzyme proteins by a tanning action. In my own experiments (Grossmann, 1962) application of tannic acid or an extract from apples which contained leucoanthocyanins significantly reduced the wilt symptoms in tomato cuttings induced by partially purified enzymes of *F. oxysporum* f. sp. *lycopersici*. However, these substances apparently were not taken up by the roots of intact plants. On the other hand, rufianic acid, 1,4-dihydroxy-anthraquinone-sulphonic acid-(2), reduced the severity of the disease even on rooted, infected plants. Under certain conditions this substance also reduced *Verticillium* wilt of tomato plants and the maceration of potato and bean tissues by *Rhizoctonia solani* Kühn (Martin, 1967; Grossmann, 1968). In this connexion it is interesting to recall that Patil and Dimond (1968b) found that rufianic acid also represses the biosynthesis of polygalacturonase by *V. albo-atrum*.

Change of Milieu

The toxicity of a compound may depend on a certain environment, for example, on the pH or on the redox status. Indian workers have shown that cystine counteracts the toxicity of *Fusarium* culture filtrates and the damage caused by the pathogen (Wheeler and Luke, 1963). These effects possibly depend on the oxidizing potential of cystine.

Earlier Zentmyer (1942) found that *Ceratocystis ulmi* (Buism.) Moreau produces *in vitro* a toxin which causes some of the typical symptoms of Dutch elm disease. In the same year Horsfall and Zentmyer (1942) reported that several compounds containing basic nitrogen such as 8-hydroxyquinoline sulphate, diaminoazobenzene, urea, malachite green and probably ammonia, counteract the toxin and are effective as chemotherapeutants against the disease. According to Feldman *et al.* (1950b), the principal toxin in filtrates of *C. ulmi* cultures is most effective at pH 4 and is inactivated progressively with increasing pH. Stem injections with alkaline chemicals such as $NaHCO_3$, $KHCO_3$ and $Ca(OH)_2$, gave good control of the disease. Even soil treatments with lime delayed the appearance of wilt symptoms (Feldman *et al.*, 1950a). Hoffman (1952) achieved good success with alkaline chemicals against oak wilt and also attributed control to neutralization of the toxin of the pathogen, *Ceratocystis fagacearum* (Bretz) Hunt.

The structure of the toxins involved in Dutch elm disease and in oak wilt are to my knowledge not yet fully known. However, some of the well-defined *Fusarium* toxins also depend on pH for their effects. Fusaric acid dissociates more and more with increasing alkalinity, especially at pH above 6·0. Consequently, its capacity to enter cells through the plasma membranes decreases and the damage to stems of tomato plants diminishes with increasing pH (Gäumann, 1957). This applies to the marticins which are also weak acids. Thus, at pH 5·4 the *dosis minima* of isomarticin for tomato plants is 8·5 mg/kg fresh weight, whereas at pH 7·5 it increases about 3-fold. In the same way peas are more damaged by a weakly acid solution of the toxin; with increasing dissociation peas become less susceptible (Kern and Naef-Roth, 1965).

It is, of course, well known that extracellular enzymes are effective only within certain pH ranges. For example, polygalacturonases from fungal parasites are usually active only in an acid milieu. A rise in pH value reduces activity and may inactivate these hydrolytic enzymes. In contrast, many pathogenic fungi produce pectin lyases which are most active in an alkaline range; their effects may be counteracted by applying substances which cause a decrease of pH of the tissues.

Alteration of the Substrate

A toxin or enzyme produced by a pathogen will damage plants only if it can affect components of the tissue or the cells. These effects will not occur if the substrate for the toxin either cannot receive it or is resistant to it for structural reasons. The host-specific toxin of *Helminthosporium victoriae* Meehan *et* Murphy (HV-toxin) is believed to act primarily on cell membranes thereby causing an increase in permeability (Wheeler and Black, 1962). The toxin consists of a polypeptide moiety and a nitrogen-containing sesquiterpene moiety called victoxinine (Pringle, 1972). The polypeptide constituent seems to be responsible for the host specificity of the HV-toxin because it combines with definite receptor sites at the cell surface. Sulphite greatly reduces toxicity although it does not affect the toxin itself. This may mean that a toxin receptor is changed and that this reduces adsorption of the toxin and, therefore, its toxicity (Pringle and Scheffer, 1964). Bisulphite and semicarbazide apparently have a similar effect (Samaddar, 1968). These compounds are known to form dissociable addition products with carbonyl groups. It is assumed, therefore, that a carbonyl group is involved in the binding of HV-toxin on the cell membrane. Hanchey (1969) recently reported that the electrolyte leakage from first leaves of HV-toxin-treated susceptible oats was suppressed by simultaneous treatment of the leaves with uranyl salts. It was not determined, however, whether this result reflected an effect of UO_2^{++} on cell membranes or the formation of a complex between uranium and the toxin itself. According to Doupnik (1968) $CaCl_2$ and, to a lesser degree $SrCl_2$ both at 0·1 M suppressed the typical HV-toxin-induced symptoms including permeability changes. Suppression by $CaCl_2$ was partially reversed by adding NaCl, KCl, or $MgCl_2$. As possible explanations Doupnik suggests, *inter alia*, a decrease of toxin uptake, maintenance of cell wall and plasmalemma permeability and stability as well as inhibition of cell wall-degrading enzymes.

The last possibility raises the point that the enzymatic degradation of pectic substances is affected strongly by calcium. Pectins are high polymer polygalacturonic acids which are partially esterified with methanol. The remaining carboxyl groups may form rigid cross connexions between the individual pectin chains through multivalent cations, particularly calcium. In this way the pectin chains may become resistant to attack by pectolytic enzymes. This may be why supplying plants with increasing amounts of calcium enhances resistance against diseases in which pectolytic enzymes contribute to pathogenesis such as black leg of Solanaceae (Stapp and Hartwich, 1957), or tomato wilt (Edgington and Walker, 1958). Treatment of plants with growth regulators, such as naphthaleneacetic acid, causes demethoxylation of the pectins and, therefore, the formation of more bridges if sufficient calcium is present. In this way the proportion of water-soluble pectins is reduced, the enzymatic degradation of pectin is decreased and in the case of tomato wilt, caused by *F. oxysporum* f. sp. *lycopersici*, resistance is increased (Edgington *et al.*, 1961).

Removal of Metabolic Blocks

Apart from changing permeability toxins may interfere with plant metabolism in other ways. Compounds structurally similar to normal metabolites may cause metabolic blocks as antimetabolites. If such blocks could be reversed with specific chemicals the effects of the toxin would be nullified and the disease would be controlled.

The wildfire toxin, produced by *Pseudomonas tabaci* (Wolf *et* Fost.) Stev. is one of the best studied bacterial phytotoxins. It causes chlorosis in tobacco and many other higher plants, and also inhibits growth of the unicellular green alga *Chlorella vulgaris* Beij. (Braun, 1955). Although the structure of the toxin has not yet been clarified, it was formerly believed to be related to the amino acid methionine. It was postulated, therefore, that the wildfire toxin acts as an antimetabolite of methionine. This was supported by the fact that another antimetabolite of methionine, namely methionine sulphoximine also causes chlorosis in tobacco and inhibition of the growth of *C. vulgaris*. Also, the toxicity of wildfire toxin and of methionine sulphoximine to *C. vulgaris* could be neutralized with methionine. However, addition of methionine did not prevent chlorosis in tobacco leaves.

This indicates that toxicity of the wildfire toxin for tobacco depends on a different mechanism. Sinden and Durbin (1968) have now established that both wildfire toxin and methionine sulphoximine inhibit glutamine synthetase prepared from pea. This enzyme catalyses the reaction

$$\text{glutamic acid} + NH_3 + ATP \overset{Mg^{++}}{\rightleftharpoons} \text{glutamine} + ADP + P_1$$

(Durbin, 1972). Inhibition of this reaction leads to a build-up of ammonia or other intermediates of nitrate metabolism which in turn induce chlorosis. By injecting large amounts of glutamine into leaves, together with the toxin or with methionine sulphoximine, accumulation of ammonia as well as chlorosis in tobacco leaves were prevented (Sinden and Durbin, 1968).

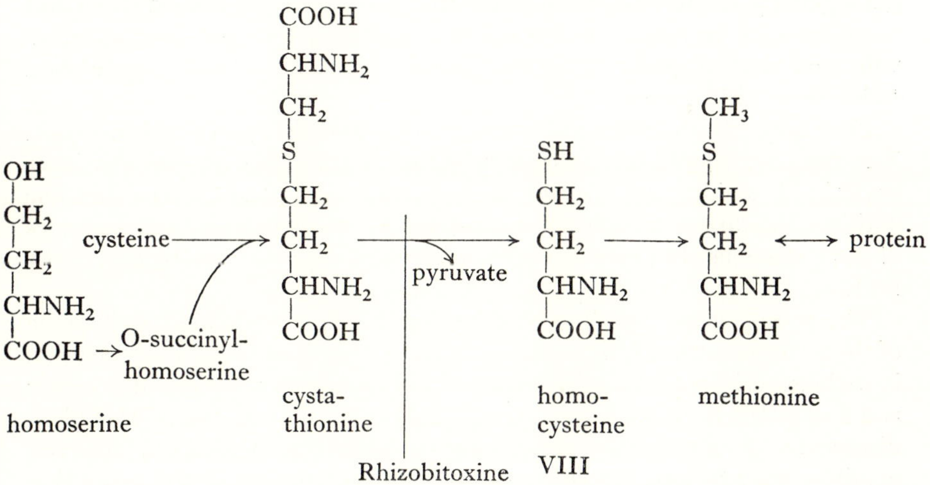

VIII

Another example is provided by rhizobitoxine which is synthesized by certain strains of the legume root-nodule bacterium *Rhizobium japonicum*

(Kirch.) Buchan. (Owens, 1969). This toxin induces chlorosis in new leaves of soybean and many other plants. Furthermore, it inhibits the growth of *Salmonella typhimurium* (Loeffl.) Cast. *et* Chalm. With this bacterium Owens and his colleagues were able to demonstrate that rhizobitoxine inhibits the enzyme β-cystathionase which catalyses the cleavage of cystathionine to form homocysteine (VIII). The inhibition of the growth of *S. typhimurium* by rhizobitoxine could be completely prevented by adding small amounts of methionine or its immediate precursor, homocysteine. Also, a toxin-induced growth inhibition of the green alga *C. pyrenoidosa* Pring. could be overcome partially by adding methionine. But attempts to prevent chlorosis in sorghum seedlings yielded ambiguous results. Chlorosis was reduced but only at methionine concentrations that bordered on being toxic themselves.

Conclusions and Prospects

The examples given in this review show that the objective of counteracting toxins by chemicals may be realized in different ways. However, the experiments have usually been done only *in vitro* or with simplified systems as in the wilting tests with cuttings. The obvious question is, of course, whether similar procedures can be developed for the practical control of plant diseases. This raises a series of problems some of which are as follows.

Because toxins and extracellular enzymes, so far as they are involved in pathogenesis, are usually produced within plant tissues the antidote should be taken up by the plant. To accomplish this different methods of application will need to be considered. From a practical point of view the application to leaves would be best.

Generally, translocation will be necessary, especially for systemic diseases. The antidote should be reasonably stable *in vivo*, but should not leave undesirable residues, at least not in food and fodder plants. Preferably, the antidote should act specifically against the toxins or extracellular enzymes of the pathogens. In particular it should not disrupt the metabolism of the host.

The prospects for success would be best if only a single toxin or enzyme of the pathogen caused all disease symptoms or at least holds a key position in pathogenesis in a particular disease—pathotoxins in the sense of Wheeler and Luke (1963). This, however, will only rarely be the case. Most plant diseases are the result of a complex interplay of different toxins, enzymes and possibly other factors. To eliminate only one component of the many that are involved may diminish the virulence of the pathogen, but is unlikely to make it avirulent.

Even under favourable conditions the toxin should be eliminated as early as possible from the chain of events comprising pathogenesis and it would seem best to aim at suppressing its biosynthesis, because merely to inactivate a toxin already present or to remove the results of toxicity does not promise permanent success so long as the pathogen remains able to produce the toxin.

Another good way could be the alteration of tissue of the host so that it cannot be attacked by toxins or enzymes of the parasite; production of large quantities of these toxins would then be of no consequence to the plant.

In recent years, highly potent systemic fungicides have been developed which kill fungal parasites within plant tissues. This obviously is the safest way of depriving pathogens of their toxicity. To me it seems, therefore, that the future of practical disease control lies more with systemic fungicides and bactericides than with substances that merely interfere with toxins or extracellular enzymes of the pathogens.

References

Bär, H. (1963). *Phytopath. Z.* **48**, 149–177.

Braun, A. C. (1955). *Phytopathology* **45**, 659–664.

Braun, R. (1960). *Phytopath. Z.* **39**, 197–241.

Brian, P. W. (1958). *Outl. Agric.* **2**, 27–32.

Byrde, R. J. W. (1963). *In* "Perspectives of biochemical plant pathology" (S. Rich, Ed.), pp. 31–41. Conn. Agric. Exp. Sta. Bull. 663.

Davis, D. and Dimond, A. E. (1953). *Phytopathology* **43**, 137–140.

Dimond, A. E. (1963). *In* "Perspectives of biochemical plant pathology" (S. Rich, Ed.), pp. 62–72. Conn. Agric. Exp. Sta. Bull. 663.

Doupnik, B., Jr. (1968). *Phytopathology* **58**, 215–218.

Durbin, R. D. (1972). *In* "Phytotoxins in plant diseases" (R. K. S. Wood, A. Graniti, and A. Ballio, Eds), pp. 19–32. Academic Press, London.

Edgington, L. V., Corden, M. E., and Dimond, A. E. (1961). *Phytopathology* **51**, 179–182.

Edgington, L. V. and Walker, J. C. (1958). *Phytopathology* **48**, 324–326.

Feldman, A. W. and Caroselli, N. E. (1951). *Phytopathology* **41**, 12.

Feldman, A. W., Caroselli, N. E., and Howard, F. L. (1950a). *Phytopathology* **40**, 8–9.

Feldman, A. W., Caroselli, N. E., and Howard, F. L. (1950b). *Phytopathology* **40**, 341–354.

Gäumann, E. (1954). *Endeavour* **13**, 199–206.

Gäumann, E. (1957). *Phytopathology* **47**, 342–357.

Gäumann, E. and Naef-Roth, S. (1956). *Phytopath. Z.* **25**, 418–444.

Gäumann, E., Naef-Roth, S., and Kern, H. (1955). *Phytopath. Z.* **24**, 373–406.

Graniti, A. (1972). *In* "Phytotoxins in plant diseases" (R. K. S. Wood, A. Graniti, and A. Ballio, Eds), pp. 1–14. Academic Press, London.

Grossmann, F. (1962). *Phytopath. Z.* **45,** 139-159.
Grossmann, F. (1968). *Neth. J. Pl. Path.* **74** (Suppl. 1), 91-103.
Hanchey, P. (1969). *Phytopathology* **59**, 1960-1962.
Hoffman, P. (1952). *Phytopathology* **42**, 11.
Horsfall, J. G. and Dimond, A. E. (1957). *Z. PflKrankh. PflPath. PflSchutz* **64,** 415–421.
Horsfall, J. G. and Zentmyer, G. A. (1942). *Phytopathology* **32**, 22–23.
Howard, F. L. (1941). *Science, N.Y.* **94**, 345.
Jost, J.-P. (1965). *Phytopath. Z.* **54**, 338–378.
Kalyanasundaram, R. and Braun, R. (1958). *Phytopath. Z.* **33**, 321–340.
Kalyanasundaram, R. and Saraswathi Devi, L. (1955). *Nature, Lond.* **175**, 945.
Kern, H. and Naef-Roth, S. (1965). *Phytopath. Z.* **53**, 45–64.
Kern, H. and Naef-Roth, S. (1966). *Phytopath. Z.* **57**, 289–297.
Martin, J. (1967). Thesis Univ. Giessen.
Matta, A. (1963). *Riv. Patol. veg., Pavia*, Ser. III, **3**, 99–106.
Oku, H. and Nakanishi, T. (1963). *Phytopathology* **53**, 1321–1325.
Owens, L. D. (1969). *Science, N.Y.* **165**, 18–25.
Patil, S. S. and Dimond, A. E. (1967). *Phytopathology* **57**, 825.
Patil, S. S. and Dimond, A. E. (1968a). *Phytopathology* **58**, 676–682.
Patil, S. S. and Dimond, A. E. (1968b). *Phytopathology* **58**, 868–869.
Patil, S. S. and Dimond, A. E. (1968c). *Phytopathology* **58**, 1063.
Pouteau-Thouvenot, M., Barbier, M., and Prévot, A.-R. (1968). *C.R. Acad. Sc. Paris* **266**, 2243–2244.
Pringle, R. B. (1972). *In* "Phytotoxins in plant diseases" (R. K. S. Wood, A. Graniti, and A. Ballio, Eds), pp. 139–154. Academic Press, London.
Pringle, R. B. and Scheffer, R. P. (1964). *A. Rev. Phytopath.* **2**, 133–156.
Samaddar, K. R. (1968). *Phytopathology* **58**, 1065–1066.
Samborski, D. J. and Forsyth, F. R. (1960). *Can. J. Bot.* **38**, 467–476.
Sinden, S. L. and Durbin, R. D. (1968). *Nature, Lond.* **219**, 379–380.
Stahmann, M. A., Clare, B. G., and Woodbury, W. (1966). *Pl. Physiol.* **41**, 1505–1512.
Stapp, C. and Hartwich, W. (1957). *Zentbl. Bakt. ParasitKde., Abt.* 2, **110**, 449–470.
Tamari, K., Ogasawara, N., and Kaji, J. (1963). *In* "The rice blast disease", pp. 35–68. The Johns Hopkins Press, Baltimore, Maryland.
van Andel, O. M. (1966). *A. Rev. Phytopath.* **4**, 349–368.
Van Sumere, C. F., van Sumere-de Preter, C., and Ledingham, G. A. (1957). *Can. J. Microbiol.* **3**, 761–770.
Wheeler, H. and Black, H. S. (1962). *Science, N.Y.* **137**, 983–984.
Wheeler, H. and Luke, H. H. (1963). *A. Rev. Microbiol.* **17**, 223–242.
Zentmyer, G. A. (1942). *Science, N.Y.* **95**, 512–513.

Discussion

Durbin: Are you aware of any *in vitro* kinetic studies with metal-requiring enzymes which show that chelating toxins such as fusaric acid act by chelation?

Grossmann: Bossi (*Phytopath. Z.* **37**, 273–316, 1960) confirmed a competitive inhibition of copper-containing polyphenoloxidase by fusaric acid but he could not prove directly that inhibition depended on a chelating of copper by fusaric acid. I am, furthermore, not aware of any kinetic studies which would prove that the effect of chelating toxins depends on chelation.

Hubbeling: Are you not of the opinion that for pectic and cellulolytic enzymes it is not the pH that is important but the relation of Ca^{++} to other ions? With *Verticillium* wilt of tomato we get heavier attack of intact plants at pH 8·0 than at a lower pH when $MgCO_3$ is added. In contrast addition of $CaCO_3$ lowers the disease index most at pH 8·0. Na^+ and NH_4^+ ions might be competitive with Ca^{++} ions.

Grossmann: Without doubt the effect of pH must be considered separately from that of calcium ions. Pectic and cellulolytic enzymes have a more or less pronounced pH optimum. The optimum for fungal polygalacturonase, for instance, usually is pH 4–5, that of pectin *trans*-eliminase, however, is pH 8 or higher. On the other hand, calcium ions may affect the degradation of pectic substances in different ways. Partially or completely demethoxylated pectin or pectate chains can be cross-linked by calcium bridges, thus becoming resistant to attack by polygalacturonases. But, calcium ions in certain ranges of concentrations stimulate the activity of *trans*-eliminases. Depending on the role of the individual enzymes in pathogenesis many interactions are possible.

Goodman: May I remind you of the work of E. D. Garber, who showed that increased buffering capacity in radish increased resistance by preventing the bacterial pathogen from changing the pH of the host tissue in which it was growing.

Grossmann: For clarity I limited my lecture to simple cases of interference with toxins and mentioned only those in which chemicals can change the milieu in ways unfavourable for a toxin or enzyme. It is also possible that a milieu favourable for a toxin or enzyme can be established only by the activity of the parasite and that this may be inhibited not only by the host tissue but also by added substances.

Main: Your paper suggests some interesting approaches to practical control of certain plant diseases. Of course, the use of inherent or genetic disease resistance is another very important practical approach. Would the methods of control you outlined, e.g. detoxification, substrate change etc., apply to genetic resistance mechanisms as well, and could they be used to predict disease resistance potential?

Grossmann: I agree that the mechanisms for elimination or inactivation of toxins may also be important in genetic resistance if the composition or the reaction of the tissue can be changed by breeding. Breeding for resistance offers also several other possibilities because natural factors of resistance aim not only at toxins and enzymes but also at invasion by and growth and sporulation of parasites.

SHORTER PAPERS

A. BACTERIAL PHYTOTOXINS

21. Isolation and the Mode of Action of the Toxin from *Pseudomonas phaseolicola**

SURESH S. PATIL AND L. Q. TAM,

Department of Plant Pathology
University of Hawaii, Honolulu, Hawaii 96822, U.S.A.

AND

P. E. KOLATTUKUDY,

Department of Agricultural Chemistry,
Washington State University, Pullman, Washington 99163, U.S.A.

Introduction

Chlorotic haloes in leaves of bean (*Phaseolus vulgaris* L.) plants infected with *Pseudomonas phaseolicola* (Burkh.) Dowson have been attributed to an extracellular toxin produced by the pathogen (Muller, 1950). When filtrates from cultures of the pathogen are injected into bean leaves they also produce chlorotic haloes (Hoitink *et al.*, 1966; Muller, 1950). Whether symptoms are caused by infection or by injection of toxin, ornithine accumulates in the chlorotic tissues (Patel and Walker, 1963; Rudolph and Stahmann, 1966). Recently, we reported that a preparation, partially purified from culture filtrates, inhibited ornithine carbamyl transferase E.C. 2.1.3.3 (OCT) from acetone powder extracts of bean leaves (Patil *et al.*, 1970). Here, we present a method for the isolation of the toxin from filtrates of culture of *P. phaseolicola* and report data in support of the hypothesis that inhibition of ornithine carbamyl transferase by the toxin may be responsible for the accumulation of ornithine *in vivo*.

Isolation of the Toxin

The toxin was isolated from shake cultures of a virulent isolate of *P. phaseolicola* (race 2) grown in peptone-salts medium (Rudolph and Stahmann, 1966), or in a synthetic medium (Watanabe, 1963). Media were sterilized in Fernbach flasks, inoculated and the flasks incubated at 18–19°C

* Published with the approval of the Director, Hawaii Agricultural Experiment Station, as Journal Series No. 1270.
This research was supported by NIH Grant No. A109477.

Culture broth (9 litres)

Centrifuge

Concentrate *in vacuo* at 70°C to 100 ml; add 9 vol methanol; filter — Residue (discard)

Concentrate filtrate to dryness; dissolve in 50 ml H_2O; add 9 vol methanol; filter — Residue (discard)

Third methanol treatment (as above); filter — Residue (discard)

Filtrate concentrated to dryness; add 20 ml H_2O — Residue (discard)

Chromatography on Sephadex G-10 column (82 × 5 cm) at flow rate 0·7 ml/min. Test fractions for halo-producing activity and OCT inhibition.

Combine *active* fractions; put on DEAE cellulose column (30 × 1·2 cm) at pH 6·5–6·8. Wash column with two bed volumes H_2O. Elute column batchwise with 0·01, 0·05 and 0·1 M NaCl (3 bed vol each).

Chromatography of concentrated salt eluates (2 ml) on Sephadex G-10 (90 × 1·2 cm) at a flow rate of 0·17 ml/min. (0·01 M salt completely elutes active fractions)/ $V_e/V_0 = 1{\cdot}9$.

FIG. 1. Isolation of *Pseudomonas phaseolicola* toxin from culture filtrates.

on a gyratory shaker. The cultures were harvested in the late log phase, centrifuged and the supernatant processed as shown in Fig. 1. Fractions from the G-10 column (last step Fig. 1) were tested for halo-inducing activity as well as for OCT inhibitory activity. Elution profiles of the two activities coincided. The active fractions were combined, concentrated and stored at 0°C before use in enzyme inhibition studies. When this material

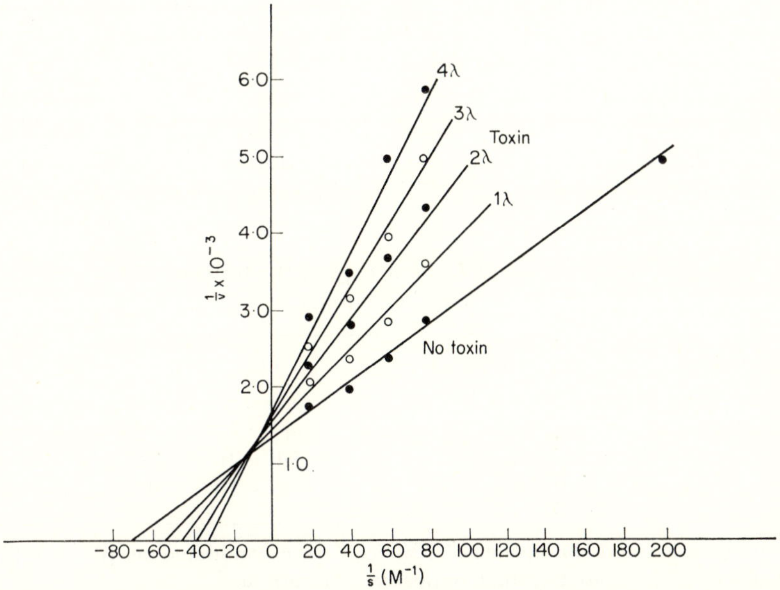

FIG. 2. Lineweaver–Burk plot for the reverse OCT reaction showing inhibition kinetics. In a final volume of 1 ml the reaction mixture contained, citrulline (25, 33, 50, 100 μmol) at pH 7; 10 μmol arsenate buffer, pH 7; and 2·8 mg enzyme protein with a specific activity of $1{\cdot}7 \times 10^{-3}$ μmol NH_4/mg protein/min. Toxin contained 203 mμmol alanine equivalents per μlitre. Incubation for 4 h at 28°C. V is the number breakdown units per h. 1 breakdown unit equals $3{\cdot}016 \times 10^{-10}$ mol ammonia.

was chromatographed on silica gel thin layer in butanol: acetic acid: water (8:2:2 v/v), it showed 2 ninhydrin positive spots. The spot with the lower R_f (0·22) which was biologically active, was chromatographed again in duplicate on silica gel thin layer in the same system. One side of the plate was sprayed with ninhydrin and the other with sulphuric acid. The ninhydrin positive spot and the sulphuric acid charred spot showed the same R_f.

OCT Inhibition Studies

OCT was partially purified from extracts of acetone powder made from 3-week-old bean plants according to a modified procedure of Kleczkowski and Cohen (1964).

First the effect of the toxin on the reverse reaction of OCT was studied. In this reaction, OCT catalyses the degradation of citrulline to ornithine and carbamyl-phosphate. The assay method was that described by Reichard (1957). The reaction mixture in a total volume of 2 ml contained 25, 33, 50, 100 μmol of citrulline (pH 7), 10 μmol of arsenate buffer at pH 7 and 2·8 mg enzyme protein with a specific activity of $1{\cdot}7 \times 10^{-3}$ μmol of NH_4 per mg of protein per min. The toxin contained 203 mμmol of alanine equivalents (Moor, 1968) per ml. The Lineweaver–Burk plot (Fig. 2) shows the kinetics of OCT inhibition at four different concentrations of toxin. The data show that the mode of inhibition of OCT for the reverse reaction is of a mixed type.

TABLE I. Comparison of K_m and V_{max} of the forward and reverse OCT reactions

Reaction*	Substrate	K_m	V_{max}
Forward reaction	Carbamyl phosphate	$1{\cdot}43 \times 10^{-4}$	2·50 μmol/10 min
	Ornithine	$5{\cdot}0 \times 10^{-4}$	2·50 μmol/10 min
Backward reaction	Citrulline	$1{\cdot}35 \times 10^{-2}$	0·037 μmol/10 min

* The forward and the backward reactions were conducted at pH 8·5 (Tris-HCl) and pH 7·0 (arsenate), their respective pH optima.

Although the reverse reaction is inhibited by the toxin the V_{max} of this reaction is considerably lower than that of the forward reaction. Data in Table I show the comparison of K_m and V_{max} of the forward and reverse reactions. The V_{max} for the forward reaction, using either substrate, is some 67-fold higher than the V_{max} for the reverse reaction. Also, Kleczkowski and Reiter (1959) have reported *in vivo* conversion of labelled ornithine to citrulline but not of labelled citrulline to ornithine. Therefore, we studied the inhibition of the forward reaction by the toxin.

The inhibition kinetics of the toxin with carbamyl phosphate as a variable substrate (ornithine at substrate saturation) is shown in Fig. 3. In order, the following were added to the reaction mixture: 50 μmol of arsenate, carbamyl phosphate (1–5 μmol), toxin 2 μl (10·9 mμmol alanine equivalents per μlitre), 0·4 mg of enzyme protein with a specific activity of 2 μmol of citrulline per mg protein per min, and 5 μmol of ornithine. Dilithium carbamyl phosphate solution was made 1 min prior to starting the assay. The reaction mixtures were incubated at 38°C.

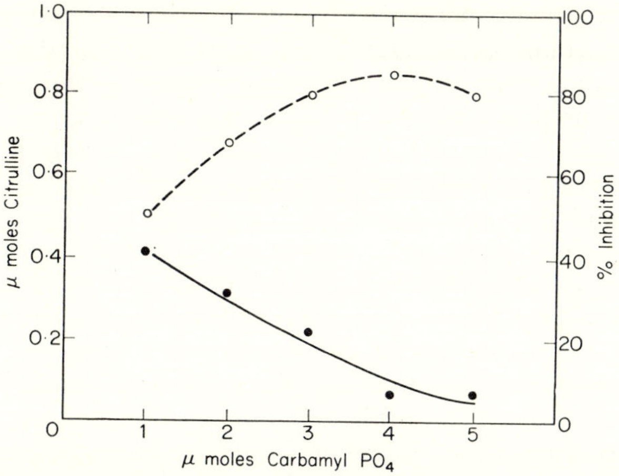

FIG. 3. Substrate concentration *vs* % inhibition for the forward reaction. Semipurified enzyme (OCT) with a specific activity of 2 μmol of citrulline per mg protein per min. In a final volume of 1 ml the assay mixtures contained: 5 μmol ornithine, carbamyl phosphate (variable), 50 μmol arsenate buffer (pH 7·5) and 0·4 mg enzyme protein. Enzyme was preincubated for 2 min at 38°C with all reactants except carbamyl phosphate. Toxin contained 10·9 mμmol alanine equivalents per μlitre 2 μlitre used per assay. Assay according to Kleczkowski and Cohen (1964). o- - - - o No toxin, ●——● Toxin.

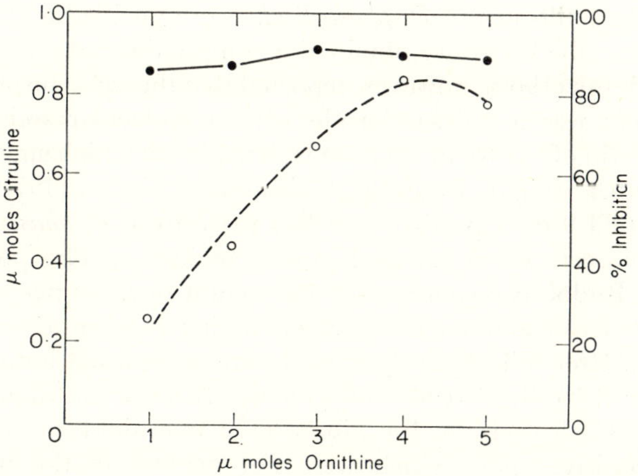

FIG. 4. Substrate concentration *vs* % inhibition for the forward reaction. All conditions exactly the same as in Fig. 3 except that ornithine was the variable substrate.

Data in Fig. 3 show that when the concentration of carbamyl phosphate is increased the inhibition of OCT by the toxin is decreased as compared to the corresponding controls. The above experiment was repeated except that ornithine was used as a variable substrate and carbamyl phosphate was at substrate saturation (5 μmol). Figure 4 shows that the degree of inhibition was the same regardless of ornithine concentration.

Enzyme inhibition studies strongly indicate that accumulation of ornithine in toxin-treated tissues is a result of OCT inhibition by the toxin. However, they do not explain how chlorosis develops in toxin-treated leaves. The inhibition of OCT may result in a citrulline deficiency in treated tissues which in turn may cause deficiencies of other intermediates of the ornithine cycle. If this is the case, toxin-treated tissues with citrulline should be protected from chlorosis. To test this, 10·0 μmol of citrulline were applied in a drop to the surface of the leaf either simultaneously or several hours after injection of the toxin. Leaves injected with toxin alone showed chlorotic haloes as usual; however, those supplied with the toxin as well as with citrulline showed no chlorosis, or at lower concentrations of citrulline showed reduced chlorosis as compared to the controls. Other naturally occurring amino acids, serine, alanine or glycine did not protect leaves from chlorosis induced by the toxin. These results indicate that OCT inhibition by the toxin is intimately related to the processes that produce chlorosis in toxin-treated tissues.

Discussion

When culture filtrates of *P. phaseolicola* were purified according to the procedure described here we obtained a compound that appeared to be a peptide. Rudolph (1969), however, reported that the toxin he obtained from *P. phaseolicola* was a polysaccharide with a molecular weight of 2100. The V_e/V_0 ratio of the toxin we have isolated is 1·9 which suggests a much lower molecular weight. Recently, Hoitink and Sinden (1970) reported a V_e/V_0 ratio of 1·9 for a partially purified toxin from *P. phaseolicola* which produces chlorosis when injected into bean leaves. The polysaccharide reported by Rudolph contained by dry weight over 80 per cent glucose. Rhamnose, fructose and to a minor extent amino and peptide nitrogen were also present. After mild hydrolysis of the toxin reported here (3% HNO_3 at 100°C for 2 h) we detected no glucose, fructose or rhamnose in the hydrolysate. Hydrolysis at 100° in N HCl overnight produced several ninhydrin positive spots which were not present in the unhydrolysed material.

The difference between our findings and those of Rudolph (1969) could be explained by postulating that *P. phaseolicola* produces two toxins, one isolated by our procedure and another by that reported by Rudolph. In fact, the "greasy looking spots" that Rudolph observes when leaves are injected with his toxin never result when our toxin is injected into bean leaves. The only symptom we observe is a discrete chlorotic halo. In the field, bean plants infected with the pathogen do not always show greasy looking spots (Patel and Walker, 1963).

The studies on the kinetics of OCT inhibition showed that the inhibition of the reverse reaction is of a mixed type. The kinetics of forward reaction is complicated because of the bisubstrate nature of the reaction. Our data (Figs 3, 4) do not show the degree of affinity of the toxin for the enzyme site(s) but they do indicate that it is the carbamyl phosphate site on the enzyme that is affected by the toxin and not the ornithine site.

The enzyme inhibition results indicate that the accumulation of ornithine in toxin-treated tissues is a result of OCT inhibition by the toxin. Although it is not clear as to how OCT inhibition causes chlorosis, it is clear that such inhibition is intimately related to the processes that lead to chlorosis. Bean plants possess all the enzymes of the ornithine cycle (Kleczkowski and Reiter, 1959) and inhibition of OCT could result in decreased arginine synthesis which in turn may affect synthesis of enzymes involved in chlorophyll synthesis. Consistent with this reasoning, the chlorotic haloes contained almost 50% less arginine compared to green tissue (Patel and Walker, 1965). Alternatively, when toxin-treated tissues were provided with citrulline no chlorotic haloes developed, indicating that alleviation of citrulline deficiency prevents chlorosis.

References

Hoitink, H. A. J., Pelletier, R. L., and Coulson, J. G. (1966). *Phytopathology* **56**, 1062–1065.

Hoitink, H. A. J., and Sinden, S. L. (1970) *Phytopathology* **60**, 1236–1237.

Kleczkowski, K., and Reiter, I. (1959). *Bull. Polo. Des. Sci.* **7**, 53–56.

Kleczkowski, K. and Cohen, P. P. (1964). *Archs Biochem. Biophys.* **107**, 271–278.

Moor, S. (1968). *J. biol. Chem.* **243**, 6281–6283.

Muller, W. H. (1950). Ph.D. thesis, Cornell University.

Patel, P. N., and Walker, J. C. (1963). *Phytopathology* **53**, 522–528.

Patel, P. N., and Walker, J. C. (1965). *Phytopathology* **55**, 889–894.

Patil, S. S., Kolattukudy, P. E., and Dimond, A. E. (1970). *Pl. Physiol.* **46**, 752–753.

Reichard, H. (1957). *Scan. J. clin. Lab. Invest.* **9**, 311–312.

Rudolph, K., and Stahmann, M. A. (1966). *Phytopath. Z.* **57**, 29–46.

Rudolph, K. (1969). *Naturwiss.* **56** (11), 569–570.

Watanabe, M. (1963). *Ann. phytopath. Soc. Japan* **28**, 201–202.

Discussion

Mirocha: How stable is the carbamyl phosphate used in the synthesis of citrulline? Was the carbamyl phosphate generated *in situ* or was the lithium salt stable enough? Also, why does the addition of citrulline reverse or prevent development of chlorosis? It does not appear that the addition of citrulline would reverse the action of the enzyme ornithine transcarbamylase.

Patil: We followed the standard practice of making the carbamyl phosphate (CAP) solution just before the beginning of the enzyme assay. The incubation was for only 10 min. It is possible that some degradation of CAP took place. We are currently investigating the degree of this breakdown under the conditions used in our assay.

We think citrulline addition prevents chlorosis because it alleviates the deficiency of citrulline caused by the inhibitory action of the toxin on ornithine carbamyl transferase of cells. Also, since the toxin is a competitive inhibitor of OCT, citrulline provided from the outside could protect the enzyme from the toxin thereby allowing the conversion of ornithine to citrulline in the normal fashion.

22. The Halo-Blight Toxin of *Pseudomonas phaseolicola:* Influence on Host–Parasite Relationship and Counter Effect of Metabolites

KLAUS RUDOLPH

Institut für Pflanzenpathologie und Pflanzenschutz der Universität, Göttingen, Germany

All work was done with highly purified preparations of the toxin of *Pseudomonas phaseolicola* (Burkh.) Dows. (Rudolph, 1969), which were colourless and produced typical halo symptoms when injected into mesophyll of young green leaves from a hypodermic needle (Rudolph and Stahmann, 1966).

Leaves of a resistant cultivar of *Phaseolus vulgaris* L., treated with toxin up to 4 days before or 1 day after inoculation with bacteria developed translucent, greasy-looking spots, typical of a susceptible reaction. Leaves treated 4 days after inoculation gave a resistant reaction and leaves treated 2 and 3 days after inoculation showed an intermediate reaction.

Estimates of numbers of bacteria in treated and untreated resistant, and in untreated susceptible leaves showed that toxin treatment allowed bacteria to multiply to the same extent in resistant as in genetically susceptible leaves (Table I). Resistance of bean leaves to *Xanthomonas phaseoli* (E. F. Sm.) Dows. var. *fuscans* (Burkh.) Starr *et* Burkh. was also decreased by the toxin, but toxin treatment had little effect on the reactions of susceptible and resistant cultivars to bean rust.

TABLE I. Effect of pre-treatment with toxin on multiplication of *Pseudomonas phaseolicola* in resistant bean leaves

Cultivar and treatment	Bacteria/cm² leaf 5 days after inoculation	Bacteria/cm² leaf 9 days after inoculation
Resistant	$10{\cdot}7 \times 10^6$	$2{\cdot}9 \times 10^6$
Resistant + toxin	$3{\cdot}4 \times 10^8$	$2{\cdot}7 \times 10^8$
Susceptible	$3{\cdot}3 \times 10^8$	$4{\cdot}7 \times 10^8$

Several investigations were carried out in order to characterize the mode of action of the toxin by treating plants with combinations of toxin and other compounds. None of the following compounds affected chlorosis and reduction of growth caused by the toxin:
metal salts (cupric acetate, cupric sulphate, cuprous chloride, ferric citrate, ferrous lactate, ferrous sulphate, ferrous tartrate, magnesium chloride), SH-compounds (mercaptoethanol, thioglycollic acid), amino acids (L-γ-aminobutyric acid, 5-aminolevulinic acid, δ-aminovaleric acid, L-arginine, L-aspartic acid, L-citrulline, L-glutamic acid, L-isoleucine, L-leucine, L-norleucine, L-norvaline, L-proline, L-valine), hormones (2,4-dichlorophenoxyacetic acid, indoleacetic acid, kinetin, naphthylacetic acid), siderochromes (ferrichrysin, ferrimycin A, ferrioxamin B), other organic compounds (carbamylphosphate, ethylenediaminetetraacetic acid, orotic acid, ribose-5-phosphate, tris(hydroxymethyl)aminomethane).

Inhibition of growth of bean callus by the toxin was not nullified by kinetin or auxins (Rudolph and Warick, 1968) as reported by other workers. However, inhibition of growth of illuminated, aerated cultures of *Euglena gracilis* Klebs. was nullified almost completely by citrulline and partially by orotic acid. Glutamic acid and dextrose had little effect whereas ornithine increased toxicity.

Only high concentrations of citrulline were effective. Typically, one tube culture of *E. gracilis* containing 50 mg sodium acetate, 100 mg peptone, 50 mg beef extract, 10–20 mg toxin and 50 ml water needed 50 mg citrulline at 28°C and 100 mg citrulline at 22°C to nullify toxicity completely.

These high concentrations suggest that the toxin does not act directly by lowering the concentration of citrulline in plant cells. It is more likely that the effects of the toxin are nullified by metabolic sequences induced by the abnormally high concentrations of citrulline that were used. But other mechanisms of toxin action which do not affect ornithine-citrulline metabolism should not be excluded (Lesemann and Rudolph, 1970).

References

Lesemann, D. and Rudolph, K. (1970). *Z. Pflanzenphysiol.* **62**, 108–115.
Rudolph, K. (1969). *Naturwiss.* **56**, 569–570.
Rudolph, K. and Stahmann, M. A. (1966). *Phytopathol. Z.* **57**, 29–46.
Rudolph, K. and Warick, R. P. (1968). *Phytopathology* **58**, 1065.

Discussion

Durbin: What concentration of citrulline is needed to protect against the toxin?

Rudolph: Depending upon the temperature, complete annulment of the toxic effect was obtained by 50–100 mg L-citrulline per 50 ml (that is $5{\cdot}7 \times 10^{-3}$ M – $1{\cdot}1 \times 10^{-2}$ M). The nutrient solution contained approximately the same absolute amounts of peptide and beef extract.

Patil: How much citrulline do you think was taken up?

Rudolph: Since the algae grew very well on peptone and beef extract with organic constituents mostly larger than citrulline, we have to assume that citrulline was taken up easily. There are no indications for a specific block for uptake of citrulline.

Kuć: Does ornithine produce halo blight symptoms and can arginine protect against the toxin of *P. phaseolicola*?

Rudolph: In tests with infiltration of leaves as well as with feeding of bean shoots we were not able to induce typical halo blight symptoms by application of nearly phytotoxic ornithine concentrations. We did not succeed in protecting against the toxin with arginine. There is a decrease of arginine in halo blight infected leaves. However, this decrease is not so marked as to explain chlorosis by absence of arginine.

Goodman: Have either Rudolph or Patil observed translocation of the toxin and the effects first recorded by Waitz and Schwartz?

Rudolph: The toxin is translocatable. When leaves are infiltrated with the toxin in confined areas, the resulting halo is often much larger. In a few cases translocation of the toxin from infiltrated leaves to developing younger leaves was observed.

Patil: In answer to Kuć, ornithine at low concentrations does cause chlorosis.

Rudolph: The ornithine-induced halo on the slide shown by Patil does not look like a typical toxin-induced halo.

Mirocha: Is the ornithine *trans*-carbamylase complex two separate enzymes?

Rudolph: No indications for a complex of two separate enzymes have been obtained.

23. Toxin production by *Xanthomonas phaseoli* var. *sojensis*

JACK D. PAXTON

Department of Plant Pathology,
University of Illinois, Urbana, Illinois 61801, U.S.A.

Xanthomonas phaseoli (E. Sm.) Dows. var. *sojensis* (Hedg.) Starr *et* Burkh. causes bacterial pustule disease of soybeans. This disease appears on soybean leaves during the warm, humid part of the summer in Illinois. The symptoms are chlorotic spots on the leaf which later develop a dark margin and finally become necrotic. At first the pustules have a distinct raised centre apparently caused by proliferation of epidermal cells.

Xanthomonas phaseoli var. *sojensis* produces a toxin in soybean leaves that causes chlorosis when put into healthy leaves. No correlation was observed between slime production by *X. phaseoli* var. *sojensis* and pathogenicity or toxin production. In this respect the pathogen appears to differ from other varieties of *X. phaseoli*. Attempts to purify and identify this toxin are under way.

Discussion

Goodman: Were washed bacteria used to inoculate leaves?

Paxton: The infiltration bacteria were not washed and the inoculum contained some culture medium. However, slime formation would probably occur in the plant before symptom development since it takes 7 days from inoculation to the appearance of symptoms.

Rudolph: In contrast to the haloes that *Xanthomonas phaseoli* causes on beans, the yellow haloes produced by *X. phaseoli* var. *sojensis* on soybean are rather small.

Paxton: These yellow haloes are rather small, however, it appears that a toxin is producing them.

24. Phytotoxic Glycopeptides of *Corynebacterium* spp.

GARY A. STROBEL

Department of Botany and Microbiology,
Montana State University, Bozeman, Montana 59715, U.S.A.

Many plant pathogenic species of *Corynebacterium* produce extracellular glycopeptides that induce wilting in plant cuttings. Three such substances isolated from *Corynebacterium michiganense* (E. Sm.) Jens. have molecular weights of *c*. 35,000, *c*. 130,000 and >200,000, and each contains amino acid residues (Rai and Strobel, 1969). Recently, Ries and Strobel (1970) isolated a phytotoxic galacto-fucosan peptide from *C. insidiosum* (McCull.) Jens. with a molecular weight of *c*. $5{\cdot}0 \times 10^6$ and the unique property of binding copper. The most highly characterized toxic glycopeptide from the plant pathogenic coryneform bacteria is the one from *C. sepedonicum* (Spieck. *et* Kotth.) Skapt. *et* Burkh. (Strobel, 1970). It has a molecular weight of *c*. 21,400. The peptide portion is about 5% of the molecule and is linked glycosidically through the —OH of threonine to a mannose residue in the toxin. Its acidic nature depends mainly on residues of 2-keto-3-deoxy-gluconic acid.

The toxic glycopeptide of *C. sepedonicum* is a vivotoxin because several milligrams of it have been isolated from potato plants infected with the pathogen. The compound from infected plants and the toxin from bacterial cultures were the same with respect to mobility by acrylamide electrophoresis, optical rotation, infrared spectra, chemical residue analyses, x-ray powder diffraction patterns and by several other criteria. There was sufficient toxin in infected plants to account for symptom production in biological assays (Strobel, 1970).

Electron microscopy by Strobel and Hess (1968) showed that the glycopeptide of *C. sepedonicum* interferes with the function and structure of plant membranes. This was corroborated in autoradiographic studies using ^{3}H-toxin, by ^3H_2O labelling, other physiological studies and more recently, electron microscopy of potato plants infected with *C. sepedonicum.*

No evidence has been obtained to support the notion that the toxin blocks movement of water through the xylem of affected plants. This led Johnson and Strobel (1970) to modify the toxin chemically to determine if it has an active site or sites. Blocking the carboxyl groups of the toxin by methylation eliminates its ability to cause collapse of young tomato stems. The modified toxin was tested with a device that measures wilt in young tomato plant hypocotyls by determining the force in mg that they can withstand in order to maintain an erect position.

This research was supported in part by NSF grant GB-12956 and is Montana Agriculture Experiment Station Paper 199, Journal Series.

References

Johnson, T. B. and Strobel, G. A. (1970). *Pl. Physiol.* **45**, 761–764.
Rai, P. V., and Strobel, G. A. (1969). *Phytopathology* **59**, 53–57.
Ries, S. M. and Strobel, G. A. (1970). *Phytopathology* **60**, 1309 (abstract).
Strobel, G. A. and Hess, W. M. (1968). *Pl. Physiol.* **43**, 1673–1688.
Strobel, G. A. (1970). *J. biol. Chem.* **245**, 32–38.

Discussion

Patil: Wood has already commented on specificity of plant pathogens in relation to their ability to produce toxins. Do you think that your toxin is involved in the specificity of your pathogen for its host plants?

Strobel: The toxin is not host specific. Some factor other than the toxin is involved in host specificity.

Barbier: Is the absorption of your peptide at 250 nm for CONH bonds only?

Strobel: That was my mistake. As you indicate the CONH absorption occurs at around 200 nm. The broad band at 250–260 nm is probably due to absorption by aromatic amino acids and a small amount of contaminating RNA.

Backman: You said that 2-keto-3-deoxy gluconic acid was the only acid methylated on the toxin molecule. Is it not true that the free carboxyl of the polypeptide is also methylated, and may also be involved in the loss of phytotoxicity?

Strobel: The digestion of the peptide from the toxin by pronase resulted in very little decrease in biological activity. Therefore, we assume that the acidic amino acids are not playing a role in the activity of the molecule, even though they may have been methylated.

Dimond: Do you have any thoughts on why the blocking of carboxyl groups results in loss of toxity?

Strobel: I suggest that you may substantially change the net charge on this compound by this alteration. This will change its adsorption by cellular structures.

Stoessl: I am uneasy about the identities of the IR and UV spectra of the toxin preparations isolated from culture and host plant. How specific do you consider X-ray diffraction patterns to be for molecules of this size?

Strobel: A comparison of X-ray diffraction pattern on molecules of this size is taken as good additional evidence for the identity of a macromolecule. However, it would not stand alone because it is not totally specific.

25. Membrane Action and Biogenesis of the Phytotoxin Syringomycin

PAUL A. BACKMAN

Department of Plant Pathology,
*North Carolina State University, Raleigh, North Carolina 27607, U.S.A.**

Pseudomonas syringae van Hall causes diseases in species of more than thirty genera of plants. DeVay *et al.* (1968) reported that a wide spectrum antibiotic, toxic to stone fruit trees, is produced by *P. syringae* in culture; they found that its production *in vitro* was related to the pathogenicity and virulence of the bacterium to peach seedlings. Sinden *et al.* (1971) found that the toxin was a polypeptide with 9 ninhydrin-reactive components and could be isolated from infected peach seedlings; the toxin was named syringomycin (SR).

Penner *et al.* (1969) found that purified SR would bind to salmon sperm or calf thymus DNA *in vitro*, and would inhibit RNA polymerase. They postulated that SR in diseased tissue could inhibit RNA synthesis and could explain the broad spectrum of antibiosis exhibited by SR. However, this study did not account for the rapid necrosis that occurs in 2–5 h when plant tissue was treated with SR.

Purified SR solutions markedly reduce the contact angle of water to a hydrophobic surface. This detergent-like property is also displayed by many of the membrane-active polypeptide antibiotics. The possibility of a site of action for SR on cell membranes was therefore investigated (Backman and DeVay, 1971).

Treatment of washed rabbit erythrocytes with SR caused a rapid leakage of haemoglobin from treated cells. A similar efflux of 260 nm absorbing materials was observed from cells of *Geotrichum candidum* Link treated with SR. Leakage of cell metabolites reached a maximum at the lethal dose for the test organism. Electron micrographs of SR-treated cells of *G. candidum* indicated that at lethal doses, massive disruption of membrane integrity occurred within 10 min and that membrane abnormalities were caused at sublethal doses.

* Present address, USDA-ARS Department of Plant Pathology and Physiology, Clemson University, Clemson, South Carolina 29631, U.S.A.

Final proof of an action on membranes was obtained by applying ^{14}C-labelled SR to freeze-dried and sectioned plant tissue; the tissue sections were then washed and coated with a photographic emulsion. After development, the microautoradiographs revealed that the radioactive SR was concentrated primarily over the plasmalemma and nuclear envelope, confirming that membranes are a primary site for binding of SR.

Studies with membrane components indicated that the site of SR binding might be membrane lipids. Cholesterol, β-sitosterol, viosterol and α-lecithin incorporated into agar media, greatly reduced the zone of inhibition around toxigenic bacterial colonies. Studies with purified toxin indicated that the reduction in inhibition was not due to an action on the producing bacterium, but rather to a probable interaction between the lipids and the toxin molecule. Sterol precursors (squalene and cholestanol) and proteins (globulins and albumins) did not reduce toxicity. It is postulated that the mechanism of reduction of SR toxicity caused by some lipids is similar to the binding of SR by cell membranes. Calcium and magnesium ions were also found to inhibit SR activity.

Serological tests of cell wall material from toxigenic and non-toxigenic isolates of *P. syringae* indicated that SR was structurally related to cell wall materials. Syringomycin appeared to react as a haptenic group of a larger cell wall antigenic molecule. The polypeptide SR could, therefore, be a product of the cell wall metabolism of pathogenic strains of *P. syringae*.

These data indicate that the phytotoxin syringomycin may be a product of the cell wall of *P. syringae*, and that its primary action is the rapid, detergent-like lysis of cellular membranes by the binding of the toxin to membrane lipids.

References

Backman, P. A., and DeVay, J. E. (1971). *Physiol. Pl. Path.* **1**, 215–233.

DeVay, J. E., Lukezic, F. L., Sinden, S. L., English, H., and Coplin, D. L. (1968). *Phytopathology* **58**, 95–101.

Penner, D., DeVay, J. E., and Backman, P. A. (1969). *Pl. Physiol.* **44**, 806–808.

Sinden, S. L., DeVay, J. E., and Backman, P. A. (1971). *Physiol. Pl. Path.* **1**, 199–214.

Discussion

Goodman: Do you ascribe fuzziness in the membrane and cell wall of the electron micrograph to the action of syringomycin?

Backman: There may be a change in the staining properties of the cell walls due to treatment with syringomycin. However, it might also be due to the quality of the reproduction of the original electron micrograph.

Durbin: What is the effect on plants of non-toxin-producing strains?

Backman: Non-toxin-producing strains are weak pathogens on peach, but on cotton, sugar beet, and bean they are non-pathogenic. On all these plants toxin-producing strains are good pathogens.

Page: Did you characterize the material absorbing at 260 nm and what substances did you use for controls ?

Backman: The material was not characterized, but the literature reports it to be nucleic acid material. The controls were untreated cells, and cells treated with base-inactivated syringomycin.

Marrè: The experiments on erythrocytes suggest the possibility that the toxin binds to some membrane protein.

Backman: Bovine serum albumin, ovalbumin, and globulin proteins did not affect the antibiosis of syringomycin, however lipids did. It appears that it is either lipid, or the lipid–protein complex that is the site of binding on the membranes.

B. FUNGAL PHYTOTOXINS

26. New Compounds from Culture Filtrates of *Fusicoccum amygdali* Del.

ALESSANDRO BALLIO, MICHELE FRAMONDINO AND GIACOMINO RANDAZZO

Laboratorio di Chimica delle Sostanze Naturali, Istituto Chimico, University of Naples, Italy

AND

CARLO G. CASINOVI, GIULIANO GRANDOLINI*, FRANCESCO MENICHINI, AND CARLO ROSSI*

Laboratori di Chimica Biologica, Istituto Superiore di Sanità, Rome, Italy

Submerged cultures of *Fusicoccum amygdali* Del. produce the highly phytotoxic metabolite, fusicoccin (I) (Ballio *et al.*, 1964). Its structure was elucidated by Ballio *et al.* (1968) and independently confirmed by Barrow *et al.* (1968).

Thin layer chromatography (silica gel G, 8% propan-2-ol in chloroform) of filtrates from cultures of *Fusicoccum amygdali* consistently shows, besides fusicoccin, small amounts of related substances with lower R_F values.

Recently, the three major by-products have been characterized as *iso*-fusicoccin (an isomer of fusicoccin), monodeacetylfusicoccin (II), and dideacetylfusicoccin (III) (Ballio *et al.*, 1970). Five other by-products which have now been purified by repeated column chromatography also have close structural relationships with fusicoccin. Three, F V/2, F V/4 and F VI, are isomers of monodeacetylfusicoccin (II). This became clear from mass spectrometry and nuclear magnetic resonance (NMR) data, and from their conversion to triacetylfusicoccin (IV) on acetylation and to dideacetylfusicoccin (III) on treatment with 0·1 N NaOH. Mass spectrometry clearly indicated that in F V/2 and F VI the acetyl group was on the glucosidic moiety (monoacetylglycosyl peak at m/e 205, deacetylaglycone peak at m/e 366), whereas F V/4 was acetylated on the aglycone moiety (aglycone

* On leave from the Istituto di Chimica Farmaceutica e Tossicologica dell'Università degli Studi, Perugia, Italy.

peak at *m/e* 408). Furthermore, F V/2 and F VI were interconvertible at slightly alkaline pH; under these conditions a third monoacetylderivative is also formed (*iso*-F VI). To locate the acetyl group more precisely, these compounds were carefully analysed by nuclear magnetic double resonance spectroscopy; this led to the assignment of structure (V) to F V/2, (VI) to F V/4, (VII) to F VI, and (VIII) to *iso*-F VI.

The other two by-products, F V/1 and F VIII, have respectively one and two oxygens less than fusicoccin; NMR shows that both contain an *iso*-propyl in place of CH_3—CH—CH_2OH and have neither *O*-acetyl nor *O*-methyl groups. Compound F VIII also lacks the *O*-*t*-pentenyl group. Studies are in progress to establish the complete structures of these compounds, knowledge of which may provide clues to metabolic pathways involved in biosynthesis of fusicoccin.

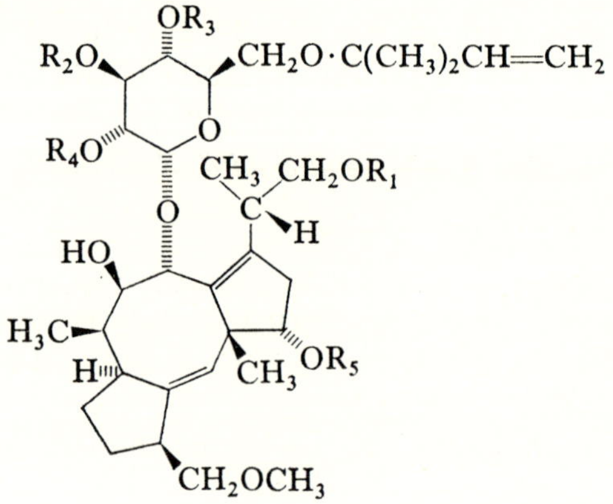

I: R_1, R_2 = $COCH_3$; R_3, R_4, R_5 = H

II: R_1 = $COCH_3$; R_2, R_3, R_4, R_5 = H

III: R_1, R_2, R_3, R_4, R_5 = H

IV: R_1, R_2, R_3, R_4, R_5 = $COCH_3$

V: R_1, R_2, R_3, R_5 = H; R_4 = $COCH_3$

VI: R_1, R_2, R_3, R_4 = H; R_5 = $COCH_3$

VII: R_1, R_2, R_4, R_5 = H; R_3 = $COCH_3$

VIII: R_1, R_3, R_4, R_5 = H; R_2 = $COCH_3$

References

Ballio, A., Chain, E. B., De Leo, P., Erlanger, B. F., Mauri., M, and Tonolo, A. (1964). *Nature* **203**, 297.

Ballio, A., Brufani, M., Casinovi, C. G., Cerrini, S., Fedeli, W., Pellicciari, R., Santurbano, B., and Vaciago, A. (1968). *Experientia* **24**, 631–635.

Ballio, A., Casinovi, C. G., Randazzo, G., and Rossi, C. (1970). *Experientia* **26**, 349–351.

Barrow, K. D., Barton, D. H. R., Chain, E. B., Ohnsorge, U. F. W., and Thomas, R. (1968). *Chem. Commun.* 1198–1200.

27. Structure—Phytotoxicity Relations with Fusicoccin Derivatives*

ALESSANDRO BALLIO, MICHELE FRAMONDINO AND GIACOMINO RANDAZZO

Laboratorio di Chimica delle Sostanze Naturali, Istituto Chimico, University of Naples, Italy

AND

ANTONIO BOTTALICO AND ANTONIO GRANITI

Istituto di Patologia vegetale, University of Bari, Italy

Sixteen derivatives of fusicoccin, prepared by chemical modification of the parent compound, have been tested for their phytotoxicity to cuttings of young tomato plants.

Except for hydrogenation of the *t*-pentenyl group, structural modification (partial or full deacetylation, acetylation, removal of the *t*-pentenyl group, removal of the glucosidic moiety) remarkably depressed or abolished phytotoxicity.

Discussion

Pringle: Hydrogenation of fusicoccin appears to increase the toxicity. Can this be correlated with changes in physical properties such as solubility?

Ballio: It is indeed possible; in the series of derivatives so far tested, with the exception of the acetylation products, we have observed a quite good linear correlation between log of the minimal dose and R_f values on thin layer chromatography with two solvent systems.

Dr. Mantle has shown that fusicoccin F(de-*t*-pentenyldideacetylfusicoccin) induces an increase of fresh weight in tomato shoots. As will be told in detail by Dr. Russi, we have observed that water absorption by etiolated pea tissue is remarkably increased not only with fusicoccin, but also with several derivatives which display a low activity in the phytotoxicity test.

Turner: How do the fusicoccin derivatives described correspond to the nomenclature A, B, C, D, E and F used by Barrow and co-workers?

* A paper on this subject has been published in *Phytopathologia Mediterranea*, **10**, 26–32, 1971.

Do the derivatives and fusicoccin A affect water absorption differently?

Ballio: Fusicoccins A, B, C, D, E and F of the British authors correspond respectively to what we call fusicoccin, isofusicoccin, monodeacetylfusicoccin, dideacetylfusiococcin, de-*t*-pentenylmonodeacetylfusicoccin and de-*t*-pentenyldideacetylfusicoccin.

Graniti: In our experiments, toxicity of fusicoccin and derivatives was assayed on tomato cuttings in a chamber at 22°C, 80% relative humidity, with very low light intensity (700 lumen/m^2). Under these conditions, the "minimal dose" of dihydrofusicoccin resulted in a little lower loss (0·6 mg/kg fresh weight), i.e. more toxic than fusicoccin (0·9 mg/kg). As already mentioned by A. Ballio, all other derivatives were less toxic than fusicoccin.

Pringle: In connexion with these examples of change in toxicity with hydrogenation, it may be interesting to note that hydrogenation of *H. carbonum* toxin decreases toxicity to about 1/500 that of the parent toxin.

Mirocha: It is interesting to note that you have observed an increase in biological activity of the dihydro derivative of fusicoccin. We find a similar relationship with the dihydro derivative of F-2 or zearalenone.

28. Investigations on the Phytotoxicity of Fusicoccins

SIR ERNST CHAIN AND PETER G. MANTLE

Biochemistry Department, Imperial College, London, U.K.

AND

BARRY V. MILBORROW

Shell Research Limited, Milstead Laboratory of Chemical Enzymology, Sittingbourne, Kent, U.K.

Fusicoccin is a terpenoid glucoside, produced by *Fusicoccum amygdali* Del. in submerged culture (Ballio *et al.*, 1964), and may be the factor causing the wilting syndrome shown by almond trees parasitized by this fungus, although this has not been established.

Fusicoccin is not species-specific and induces a wilt of plant shoots when introduced into the transpiration stream through a cut surface. Intact leaf and root surfaces were penetrated only very slowly. Under conditions of moisture tension the uptake by tomato shoots of small amounts of fusicoccin (*c.* 2 μg/g fresh weight) induced accelerated evapotranspiration which, being in excess of water uptake, resulted in a loss of turgor. This was not caused by impaired vascular flow. Protoplasmic streaming was not inhibited by fusicoccin indicating that cells are not killed before or during the initial wilting process. Leakage of cytoplasmic solutes, whether ions or pigments, was negligible. This suggests that no gross damage to cytoplasmic membranes occurred, and the absence of any observable ultrastructural changes in tomato leaf tissues which had wilted slightly in response to fusicoccin supported this conclusion. The semipermeable integrity of cytoplasmic membranes was not impaired as was shown by the ability of fusicoccin-wilted tomato shoots to regain full turgor during immersion in water for one hour. When these turgid tissues were returned to drying conditions they wilted again. This cycle was repeated several times, demonstrating that the recovery from the wilted conditions was not due to metabolism or leaching out of the fusicoccin. High concentrations of fusicoccin (10 and 100 μg/ml) respectively elevated the respiration rate of tomato leaf disks and suppressed growth of plants in submerged culture. Respiration was not stimulated by lower concentrations of fusicoccin not exceeding 1 μg/ml which, nevertheless, were well able to induce wilting when applied via the transpiration stream.

Fusicoccin caused an increase of stomatal aperture, confirming the observations of Turner and Graniti (1969), but tomato shoots in which the stomata had been closed by an application of abscisic acid still wilted in response to fusicoccin treatment in the same way as shoots which had been treated with fusicoccin alone. This separates any effect on stomatal aperture from involvement in the primary process by which the wilting syndrome is initiated, and which has been the concern of the present investigation (Chain *et al.*, 1971). The stomatal guard cells, by virtue of their anatomical position, are likely to be among the last cells in a leaf to receive a substance introduced into the transpiration stream. Thus, a wilt-inducing agent might first be expected to relax the turgor of cells surrounding the stomata, thus allowing the guard cells to expand slightly and exhibit an increased stomatal aperture.

Progressive de-acetylation of fusicoccin reduced its potency as a wilting agent. However, tomato shoots showed a novel response after taking up small amounts of desacetyl-desisopentenyl-fusicoccin. Their fresh weight increased by 20% within a few hours and during this time the shoots apparently became more plastic, bending over at right angles while remaining fully turgid. The leaves became in-rolled upwards suggesting different turgidity of upper and lower surfaces such as would be expected in leaves the terminal vessels of which are situated nearer to the abaxial surface. This implied that desacetyl-desisopentenyl-fusicoccin possessed an auxin-like property, which, however, was not observed under conditions of water stress in response to fusicoccin solutions over a similar range of 0·2–10 μg/ml. After concentrations of desacetyl-desisopentenyl-fusicoccin in tissues, calculated from the volume of solution taken up, had reached about 5 μg/g fresh weight a fusicoccin-like wilting response was elicited. It is not yet clear whether this fusicoccin derivative has two distinct effects on plant tissues, one auxin like and the other wilt-inducing, or whether one physiological property depends for its forms of expression on the concentration of the derivative in the tissues.

Ophiobolin, the phytotoxicity of which was first investigated by Orsenigo (1957) and which has a close structural similarity to fusicoccin, caused a loss of pigment from beetroot disks within fifteen minutes, indicating irreversible damage to cytoplasmic membranes which does not occur as a response to fusicoccin. Rice shoots wilted in fusicoccin solutions while remaining green, whereas those in ophiobolin solutions became chlorotic within 24 h without wilting. Thus the modes of action of these structurally similar substances appear to be quite distinct. It is concluded that fusicoccin though not primarily toxic (*sensu stricto*) to cells, induces death by desiccation. The ability of the cells to retain water is impaired but the precise mechanism by which this is effected remains obscure.

References

Ballio, A., Chain, E. B., DeLeo, P., Erlanger, B. E., Mauri, M., and Tonolo, A. (1964). *Nature, Lond.* **203**, 297.

Chain, E. B., Mantle, P. G., and Milborrow, B. V. (1971). *Physiol. Pl. Path.* **1**, 495.

Orsenigo, M. (1957). *Phytopath. Z.* **29**, 189.

Turner, N. C. and Graniti, A. (1969). *Nature, Lond.* **223**, 1070.

Discussion

Graniti: Many results obtained by Mantle agree with our data and we are pleased to have them confirmed.

Here I would like to comment further in the light of our results from recent experiments.

Fusicoccin is readily absorbed by young bean plants through intact roots (Graniti and Russo, 1970, *Phytopath. medit.* **9**, 75–86). The toxin (6·8 μg/ml) was added to the flasks of Knop's solution in which the plants were growing. In a greenhouse at about 25°C and 80% RH, with 3000 lux maximum daylight, the treated plants wilted within one day.

Electron microscopic examination of sections through the mesophyll of wilting leaves showed that cell collapse occurred after pseudoplasmolysis. Among cellular organelles, the chloroplasts were most damaged. The observed changes can be explained in terms of increase of cell wall plasticity through an auxin-like mechanism and subsequent disruption of the permeability of cell membranes.

These results show that fusicoccin can be absorbed by intact plant surfaces, provided they are not protected by a heavy cuticle. In fact, spraying fusicoccin solutions on leaves affects only stomata if the cuticle is thick, but produced wilt and necrosis in plants with thin cuticle (Graniti and Turner, 1970, *Phytopath. medit.* **9**, 160–167).

29. Fusicoccin—a Phytotoxin that Opens Stomata

NEIL C. TURNER

The Connecticut Agricultural Experiment Station, New Haven, Connecticut 06504, U.S.A.

Fusicoccum amygdali Del., the pathogen of almond (*Prunus amygdalus* St.) and peach (*P. persica* (L.) St.) not only causes cankers around infected buds and nodes, but also causes the leaves at the distal portion of the infected shoot to wilt and die. The pathogen was not found in the wilted tissues, but a phytotoxin, fusicoccin, was isolated from the actively growing mycelium and found to produce symptoms in almond and peach cuttings closely resembling those that follow infection by the pathogen (Graniti, 1964).

The transpiration of tomato (*Lycopersicon esculentum* Mill. cv. "Super Marmande") cuttings in an illuminated room maintained at 30°C and 32 to 42% relative humidity, was increased from 0·5 g/h to 1·2 g/h within 3 h of being placed in 10^{-5} M fusicoccin (Graniti, 1964). A modification of Hygen's drying technique indicated that fusicoccin increases both stomatal and cuticular transpiration (Procacci and Graniti, 1966).

Subsequently, the stomata in tomato and a number of other species were found to open when 10^{-5} M fusicoccin was either absorbed through the petiole or painted on to the leaf (Turner and Graniti, 1969; Graniti and Turner, 1970). Changes in stomatal resistance were observed with a ventilated diffusion porometer (Turner and Parlange, 1970). The porometer quickly indicates the resistance to diffusion of water vapour through the leaf epidermis; changes in the width of the stomatal pore were occasionally measured on replicas obtained from silicon rubber impressions (Sampson, 1961). Thus, when dogwood (*Cornus florida* L.) cuttings were placed in 10^{-5} M fusicoccin, the resistance of the abaxial leaf surface to diffusion of water vapour first increased, but after 1 h the resistance was already less than that in the controls and after 3 h dropped to only 30% of that in the controls. Further, in young bean (*Phaseolus vulgaris* L. cv. "Pinto") plants painted with 10^{-5} M fusicoccin, the resistance of both leaf surfaces was reduced to one-third that in the controls within 3 h and this difference was maintained for 18 h.

Fusicoccin not only opens stomata in the light, but also does so in the dark. The stomata of bean plants kept in the dark increased from a mean width of 0·6 μm in the controls to 1·8 μm in leaves painted with fusicoccin; the epidermal resistances were 15·6 sec cm^{-1} and 0·8 sec cm^{-1}, respectively. However, the time required for fusicoccin to open stomata more widely than the controls was greater in the dark than in the light (Graniti and Turner, 1970).

Leaf water potentials, measured with a pressure chamber (Scholander *et al.*, 1964; Turner *et al.*, 1971) were decreased in leaves treated with fusicoccin (Turner and Graniti, 1969). Bean leaves painted with 10^{-5} M fusicoccin first increased in potential from −3·3 to −2·1 bars as the stomata closed but then decreased as the stomata opened, dropping to −11·4 bars after 6 h. The decrease in leaf water potential after stomatal opening is probably the major cause of the wilting observed in diseased tissue. However, wilting has been observed in leaves treated in the dark, suggesting that factors additional to a rapid decrease in potential may bring about loss of turgor.

Faster evaporation not only occurs in leaves adequately supplied with water, but also from excised shoots (Procacci and Graniti, 1966; Turner, 1970). In a ventilated room at 26 ± 1°C and 80 ± 5% relative humidity, lucerne (*Medicago sativa* L.) shoots sprayed with 10^{-5} M fusicoccin 3 h prior to cutting, dried to 50% of their initial fresh weight in 8 h compared to 28 h for the controls; the time to dry to a quarter of the initial fresh weight was 44 and 88 h respectively. Fusicoccin also speeded the drying of lucerne shoots in the field, even to a moisture content suitable for storing as hay (Turner, 1970). The faster drying has been shown to be due, in part, to the slower closure of stomata previously opened with fusicoccin. But the faster drying rate continued after the stomata had closed to an aperture less than 1 μm, the smallest that can be microscopically detected. Procacci and Graniti (1966), therefore, suggested that fusicoccin increased cuticular transpiration. However, fusicoccin did not change the water loss through the cuticle in dogwood leaves, grapes (*Vitis vinifera* L.) or grapes from which the waxy "bloom" had been removed by chloroform. Therefore, the increased rate of drying of treated lucerne and tomato may be caused by the failure of the treated stomata to close as tightly as normal in desiccated leaves.

Having shown that fusicoccin opens the stomata in both the light and in the dark, I now speculate on its mechanism of action. It is known that stomata open as the guard cells increase in turgor relative to the subsidiary cells. Low concentrations of fusicoccin cause leaf tissue to absorb water (Ballio *et al.*, 1968), possibly because of increased cell wall plasticity (Rasi Caldogno *et al.*, 1969); any differential absorption of water by the guard cells greater than that by the subsidiary cells will open the stomata. Fusicoccin also changes cytoplasmic permeability, particularly to potassium (Ballio *et al.*, 1968), an electrolyte involved in stomatal changes (Sawhney and Zelitch, 1969). Since Ballio *et al.* (1968) showed that an increase in the absorption of water occurred within 1·5 h whereas changes in permeability to potassium only occurred after 7 h, the rapid opening of stomata by fusicoccin probably follows changes in water absorption and not changes in permeability to potassium ions (Graniti and Turner, 1970).

References

Ballio, A., Graniti, A., Pocchiari, F., and Silano, V. (1968). *Life Sci.* **7**, Ser. II, 751–760.

Graniti, A. (1964). *In* "Host–parasite Relations in Plant Pathology" (Z. Király and G. Ubriszy, Eds), pp. 211–217. Budapest.

Graniti, A. and Turner, N. C. (1970). *Phytopath. medit.* **9**, 160–167.

Procacci, R. and Graniti, A. (1966). *First Cong. Mediterranean Phytopathol. Union*, Bari, pp. 87–93.

Rasi Caldogno, F., Pennacchioni, A. and Lado, P. (1969). *Giorn. bot. ital.* **103**, 629–630 (abstract).

Sampson, J. (1961). *Nature, Lond.* **191**, 932–933.

Sawhney, B. L. and Zelitch, I. (1969). *Pl. Physiol.* **44**, 1350–1354.

Scholander, P. F., Hammel, H. T., Hemmingsen, E. A., and Bradstreet, E. D. (1964). *Proc. natn. Acad. Sci. U.S.A.* **52**, 119–125.

Turner, N. C. (1970). *Agron. J.* **62**, 538–541.

Turner, N. C. and Graniti, A. (1969). *Nature, Lond.* **223**, 1070–1071.

Turner, N. C. and Parlange, J.-Y. (1970). *Pl. Physiol.* **46**, 175–177.

Turner, N. C. DeRoo, H. C., and Wright, W. H. (1971). *Conn. Agric. Exp. Station, Special Soils Bull.* No. xxxiii.

Discussion

Schlosser: Did you consider that alterations in metabolism, such as an interference with energy production, could also lead to opening of the stomata, in which case the stomatal opening would be only a secondary effect?

Turner: Present evidence suggests that stomatal opening is the energy requiring process and that closure appears to be the "relaxed" state. Any reduction in energy production by fusicoccin would, therefore, lead to stomatal closure and not the opening observed in the present studies. For stomatal opening to be a secondary effect, fusicoccin would have to stimulate energy production by the plant. I have not studied the effect of fusicoccin on energy production, but I believe that Dr. Marrè and his co-workers have some unpublished studies in this field; perhaps he can elaborate on this point.

Marrè: The present evidence is rather against the idea that responses of the leaf to fusicoccin depend on energy shortage. Fusicoccin-induced water uptake stimulation is strongly inhibited by uncouplers such as 2–4 dinitrophenol and appears, therefore, as an energy dependent process. Moreover, in the case of the pea internode segments, we found ATP levels at least as high as in the controls.

Kuć: Have you observed wilting on almond trees without cankers? Do cankers appear before wilting or wilting before canker formation? Does fusicoccin cause canker formation?

Turner: Professor Graniti is better placed to answer this question.

Graniti: Infection by *Fusicoccum amygdali* through a bud or a leaf scar of an almond shoot is followed within a few days by wilting of leaves located at some distance from the infection node, where the fungus growth is still very limited. These systemic symptoms develop during the early stages of the disease, while both bark and xylem are free of gums which are soon produced in tissues adjacent to the infected area. Clearly, therefore, wilting does not result from blocking of xylem vessels, but from the action of the toxin itself on leaf tissue.

Although gummosis is not entirely effective in stopping the fungus advance, it hinders the spread of toxic metabolites to distant uninvaded organs through conductive or parenchymatous tissues. When gummosis sets in, therefore, the pathogen behaves as a local parasite and its action becomes more and more limited to the invaded tissues, until corky layers are formed all round the lesion. Cankers are thus developed.

These events seem to change an initially-toxigenic, potentially-systemic disease into a localized and chronic one.

Turner: Before the discussion is closed, I understand that Dr. Graniti has some recent results to present.

Graniti: I should like to mention some data recently obtained in our laboratory by Dr. Bottalico about the influence of light on fusicoccin toxicity to tomato cuttings.

The experimental conditions were the following: tomato plants grown in a greenhouse in December; toxin concentration 6·8 μg/ml (10^{-5} M fusicoccin); wilting tests carried out in a room at 22°C, 80% RH, illuminated with fluorescent daylight lamps.

Toxicity of fusicoccin as estimated by the Gäumann test (Graniti, 1964). *Phytopath. medit.*, **3**, 125–128), decreased by increasing light intensity, as shown by the following figures:

Light intensity (lumens/m^2)	*Minimal dose* (mg/kg fresh weight)
0 (dark)	0·11
700	0·89
1000	1·62
1500	2·92
2000	7·34
4000	10·20

30. Effects of Victorin and Fusicoccin on Stomatal Behaviour in Oats (*Avena sativa* L.)

NEIL C. TURNER

The Connecticut Agricultural Experiment Station, New Haven, Connecticut 06504, U.S.A.

Victorin, the host specific toxin of *Helminthosporium victoriae* Meehan *et* Murphy, (Luke and Wheeler, 1955) and fusicoccin, the non-specific toxin produced by *Fusicoccum amygdali* Del., (Graniti, 1964), have terpenoid moieties (Ballio *et al.*, 1968; Owens, 1969) and both hasten the drying of excised plants (Wheeler and Hanchey, 1968; Turner, 1970). Fusicoccin increases the transpiration rate of cuttings (Graniti, 1964) by opening stomata (Turner and Graniti, 1969; Graniti and Turner, 1970), whereas victorin decreases transpiration and causes a transient closure of stomata (Wheeler, 1969). The effects of victorin on transpiration and stomatal behaviour in blight susceptible (cv. "Park") and blight resistant (cv. "Russel") oat leaves were studied and the effects of fusicoccin and victorin on stomata have been compared. Stomatal changes were followed with a ventilated diffusion porometer (Turner and Parlange, 1970), which measures the resistance to diffusion of water vapour through the epidermis.

Fully-expanded leaves of oats were cut and quickly placed into beakers containing solutions of victorin of 0, 0·00012, 0·0012, 0·012, 0·12, 1·2 and 12 units/ml (Luke and Wheeler, 1955). They were kept in a room at 27°C, 55% RH and 2500 ft-candles illumination. The transpiration of susceptible and resistant oat leaves with cut ends in distilled water did not differ significantly after 3 h, but victorin at 1·2 units/ml and greater reduced transpiration in the susceptible variety by 60%; transpiration in the resistant variety was unaltered at any concentration. In the blight susceptible oats, all concentrations of victorin caused transient closure of the stomata; this occurred earlier with the higher concentrations of victorin. Thus, maximum closure occurred after 1 h at 1·2 units/ml, after 3 h at 0·012 units/ml and after 6 h at 0·00012 units/ml. In the resistant cultivar an increase in resistance above that in the control was observed after 3 h at 1·2 units/ml, at 12 h at 0·012 units/ml, and only by extending the experiment to 24 h, was an increase observed at 0·00012 units/ml. Water soaking of tissue between the vascular bundles was observed after 3 h in blight susceptible leaves given 1·2 units/ml of victorin. Moreover, after 24 h susceptible leaves given 1·2 and 0·012 units/ml of victorin and resistant leaves given 1·2 units/ml of victorin were wilted and chlorotic.

A comparison of fusicoccin and victorin painted on to leaves of susceptible and resistant oats under the same environmental conditions, showed that 1·2 units/ml of victorin consistently increased epidermal resistance of the blight susceptible variety; a transient increase in epidermal resistance of the blight resistant variety was observed after 0·5 h. At 10^{-5} M, fusicoccin opened stomata in both cultivars.

Thus, both absorption and painting of leaves have shown that victorin causes stomata to close. The absorption studies showed that stomatal closure was transitory in blight susceptible leaves and that the transient closure was earlier, the greater the concentration of victorin. Blight resistant leaves were unaffected when painted with victorin and only slowly affected when the toxin was absorbed. Fusicoccin, in contrast to victorin, opened the stomata of oats as of other species (Turner and Graniti, 1969). Moreover, its action was similar in both blight susceptible and resistant cultivars. Clearly, therefore, the mechanism of action of the toxins in the leaf differs in the two diseases (Turner, 1972).

References

Ballio, A., Brufani, M., Casinovi, C. G., Cerrini, S., Fedeli, W., Pellicciari, R., Santurbano, B., and Vaciago, A. (1968b). *Experientia* **24**, 631–635.

Graniti, A. (1964). *In* "Host–parasite relations in Plant Pathology" (Z. Király and G. Ubriszy, Eds), pp. 211–217. Budapest.

Graniti, A., and Turner, N. C. (1970). *Phytopath. medit.* **9**, 160–167.

Luke, H. H. and Wheeler, H. E. (1955). *Phytopathology* **45**, 453–458.

Owens, L. D. (1969). *Science* **165**, 18–25.

Turner, N. C. (1970). *Agron. J.* **62**, 538–541.

Turner, N. C. (1972). *Am. J. Bot.* **59**, (In press).

Turner, N. C. and Graniti, A. (1969). *Nature, Lond.* **223**, 1070–1071.

Turner, N. C. and Parlange, J.-Y. (1970). *Pl. Physiol.* **46**, 175–177.

Wheeler, H. (1969). *Phytopathology* **59**, 1057 (abstract).

Wheeler, H., and Hanchey, P. (1968). *A. Rev. Phytopath.* **6**, 331–350.

Discussion

Scheffer: My experience has been that concentrations of *Helminthosporium victoriae* toxin such as you have used cause no visible effects on resistant oat plants. Damage to such plants at the higher concentrations you used indicates that there were too many impurities in your toxin preparation. I would like to see some of your experiments repeated with a cleaner preparation of the toxin.

Page: With reference to our observations on the influence of adenine in desiccation of potato leaves, and its known hormonal effect, it might prove useful to try adenine and purine analogues in your work.

Can you ensure that there is no effect from shading when your apparatus is in position on a leaf?

Turner: Since the ventilated diffusion porometer is usually clamped on to the leaf for less than 30 seconds, insufficient time elapses for changes in stomatal aperture to occur.

Graniti: Have you painted leaves of plants other than oats with victorin?

Turner: I have not used victorin on any plants other than oats.

31. The Measurement of Water Status in Healthy and Diseased Plants

NEIL C. TURNER

The Connecticut Agricultural Experiment Station, New Haven, Connecticut 06504, U.S.A.

The many references to wilting reported at this Institute clearly indicate the widespread effect of phytotoxins on the transpiration and water status of diseased plants. Wilting is frequently sufficient as an index of phytotoxic activity, but this gives little indication of the mechanisms whereby toxins affect plant hydration. Toxins may cause wilting by irreversible damage to the permeability of the cell membranes, by reversible permeability changes, or by changing the water and vapour flow through the plant. Some techniques which can be used to identify and measure the effect of phytotoxins on plant hydration are briefly reviewed below.

The energy status of the water in the normal plant cell is usually expressed as the total water potential, ψ (Taylor and Slatyer, 1962; Taylor, 1968). Three major components combine to determine the total water potential of a cell, the osmotic potential (π) due to the presence of solutes within the cell, the matrix potential (ψ_m) arising from the imbibitional forces of the colloids in the cell and the forces of capillarity in the cell wall, and the turgor potential (P) developed in the cell in response to the osmotic potential and tensile strength of the cell wall. The ψ_m is usually very small relative to π and P and can be disregarded. Thus the two major components of the water potential usually combine:

$$-\psi = -\pi \mp P \tag{1}$$

The signs indicate the normal condition of each component when referred to pure free water. Equation 1 may be more familiar in the earlier terminology as:

$$DPD = OP - TP \tag{2}$$

where DPD is the diffusion pressure deficit, OP is the osmotic pressure and TP is the turgor pressure. Wilting occurs when ψ becomes smaller than π, allowing P to become zero or negative. The turgor pressure of plant tissue is usually calculated from measurements of ψ and π on the same tissue.

A large number of methods for the measurement of total water potential are available. Only the more commonly used methods are described. For a comprehensive review of the methods and sources of error see Barrs (1968); Boyer (1969a) has a less extensive but useful review.

Measurement of the Total Water Potential

(1) *Thermocouple psychrometry.* This is indisputably the most accurate method used at present. Two versions of the thermocouple psychrometer have been described (Spanner, 1951; Richards and Ogata, 1958). Both methods involve the isolation of a sample in a sealed chamber, and the measurement of the vapour pressure (or relative humidity) when equilibrium has been reached between the gas phase and the sample; the thermocouple measures the wet-bulb depression of the temperature. One important limitation of the technique is that strict temperature control is essential.

(2) *Pressure chamber technique.* This method, which is not temperature sensitive and is readily used in the field or laboratory, uses the pressure chamber described by Scholander *et al.* (1964, 1965). An excised leaf or branch is quickly placed into the chamber with the cut end protruding through the gas-tight seal in the lid of the chamber. Gas pressure from a cylinder of nitrogen is applied to the leaf or twig until the meniscus of the xylem sap returns to the cut surface. The balancing pressure is closely correlated with the leaf water potential in most species (Boyer, 1967), but can be sufficiently in error to warrant calibration against the thermocouple psychrometer whenever a new species is used (Boyer, 1967; Kaufmann, 1968). Although initially used with leafy twigs, the technique has been successfully used with other species, e.g. corn, sorghum, tobacco, pepper, wheat (DeRoo, 1969, 1970; Fischer, in press; Turner *et al.*, 1971).

(3) *Water content.* The water saturation deficit (*WSD*) or relative water content (*RWC*) can be simply measured without recourse to expensive equipment except an accurate balance. Both require the measurement of the field-fresh weight (*FW*), the fully turgid weight (*TW*) and the oven-dry weight (*DW*) of a plant sample. Then:

$$WSD = \frac{(TW - FW)}{(TW - DW)} \times 100$$

$$RWC = \frac{(FW - DW)}{(TW - DW)} \times 100$$

When used to determine the water potential, by calibration against the thermocouple psychrometer, the technique suffers from the disadvantage that small changes in *RWC* produce large changes in leaf water potential. This can lead to a considerable error in the estimation of water potential.

Measurement of Osmotic Potential

(1) *Determination of the freezing point.* The tissue to be sampled is quickly frozen to break the cell membranes and the cell sap expressed under pressure. Currier (1944) describes a method for determination of the freezing point depression of small volumes of cell sap. Commercial cryoscopes are also available.

(2) *Determination of the vapour pressure.* This technique also uses expressed cell sap from quickly frozen samples of tissue. In this case the change in vapour pressure due to the presence of solutes in the cell sap is measured. The thermocouple psychrometer described above can be used for this purpose. Vapour pressure osmometers are also available commercially.

(3) *Pressure chamber technique.* Scholander *et al.* (1964, 1965) showed that the pressure chamber can be used to determine the osmotic potential in living tissue. The chamber is used to expel liquid progressively from the xylem; the volume of liquid expelled (v) is then plotted as a function of the pressure (p) required to expel each increment of water. The plot of $1/p$ against v will be linear when the turgor pressure is zero; interpolation of the curve to the point where the applied pressure just causes zero turgor, corresponds to the π of the cells (Boyer, 1969a).

Other methods for determining π are reviewed by Barrs (1968).

Use of Pressure Chamber for Determination of Membrane Damage

The pressure chamber (Scholander *et al.*, 1964, 1965) can also be used to indicate whether membrane permeability has been affected by a plant toxin. The chamber is used to express xylem sap under pressure as described for the determination of osmotic potential. The pressure-volume plot is compared for healthy and diseased tissue; if the two have a similar slope the cell permeability is unaffected, but if the slopes differ, the toxin has damaged the cell membrane. Hammel and Waggoner used this technique to show that *Ceratocystis ulmi* (Buism.) Moreau did not change the permeability of elm leaves (Dimond, this volume, p. 292).

Resistance to Water Transport in the Plant

In the transpiring plant, water movement can be considered to occur along a thermodynamic continuum between the soil and atmosphere. The difference in water potential (ψ) between the atmosphere and soil acts as the driving force for water movement, but in the continuum the water encounters a series of resistances (r). The rate of water movement, T, is given by:

$$T = \frac{\psi_{\text{soil}} - \psi_{\text{root surface}}}{r_{\text{soil}}} = \frac{\psi_{\text{root surface}} - \psi_{\text{xylem}}}{r_{\text{root}}} =$$

$$\frac{\psi_{\text{xylem}} - \psi_{\text{leaf}}}{r_{\text{xylem}} + r_{\text{leaf}}} = \frac{\psi_{\text{leaf}} - \psi_{\text{air}}}{r_{\text{stomata}} + r_{\text{boundary layer}}}$$

An increase in the resistance to water flow will either decrease the transpiration or increase the potential gradient in the appropriate segment of the continuum. Provided a phytotoxin does not break down the osmotic system of the plant, measurement of the water potential or the resistance to flow can be used to indicate the affected portion of the pathway and determine the magnitude of the effect. The interchangeability of the two measurements is demonstrated in Fig. 1, taken from data by Turner and Graniti (1969) with fusicoccin. Fusicoccin changes the stomatal resistance of treated plants; concurrent measurements of stomatal resistance and leaf water potential showed a remarkably similar relative change in both parameters.

Techniques for the measurement of water potential have been described. The following techniques can be used to measure plant resistances.

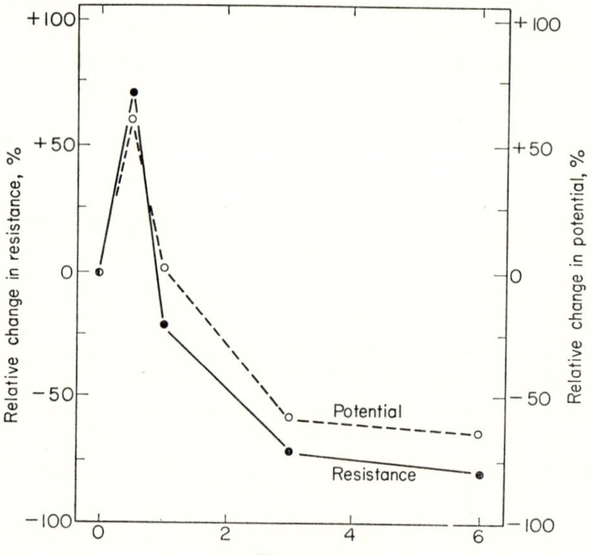

FIG. 1. Relative changes with time of stomatal resistance and leaf water potential measured in bean leaves after treatment with 10^{-5} M fusicoccin (see Turner and Graniti, 1969).

(1) *Stomatal resistance.* One method that has found wide acceptance is the portable diffusion porometer. A small chamber is clamped on to a leaf and the rate of diffusion of water vapour from the leaf into the chamber is measured over a known humidity range with a lithium chloride sensor (Wallihan, 1964). Several modifications and calibration procedures have been published (Van Bavel *et al.*, 1966; Kanemasu *et al.*, 1969; Kramer, 1969; Byrne *et al.*, 1970; Turner and Parlange, 1970). Other methods of measurement have recently been reviewed by Meidner and Mansfield (1968) and Barrs (1968).

(2) *Leaf, stem and root resistances.* Resistances to transpiration within the plant can be calculated from the measured water flow through either the roots, stems, or leaves, or any combination of the three, under pressure. Jensen *et al.* (1961) described a chamber for this type of study. Dimond (1966) also used this technique to confirm calculations of the resistance from measured xylem vessel radii.

Boyer (1969b) has described, in a preliminary report, a simple method of measuring the plant resistances from measurements of the recovery of water potential within the leaf after an induced deficit. The recovery of the leaf water potential with the plant in the soil, the plant alone, the stem and leaf alone, and the leaf alone, are followed, and the resistances within the plant calculated from the half-time for recovery.

Potentials within the stem can also be measured from the leaf water potentials, measured with either the pressure chamber or the thermocouple psychrometer, if the leaf transpiration is prevented by covering with polyethylene and aluminium foil. If the water movement through the stem is known, the resistances within the stem can be calculated from the measured potentials (Begg and Turner, 1970).

For a more comprehensive review of water and water movement in plants the reader is referred to recent books by Slatyer (1967) and Kramer (1969).

References

Barrs, H. D. (1968). *In* "Water Deficits and Plant Growth" (T. T. Kozlowski, Ed.), Vol. 1, pp. 235–368. Academic Press, New York.

Begg, J. E. and Turner, N. C. (1970). *Pl. Physiol.* **46**, 343–346.

Boyer, J. S. (1967). *Pl. Physiol.* **42**, 133–137.

Boyer, J. S. (1969a). *A. Rev. Pl. Physiol.* **20**, 351–364.

Boyer, J. S. (1969b). *Science*, **163**, 1219–1220.

Byrne, G. F., Rose, C. W., and Slatyer, R. O. (1970). *Agr. Meteorol.* **7**, 39–44.

Currier, H. B. (1944). *Pl. Physiol.* **19**, 544–550.

DeRoo, H. C. (1969). *Agron. J.* **61**, 969–970.

DeRoo, H. C. (1970). *Tob. Sci.* **14**, 105–106.

Dimond, A. E. (1966). *Pl. Physiol.* **41**, 119–131.

Fischer, R. A. (In press). *In* "Plant Response to Climatic Factors." *Proceedings of the Uppsala Symposium, Sept. 1970* UNESCO.
Jensen, R. D., Taylor, S. A., and Wiebe, H. H. (1961). *Pl. Physiol.* **36**, 633–638.
Kanemasu, E. T., Thurtell, G. W., and Tanner, C. B. (1969). *Pl. Physiol.* **44**, 881–885.
Kaufmann, M. R. (1968). *Forest Sci.* **14**, 369–374.
Kramer, P. J. (1969). "Plant and Soil Water Relationships: A Modern Synthesis." McGraw-Hill, New York.
Meidner, H. and Mansfield, T. A. (1968). "Physiology of Stomata." McGraw-Hill, London.
Richards, L. A. and Ogata, G. (1958). *Science* **128**, 1089–1090.
Scholander, P. F., Hammel, H. T., Hemmingsen, E. A., and Bradstreet, E. D. (1964). *Proc. natn. Acad. Sci. U.S.A.* **52**, 119–125.
Scholander, P. F., Hammel, H. T., Bradstreet, E. D., and Hemmingsen, E. A. (1965). *Science* **148**, 339–346.
Slatyer, R. O. (1967). "Plant–Water Relationships." Academic Press, New York.
Spanner, D. C. (1951). *J. exp. Bot.* **11**, 145–168.
Taylor, S. A. (1968). *In* "Water Deficits and Plant Growth" (T. T. Kozlowski, Ed.), Vol. 1, pp. 49–72. Academic Press. New York.
Taylor, S. A. and Slatyer, R. O. (1962). *UNESCO Arid Zone Res.* **16**, 339–349.
Turner, N. C. and Graniti, A. (1969). *Nature* **223**, 1070–1071.
Turner, N. C. and Parlange, J.-Y. (1970). *Pl. Physiol.* **46**, 175–177.
Turner, N. C., DeRoo, H. C., and Wright, W. H. (1971). *Conn. Agric. Exp. Station, Special Soils Bull.* No. xxxiii.
Van Bavel, C. H. M., Nakayama, F. S., and Ehrler, W. L. (1965). *Pl. Physiol.* **40**, 535–540.
Wallihan, E. F. (1964). *Pl. Physiol.* **39**, 86–90.

32. Effect of Fusicoccin on Abscission, Cellulase Activity and Ethylene Production in Citrus Explants*

ALBERT W. FELDMAN

University of Florida Citrus Experiment Station, IFAS, Lake Alfred, Florida, U.S.A.

AND

ANTONIO GRANITI AND LORENZO SPARAPANO

Istituto di Patologia vegetale, University of Bari, Italy

The effect of fusicoccin (Ballio *et al.*, 1968), a phytotoxic carbotricyclic diterpene glucoside produced by *Fusicoccum amygdali* Del. on abscission, ethylene production and cellulase activity in citrus explants was determined. Activity of fusicoccin was compared with 8 accelerators or inhibitors of abscission.

Abscission was retarded for 65–72 h with fusicoccin at 0·1 μg/explant and in this respect fusicoccin was more effective than IAA, naphthalene-acetic acid, and kinetin. IAA often accelerated abscission in explants from 2-month-old leaves. Fusicoccin was less effective than either cycloheximide, an inhibitor of protein synthesis, or 2,4-D both of which inhibited abscission for 120 h at 0·025 μg/explant.

Use of 0·4 μg/explant of either fusicoccin or 2-chloroethylphosphonic acid, a compound that readily releases ethylene in plant tissues, increased ethylene evolution sevenfold within 24 h; this continued for another 48 h and was associated with an increase in cellulase activity. Ethylene evolution was also rapidly increased by 2,4-D and less rapidly by cycloheximide, but these substances decreased cellulase activity and retarded abscission.

Increase in abscission was associated with an increase in cellulase activity except for fusicoccin-treated explants in which inhibition of abscission appears to be caused by several interacting factors, including toxicity to, and impairment of the sensitivity of the tissues in the separation zone to stimulation of abscission by ethylene.

* A paper on this subject has been published in *Physiological Plant Pathology*, **1**, 115–122, 1971.

References

Ballio, A., Brufani, M., Casinovi, C. G., Cerrini, S., Fedeli, W., Pellicciari, R., Santurbano, B., and Vaciago, A. (1968). *Experientia* **24**, 631–635.

Discussion

Wood: How is it possible for cellulase to cause abscission?

Feldman: Cellulase activity has been shown to contribute to cytolysis and cell disintegration in the abscission zone and there is now considerable evidence to indicate that the activity of this enzyme is mediated by ethylene. (R. H. Horton and D. J. Osborne (1967). *Nature* **214**, 1086–1088; A. Ratner, R. Goren, and S. P. Monselise (1969). *Pl. Physiol.* **44**, 1617–1623.)

Morpurgo: I should like to refer to experiments done in collaboration with Dr. Bandiera following a suggestion by Professor Marrè. We have found that fusicoccin can replace auxin in media supporting independent growth of carrot callus. But it does not support growth of callus of *Nicotiana tabacum* and it does not stimulate callus formation from seeds of *N. tabacum*, and *Solanum lycopersicum*. But it does enhance growth of plantlets until the development of numerous leaves when the plantlets suddenly wilt.

33. Growth Stimulation of Etiolated Pea Tissue by Fusicoccin and Some Derivatives*

ALESSANDRO BALLIO

Laboratorio di Chimica delle Sostanze Naturali, Istituto Chimico, University of Naples, Italy

AND

FRANCESCO POCCHIARI, SERENA RUSSI AND VITTORIO SILANO

Laboratori di Chimica Biologica, Istituto Superiore di Sanità, Rome, Italy

A previous investigation described a remarkable effect on water balance and transpiration of tomato seedling cuttings by fusicoccin, the phytotoxic glucoside produced in submerged culture by *Fusicoccum amygdali* Del. (Ballio *et al.*, 1968).

Similar effects on etiolated pea tissue by fusicoccin and a series of related compounds have now been observed. Whereas the dry weight of cuttings of etiolated pea seedlings is unaffected by fusicoccin, the wet weight increases fourfold in the presence of 1 μg/ml of fusicoccin. This effect is strictly dependent on the oxygen tension in the gas phase.

The fusicoccin derivatives which were tested caused either similar, lesser or no effects. This indicates that certain structural features of fusicoccin are not essential for producing an increase of water absorption in cuttings of etiolated pea seedlings.

Because during this investigation evidence was obtained that the apical section of the first distal internode is the part most sensitive to fusicoccin, quantitative measurements were made of its effect on isolated internodes of etiolated pea seedlings. Increase of the length and wet weight of internodes was stimulated by fusicoccin. This effect too is influenced by the oxygen tension in the gas phase.

An early effect induced by fusicoccin is an increase in the plasticity of the tissue as estimated by measuring with an appropriate device the irreversible deformation of etiolated pea internodes subjected to a weight.

* A paper on this subject has been published in *Physiol. Pl. Path.* **1**, 95–104 (1971).

These results, as well as those of another physiological, and biochemical investigation (Marrè *et al.*, this vol. p. 417), suggest that fusicoccin can be considered in some respects as an auxin-like substance.

Reference

Ballio, A., Graniti, A., Pocchiari, F., and Silano, V. (1968). *Life Sci.* **7**, 751–760.

34. On the Relation between the Growth Promoting and the Toxic Action of Fusicoccin

ERASMO MARRÈ, PIERA LADO
AND FRANCA RASI CALDOGNO

Laboratorio di Fisiologia Vegetale, Istituto di Scienze Botaniche, Università di Milano and Centro di studio del C.N.R. per la biologia cellulare e molecolare delle piante, Milano, Italy

Fusicoccin, the toxic agent produced by *Fusicoccum amygdali* Del. (Graniti, 1962; Ballio *et al.*, 1968) induced rapid wilting of leaves, followed by dehydration, probably as a consequence of increased transpiration through stomata. Fusicoccin also stimulates water uptake and cell enlargement in excised leaf fragments and etiolated pea internode segments. Recent work in our laboratory on the characterization of the growth promoting action of fusicoccin is summarized below.

Experiments on Pea Internode Segments

In the pea internode segment test, cell enlargement induced by fusicoccin is accompanied by irreversible stretching of the cell wall. Optimum concentration of fusicoccin ($1{\cdot}4 \times 10^{-5}$ M) induces an effect on cell enlargement about twice that induced by IAA at its optimum concentration (10^{-5} M). Concentrations of fusicoccin and IAA giving the same effects on cell elongation also have similar effects on increase in cell volume. An important difference between the two dose/response curves is that whereas IAA at concentrations above the optimum rapidly becomes inhibitory, this does not happen with concentrations of fusicoccin up to 10^{-4} M.

Inhibitors of glycolysis (10^{-2} M NaF), of RNA and protein synthesis (actinomycin D, cycloheximide) and uncouplers of phosphorylation (2,4 DNP) all severely decreased the effects of fusicoccin and IAA on cell elongation. At optimum concentrations IAA stimulated growth by *c.* 200%, respiration by *c.* 20%, and increased glucose-6-phosphate and pyruvate concentrations by *c.* 20% and *c.* 100% respectively. Fusicoccin at optimum concentration induced larger responses of all of these parameters.

10^{-5} M IAA decreased slightly but consistently the rate of dissimilation to CO_2 of C–1 labelled glucose. Here again, fusicoccin induced a similar but much larger effect.

Experiments on Leaf Disks

In tomato leaf disks fusicoccin (10 γ/ml) significantly stimulated water uptake whereas IAA, over a wide range of concentrations, did not affect this process.

No significant differences in osmotic pressure were found between fusicoccin-treated and control leaf disks. Even in the presence of 0·03 M mannitol, water uptake in the presence of fusicoccin was much larger than that of controls in water. This suggests that fusicoccin acts by increasing cell plasticity rather than by increasing the osmotic potential.

The data on the physiology and the biochemistry of the effects of fusicoccin on the pea internode segments are interpreted as showing that a consistent part of the mechanism of action of fusicoccin and one of the natural auxins on growth and metabolism are in common even if the results suggest a different primary site of action for the two classes of compounds.

In leaf disks stimulation of water and uptake and cell enlargement by fusicoccin seems to depend on an effect of the toxin in increasing cell wall plasticity presumably by the same mechanism involved in promotion of cell enlargement in pea internodes.

This supports the hypothesis that even the wilting and the dehydration of fusicoccin-treated whole leaves depends on modifications of the plasticity of the walls of leaf cells (and possibly of the stomatal cells), so as to prevent their normal regulation of water evaporation from internal tissues.

Discussion

Mantle: At what concentration of fusicoccin do you observe the cell extension effects of this substance?

Marrè: In the case of the pea internode segments a significant effect is already clear at 5×10^{-8} M, and the optimal concentration is from 10^{-5} M to 10^{-4} M. Using higher concentrations is made difficult by the poor solubility of the compound. In the case of leaf disks the optimal concentration is around 10^{-5} M.

Mantle: Do you consider that the cell extending property of fusicoccin is directly involved in the process of wilting in which it has been suggested that stomatal opening in excess is the contributory factor?

Marrè: It seems likely that a change of plasticity of the cell walls of the stomatal cells, or, alternatively, a change of the water balance between stomatal cells and neighbouring tissues, due to a differential effect of the toxin on the water uptake capacity of the various leaf tissues, might influence the mechanisms of opening of the stomata, and thus transpiration.

Matta: What is the effect of fusicoccin on IAA content?

Marrè: As far as I know there are no direct data in this regard. However, the hypothesis that fusicoccin stimulates cell enlargement by increasing the level of active auxin seems to me quite unlikely. In fact, the effect of fusicoccin on this process, in the pea internodes, is twice as much as that of IAA at optimal concentration; and when both fusicoccin and auxin (optimal) are present, the response is equal to that to fusicoccin alone. Also, in the case of metabolic responses (such as O_2 uptake stimulation, increase in the glucose-6-P and pyruvate levels, etc.) we found that the effects of fusicoccin are about double those of auxin at the optimal concentration. Moreover, the fact that fusicoccin markedly stimulates water uptake in leaf disks, whereas none of the auxins we tested (IAA, 2-4 D, pClPAA) was active over a wide concentration range, makes it difficult to explain the effect of fusicoccin on cell enlargement as due to change of the endogenous auxin level or activity.

Goodman: Has anyone in either the Italian or British group done experiments to measure plastic and elastic wall "stretch" as caused by fusicoccin or IAA?

Marrè: Preliminary experiments carried out by the Rome group and ourselves indicate that both the plastic and the elastic compounds are affected in the same way by fusicoccin and by auxin. These results, however, are qualitative rather than quantitative.

Mantle: What techniques were used to measure cell wall stretching and osmotic pressure?

Marrè: Osmotic pressure was calculated from measurements of the freezing point of the cell sap. Irreversible stretching of the cell wall in the internode segments was evaluated by measuring the residual increase of fresh weight after plasmolysis in a hypertonic mannitol solution.

Mantle: Would the concentrations of fusicoccin used wilt the plant in 4–6 h?

Also, since the symptom is reversible, could the effect on the cell wall be reversible?

Marrè: Although we did not run experiments under strictly comparable conditions, the concentrations stimulating water uptake (optimum around 10^{-5} M) are of the same order as those inducing rapid wilting of whole leaves.

As to the point of reversibility, if you accept (what is far from being demonstrated) that wilting is a rather remote consequence of structural changes of the cell wall, and that you can observe wilting only under a condition of water stress, then you might get at the same time reversible wilting and irreversible stretching of the cell wall. When the water stress is removed, wilting would disappear, even if the stretching remained.

35. Role of a Toxin Produced by *Myrothecium roridum* in Myrothecium Leaf Spot of Red Clover

FELIX L. LUKEZIC AND BARRY M. CUNFER

Department of Plant Pathology, The Pennsylvania State University, University Park, Pennsylvania 16802, U.S.A.

Myrothecium leaf spot usually begins as small, dark, circular, water-soaked lesions that later become necrotic and irregular as they enlarge. At a later stage, fungal sporodochia are formed on both surfaces of the lesion. However, when leaves are inoculated with sporodochial washings, numerous minute necrotic spots often quickly appear. Many of these do not enlarge, and no fungal growth is associated with them, suggesting that a toxic mechanism may be involved. A phytotoxic metabolite obtained from the dark green fluid associated with sporodochia grown on autoclaved red clover leaves was purified and partially characterized. The compound produces necrosis on red clover leaves and other species. It is stable after storage for 30 days at 5° and 30°C and retains its activity after autoclaving. The substance is neutral, aliphatic, non-proteinaceous, and contains a carbonyl bond. It plays a part in pathogenesis by creating injuries on which spores may germinate and then penetrate the leaf. The toxin may also be more directly involved in the disease because it is extracted from infected red clover leaves with petroleum ether.

The infrared spectrum and phytotoxicity of the compound are similar to necrocitin a antifungal, phytotoxic metabolite isolate from *Myrothecium roridum* Tode *ex* Fr. culture filtrates.

References

Cunfer, B. M., Graham, J. H., and Lukezic, F. L. (1969). *Phytopathology* **59**, 1306–1309.

Cunfer, B. M. and Lukezic, F. L. (1970). *Phytopathology* **60**, 341–344.

Pawar, U. H., Deshmukh, P. V., and Thirumalachar, M. J. (1965). *Hindustan Antibiotics Bull.* **8**, 59–63.

36. Role of Toxin in the Leaf Spot Disease of Alfalfa caused by *Leptosphaerulina briosiana*

LEIF SUNDHEIM

The Agricultural College of Norway, 1432 Vollebekk, Norway

Leptosphaerulina briosiana (Poll.) Grah. *et* Luttr. causes leaf spots on several forage legumes. In red and white clover, black, necrotic spots appear on leaves and the disease is named pepper spot. In alfalfa (*Medicago sativa* L.), necrotic spots develop with a distinct chlorotic halo around the margins. This halo only develops at high light intensities. In the field it occurs mainly on the top leaves of infected plants whereas the halo is usually absent from the lower leaves. In inoculation experiments under controlled environment conditions chlorosis developed at 2000 but not at 400 ft-candles.

Histological studies of sections of naturally and artificially inoculated leaves indicated that mesophyll cells were affected ahead of the hyphae. Unsuccessful attempts were made to extract biologically active materials from infected leaves.

The fungus was grown in a liquid medium on a shaker at 25°C for 10 days and the mycelium was harvested by filtration. The cell free filtrate was tested for biological activity by placing 0·1 ml in rubber wells placed on leaflets. Puncture of the epidermis facilitated entry of the filtrate into the leaves.

Cell-free filtrates from cultures of the fungus caused spots that were visible 12 h after application. The spots enlarged rapidly in size and in 48 h they were 2–5 mm in diameter. The lesions were similar in shape and size to the necrotic spots typical of the disease. However, they lacked the chlorotic margins of the natural lesions. Filtrates from non-inoculated cultures had no effect on leaves.

The culture filtrate was evaporated to dryness under vacuum at 35°C. The residue was taken up in water, centrifuged at 3500 g for 30 min and the supernatant solution tested for biological activity. The toxic principle was water soluble but insoluble in ethanol, acetone and ether. Its toxicity was not lost by heating to 100°C for 10 min.

In further attempts at purification a precipitate made by adding lead acetate was removed from solution by centrifugation and the supernatant was passed through an anion exchange column. The eluted substance was concentrated, spotted on paper and descending chromatograms were run with a pyridine–butanol–water solvent.

Fractions were eluted from the paper with water and tested for activity on leaves. The active fraction corresponded to a spot on the paper which reacted strongly with ninhydrin indicating the presence of free amino groups. Thus the toxin could be a peptide containing molecule. No reactions for phenols, sugars, aldehydes or indoles were obtained in the area on the chromatogram containing the active principle.

In summary, the toxic principle from culture filtrates of *L. briosiana* can produce some but not all of the symptoms of the disease which this fungus causes on *Medicago sativa*.

Reference

Sundheim, L. and Wilcoxson, R. D. (1965). *Phytopathology* **55**, 546–553.

Discussion

Lukezic: Did you observe any effect of light on the activity of the culture filtrate ?

Sundheim: High light intensities are needed for symptom development both in artificial inoculation experiments and under natural conditions in the field. Whether light affects the symptoms caused by culture filtrate has not been studied.

37. The *Cercospora beticola* toxin

ECKART SCHLOSSER

Department of Crop Production and Protection,
Faculty of Agricultural Sciences, American University of Beirut,
Lebanon

The sugar beet leaf spot disease, caused by *Cercospora beticola* Sacc., is widespread where warm and humid conditions prevail. Numerous spots develop on the outer leaves and often coalesce to form larger necrotic areas, finally resulting in dead, dried leaves. The necrotic leaf spots are often surrounded by a halo of yellow-green tissue which is free of fungal mycelium. It appears that before the fungus colonizes tissue it preconditions it by secreting one or more toxic compounds which act in advance. On the basis of this observation I tested the hypothesis that *C. beticola* can produce toxins which might play a rôle either in the symptom development or in pathogenesis in this host–parasite combination.

In vitro the fungus produces a non-host specific, yellow coloured toxin which was partially purified and characterized. It is synthesized by all strains of the pathogen that were tested and is apparently the only toxin produced. The toxin has antibacterial but not antifungal properties. At low concentrations (1–10 μg/ml) the toxin inhibits root formation of tomato cuttings and induces necrotic lesions on sugar beet leaves similar to those caused by the fungus. The level of toxin synthesis by *C. beticola* strains appears to be linked with the difference in their potential as parasites. The toxin forms rather stable complexes with Fe^{+++} and Co^{++} ions which indicates that it might interfere with physiological processes in which these ions are involved. All attempts to demonstrate toxin production *in vivo* have failed so far.

Discussion

Mussell: Does infection with sugar beet yellows virus break tolerance to both the pathogen and the toxin or only to the pathogen?

Schlosser: Sugar beet yellows breaks down the tolerance to the pathogen. The reaction towards the toxin is not known.

Scheffer: What concentrations of the toxin are required to damage tolerant and susceptible plant cells?

Schlosser: For susceptible plant cells 5–20 $\mu g/ml$ were sufficient to cause necrosis. With resistant cells 200–500 $\mu g/ml$ were required to induce visible damage.

Wood: Did cultures stored for a long time retain pathogenicity?

Schlosser: There may have been changes in pathogenicity although we did not notice any striking effects. Cultures which have been in continuous subculture for five years readily infected susceptible sugar beet varieties with no differences from cultures freshly isolated.

38. Phytotoxins in the Brown Spot Disease of Tobacco Caused by *Alternaria tenuis**

CHARLES E. MAIN AND R. W. PERO†

Department of Plant Pathology, North Carolina State University, Raleigh, North Carolina 27607, U.S.A.
Present address of junior author:

Brown spot is a foliar disease of tobacco caused by *Alternaria tenuis* Nees. Lesions consist of invaded, necrotic centres surrounded by non-invaded chlorotic and green pre-halo zones beyond which is healthy tissue. The gradient of biochemical and physiological alterations within and near these lesions have been proposed as a model system to study lesion development and the nature of tolerance to this disease (Main, 1969).

The chlorosis indicates production of fungal metabolites that diffuse outwards from the invaded centre. Three dibenzopyrone phytotoxins, alternariol monomethyl ether, alternariol and altenuene, have been isolated from cultures of pathogenic isolates of *A. tenuis* and characterized by co-chromatography, melting point, infra-red and ultraviolet absorption spectra, proton magnetic resonance and mass spectrum techniques. All three compounds induced chlorosis when bioassayed by injection into tobacco leaf laminae. Chlorosis intensity was related to toxin concentration between 25–1000 μg/ml. Necrosis did not occur even at concentrations above 1000 μg/ml. Alternariol monomethyl ether was detected in leaf tissue one hour after injection, but was not recovered after 72 h indicating that it may be metabolized or degraded by the leaf as chlorosis developed (Pero and Main, 1970). The toxin may be a precursor giving rise to polyphenols because free phenols decreased in the lesion compared to their concentrations in healthy tissue. Acetone powder extracts of halo and pre-halo tissues have been shown to oxidize tyrosine, dihydroxyphenylalanine and chlorogenic acid resulting in the formation of brown pigments (Spurr and Main, 1969).

* Paper No. 3415 of the Journal Series of the North Carolina State University Agricultural Experiment Station, N.C. 27607, U.S.A.

† National Institute of Environmental Health Sciences, Cell Biology Section, Research Triangle Park, North Carolina 27709, U.S.A.

Attempts to detect the phytotoxins within the natural lesion have been unsuccessful. Freeze-dried tissues of model lesions were extracted with hot acetone, absorbed on to Silica gel G, chromatographed in columns using a tetrahydrofuran (THF): benzene elution series, and analysed for the three toxins by thin-layer (TLC) and gas chromatography (GC) methods (Pero and Owens, 1971). Certain fluorescent components of tobacco tissue separated with R_F values similar to those of the purified toxins. However, preparative TLC of the extracts in the appropriate solvent systems (alternariol monomethyl ether, 5:95 v/v ethanol in benzene; alternariol, 20:80 v/v/THF in benzene; and altenuene, 50:50 THF in benzene), cochromatography with standard phytotoxins, and GC analysis using an OV-17 stationary phase, showed that concentrations of the toxins in the lesions were less than 1 μg/g leaf tissue. The details of methodology are presented elsewhere (Pero *et al.*, 1970).

A fourth water-soluble phytotoxin has been isolated and appears to be a large molecular weight polysaccharide. Chlorosis of tobacco leaf lamina occurred at 0·5 μg/ml indicating high activity. Necrosis occurred rapidly at higher concentrations, and followed chlorosis at lower concentrations. The toxin was not antibiotic to *Chlorella pyrenoidosa* Pring. or several bacterial species, and did not inhibit germination of *A. tenuis* spores at concentrations of 1000 μg/ml.

Varietal tolerance to the brown spot disease is related to the extent of chlorotic halo surrounding the necrotic lesion. During commercial flue-curing of tobacco, the halo and pre-halo zones become necrotic and contribute to the disease damage which affects quality and marketability of the cured leaf. The extent of halo formation is influenced by varietal tolerance, toxin production by *A. tenuis*, leaf age and environmental factors. The model lesion system should provide an experimental approach to elucidate the role of each of the parameters involved.

References

Main, C. E. (1969). *Phytopathology* **59**, 1040.

Pero, R. W. and Main, C. E. (1970). *Phytopathology* **60**, 1570–1573.

Pero, R. W. and Owens, R. G. (1971). *Anal. Biochem.* (In Press.)

Pero, R. W., Owens, R. G., Dale, S. W., and Haven, D. (1971). *Biochem. biophys. Acta* **230**, 170–179.

Spurr, H. W., Jr. and Main, C. E. (1969). *Phytopathology* **59**, 1051.

Discussion

Barbier: You have described the isolation of three phenolic phytotoxins from *Alternaria tenuis*. What about the so-called "Tentoxin"?

Main: Tentoxin is a cyclic-tetrapeptide (MW 414) described by Dr. George Templeton of The University of Arkansas. We have not detected tentoxin in culture filtrates of our *Alternaria tenuis* isolates. Two of our phenolic phytotoxins, e.g. alternariol (M.W. 258) and alternariol monomethyl ether (M.W. 272), have been described previously. The third toxin, altenuene (M.W. 292), is a new compound with a hydroxyl group at the 3′ position and only one double bond located 1′–6′ in the primed ring. All three are dibenzopyrone derivatives.

Mirocha: There is a great deal of similarity between the metabolites, i.e. alternariol, the methoxy ether and the H.I.F. compound produced by an isolate of *Alternaria* of the *tenuis* group, that we are studying. The characteristics of the H.I.F. material appears similar to those of our mycotoxin which is active against mammalian cells.

Stoessl: The biological activity of hydrogenation products has been commented upon earlier in this symposium. You have isolated a somewhat unusual compound in altenuene. I want to ask whether such hydrogenations of aromatic nuclei are common in *Alternaria* species since we in our Institute obtained some tetrahydro-anthraquinones from *A. solani* some time ago.

Main: The biosynthetic relationship of altenuene to other *Alternaria* metabolites is not known. However, Gatenbeck (*Acta. chem. Scand.* **19**, 65–71, 1965) proposed alternariol synthesis as head-to-tail condensation of malonate units into polyketo chains which cyclize and aromatize immediately. The presence of a hydrogenated ring precursor has not been noted. Earlier work by Thomas (*Biochem. J.* **80**, 234–240, 1961) supports these findings. In fact, Thomas, and others (*J. agr. Food Chem.* **18**, 260–267, 1970) suggest that structural modifications of alternariol were probably the result of enzyme induction due to the presence of alternariol itself.

39. Production of Phytotoxins by *Phytophthora nicotianae* B. de Haan var. *parasitica* (Dast.) Waterh.

ALESSANDRO BALLIO AND LUCIANA GIANANI

Laboratorio di Chimica delle Sostanze Naturali, Istituto Chimico, University of Naples, Italy

AND

RAFFAELE BORRELLI, ANTONIO BOTTALICO AND ANTONIO GRANITI

Istituto di Patologia vegetale, University of Bari, Italy

In *Phytophthora* rots of citrus, death of cells occurs beyond the region of advancing hyphae. This suggests that the pathogen acts by secreting toxins.

On the basis of previous research, phytotoxins were produced *in vitro* by growing an isolate of *P. nicotianae* B. de Haan var. *parasitica* (Dast.) Waterh. on a synthetic medium containing sucrose, L-asparagine, mineral salts, and vitamins (Graniti, 1969). Shaken flasks and 20 or 35 litre stirred fermenters were used. Phytotoxicity of culture filtrates reached its maximum after shaken cultures had grown for 8 days at 29°C or when submerged cultures had grown for 60–90 h at 27°C in stirred fermenters.

Diluted filtrates taken up by tomato cuttings readily caused marginal and interveinal necrotic spots on leaves.

Treatment of culture filtrates with 70% ethanol removed material that was inactive; the supernatant solution retained full activity after dialysis, and also after extraction with organic solvents. This indicated that hydrophilic, macromolecular substances were responsible for toxicity. Consequently, column chromatography on Sephadex G 25 "Fine" was chosen for the first stages of purification. Further fractionation on DEAE-Sephadex gave two groups of active fractions which caused different symptoms on leaves of tomato cuttings.

Similar results were obtained when CM-Sephadex was used for the second step in purification. It thus appears that more than one phytotoxin is present in filtrates of cultures of *P. parasitica*.

Studies are in progress to elucidate the chemical nature of the active compounds; there is good evidence that at least one of them is a polysaccharide.

References

Graniti, A. (1969). *In* "Proc. First Intern. Citrus Symposium" (H. D. Chapman, Ed.), Vol. III, pp. 1187–1190. University of California, Riverside, U.S.A.

Discussion

Grossmann: Are you able to exclude the possibility that your toxins contain proteins. I believe the *Phytophthora infestans* produces toxins that may be proteins.

Graniti: Professor Ballio is better placed to answer this question.

Ballio: As reported in the slides referring to the chromatographic fractionations of the active material, we have always followed the absorbancy at 260 nm during elution. Results so far obtained indicate the possibility that one of the purified toxins is a glycopeptide.

Mirocha: Will you please comment on correlations and similarities between symptoms caused by fractions from culture filtrates on the tomato cuttings used in your bioassays and those found in citrus plants, the natural host of the pathogen?

Graniti: Evidence that some toxic material spreads in tissues of citrus bark infected by *Phytophthora nicotianae* var. *parasitica* was easily obtained. Samples of bark were removed at the margin of the lesion and placed in a plasmolysing solution containing a vital stain. In the region behind the advancing hyphae, the host cells appeared to have lost their ability to plasmolyse. The hyphae themselves appeared to be embedded in dead tissue.

The active fractions from culture filtrates of the pathogen caused the appearance of necrotic spots on leaf blades when supplied to tomato cuttings through their vascular system.

Therefore, if a toxin is involved, its effect is probably necrotic.

40. Toxin production by *Phytophthora megasperma* Drechsl. var. *sojae* Hild.

JACK D. PAXTON

Department of Plant Pathology,
University of Illinois, Urbana, Illinois 61801, U.S.A.

Phytophthora root and stem rot of soybeans is characterized by a rapid wilting of infected plants. Inoculated susceptible plants, held under 100% relative humidity at 21°C, wilt within 24 h. The margins of affected leaves often collapse and then become dry through loss of water.

Phytophthora megasperma Drechsl. var. *sojae* Hild. race 3 was cultured in soybean broth (5 soybeans/100 ml water). In this medium enough toxin accumulates after 7 days to cause wilting of one-week-old seedlings of cultivar Harosoy 63 soybeans. Wilting also occurs in plants without roots indicating that the toxin effect is not primarily on the root system.

This toxin is water soluble, stable at 121°C for 20 min and does not pass through dialysing membranes. Attempts are now under way to purify and identify it.

Discussion

Graniti: Do the pinching symptoms at the stem ends depend on the concentration of culture filtrate? Do they appear only when undiluted filtrates are used?

Paxton: I think that the pinching symptoms shown are an artefact of culture conditions since they are not part of the syndrome.

I did not test for concentration dependence.

41. A host-specific toxin produced by *Cochliobolus heterostrophus*

VIGGO SMEDEGÅRD-PETERSEN

Department of Plant Pathology,
The Royal Veterinary and Agricultural University,
DK-1871 Copenhagen V, Denmark

Certain isolates of *Cochliobolus heterostrophus* Drechsl. which induce blight reactions on corn produce a host-specific pathotoxin. Blighting symptoms are characterized initially by a diffuse, generalized water soaking and chlorosis of leaf tissue apparent 2–3 days after inoculation, then rapid necrosis of the affected tissue. Filtrates were obtained from cultures grown on Fries' modified medium with 2% yeast extract. Toxin production was measured by dilution end point methods using a seedling bioassay measuring inhibition of root elongation and detached leaves floated on culture filtrates. Minimum pH of filtrates was correlated with maximum mycelial growth and occurred 9–11 days after inoculation. Maximum toxin production occurred 4–6 days later. Culture filtrates and extracts from diseased plants induced all disease symptoms and exhibited the same specificity as the pathogen to susceptible and resistant lines of corn. *Sorghum vulgare* Pers. and *S. vulgare* var. *sudanese* (Piper) Hitch. were susceptible to the isolates and their toxins whereas eight other gramineous species were resistant to both. Toxin production is under genetic control in that ascospore progeny with greater or less capacity to produce toxin than parental isolates were obtained from genetic crosses. Certain of the chemical characteristics of the pathotoxin have been determined.

References

Smedegård-Petersen, V. and Nelson, R. R. (1969). *Can. J. Bot.* **47**, 951–957.

Discussion

Paxton: Is toxin production correlated with the virulence of your isolates?

Smedegård-Petersen: There is a correlation between toxin production and pathogenicity because the toxin affects only hosts of the pathogen. Correlation between virulence and toxin production was not tested.

Kuć: Do you observe symptoms when you inoculate a resistant plant or a non-host with the fungus?

Smedegård-Petersen: Resistant plants reacted with either no symptoms or with so few and as tiny flecks that they were considered highly resistant.

Durbin: When did you harvest ascospore cultivars?

Smedegård-Petersen: We found a relationship between pH of culture filtrate and toxin yield which suggested that maximum toxin yield would be attained approximately 5 days after the pH of growth medium reached a minimum.

42. Hydrogen Cyanide Production by Virulent Isolates of a Psychrophilic Fungus in Their Interaction with Alfalfa, Winter Wheat, and Different Species of Grass

JACK B. LEBEAU

Canada Department of Agriculture,
Research Station, Lethbridge, Alberta, Canada

Winter crown rot of forage crops, winter wheat and turfgrass on the Canadian prairies is caused by a sterile psychrophilic basidiomycete. The pathogen has been isolated from grass as far north as Alaska. No fruiting structures have been discovered but the presence of hyphal clamp connections and the low temperature requirements for growth and pathogenesis have led to the description of this fungus as a low-temperature basidiomycete. The aestival stage of this basidiomycete is not known but much has been learned recently of its physiological properties and its interaction with the host during the winter.

Lebeau and Dickson (1953) reported that HCN was liberated by the fungus cultured on both natural and synthetic media. Ward *et al.* (1961) grouped isolates into three categories A, B and C on the basis of their cultural appearance, virulence, and ability to liberate HCN in culture and in the host plant. Type A isolates grew slowly and produced no HCN in culture but released large amounts of HCN in interactions with host plants and were highly virulent. Type B isolates grew rapidly and produced large quantities of HCN in culture but smaller amounts in host tissue and were less virulent than type A. Type C isolates grew rapidly but did not release HCN in culture or in conjunction with the host and were not pathogenic.

The pathogen was first isolated from severely damaged turfgrass in central Alberta by Broadfoot and Cormack (1941). A short time later it was obtained from alfalfa and other legumes. Cormack (1952) demonstrated the wide host range of the fungus and found that it attacked cultivated and native species of many genera of monocotyledons and dicotyledons. Rhubarb and chives appeared to be immune and bromegrass and the Siberian strain of alfalfa (*Medicago falcata* L.) were highly resistant.

Pathogenesis has been repeatedly demonstrated by Lebeau (1966) to be associated with the accumulation of toxic amounts of HCN in host tissue.

Mass invasion by mycelium and disorganization of host cells were not observed until 75 days after the first positive tests for HCN in the tissues were obtained (Lebeau *et al.*, 1959).

Colotelo and Ward (1961) suggested that production of HCN in alfalfa plants infected by the low-temperature basidiomycete depended on β-glucosidase of the fungus acting on cyanogenic substrates of the host. They believed that resistance of plants is related to their content of cyanogenic glucosides. This explanation of pathogenesis seemed logical but no consistent evidence has been obtained to support it.

Recent studies show that resistance to this pathogen breaks down at low temperatures. Under these conditions the host may produce precursors that release HCN when acted upon by the fungus.

Foliage sprays of sodium tetraborate provided remarkable control of the disease in alfalfa and virtually eliminated accumulation of HCN in crown-bud tissue (Lebeau and Atkinson, 1967). The borate ion may interfere with synthesis of a cyanogenic compound in the host, or may exclude cyanogens of the host from reactions involving the pathogen. Regardless of the mechanism of control, however, inhibition of cyanide production by sodium tetraborate must be regarded as further evidence that HCN is the active toxin in winter crown rot.

References

Broadfoot, W. C. and Cormack, M. W. (1941). *Phytopathology* **31**, 1058–1059.
Colotelo, N. and Ward, E. W. B. (1961). *Nature, Lond.* **189**, 242–243.
Cormack, M. W. (1952). *Can. J. Bot.* **30**, 537–548.
Lebeau, J. B. (1966). *Crop. Sci.* **6**, 23–25.
Lebeau, J. B. and Atkinson, T. G. (1967). *Phytopathology* **57**, 863–865.
Lebeau, J. B., Cormack, M. W., and Moffatt, J. E. (1959). *Phytopathology* **49**, 303–305.
Lebeau, J. B. and Dickson, J. G. (1953). *Phytopathology* **43**, 581–582.
Ward, E. W. B., Lebeau, J. B., and Cormack, M. W. (1961). *Can. J. Bot.* **39**, 297–306.

Discussion

Durbin: Where did the three isolates you used come from? What evidence is there that the three are the same fungus?

Lebeau: The three types of isolates A, B and C, of this psychrophilic basidiomycete were obtained from grasses and legumes in the Canadian prairies. We are not certain that they are the same fungus but they are sterile with hyphal clamp connections and were grouped together on the basis of their cultural appearance, virulence, and ability to liberate HCN in culture and in the host plant.

Mussell: What is the temperature range for growth of the low temperature basidiomycete *in vitro*?

Lebeau: The temperature range for growth *in vitro* for this fungus is approximately −1° to 24°C.

Paxton: I would like the relationship between temperature and disease development clarified. Does the plant become more resistant at higher temperatures? Removal of snow cover, which would lower soil temperatures, tends to control this disease. Why should this occur?

Lebeau: It is true that the host becomes more resistant at higher temperatures. Removal of snow lowers soil temperature but it also reduces the relative humidity required for the development of the pathogen.

Grossmann: Is it possible by treatment with cyanides to predispose plants to infection by the normally avirulent strain?

Lebeau: We have done no work on predisposition of the host by exposure to HCN. Therefore, we do not know whether this treatment would make avirulent isolates virulent.

Goodman: Does HCN reproduce the symptoms of the disease?

Lebeau: Yes, treating the plants with HCN produces typical symptoms of the disease.

Main: The relative level of different gases in the ambient atmosphere are known to influence development of a number of diseases. Have you any information on the levels of O_2, CO_2, etc. under snow cover and how these influence the host, the pathogen, or their interaction?

Lebeau: We have no information on the effects of other gases that may be present under the snow.

43. On the origin of Hydrogen Cyanide in Cultures of a Psychrophilic Basidiomycete

EDMUND W. B. WARD, DENIS THORN
AND A. N. STARRATT
Canada Department of Agriculture, Research Institute, London, Ontario, Canada

Hydrogen cyanide production in cultures of a psychrophilic basidiomycete was first investigated on the assumption that it occurred by hydrolysis of a typical cyanogenic glycoside. No evidence for this assumption was obtained. Evidence was obtained for the origin of cyanide-carbon from carbon-2 of glycine, an amino acid from which it would be theoretically impossible to derive a typical cyanohydrin (Ward and Thorn, 1966). A claim that HCN arises from isoleucine in this organism (Stevens and Strobel, 1968) could not be substantiated. Steam distillates from cultures incubated with isoleucine-ul-^{14}C were radioactive, but the radioactivity remained in the distillate after removal of the HCN as insoluble AgCN. The bulk of the radioactive material was extracted into ether after acidification of the distillate; chromatography indicated that it consisted of organic acids. Steam distillates from cultures incubated with isoleucine-2-^{14}C contained the equivalent of less than 1% of the radioactivity obtained with isoleucine-μ-^{14}C. Carbon-2 of isoleucine thus not only failed to give cyanide carbon but made a proportionately minor contribution to the total radioactivity recovered in steam distillates.

References

Stevens, D. L., and Strobel, G. A. (1968). *J. Bact.* **95**, 1094–1102.
Ward, E. W. B. and Thorn, G. D. (1966). *Can. J. Bot.* **44**, 95–104.

Discussion

Strobel: I appreciate your comments on your work on the origin of cyanide in a psychrophilic basidiomycete. Would you care to comment on our observations on the occurrence of two specific β-glucosidases and an aliphatic nitrilase that act on cyanogenic compounds that we have demonstrated in cultures of this fungus?

Ward: I do not feel that the production of two enzymes by the fungus is in itself evidence that the fungus also produces the substrates for these enzymes. It should be noted that the fungus is a parasite of plants which do produce these substrates.

Durbin: Are you [Ward] and Strobel working with the same fungus?

Ward: The isolate of the fungus in both cases is the W_2 isolate obtained from Dr. J. B. Lebeau, Lethbridge, Canada.

44. Toxic Proteins Secreted by Cotton Isolates of *Verticillium albo-atrum*

HARRY W. MUSSELL

Boyce Thompson Institute for Plant Research, Inc.,
Yonkers, N.Y. 10701, U.S.A.

Verticillium albo-atrum Rke. *et* Berth. causes disease symptoms on all aerial parts of infected plants while confined almost exclusively to vascular tissue. The appearance of symptoms some distance away from the pathogen suggests that toxins secreted by it may play a role in disease development. Previous workers were unable to demonstrate production of small molecular weight phytotoxins by *V. albo-atrum* (Green, 1954; Pegg, 1965; Stoddart and Carr, 1966); I therefore decided to investigate whether proteins secreted by the pathogen caused some disease symptoms.

Proteins were harvested from 12-day cultures of the severe defoliating patho-type of *V. albo-atrum* (isolates T-1, V3H and V19) on modified Czapek-Dox liquid medium containing 1% glucose. Preliminary experiments indicated that precipitation of these proteins with ammonium sulphate altered both their enzymic character and phytotoxicity. Because of this, the proteins were concentrated and desalted by ultrafiltration.

Proteins from 3 litres of culture filtrates were concentrated to 20 ml and fractionated on Sephadex G-75. Fractions were pooled to represent molecular weight ranges of over 100,000 (I); 100,000–45,000 (II); 45,000–25,000 (III); 25,000–10,000 (IV); 10,000–1500 (V); 4500–2500 (VI) and under 2500 (VII). They were desalted and applied to cuttings from 8-week-old cotton plants in deionized water at concentrations equal to and 1/100th of their original concentration in culture filtrates. Fractions I, II, III, and V were found to be phytotoxic.

At the higher concentration, Fractions I, II, and III caused a severe persistent wilt within 6 h, and Fraction V induced severe necrosis and wilting of cuttings 24–48 h after application. As pointed out by Green (1954), wilting is an unreliable index of phytotoxicity and may be caused by mechanical occlusion of vascular tissue. This probably happened with Fractions I, II, and III because these cuttings wilted at rates close to those of cuttings deprived of water. The wilt caused by Fraction V, however, was probably due to its toxicity, because treated plants appeared normal for 24 h, then wilted, and did so irreversibly. The active material in this fraction was probably a phytotoxin secreted by the pathogen.

At the lower concentration, Fractions I, II, and III caused chlorosis and necrosis of leaves indistinguishable from that seen in *Verticillium* wilt. These symptoms appeared in 24 h, and became progressively more prominent during the next 24 h. Seventy-two hours after applying the fractions, many leaves were either necrotic or had abscissed. The material causing abscission seemed to be concentrated in Fraction I whereas Fraction II induced most chlorosis.

The materials in these fractions were identified as proteins from their molecular weight, high optical density at 280 mμ, positive reaction with Lowry reagents, and heat lability. Boiling the fractions for 5 min before use resulted in total loss of phytotoxicity.

From these experiments, it seems that most if not all of the symptoms of *Verticillium* wilt in cotton are caused by proteins secreted by the pathogen.

References

Green, R. J., Jr. (1954). *Phytopathology* **44**, 433–437.
Pegg, G. F. (1965). *Nature* **209**, 1228–1229.
Stoddart, J. L. and Carr, A. J. H. (1966). *Ann. appl. Biol.* **58**, 81–92.

Discussion

Matta: You mentioned lack of ethylene production from *Verticillium* in culture. Is ethylene produced by the plant in the first phases of infection?

Mussell: We have not been able to detect significant quantities of ethylene evolved by infected cotton using bioassays. However, Wise and DeVay have recently reported that the infected cotton plants do evolve more ethylene than healthy plants.

Durbin: Would you give some details on the bioassay system used to detect ethylene production by infected cotton plants?

Mussell: We have used both tomato and pea seedlings looking predominantly for the epinastic responses reported in the literature.

Wood: I am surprised that you used a medium containing glucose as the main carbon source for the production of pectic enzymes. We have usually found that these enzymes are produced much more freely in the presence of their specific substrates.

Mussell: Pectic enzyme production by *V. albo-atrum* does not respond to the presence of substrate as well as has been reported for other fungi. Patil and Dimond and myself and Green have both observed that this fungus produces more PG when growing on glucose than on pectic media.

Goodman: Do you get reasonably good growth of your isolates on the glucose-salts medium?

Mussell: Yes, we get very good growth of *V. albo-atrum* on our modified Czapek-Dox medium containing glucose. The figures vary for different isolates of the fungus, but an average figure would be from 75 to 90 mg dry weight of mycelium in 12 days.

Page: Our observations on the inhibition of pectic enzyme induction by glucose in cultures of *Fusarium* spp. confirm similar observations made by Wood.

Mussell: I have done parallel studies using *F. oxysporum* f. sp. *lycopersici* and *V. albo-atrum* and, under identical conditions, *Fusarium* produced more PG on media containing pectin and pectin derivatives, while *V. albo-atrum* consistently produced more PG on glucose.

Patil: Page stated that glucose represses synthesis of polygalacturonase in cultures of *F. oxypsorum* f. sp. *lycopersici*. In contrast *V. albo-atrum* produces pectic enzymes abundantly on glucose media. In pectate media, production is much less, and adding pectate to glucose media does not significantly increase production of these enzymes by this organism.

45. Detection of O^{18} in Ophiobolins Isolated from Cultures of *Cochliobolus miyabeanus* grown in O^{18} Atmosphere

ANTONIO SCALA

Istituto di Chimica Organica dell'Università-Milano, Italy

Cochliobolus miyabeanus (Ito and Kurib.) Drechs. is a fungus which produces the phytotoxins ophiobolin A and B (Fig. 1) (Canonica *et al.*, 1966a, b). The methods used to grow *C. miyabeanus* in an atmosphere enriched with O^{18} and to determine which oxygen atoms of the ophiobolin molecules derive from the atmosphere are described below.

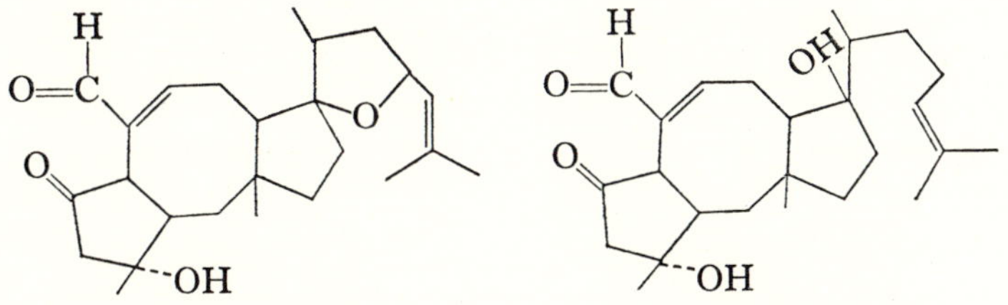

Ophiobolin A　　　Ophiobolin B

FIG. 1. Structure of Ophiobolin A and Ophiobolin B.

We had to solve different kinds of problem. The most important was to keep an atmosphere constant for at least 80 h, and to use as little O^{18} as possible keeping in mind the reliability of the results and the cost of the experiment. We had found that *C. miyabeanus* did not produce ophiobolins either in pure oxygen or in air containing less than 10–15% oxygen. Furthermore, a large amount of carbon dioxide is produced and a large amount of oxygen is utilized during incubation of the fungus in spite of the use of resting cells. For these reasons we had to design an apparatus by which for three days we could keep an atmosphere free from carbon dioxide and with percentages of oxygen and nitrogen as close as possible to those of natural air.

Because the micro-organism utilizes about 1500 ml of oxygen during incubation, it was necessary to fill a container from which the gas was continuously supplied to the incubation flask. Also, it was important to avoid too large a concentration of oxygen in the atmosphere around the fungus.

A diagram of the apparatus used is shown in Fig. 2. The 750 ml container for 150 ml of culture is a modified Warburg flask; 15 ml 50% KOH are placed in the central well to absorb carbon dioxide produced during incubation. The flask has a connexion for the vacuum pump. The oxygen reaches the flask through a flexible tube and then through a glass tube which ends a few millimetres above the liquid surface. A sterile cotton plug prevents gas from returning to the container. The break-seal flask contains 100 ml

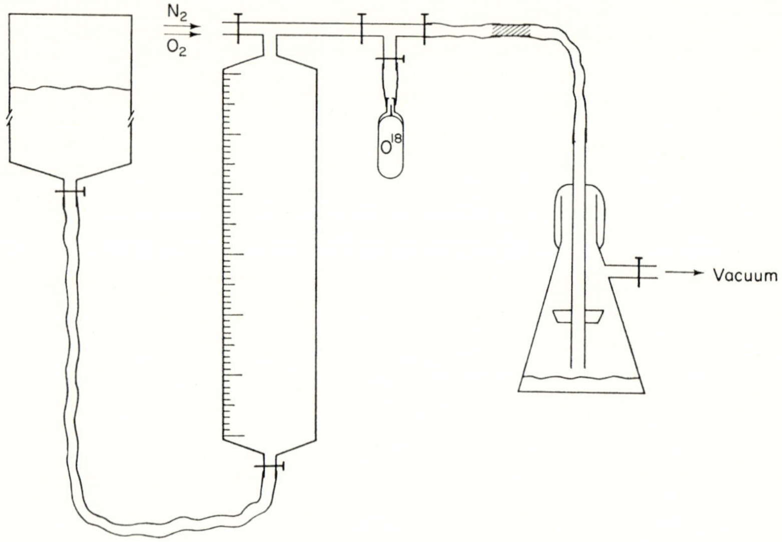

FIG. 2. Diagram of apparatus for growing *Cochliobolus miyabeanus* in an O^{18} enriched atmosphere.

of 99% O^{18} 1% being O^{16}. On breaking the seal, the O^{18} is mixed with the two litres of O^{16} contained in the cylinder. The flask is filled with nitrogen at 0·8 atm while the oxygen is kept continuously under slight pressure.

The active metabolism of the fungus keeps a continuous small negative pressure in the flask, allowing the oxygen to flow from the cylinder to the Warburg flask. In this way the composition of the atmosphere over the culture after three days was 20·4% oxygen and 1% carbon dioxide.

The O^{18}/O^{16} ratio was 0·0455 as determined by mass spectrometry. From a culture grown in these conditions we obtained, after methylene chloride extraction of the mycelium and purification by absorption chromatography, 35 mg of ophiobolin A and 180 mg of ophiobolin B. The oxygen atoms of ophiobolins may originate either from the water or from the atmosphere according to the cyclization mechanism of the acyclic precursor.

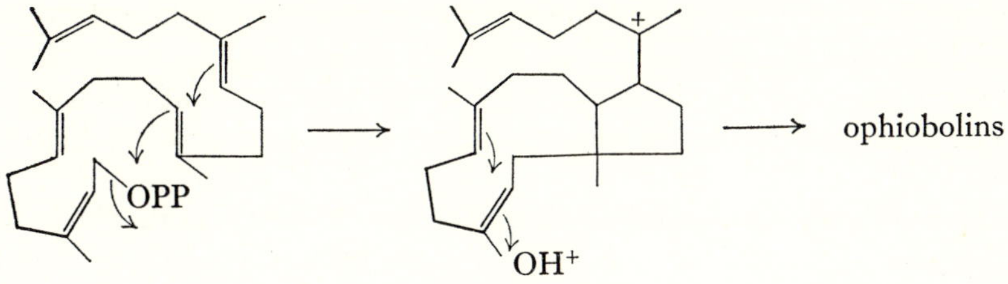

FIG. 3. Hypothetical mechanism proposed by Nozoe *et al.* (1965) for origin of the OH group.

According to the hypothetical mechanism proposed by Nozoe *et al.* (1965) (Fig. 3), the hydroxy group at position 3 originates from atmospheric oxygen. However, in the mechanism shown in Fig. 4, the OH originates from the medium. It was easy to hypothesize that the two carbonylic oxygens derive from the atmosphere but that they slowly exchange with the water. Alternatively, the oxygenated functions at positions 3 and 14 of ophiobolins might derive either from the atmosphere or from the water.

Two general methods can be used to determine the O^{18} enrichment of a compound. Either carbon dioxide obtained by pyrolysis of a compound or the compound itself can be checked by mass spectrometry for the O^{18}

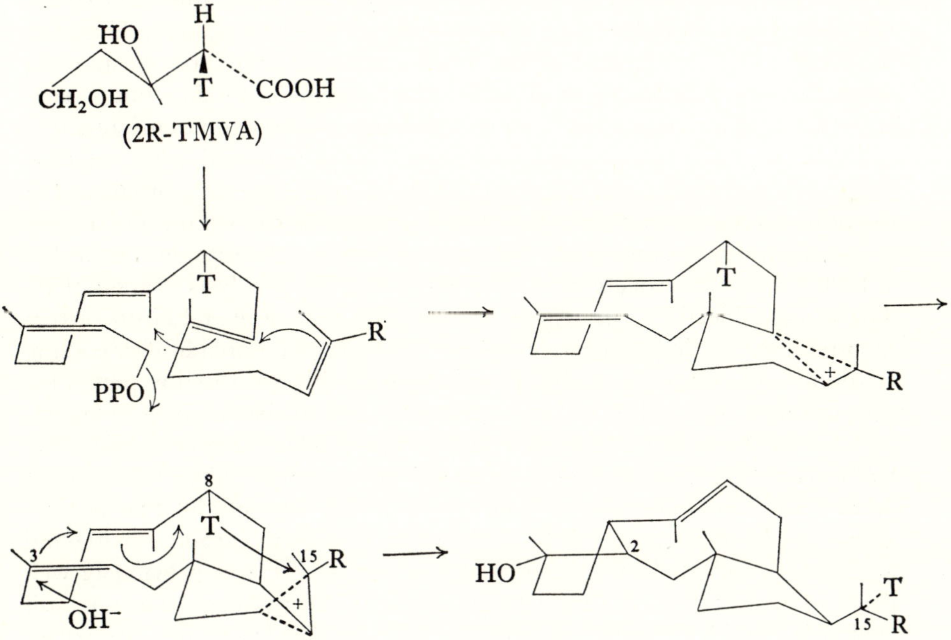

FIG. 4. Origin of the OH group.

content. We used the second method to determine the ethereal oxygen of ophiobolin A.

One of the most prominent peaks in the mass spectrum of this compound can be found at *m/e* 165. By high resolution mass spectrometry, the composition $C_{11}H_{17}O$ can be attributed to this peak, which can originate only from the part of the molecule containing the oxygen bridge (Fig. 5). The intensity of the peak at *m/e* 167 has to be 0·88% of the peak at *m/e* 165, as calculated

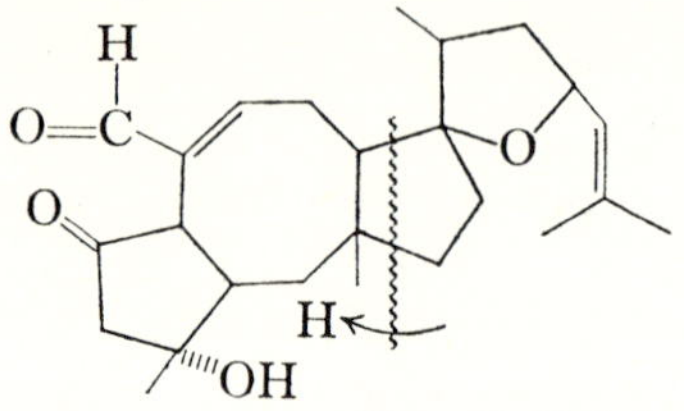

⟶ $C_{11}H_{17}O$

found m/e 165·1267
calcd. m/e 165·1279

FIG. 5. Origin of $C_{11}H_{17}O$ of Ophiobolin A.

from the relative abundance of the isotopes. In the ophiobolin A spectrum the peak at *m/e* 167 is about 1·5% as intense as the peak at *m/e* 165, in good agreement with the calculated value. In the mass spectrum of O^{18} enriched ophiobolin A, the peak at *m/e* 167 is 5·2% as intense as the peak at *m/e* 165. From this finding an O^{18} enrichment of 3·7% is calculated in good agreement with the theoretical value of 4·35%. In this way we established that the ethereal oxygen of ophiobolin A derives from the atmosphere. Unfortunately, the O^{18} enrichment of the other oxygens of ophiobolins cannot be determined by a similar method. In fact the mass spectrum does not present any other peak intense enough for the calculation of the isotopic enrichment.

Therefore, pyrolysis of the compound by a modified Unterzaucher method (Unterzaucher, 1940), was used to determine the O^{18} enrichment of the other oxygens of ophiobolins. The reaction was carried out in conventional apparatus for microdetermination of oxygen; the sample was heated at 1200°C in a stream of argon which ensured that the pyrolysis was complete. The gas produced by the reaction is transformed completely into carbon monoxide by passing over active carbon. The carbon monoxide after cooling at room temperature is oxidized to carbon dioxide with iodine pentoxide supported on silica gel. It is very important to use an absolutely dry reagent. Miller and Anderson (1959) have shown that under these conditions one half oxygen of carbon dioxide is derived from the enriched sample and the other half from iodine pentoxide. So that only one of the two oxygen atoms of carbon dioxide will be enriched in O^{18}. Both carbon dioxide and argon were collected in a liquid nitrogen trap which was previously well dried to avoid traces of water. After completion of pyrolysis the trap, still cooled in liquid nitrogen, was evacuated at 0·5 mmHg to remove argon. The system was then connected to a mass spectrometer and kept at −190°C until the disappearance of peaks of argon and of water possibly contained in the instrument. Liquid nitrogen was now substituted by dry ice in acetone. In this way the temperature reached –70°C and carbon dioxide evaporated from the trap while traces of water possibly present still remained in the trap. Argon has to be removed completely because its mass spectrum shows a peak at *m/e* 40 which could cause problems. Elimination of water is also important because at room temperature in the mass spectrometer a rapid exchange with the carbon dioxide can occur.

Under the above conditions the O^{18} enrichment can be determined by calculating the ratio between the peaks at *m/e* 44 which derives from carbon dioxide containing two O^{16} and at 46 which derives from carbon dioxide containing one O^{18} and one O^{16}, and utilizing the expression:

$$N = (x - 0{\cdot}00204)\,T/(0{\cdot}0455 - 0{\cdot}00204);$$

where N is the number of O^{18} enriched atoms per molecule, T is the total number of oxygen atoms per molecule, 0·0455 is the O^{18}/O^{16} ratio for oxygen in the experiment, $x = R^1/1 + R^1$, $R^1 = R - 0{\cdot}00204$, 0·00204 is the natural O^{18}/O^{16} ratio and R the observed ratio between the abundance of the peaks at *m/e* 46 and 44 respectively.

Figure 6 shows the analysed compounds and the number of enriched oxygen atoms present in each molecule.

The fractional value of enriched oxygen we found for ophiobolin A can be attributed to a partial exchange of the two carbonyls with the medium during incubation and isolation of the compound.

Ophiobolin A was dehydrated in alkaline medium to yield anhydro-ophiobolin A. Under these conditions the hydroxy group in position 3 is eliminated and the two carbonyls exchange with the medium to a large extent. In fact, dehydration of ophiobolin A in the presence of water enriched with O^{18} yielded anhydro-ophiobolin A in which only 75% of the carbonyl oxygens had exchanged with the medium.

Therefore, the value of 1·1 O^{18} enriched atoms which we found for anhydro-ophiobolin A is in good agreement with the results directly obtained from the mass spectrum of ophiobolin A showing that only ethereal oxygen is enriched.

Ophiobolin A with excess of hydrazine followed by acidic treatment yields the pyridazine derivative shown in Fig. 6 containing only one enriched oxygen atom.

From these results we concluded that in ophiobolin A both carbonyl oxygen and ethereal oxygen derive from the atmosphere while the hydroxylic oxygen originates from the water. The same analysis performed on ophiobolin B demonstrated that in this compound also only the oxygen atom at position 3 originates from the medium.

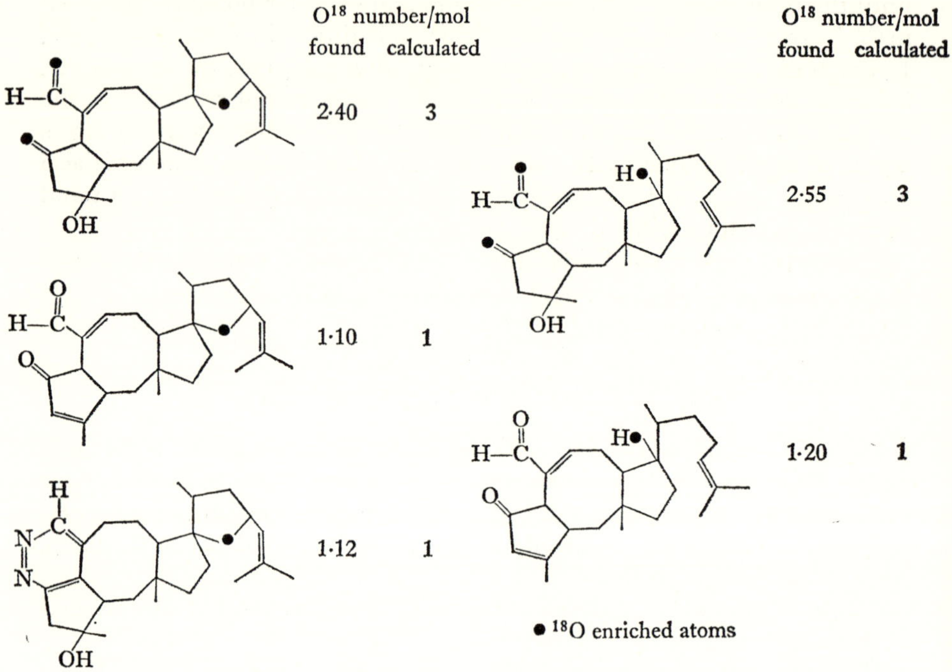

FIG. 6. Number of enriched oxygen atoms in different products analysed.

All the results agree with the hypothesis that the cyclization of geranylfarnesylpyrophosphate occurs with the introduction of an OH^- at position 3 as reported in Fig. 4.

The use of O^{18} to determine the origin of the oxygen atoms of a molecule is usually considered an expensive and sophisticated technique but as this paper demonstrates, the use of simple expedients makes methods easy and inexpensive for certain biological experiments.

References

Canonica, L., Fiecchi, A., Galli Kienle, M., and Scala, A. (1966a). *Tetrahedron Lett.*, 1211.

Canonica, L., Fiecchi, A., Galli Kienle, M., and Scala, A. (1966b). *Tetrahedron Lett.*, 1329.

Nozoe, S., Morisaki, M., Tsuda, K., Itaka, Y., Takahashi, N., Tamura, S., Ishibashi, K., and Shirasaka, M. (1965). *J. Am. chem. Soc.* **87**, 4968.

Unterzaucher, J. (1940). *Ber. Deut. Chem. Ges.* **73**, 391.

Miller, W. G. and Anderson, L. (1959). *Anal. Chem.* **31**, 1669.

Discussion

Kuć: Do you know how molecular oxygen, possibly via an oxygenase, is incorporated into the molecule and how the ether is formed?

Scala: We do not know how molecular oxygen is incorporated into ophiobolin B. We have noticed that tritium label of the isopropylidene group in ophiobolin A biosynthesized from 2T-MVA is lower than in ophiobolin B. This loss of radioactivity corresponds to the exchange of one hydrogen atom of one of the C-19 methyls. This finding can be rationalized with the following cyclization mechanism for the formation of the oxygen ring:

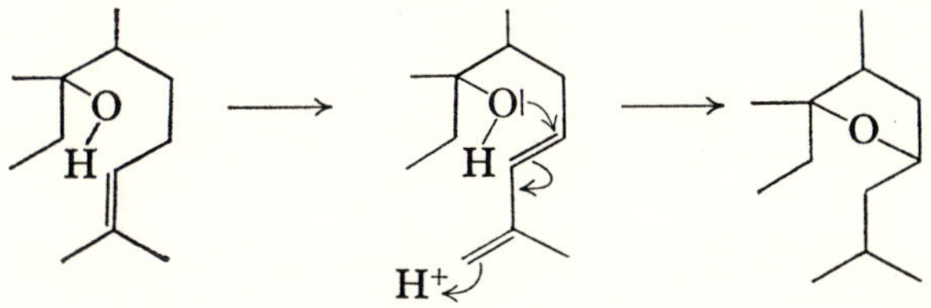

C. SHORTER PAPERS ON MISCELLANEOUS TOPICS

46. Host-Specific Toxins as Determinants of Successful Colonization by Fungi

OLEN C. YODER

Department of Botany and Plant Pathology, Michigan State University, East Lansing, Michigan 48823, U.S.A.

Host-specific toxins are known to cause the major symptoms of certain infectious plant diseases (Scheffer and Yoder, p. 251). We may now ask whether or not toxins are required for initial colonization of plant tissues by fungi. In this question lies the heart of the basic problem of host range limitation. All plant pathogens are more or less host-specific, and some are limited to only one genotype of one species. The special relationship between pathogen and susceptible host, as distinct from an aborted attempt to colonize a resistant host, can be detected very early in the infection process. Therefore, any factor which determines host specificity must be present and functional during the initial interactions between host and parasite.

In general, spores of fungal pathogens germinate, form appressoria, and penetrate the surface of susceptible and resistant leaves. Growth in resistant leaves stops after penetration whereas growth in susceptible leaves continues with little evidence of damage to the host cells for some time. Fungi which produce host-specific toxins follow the usual pattern of invasion; thus they should be good models for the study of the molecular basis of disease initiation. This report deals with the role of *Helminthosporium victoriae* Meehan *et* Murphy (HV) toxin in establishing the parasitic relationship between *H. victoriae* and its host, oats (*Avena sativa* L.). Comstock and Scheffer (1970) have reported similar data for *H. carbonum* Ullstr., a pathogen of maize (*Zea mays* L.).

The first evidence that HV-toxin is involved in disease initiation was reported by Nishimura and Scheffer (1965). Resting spores of pathogenic *H. victoriae* contain toxin whereas spores of non-pathogenic isolates or mutants do not. When toxin-containing spores germinate on the leaf surface, toxin is released (Nishimura and Scheffer, 1965). Toxin placed on a leaf surface can diffuse into the tissue within 15 min (Yoder and Scheffer, 1969). Thus, limited zones of host tissue are exposed to toxin at the beginning of the host–parasite interaction. The earliest known response of susceptible tissue to toxin is loss of electrolytes which occurs almost at once (Scheffer and Samaddar, 1970). Similarly, the earliest observed host response to infection is loss of electrolytes which is evident 6 h after inoculation (Yoder and Scheffer, 1969). The delayed effect of infection, as compared to toxin, can be attributed to time required for spore germination and toxin production. The fungus responds to the toxin-produced changes in host cells, making more growth in susceptible than in resistant tissues within 12 h after inoculation.

The requirement of toxin for successful colonization was examined further by use of two fungi not pathogenic to oats: *H. carbonum* (a pathogen of maize), and a non-pathogenic mutant of *H. victoriae* which did not produce toxin. Spores of these fungi germinated on the oat leaf surface, penetrated the epidermis and stopped growing. Leaves, susceptible and resistant to *H. victoriae*, were resistant to these fungi. When a small amount of HV-toxin was added, the spores germinated, penetrated, and developed throughout susceptible leaves just as did pathogenic *H. victoriae*. Toxin did not allow *H. carbonum* and the *H. victoriae* mutant to colonize resistant oat leaves. These data support the hypothesis that toxin is required for successful infection of susceptible tissues (Yoder and Scheffer, 1969).

If HV-toxin conditions susceptible cells for colonization by *H. victoriae*, what then determines resistance? All evidence indicates that resistance to the fungus is the same as insensitivity to the toxin (Scheffer and Yoder, p. 251). Toxin does not condition resistant oat cells, and there is no colonization. Since insensitivity to toxin is a constitutive characteristic (Samaddar and Scheffer, 1970), the fate of infection appears to be determined by factors present or absent in cells before fungal penetration occurs. However, possible contributions of fungus-induced host responses have been considered. Visible host responses such as thickened epidermal cell walls and hypersensitive-type reactions could not be correlated with resistance because: (1) resistance was usually expressed without visible responses; (2) fungal growth was not consistently checked in cells which showed such visible responses; (3) such responses occurred too late to account for resistance (Yoder and Scheffer, 1969).

Another response of oat tissues to infection is production of phytoalexin (PA). Nishimura and Scheffer (1965) described the following characteristics of this response: (1) susceptible and resistant leaves produce equal amounts of PA after inoculation; (2) both pathogenic and non-pathogenic isolates induce equal amounts of PA; (3) spores of both pathogenic and non-pathogenic isolates are equally sensitive to PA; (4) PA is detected only after fungal growth in the resistant plant had stopped; (5) toxin does not interfere with PA production by susceptible oats; and (6) toxin, without spores, does not induce PA. These considerations indicate that the PA produced by oats has no significant role in initial resistance to *H. victoriae*. However, the usual resistance responses are another indication of the similarity between disease caused by *H. victoriae* and many other infections (Scheffer and Yoder, p. 251).

It is sometimes assumed that disease caused by *H. victoriae* differs from most diseases in that *H. victoriae* acts as a saprophyte (Hadwiger and Schwochau, 1969). The fact that toxin released by germinating spores changes host cell physiology before the fungus penetrates the tissue supports this assumption. However, closer examination shows no evidence that *H. victoriae* is more or less saprophytic than are other plant pathogens. The capabilities of *H. victoriae* were compared with those of a known saprophyte, *Neurospora tetrasperma* Shear *et* Dodge, and four fundamental differences were observed. *H. victoriae* attached to leaf surfaces, penetrated leaves, established a compatible relationship with its susceptible host, and induced host responses in resistant plants. *N. tetrasperma* did none of these things. Fungi with pathogenic potentials, such as *H. victoriae*, appear to have the inherent ability to attach to and penetrate surfaces. The saprophyte *N. tetrasperma* has no mechanism for attachment and penetration and can colonize only dead tissue (Yoder and Scheffer, 1969).

An attempt was made to determine whether or not susceptible cells are dead when hyphae of *H. victoriae* are growing among them. Since it is difficult to define cell death, several functions were evaluated 12 h after inoculation: (1) cells were not visibly affected as determined by light microscope observation; (2) plasma membrane characteristics were changed as early as 6 h after inoculation, but this is not proof of cell death; (3) dehydrogenase activity was not affected; (4) protein synthesis was only partially inhibited (this could be secondary to the membrane effect); and (5) phytoalexin synthesis was just beginning at 12 h and increased after that time. Thus, by the time hyphae were well established in the tissues, host cells were not "dead", as indicated by their metabolic activity and ability to carry on new syntheses.

Acknowledgement

This work was supported in part by the National Science Foundation (GB 6560) and by the Michigan Agricultural Experiment Station (Journal Article No. 5240).

References

Comstock, J. C. and Scheffer, R. P. (1970). *Phytopathology* **60**, 1288 (Abst.).
Hadwiger, L. A. and Schwochau, M. E. (1969). *Phytopathology* **59**, 223–227.
Nishimura, S. and Scheffer, R. P. (1965). *Phytopathology* **55**, 629–634.
Samaddar, K. R. and Scheffer, R. P. (1970). *Pl. Physiol.* **45**, 586–590.

Scheffer, R. P. and Samaddar, K. R. (1970). *In* "Recent Advances in Phytochemistry", Vol. 3, pp. 123–142. Appleton-Century-Crofts, New York.
Yoder, O. C. and Scheffer, R. P. (1969). *Phytopathology* **59**, 1954–1959.

Discussion

Mussell: What levels of toxin were required to make the non-pathogenic mutant a successful pathogen?

Yoder: A toxin preparation which gave complete inhibition of susceptible seedling root growth at 0·016 μg/ml permitted growth of the non-pathogenic mutant in susceptible leaves at a concentration of 0·16 μg/ml. More dilute solutions did not aid the non-pathogen in initial colonization of susceptible tissue.

Wood: Take the case of an avirulent strain of *H. victoriae* that does not produce toxin and a virulent one that does, on oats susceptible to the toxin. Both infect cells and presumably kill them in the earliest stages of infection. But the avirulent race does not continue to do so. First I should like to know whether the avirulent strain can continue to grow on dead oat tissue; because if it can, why does it not do so after it has killed the cell it penetrates? If it can kill one cell why does it not kill many more? Are you not neglecting the role of the host cell in restricting the growth of the pathogen?

Next, the *H. victoriae*/oats system seems to be simpler than many others. Is it envisaged that highly specific toxins will explain the complex pattern of resistance and susceptibility in the *Colletotrichum lindemuthianum*/bean complex?

Lastly, and a little teleologically, if *H. victoriae* is as specific as you claim what did it do before Victoria type oats appeared in N. America?

Yoder: There is evidence indicating that susceptible tissue is colonized by the virulent strain before the cells are killed. The avirulent strain penetrates the cell wall and stops growing; this cell does not appear to die for some time, either. Ultimately, cells do die and they are not colonized by the avirulent strain (which presumably can grow on artificially killed oat tissue). Obviously, these dead cells are not a suitable substrate for the fungus. Why? Any answer must accommodate this fact: resistance to the fungus cannot be genetically separated from insensitivity to the toxin. (See Scheffer's comment, p. 462). We hope to understand key events which determine whether or not an infection will be successful. In this case, insensitivity to toxin appears to determine resistance. Events which occur after that, such as host cell responses, may be incidental or secondary to the controlling mechanisms which determine resistance or susceptibility. I do not wish to neglect the role of the host cell in resistance but rather to emphasize the evidence, genetic and otherwise, which helps us to understand this particular case.

As yet, chemical determinants of pathogenicity have been described for only 8 fungi (see Scheffer's paper). It seems likely, however, that there is a chemical basis for all plant diseases, although this could take various forms. There are no available data, to my knowledge, that give a satisfactory chemical explanation of resistance and susceptibility in such complex cases as *C. lindemuthianum*/bean.

We cannot answer the question on *H. victoriae* survival, but we can speculate. Perhaps *H. victoriae* uses mechanisms other than HV-toxin to grow and reproduce in the absence of susceptible oats. HV-toxin-producing strains of *H. victoriae* have been found in parts of the world where susceptible oats have never been grown. We would expect, from evolutionary theory, that useless appendages are not maintained. Why would *H. victoriae* maintain a gene for production of toxin if toxin is of no value to it? It is possible that HV-toxin affects certain other species (to the advantage of the fungus), in such subtle ways that the effects have not been detected. We know that there are selections of oats which tolerate relatively high levels of toxin, yet toxin is responsible for pathogenicity, as shown by the genetic evidence.

Meyer: When you add the toxin to a leaf with an avirulent strain of *H. victoriae* you induce disease. What happens when you inoculate leaves with other *Helminthosporia* or any other fungus after treatment with the toxin, because if, in this case, you do not induce disease, specificity of *H. victoriae* should then be shared with some other factor of the parasite?

Yoder: *H. carbonum*, pathogenic to maize but not to oats, will colonize susceptible oats in the presence of HV-toxin.

Matta: The occurrence of the same level of phytoalexins in both resistant and susceptible oat–*H. victoriae* combinations appears to be rather peculiar when compared with other systems characterized by phytoalexin production. This would indicate that in any case *H. victoriae* is unable to establish itself as an effective parasite. On the other hand the toxin, superimposed on earlier events in infection as a determinant of pathogenicity (but not of parasitism), kills the cells permitting the saprophytic development of the fungus.

Yoder: Phytoalexin production by both susceptible and resistant varieties has been reported for other disease systems. There are several indications that phytoalexin does not play an important role in initial resistance of oats to *H. victoriae* (see text). I have examined the possibility that *H. victoriae* develops as a saprophyte in dead host tissue. Comparison of *H. victoriae* with a known saprophyte and data on the metabolic conditions of susceptible cells after colonization by *H. victoriae* (see text) indicate that *H. victoriae* attacks living tissue just as other pathogens do. The fact that susceptible cells produce a phytoalexin after colonization by *H. victoriae* implies that cells are not dead and that this phytoalexin is not responsible for limitation of fungal growth.

Scheffer: Resistance to these infections is equivalent to lack of sensitivity to toxins. Many lines of evidences bear on this. The fact that the invaded cell of the resistant variety later becomes unsuitable as a substrate for the fungus (because of phytoalexins etc.) seems to be incidental to an understanding of this type of resistance.

The main point of Yoder's talk was the similarity of these cases to the usual case of fungal invasion of leaves. I would like to throw the question back to Matta: In what way do these infections differ from the usual, except that host-selective compounds are known? How are they more saprophytic than usual for fungal pathogens?

Schlosser: Is there any evidence for or against the involvement of an enzyme which breaks down victorin in the first 12 h after infection in which susceptibility or resistance is determined?

Yoder: Both susceptible and resistant oats inactivate toxin equally. There is no direct evidence that the inactivation is enzymatic. This was discussed in more detail by Scheffer in a previous presentation.

Feldman: I was interested in your comment that disease could be produced with a non-pathogen and toxin but not with a saprophyte and toxin. May I ask how one distinguishes a non-pathogen from a saprophyte?

Yoder: I prefer the term "potential pathogen" over "non-pathogen". This describes an organism which is pathogenic to one or more genotypes but is non-pathogenic to most plants. Fundamental differences between the behaviour of a potential pathogen and a saprophyte are described in the text.

Rudolph: You have stated that phytopathogenic fungi, non-pathogenic to oats, could be made to penetrate host tissue by adding the toxin at the same time. From these observations you concluded that only nutrients have to be made available to induce a susceptible reaction of oats towards phytopathogenic fungi. My question is: How many different fungi, which are normally non-pathogenic to oats, did you try? How broad was your spectrum? I feel this questionable to draw such general conclusions, if only few species have been tested.

Yoder: The conclusion drawn from these data is that toxin is required for successful colonization of susceptible tissues. The question of fungal nutrition has not been examined. Fungi used in these experiments are described in the text.

Paxton: Gerdemann and Chamberlain showed that heat treatment of soybeans reversibly induces susceptibility to potential pathogens of other crops but not to strict saprophytes.

Hilu studied *Helminthosporium turcicum* attack on corn leaves and found equal growth on resistant and susceptible leaves for the first three days.

We have found phytoalexin is first detected 3 days after inoculation of the resistant corn leaf.

Does the amount of toxin applied to the leaf along with a non-toxin producing strain of *H. victoriae* determine the amount of invasion by the fungus? If toxin is responsible for disease development, a strict correlation should exist between amount of toxin applied and fungal invasion of the host. Why do these leaf lesions become limited rather than completely kill the host?

Yoder: A threshold level (0·16 μg/ml) of toxin is required for initial colonization of susceptible oats by a non-pathogen. The objective of this study was to investigate the requirement of toxin for successful establishment of the fungus in tissues. Limitation of lesion size is a later event and was not investigated. However, either spreading or limited lesions can develop, depending on conditions of inoculation and incubation.

Mirocha: I would like to refer to the comment that if nutrients were supplied to an avirulent strain of the fungus it would grow or proliferate in the host. My experience with rust fungi has led me to believe that soon after infection (within 4 h) hydrolysis of starch occurs and glucose is available. I would be surprised if *H. victoriae* would not do the same because *H. sativum* has β-amylase as a constitutive enzyme. This may not mean that because glucose is available the fungus will utilize it. I propose that the restriction in growth is due to another factor, as yet unknown, rather than lack of nutrients.

Page: There are strains of *Fusarium oxysporum* that are parasites but cause no visible symptoms in some plants. There are strains of *F. oxysporum* in the soil that are saprophytes. Also, *formae speciales* of *F. oxysporum* may be highly specific with respect to a host and would be non-pathogenic on a non-host.

47. Suppression of the Resistance Response as an Active Mechanism for Susceptibility in the Potato–*Phytophthora infestans* Interaction

J. L. VARNS AND JOSEPH KUĆ

Department of Biochemistry, Purdue University, Lafayette, Indiana, U.S.A.

A consistent response of tubers to infection by incompatible races of *Phytophthora infestans* (Mont.) de Bary (hypersensitive resistance) was demonstrated with eleven cultivars and six races of the fungus. The response included rapid necrosis of host cells and the accumulation of two terpenoids,

TABLE I. Treatments received by potato slices.

Abbreviation	Treatment
R, S	Inoculated with an incompatible race (race 4), or compatible race (1.2.4) and incubated 72 h
R 12 h S, S 12 h R	Inoculated, incubated 12 h, reinoculated and incubated 72 h
Aged 12 h R, 12 h S	Aged 12 h, inoculated and incubated 72 h
R 24 h S, S 24 h R	Inoculated, incubated 24 h, reinoculated and incubated 72 h
Aged 24 h R, 24 h S	Aged 24 h, inoculated and incubated 72 h
R 24 h homog, S 24 h homog	Inoculated, incubated 24 h, treated with sonicated homogenate of fungus and incubated 72 h
Aged 24 h homog	Aged 24 h, treated with sonicated homogenate of fungus and incubated 72 h
S 36 h R	Inoculated, incubated 36 h, reinoculated and incubated 72 h
Aged 36 h R	Aged 36 h, inoculated and incubated 72 h

S—Susceptible; R—Resistant.

rishitin and phytuberin. Sonicates or extracts of the fungus elicited the above response with all cultivars including those lacking R genes for resistance. A compatible or susceptible interaction was characterized by the suppression of both hypersensitive collapse of cells and accumulation of rishitin and phytuberin (Tables I, II). An alteration of cellular response in

TABLE II. Regulation of terpenoid accumulation in Kennebec potato slices by *Phytophthora infestans*

Treatment	Rishitin	μg/g/ dry wt* Phytuberin	Estimation‡ unknown
Fresh tissue	—†	—†	—
Aged 36 h	—	—	—
Aged 72 h	—	—	+—
R	118	5	+—
S	—	—	+++
R 12 h S	217	20	+—
S 12 h R	42	3	+++
Aged 12 h R	130	6	+—
Aged 12 h S	—	—	++
R 24 h S	334	23	+—
S 24 h R	56	2	+++
Aged 24 h R	238	4	+—
Aged 24 h S	—	—	++
S 24 h homog	52	2	+++
Aged 24 h homog	415	3	+—
R 24 h homog	400	20	+
S 36 h R	9	2	+++
Aged 36 h R	76	5	+—

* Analyses by thin layer and gas–liquid chromatography of combined first and second mm of slices.

† Compound not detected.

‡ Relative estimation based on colour reaction after spraying silica gel G plates with vanillin sulphuric acid reagent.

— = none detected; +— = trace; +++ = maximum colour development and size of spot.

the host during the compatible interaction suppressed the ability of the host to respond normally to subsequent inoculation with an incompatible race. Conversely, once the host was inoculated with an incompatible race, suppression from subsequent inoculation with a compatible race did not develop. A compatible interaction also suppressed the hypersensitive host response to sonicates of the fungus. An unknown compound, present in trace quantities in aged tuber slices or incompatible interactions, increased markedly in compatible interactions or after inoculation with a compatible race followed by an incompatible race. A 12-h period after inoculation either established suppression or elicited the hypersensitive host response. The authors suggest that compatible races of the fungus suppress the normal host response to infection and thereby develop in the tissue and cause disease.

Discussion

Yoder: How does the time required for a host response in the incompatible interaction compare with the time after inoculation when the fungus stops growing in the tissue?

Kuć: According to the work of Tomiyama, the growth of the fungus is not retarded for at least 2·5 h after hypersensitive death of the host cells. Approximately 12 h after inoculation, growth of the fungus was still evident but abnormal. In our observations, 24 h after inoculation growth of the fungus was stopped in the incompatible interaction. Rishitin and phytuberin could be quantitatively determined in the top 2 mm of infected slices 24 to 30 h after inoculation and detected in trace quantities 12–18 h after inoculation. According to Tomiyama, the concentration around the sites of penetration is considerably higher than in the top 1 mm; therefore, accumulation of the compounds probably starts at the time of or shortly after hypersensitive collapse of the cells.

Yoder: Can homogenates of both compatible and incompatible races of the pathogen induce the formation of rishitin and phytuberin?

Kuć: Yes, sonicated homogenates of the fungus induced the accumulation of rishitin and phytuberin in all cultivars even those lacking R genes for resistance.

Feldman: Have you determined the ED_{50} of phytuberin and rishitin?

Kuć: The ED_{50} for rishitin and phytuberin is approximately 2×10^{-4} M.

Matta: I was pleased to hear the convincing evidence that susceptibility depends on suppression of resistance responses in the *Phytophthora infestans*–potato tuber system. Similar results have also been obtained with vascular fungi. Beckman found that the respiratory increase induced in radish seedlings by a mixture of saprophytes is prevented in the presence of the pathogenic form of *Fusarium oxysporum.* In our laboratory we have observed that the sudden increase of phenolics that is associated with the protection induced by avirulent forms of *F. oxysporum* is suppressed just by addition of virulent conidia to the inocula of the avirulent conidia.

Kuć: I believe the mechanism for susceptibility in a number of host–parasite interactions is dependent upon the suppression of a non-specific resistance response. Your work and the examples you cite support this hypothesis.

Main: It has been demonstrated that polygenic or field resistance varies greatly between potato varieties and breeding lines once the dominant R gene for resistance is overcome by the appropriate race of *P. infestans.* In your various R gene fungus race combinations, do you find varietal differences in the total amount of rishitin or phytuberin detected with time related to varietal response? Do you think either of these two compounds are involved in polygenic resistance, or are they specific to the dominant R gene: incompatible race phenomenon you have described?

Kuć: We have not observed differences in phytuberin and rishitin accumulation in the hypersensitive response that we can relate to varietal or race differences. I do not believe resistance is based on a single mechanism of response. Just as with animals, I believe several mechanisms or orders of resistance are found in plants. The hypersensitive collapse and associated accumulation of rishitin and phytuberin may be one of them; however, even the magnitude of the response to injury appears to be a resistance mechanism. After cutting, potato tubers accumulate large quantities of α-solanine and α-chaconine. Both compounds are inhibitory to the growth of *P. infestans* and may constitute another resistance mechanism. Numerous quantitative genes may influence the rapidity or magnitude of a response as they do fruit colour or size.

48. Multiple Infection and Symptom Mitigation in Vascular Wilt Diseases

J. A. MEYER AND H. MARAITE

Laboratoire de Phytopathologie de la Faculté des Sciences Agronomiques, Université Catholique de Louvain, Heverlee, Belgium

Muskmelon plants were inoculated simultaneously with strains of *Fusarium oxysporum* f. sp. *melonis* differing in virulence and in that some were deficient for biotin. It was shown that a plant could become infected by more than one strain after inoculation with a mixture, and that symptoms appear more slowly in plants inoculated with a virulent *and* less virulent strain than in plants inoculated only with a virulent strain. The results that were obtained were best explained in terms of competition between strains at sites of penetration.

Discussion

Matta: If competition for the substrate is the mechanism of the observed protection, I do not see why the total quantity of mycelium of the two strains in the stem would be decreased.

I would not exclude even in your case that the presence of the less virulent strains could elicit a more effective defence reaction by the plant tissue.

Meyer: Effectively, the quantity of mycelium within the stem decreased with the mitigation of symptoms and the competition for the substrate is the best hypothesis to explain my results though I do not exclude any other possibilities.

Molinas: What does Meyer think about the work of Grente *et al.* with simultaneous inoculations of virulent and avirulent strains of *Endothia parasitica* into *Castanea* in Italy and W. of France ? These workers postulated an extranuclear factor that can be transmitted from the avirulent to the virulent strain.

Meyer: The hypothesis of Grente is fascinating but, in our case, the strains did not show any change in pathogenicity after simultaneous passage through susceptible plants.

Kuć: Undoubtedly different mechanisms can explain cross protection in different host–pathogen interactions. In the *Colletotrichum lindemuthianum* bean interaction I doubt if cross protection can be explained by competition since the heat inactivated fungus can be used to cross protect.

Page: Are biotin requiring strains of *Fusarium oxysporum* common? I have not heard of them in N. America.

Meyer: Biotin requiring strains of *Fusarium oxysporum* are very common among our isolates of forma *melonis* (Meyer and Maraite, *Nature* **216**, 5110: 70–71, 1967); this has also been noted by Armstrong, J. K. and J. M. for a strain of *F. oxysporum* isolated from cowpea (*Phytopathology* **46**, 7 (1956) abst.).

Graniti: Grente and Sauret (1969, *C. R. Acad. Sci.*, Paris **268-D**, 3173–3176) have recently shown that the reduced pathogenicity (what they called "hypovirulence exclusive") of the white strains of *Endothia parasitica* is controlled by some infectious cytoplasmic factors. The viral nature of these factors has been suggested, but not yet proved.

49. Amino Acid Antimetabolites as Causal Agents in Plant Disease*

SHREVE S. WOLTZ

Agricultural Research and Education Centre, University of Florida, Bradenton, Florida 33505, U.S.A.

Frenching of tobacco, the first recorded disease of the type under discussion, was first described in 1688 in a letter by John Clayton to the Royal Society (Lucas, 1958) and it has since been studied extensively by many workers. Steinberg (1952) has most closely approached a solution to the problem by producing the frenching syndrome with isomers of isoleucine and with diffusates from *Bacillus cereus* Frankl. *et* Frankl. cultures. Woltz and Jackson (1961) demonstrated that symptoms of yellow strapleaf disease of chrysanthemum could be produced by applying certain isomers of isoleucine around roots of chrysanthemum plants. They found that soils associated with yellow strapleaf also caused frenching of tobacco.

Woltz (1963) found that leucine and valine prevented symptom production when applied with DL-alloisoleucine to the root zones of chrysanthemum plants. The branch-chained amino acids, leucine, isoleucine, and valine act as metabolic antagonists among themselves and their analogues are effective as antimetabolites. Analogues of leucine were sprayed on leaves and drenched around roots of chrysanthemum plants naturally affected with yellow strapleaf; DL-2 methyl leucine was the most effective in causing stunted, deformed plants to grow more normally (Woltz, 1967).

A potent antimetabolite of leucine, 1-amino-2-nitrocyclopentane-1-carboxylic acid (ANCPA) produced by certain strains of *Aspergillus wentii* Wehmer, has been studied extensively by Brian *et al.* (1965), Broadbent and Radley (1966), and Burrows *et al.* (1965); it causes frenching when applied to tobacco plants. Woltz and Littrell (1968) found that severe yellow strapleaf could be produced by filtrates from cultures of *A. wentii* or by ANCPA and that plants of *Chrysanthemum morifolium* Ram. grown aseptically in vermiculite and a nutrient solution developed typical symptoms after the substrate was contaminated with *A. wentii*, but not when *B. cereus* was used. Plants were only 22% as high as control plants 46 days after *A. wentii* had been added.

* Florida Agricultural Experiment Stations Journal Series No. 3641.

ANCPA induced more yellowing of leaves, but a quicker recovery from growth retardation than did a mixture of DL-isoleucine and DL-alloisoleucine. The severity of both syndromes was markedly reduced by simultaneous application of L-leucine. L-valine prevented many of the effects of isoleucines but did not counteract ANCPA.

Less ANCPA is required to produce symptoms when it is injected into stems of seedlings than when it is sprayed on foliage or drenched around the roots. Injection permits quantitative studies and eliminates differences in rates of absorption of metabolite and antimetabolite.

Attempts to produce yellow strapleaf by contaminating autoclaved soil with *A. wentii* were unsuccessful.

The effects of spraying plants with filtrates from *A. wentii* cultures have also been studied (Woltz, 1969). Plants exhibited three types of reaction: (a) severely affected with long-lasting effects; (b) mildly affected with transitory symptoms; and (c) little or no effect. The symptoms produced were similar to those of frenching and yellow strapleaf. Stems of watermelon plants split open longitudinally following foliar applications of a culture filtrate. Flowers of *Ipomoea purpurea* Lam. had narrowed and twisted petals and were polypetalous instead of gamopetalous. In less severe responses the corolla-lobe segments became more prominent. About half of a number of tomato fruit 1–2 cm in diameter, sprayed with ANCPA developed symptoms of the physiological disorder, cat-face. The effect was specifically a failure of fruit wall to develop normally with resultant cracking and splitting. Injection of ANCPA into peduncles caused all fruit to develop cat-face symptoms.

It is likely that normal protein synthesis is altered by antimetabolites of leucine. Plants not affected by them may produce more leucine and thus escape damage or, as for azetidine-2-carboxylic acid an analogue of proline, the transfer RNA of susceptible plants may accept the "activated" analogue which is then incorporated into protein in the place of natural amino acids (Fowden, 1963; Peterson and Fowden, 1963). Plants susceptible to azetidine-2-carboxylic acid contain amino acid synthetases able to activate the analogue whereas resistant plants do not.

Structural analogues of certain amino acids produce syndromes characteristic of natural diseases in many species when sprayed on leaves, injected into plants or applied to roots. It has been demonstrated that the analogues often act as antimetabolites of related amino acids. Antagonists of L-leucine, in order of increasing effectiveness include: L-isoleucine, D-isoleucine, L-alloisoleucine and ANCPA. Symptoms of frenching and yellow strapleaf are produced by each of these antimetabolites; the effect of L-alloisoleucine is reversed by DL-valine.

The development of physiological diseases of the type under discussion depends on active plant growth. Nitrogen nutrition is closely linked to symptom development, deficiency preventing adequate growth for symptom expression, and excess preventing the expression. Prevention by large amounts of nitrogen may depend on production of greater amounts of natural amino acids which overcome competitive inhibition by the analogues.

Frenching, yellow strapleaf and similar diseases develop under specific soil conditions. The factors that have been found to predispose to the development of such diseases include warm soils that contain relatively large amounts of moisture and organic matter as well as adequate soil nutrients. All evidence indicates that the causal micro-organism(s) is non-parasitic and releases a phytotoxin into the soil environment.

References

Brian, P. W., Elson, G. W., Hemming, W. G., and Radley, M. (1965). *Nature* **207**, 998–999.

Broadbent, D. and Radley, M. E. (1966). *Ann. Bot.*, N.S. **30**, 763–777.

Burrows, B. F., Mills, S. D., and Turner, W. B. (1965). *Chem. Comm.* **5**, 75–76.

Fowden, L. (1963). *J. exp. Bot.* **14**, 387–398.

Lucas, G. B. (1958). The Scarecrow Press, Inc. New York, Chap. 29, p. 316–358.

Peterson, P. J. and Fowden, L. (1963). *Nature* **200**, 148–151.

Steinberg, R. A. (1952). *Pl. Physiol.* **27**, 302–308.

Woltz, S. S. and Jackson, C. R. (1961). *Pl. Physiol.* **36**, 197–201.

Woltz, S. S. (1963). *Pl. Physiol.* **38**, 93–99.

Woltz, S. S. (1967). *Proc. Fla State Hort. Soc.* **70**, 459–461.

Woltz, S. S. and Littrell, R. H. (1968). *Phytopathology* **58**, 1476–1480.

Woltz, S. S. (1969). *Proc. Fla. State Hort. Soc.* **72**, 165–167.

50. Ultrastructure of Fungus-Infected Onion Roots and Fungus Exudate-Treated Roots

WILFORD M. HESS AND DARRELL J. WEBER

Department of Botany, Brigham Young University, Provo, Utah 84601, U.S.A.

Two cultivars of onion were infected by a virulent isolate of the onion pink root fungus *Pyrenochaeta terrestris* (Hansen) Gorenz *et al.* (Hess, 1969). Ultrastructural observations of infected roots of both cultivars indicated that the fungus penetrated primarily or entirely by dissolution of host cell walls rather than by pressure of the fungus. Membrane fragments, lipid bodies, and nuclei were still recognizable in host cells in advanced stages of infection, but in general, the organelles of the onion cells were completely disrupted.

When the fungus was grown on paper consisting of gymnosperm cells the paper was rapidly utilized as a carbon source. Cells from healthy and infected onion roots were fixed by glutaraldehyde–acrolein–osmium techniques of Hess (1966). Figure 1 shows the normal appearance of cell wall and cytoplasm. Figure 2 shows a fungus cell penetrating an onion cell wall; here the tissue was processed by the techniques of Albersheim *et al.* (1960) which involves the staining of methyl esters of pectin by hydroxylamine and ferric chloride. Lead citrate post section staining (Reynolds, 1963) was used to enhance the contrast. The photograph shows the dissolution of the cell wall at the point of penetration.

Exudates from fungi growing on excised roots or a synthetic medium containing cellulose were assayed with onion seedlings. The organelles of cells of onion roots treated with these exudates were completely disrupted and cell walls were partially dissolved. Figure 3 shows a portion of a cell wall of an exudate-treated cell. It is evident in Figure 3 that the plasma membrane is disrupted and is no longer associated with the cell wall. The cell walls of exudate-treated cells were more severely affected than were those of tissue infected by the fungus. Figure 4 shows cytoplasm from a portion of an exudate-treated onion root cell which resemble cells from infected tissue except for the disruption of nuclei in cells treated with the exudate.

Isolation procedures for characterizing a toxic product indicate that it is heat stable and that toxicity is lost by dialysis. Purification of the product by

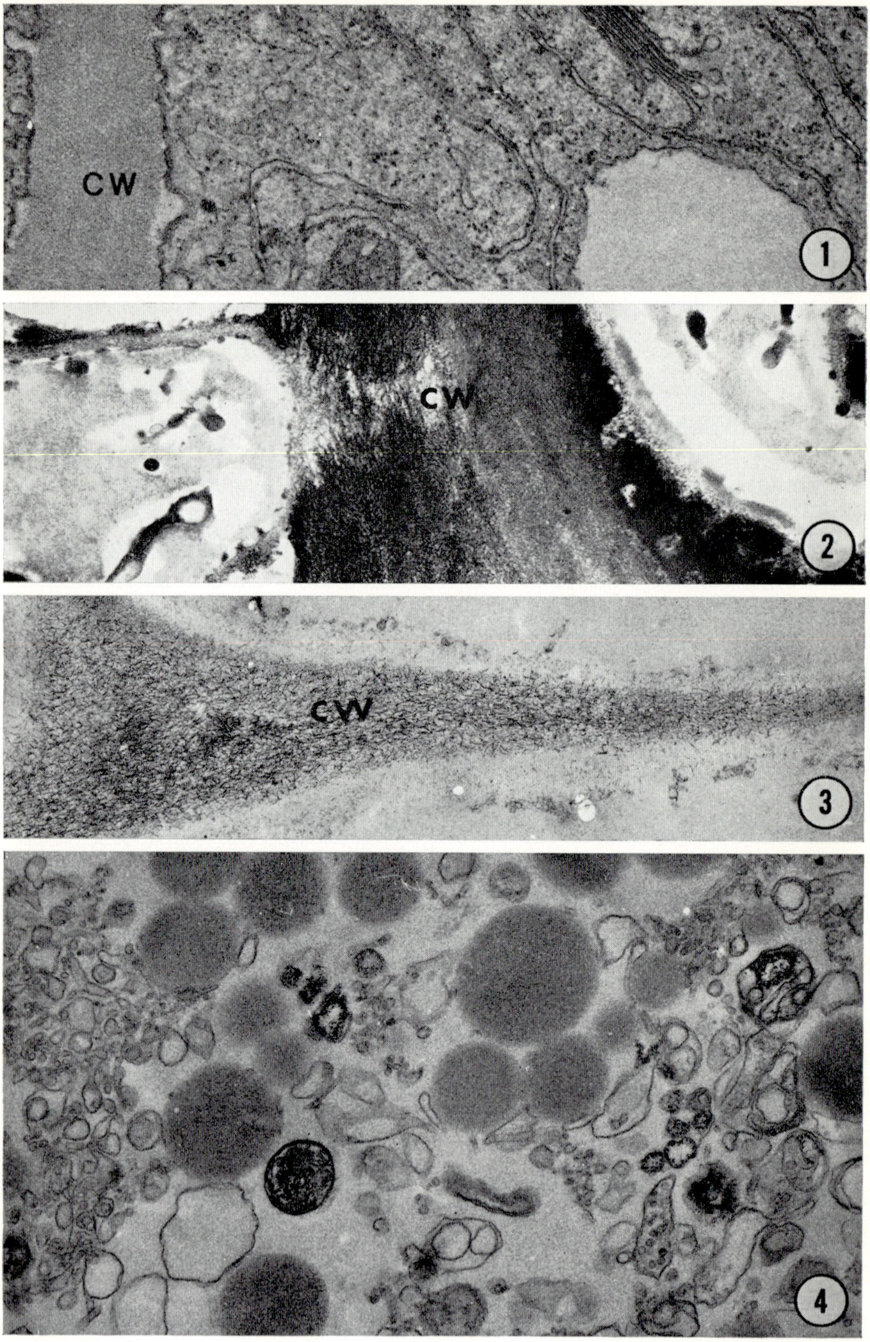
CW
1
CW
2
CW
3
4

Sephadex column chromatography showed that it is a small molecular weight compound.

Acknowledgement

Supported in part by NSF grant GB-14161 and NIH Career Development Award No. 1-K4-GM-29, 976-01 to W. M. Hess and National Air Pollution Control Administration, Consumer Protection and Environmental Health Service Grant No. AP 01135 to D. J. Weber.

References

Albersheim, P., Mühlethaler, K., and Frey-Wyssling, A. (1960). *J. Cell Biol.* **8**, 501–506.

Hess, W. M. (1969). *Am. J. Bot.* **56**, 832–845.

Hess, W. M. (1966). *Stain Technol.* **41**, 27–35.

Reynolds, E. S. (1963). *J. Cell Biol.* **17**, 208–213.

Discussion

Mussell: What fractionation methods were used?

Hess: The toxin or toxic product was purified by a combination of Sephadex G-25 column chromatography, ion exchange columns (Dowex 50 and 2), and by heat precipitation. The toxic product passes through dialysis tubing, but this was not used as a purification procedure.

Goodman: How long does it take the fungal exudate to cause the extensive root cell degradation shown?

Hess: For susceptible hosts the degradation can take place within 3–5 days. For resistant hosts it will commonly take longer if it takes place at all.

Kuć: Is the darkening of the cell wall material after treatment with lead citrate due to pectin or pectic substances?

Hess: I do not know.

Matta: Do you have any idea of the nature of the electron dense material around the infection hypha in the cell wall?

FIGS 1–4. Onion root cells. FIG. 1. Portion of a non-treated cell showing cytoplasm and cell wall (CW), (×23,500). FIG. 2. Hyphal cell penetrating host cell wall (CW), (×23,500). FIG. 3. Portion of a cell wall (CW) from a fungus exudate-treated cell (×23,500). FIG. 4. Portion of the cytoplasm from a fungus exudate-treated cell. (×23,500).

Hess: We have not conducted histochemical investigations to date to determine the nature of this material. However, it is thought to be a stainable breakdown product caused by degradation of the host tissue by the fungus.

Kuć: As a compromise in the debate as to whether penetration is by mechanical pressure or chemical action I offer our work with *Venturia inaequalis*. We believe this fungus penetrates by both chemical action and mechanical pressure.

51. Qualitative and Quantitative Changes Induced in the Polyphenoloxidases and Peroxidases of Plants by Several Ecological Factors

HENRI MARAITE

Laboratoire de Phytopathologie de la Faculté des Sciences Agronomiques, Université Catholique de Louvain, Heverlee, Belgium

Infection of plants by micro-organisms usually increases the activity of host polyphenoloxidase (PPO) and peroxidase (PO). These changes are associated with the more general metabolic alterations that are induced by pathogens and are often considered to be involved in the mechanisms of resistance (Kosuge, 1969). Mechanical injury (Novacky and Wheeler, 1970), treatments with growth regulators (Galston *et al.*, 1968) or increase in temperature (Akazawa and Uritani, 1961) also alter oxidase activities, which also change during differentiation of plant tissues (Alvarez, 1968). The variation in activity thus seems to be non-specific.

Separation of proteins in plant extracts by polyacrylamide gel disk electrophoresis now makes it possible to study these changes at the iso-enzyme level (Shannon, 1968).

The work described below demonstrates that colonization of muskmelon plants (*Cucumis melo* L.) by the vascular pathogen *Fusarium oxysporum* Schlecht. f. sp. *melonis* Snyd. *et* Hans. (Maraite, 1969) induces metabolic alterations that include quantitative and qualitative changes in the PPO and PO activities of extracts from infected plants.

During the first stage of pathogenesis, there is an increase of the PPO and PO activities in different parts of the plant, even in those not containing the pathogen. These changes occur before leaves show any internal or external symptoms. Zymograms of extracts from inoculated plants show a slight increase in the intensity of the original isoenzyme bands and the appearance of some new ones. Similar changes also appear in extracts from older, uninoculated plants. For PPO and PO, therefore, plants in the early stages of the disease, are similar to older plants that have aged naturally.

In the second phase of disease, characterized by exponential increase of the parasite in the plant, vascular browning and the disintegration and death of colonized tissues, there is a still more rapid increase of the PPO and PO activities. In the PO zymograms of extracts from infected plants the changes are mainly quantitative compared with uninoculated plants, with an increase in the intensity of certain isoenzyme bands. In the PPO zymograms, however, some plant isoenzyme bands are more intense, others are less intense, and new bands are present. From (1) similarities between the electrophoretic mobilities of these new PPO isoenzymes, the bands which become more intense, and the isoenzyme bands of the pathogen detected after colonization of autoclaved muskmelon tissues, (2) the correlation between the increase of intensity of these bands and colonization by the pathogen, I conclude that the pathogen is directly involved in the increase of PPO activity and the alteration of the isoenzyme pattern observed in the infected plant, by the production of its own enzymes. Also, liberation of latent plant oxidases by tissue-degrading enzymes of the pathogen is possible (Deverall and Wood, 1961).

A soil drench with the systemic fungicide benomyl induces in muskmelon plants an increase in the PO and a slight decrease in PPO activity. In leaves of treated plants showing symptoms of phytotoxicity (marginal chlorosis and necrosis) PPO and PO activities are very high. No qualitative alterations were detected in PPO and PO zymograms.

References

Akazawa, T. and Uritani, I. (1961). *Phytopathology* **51**, 668–676.

Alvarez, M. R. (1968). *Am. J. Bot.* **55**, 619–625.

Deverall, B. J. and Wood, R. K. S. (1961). *Ann. appl. Biol.* **49**, 473–487.

Galston, A. W., Lavee, S., and Siegel, B. Z. (1968). *In* "Proc. Inter. Conf. Plant Growth Subst. 6th. Ottawa 1967" (F. Wightman and Setterfield, Eds). Runge Press, Ottawa.

Kosuge, T. (1969). *A. Rev. Phytopath.* **7**, 195–222.

Maraite, H. (1969). Pathogénie du flétrissement du melon causé par *Fusarium oxysporum* f. sp. *melonis*. Ph.D. thesis, University of Louvain, Leuven.

Novacky, A. and Wheeler, H. (1970). *Phytopathology* **60**, 467–471.

Shannon, L. M. (1968). *A. Rev. Pl. Physiol.* **19**, 187–210.

Author Index

Numbers in italics indicate the pages on which references are listed in full

C

D

E

F

L

Q

R

S

T

U

V

Subject Index

B

D

G

H

I

J

K

L

M

R

T

X

Y

Z